# Sustainable Organic Farming

# Sustainable Organic Farming

Edited by **Wendel Mason**

SYRAWOOD
PUBLISHING HOUSE
New York

Published by Syrawood Publishing House,
750 Third Avenue, 9th Floor,
New York, NY 10017, USA
www.syrawoodpublishinghouse.com

**Sustainable Organic Farming**
Edited by Wendel Mason



International Standard Book Number: 978-1-68286-329-9 (Hardback)



Printed in China.

# Contents

# Preface

The main aim of this book is to educate learners and enhance their research focus by presenting diverse topics covering this vast field. This is an advanced book which compiles significant studies by distinguished experts. This book addresses successive solutions to the challenges arising in the area of application, along with it; the book provides scope for future developments.

Intensive farming techniques to increase yield have considerable impact on the environment. Sustainable organic farming is an attempt to curb such harmful effects by introducing microbial fertilizers instead of synthetic chemicals and limiting the use of pesticides and plant growth regulators, which help reduce the burden caused due to intensive farming techniques. This book attempts to assist those interested in this field with the help of researches and case-studies by experts from around the globe on topics such as transgenic crops in sustainable development, genetic modification as a route for delivery of sustainable crop protection, ecosystems and environment, etc. This book will prove to be an essential guide for both academicians and those who wish to pursue this discipline further.

It was a great honour to edit this book, though there were challenges, as it involved a lot of communication and networking between me and the editorial team. However, the end result was this all-inclusive book covering diverse themes in the field.

Finally, it is important to acknowledge the efforts of the contributors for their excellent chapters, through which a wide variety of issues have been addressed. I would also like to thank my colleagues for their valuable feedback during the making of this book.

**Editor**

# Optimal food plan for rural households' food security in Kwara State, Nigeria: The goal programming approach

**O. A. Omotesho and A. Muhammad-Lawal***

Department Agricultural Economics and Farm Management, University of Ilorin, P. M. B. 1515, Ilorin, Nigeria.

**This study developed optimal food consumption plan for the rural households in Kwara State, Nigeria. A two-stage simple random sampling technique was employed in collecting data used for the study between October, 2005 and January, 2007. The main tools of analysis used in this study are the Food Security Index and the Linear Goal Programming (LGP) Model. This study reveals that 65.45% of the rural households are food insecure. Rice, maize, cowpea, gari, fish and palm oil consumed at 4.17, 18.59, 2.92, 2.66, 8.53 and 3.10 Kg per week respectively constituted the least-cost food plan for the rural households. With an average household size of eight male adult equivalents, this plan has a cost implication of ₦73.73 per person per day. This shows that the cost of meeting this least-cost food plan was 38% lower than the one U.S. Dollar World Bank poverty line per person per day. This study recommends the need for the rural households to be educated on the nutritional implication of the consumption of various food items.**

**Key words:** Goal programming, optimal combination, food plan, rural households, food security, Kwara State, Nigeria.

## INTRODUCTION

### Background to the study

Food security has been described as an important aspect in any consideration of the sustainability of the wealth of a nation. This is in view of its role as a critical factor in economic development, peace and stability (Akanji, 1993; World Food Summit, 1996; Onyido, 1997; Osundare, 1999; Adegboye, 2004). Food is of high importance in matters of human wellbeing and economic productivity. Hence, the need for food in Nigeria, like other parts of the world, has become a policy issue. Besides, the nutrients contained in food are necessary for proper body functions (Olayide, 1982). To be food secure, sufficient resources are usually required to produce or purchase adequate food. However, this does not guarantee good nutrition and health as we can see from the diet-related health problems among even more affluent population groups. Even though eating well is vital for a healthy and active life, many people in virtually all countries do not eat well because of poverty and a lack of nutrition education. To have food security and be adequately nourished, an individual needs an understanding of what constitutes an appropriate diet for good healthy condition as well as the resources, skills and motivation to make good food choices (FAO, 2005a).

Developing policies and interventions to increase food security therefore requires an understanding of each of these factors, their inter-relationships and their relevance to particular groups of people (FAO, 2004). It can therefore be seen that the task of providing nutrition education especially in the low income countries must be accorded the necessary priority. Since it is important for the people to know how best to use their resources to ensure nutritional wellbeing, nutrition-education is therefore a key for developing the skills and motivation needed for the people to eat well (FAO, 2005a). Bliss and Stern (1978) observed that small increases in food consumption can bring about substantial improvement in productivity.

Goldsmith et al. (1967) and de Ville de Gouyet et al. (1979) observed that given most national food patterns, protein needs can be met if the caloric supply is adequate and the diet corresponds to the national average. This may explain why most studies on food security focus more on the calorie availability and consumption (Feleke et al; 2003, Makinde, 2000). An analysis of trends in the

*Corresponding author. lawaz71@yahoo.com.

supply of dietary energy can therefore help to reveal patterns that provide a basis for assessing the adequacy of the food supply, from which nutritional inferences can be drawn. Carbohydrates, fat and protein comprise the three principal sources of energy in the human diet. Individual foods contain different proportions of the three principal macronutrients. Animal products (meat and dairy) are rich sources of protein and fat, while cereals, fruits and vegetables contain a large proportion of carbohydrate. The quantity and quality of each component is particularly important when analyzing nutritional adequacy (Kennedy, 2001).

In a study on food security, Olayemi (1998) gave the threshold for food security as the ability of a household to meet 2260 Kcal of energy and 65 g crude protein per capita below which a household was classified as food insecure. This is however at variance with Olayide (1982) who gave daily consumption of 2470 kcal of energy and 65 g of protein per capita as the bench mark for food security. In the views of Joseph and Ajayi (2002), the recommended minimum nutrient requirements to be consumed per day per capita includes the 2191 Kcal and 65-86 g crude protein out of which at least 35g or (40%) must be animal protein. Unlike vitamins and minerals, which sometimes perform unique functions to meet the body's needs, fats, carbohydrates, and proteins substitute for one another to some extent to meet the body's energy needs (National Academies of Sciences, 2002). On the basis of the analysis of the recommended daily calorie distribution by the National Academies of Sciences (2002) and Honfoga and van den Boom (2003), 45 to 65% of the calorie supply should be made from carbohydrates including cereals, roots/tubers and fruits; 20 to 35% form fats and oils including lipids from oil crops and animal products; and 10 to 35% from protein including animal products and pulses.

Unlike nutrition related studies in the livestock industry, very few studies appear to have been done in the general area of human dietary combination and optimization. Many nutrition-related articles, especially in Africa, report results of studies of the nutrient composition of food without indicating the optimal levels desirable for human consumption, despite awareness about daily recommended levels required of the different nutrients (Ogunyinka and Norman, 2003). In view of this, there is hardly any study that identifies the optimal food consumption plan that will meet the food security need of the rural households at least costs from the locally available food items. Developing optimal food plan for the food insecure households may be needed to reduce the problems associated with ignorance on nutritious and low cost food items faced by some poor rural households. Furthermore, such plan can be useful in determining the costs of living in the rural areas of Kwara State. This study also holds implications for the poor, many of whom are vulnerable to specific health hazards because of their lack of knowledge about the appropriate ( that is, healthy) composition and level of nutrient intake. The results of this study will therefore indicate the least cost diet from among the food items consumed in the rural areas.

Goal programming is an advanced extension of linear programming found to be effective in handling multi objective problems (Njiti and Sharpe, 1994; Adejobi, 2004). It is one of the most widely used techniques for dealing with multiple criteria decision making in management with relevant application to similar farm planning problems (Romero and Rehman, 1985). This is in line with Habeeb's (1989) observation that goal programming has been regarded as the best mode for feasible allocation of resources especially in the Nigerian agricultural sector which have multiple and conflicting objectives.

Goal programming is presented as a decision-making framework that accommodates multiple objectives and explicitly determines how spatial characteristics should be traded off in cases where different, and perhaps opposing, characteristics are important for achieving the distinct management objectives. Goal programming provides a flexible way for decision-makers to incorporate the multiple objectives into a single composite objective. This decision-making framework translates a number of individual objectives into a single overall composite objective and measures how far the results of each strategy fall from the composite objective (Marshall and Homans, 2001).

Under the goal programming framework, a management objective function is created that minimizes a weighted aggregation of the distance of each objective from some pre-specified target level for that objective. In order to standardize the measures of distance across objectives that are measured in different units, this distance is measured as relative deviation from the target level (Marshall and Homans, 2001).

Goal programming model is of wide application as a technique for multiple objective decisions making which has gained increasing amount of attention from management scientists, production mangers and practicing managers with whom it has been formulated to solve operation management problems relating to planning decision making and control (Adewumi, 2002). Goal programming model was applied to a management scenario with multiple objectives in order to explore how important different characteristics of absolute and relative location are at contributing to individual objectives and how that information can be used to create a ranking system with respect to an overall objective (Marshall and Homans, 2001). It can be used to develop optimal level of production activities which will satisfy other concurrent goals (Fiske et. al., 1994). Even though application of goal programming techniques in agriculture is limited in developing countries, Adewumi (2002) reported that Monarchi et al. (1976), Wheeler and Russel (1977), Barlet and Clawson (1978), Bazara and Bouzaer (1981), Barnett et al. (1982), Marton and Sanchohez (1982),

**Table 1.** Sample distribution.

| Zones | Villages distribution | Sampled villages | Households sampled | No of households with valid responses |
|---|---|---|---|---|
| A | 217 | 4 | 68 | 64 |
| B | 237 | 5 | 85 | 84 |
| C | 483 | 10 | 170 | 166 |
| D | 311 | 6 | 102 | 97 |
| Total | 1258 | 25 | 425 | 411 |

Source: Field Survey, 2005/07.

Sandiford (1986) and Aromolaran (1992) have applied goal programming model in one form of agricultural planning decision or the other.

Two major variants of goal programming model include the Lexicographic goal programming model and the weighted goal programming model (Romero and Rehman, 1985). The main area of difference between them is that in the Lexicographic Goal Programming Model, priorities are assigned to objectives in a pre-emptive fashion whereas the Weighted Goal Programming assigns equal priorities to all objectives but attaches weight, which reflect the relative contribution of each objective (Adewumi, 2002; Adejobi, 2004). For the purpose of this study, a weighted goal programming model was used to develop an optimum plan for rural faring household food security. This is because rural households are assumed to attach equal priorities to food security goals.

## Objectives of the study

The main objective of this study is to analyse the level of food consumption of rural households.
The specific objectives of the study are to:

1. Determine the level of rural households' food consumption.
2. To determine the food insecurity status of the rural households.
3. Develop optimal food consumption plan for the rural households' food security.

## METHODOLOGY

### Study area

This study was conducted in Kwara State, Nigeria. Kwara state lies within the North Central geopolitical zone of Nigeria. It has a land area of about 32,500 square kilometres. According to the 2006 National Population Census figure, Kwara State has a total population of 2,371,089. This is made up of 1,220,581 males and 1,150,508 females. The State is bounded in the north by Niger State, in the South by Osun and Ondo States, in the East by Kogi State and in the West by Oyo State. Kwara State shares an international boundary with the Republic of Benin (Taiwo, 2005). Kwara State is located between latitudes 7°45'N and 9°30'N and longitudes 2°30'E and 6°25'E. The topography is mainly plain lands to slight gentle rolling. The annual rainfall ranges between 1000mm and 1500 mm. Average temperature ranges between 30 and 35°C (KWADP, 1996).

There is a total of One thousand, two hundred and fifty eight rural communities in Kwara State (National Population Commission, 1991). More than 90% of the rural populations who constitute about 80% of the total population are engaged in farming (Yusuf, 2000). Kwara State is divided into four zones. This classification is in consonance with agro-ecological and cultural characteristics of the areas. The zones are as follows: Zone A: Baruteen and Kaiama LGAs; Zone B: Edu and Patigi LGAs; Zone C: Asa, Ilorin East, Ilorin South, Ilorin West and Moro LGAs; and Zone D: Ekiti, Ifelodun, Irepodun, Isin, Offa, Oke-Ero and Oyun LGAs.

### Sampling technique

This study was carried out between October, 2005 and January, 2007. A two-stage simple random sampling technique was employed for this study. The first stage comprised random selection of twenty five villages which constitute about two percent of the total villages in the study area. The selected villages were distributed across the four agro-ecological zones in proportion to the distribution of the villages in each zone. This was done by selecting two percent of the total villages in each zone. This sampling frame for the villages was obtained from the 1991 National Population Commission village listing.

The second stage involves random selection of seventeen households from each of the selected villages. Information on the number of the total households in each village was obtained from the village council. This was used in preparing the list of the households that form the sampling frame from which the random selection of the households was carried out. Every other household was thereafter randomly selected from the list. A total of four hundred and twenty five rural households were selected for this study. Responses from a total of four hundred and eleven rural households were however found useful for this study. This is as distributed in Table 1.

### Method of data collection

Data used for this study were collected over a period of sixteen months, between October, 2005 and January, 2007. The primary source was through the use of a well structured questionnaire administered by trained enumerators.

### Analytical techniques

#### Food security index

This was used to determine the level of food insecurity among the

**Table 2.** Structure of the weighted goal programming model for the rural households' food security.

| Goals | Goal Statement: Achievement of | Objective function to Minimize | Deviational variable in objective function | Priority level | Weight |
|---|---|---|---|---|---|
| Meeting the minimum vegetable protein consumption | 171 kcal per day per adult equivalent from vegetable protein | Under achievement | $d_i^-$ | 1 | 5 |
| Meeting the minimum animal protein requirement | 199.5 Kcal per day per adult equivalent from animal protein | Under achievement | $d_i^-$ | 1 | 5 |
| Meeting the minimum fat requirement | 494 Kcal per day per adult equivalent from fat | Under achievement | $d_i^-$ | 1 | 5 |
| Meeting the minimum carbohydrates | 1605.5Kcal per day per adult equivalent from carbohydrates | Under achievement | $d_i^-$ | 1 | 5 |

rural households. Food security equation used by Feleke et al. (2003) was adopted for this study. The equation is stated as follows:

$$C^* = C_j - y_j$$

Food security indicator for this study is defined by calorie consumption Ci and calorie requirement $y_j$ of the $j^{th}$ household. For food secure households, $C^* \geq 0$ whereas, for food insecure households $C^* < 0$

For the purpose of this study, calorie requirement was based on the daily-recommended caloric intake of 2470 kcal per adult equivalent (Olayide, 1982).

**Linear goal programming model**

This was used to develop optimal food plan for rural households in Kwara State. Following Manyong and Degand (1995), the model was used as follows:

$$\text{Minimize } Z = \sum_{i=1}^{g}(W_i.n_i + W_i.\ p_i)$$

Subject to

$$\sum_{i=1}^{g} a_{ij}X_j + d_i^- - d_i^+ = a_i$$

$$\sum_{j=1}^{m} a_{kj}X_j \leq bk$$

$$X_j,\ d_i^-,\ d_i^+ \geq 0$$

Where Z = Objective function. It includes the total cost per kilogram in grain of food items available to the households as well as the positive and negative deviations for the goals.
$d_i^-$ = negative deviation if $a_i$ is under-achieved
$d_i^+$= positive deviation if $a_i$ is over - achieved
$W_i$ = weight attached to deviation $d_i^-$or $d_i^+$
$a_{ij}$ = matrix of the marginal contribution of $X_j$ to satisfying $a_i$.
$X_j$ = $j^{th}$ food item consumption measured in kilogram. In line with United State Department for Agriculture (USDA, 2007), food items consumed by at least twenty five percent of the sample were included in this analysis.
$a_{kj}$ = coefficient of use of $b_k$
$b_k$ = available level of resource k. This was determined by the average food consumption in Kilogram;
m = number of resource(s) which in this study equals 1;
$a_i$ = aspiration levels for the $i^{th}$ goal. i is the total food consumption and it measured in Kg.
g = number of goals. Given that household calorie requirement is estimated at 2470 kcal per day per adult equivalent, there were four goals to be achieved.

These according to the National Academy of sciences (2002) are given as follows:

65% of calorie estimated at 1605.5 Kcal per day per adult equivalent from carbohydrates.
20% of calorie estimated at 494 Kcal per day per adult equivalent from fat.
15% of calorie estimated at 370.5 Kcal per day per adult equivalent from protein. Since protein comprises both animal and vegetable sources, energy supplied from protein was estimated at 171 kcal per day per adult equivalent from vegetable protein; and
The remaining 199.5 Kcal per day per adult equivalent from animal protein.

Matrix of the model for rural households' food security is as presented in Table 2.

**Limitations to the study**

There is a general lack of accurate data on the number of households in the rural areas of the study area. In developing the sampling frame, this study relies on information given by the members of the village councils. The accuracy of such data however depends on the ability of the council to recollect all the households in the village. In preparing food with least cost to the rural households, the study assumed that the prices of the food items remain constant. Besides, the study assumes that the food items are available in the rural households all year round.

**Table 3.** Distribution of food consumption in the rural households in Kwara State.

| Food items | Percentage calorie (%) | Quantity consumed (Kg) | Value (₦) |
|---|---|---|---|
| Rice | 11.46 | 5.77 | 646.24 |
| Sorghum | 9.26 | 4.66 | 199.72 |
| Maize | 10.47 | 5.12 | 196.92 |
| Cowpea | 4.81 | 2.56 | 216.62 |
| Gari | 7.11 | 3.68 | 184.00 |
| Lafun | 9.18 | 4.75 | 158.32 |
| Yam | 13.33 | 21.33 | 888.82 |
| Elubo | 5.79 | 3.19 | 348.00 |
| Meat | 1.42 | 1.11 | 555.00 |
| Fish | 0.77 | 1.03 | 257.50 |
| Egg | 0.14 | 0.26 | 62.40 |
| Milk | 0.04 | 0.13 | 119.99 |
| Palm oil | 6.55 | 1.28 | 223.26 |
| G/nut oil | 2.66 | 0.52 | 104.00 |
| Orange | 1.39 | 6.13 | 559.79 |
| Okra | 0.08 | 0.45 | 60.00 |
| Tomato | 0.14 | 1.19 | 96.22 |
| Onion | 0.12 | 0.68 | 57.63 |
| Vegetable | 0.15 | 0.54 | 17.82 |
| Melon | 0.03 | 0.21 | 80.00 |
| Pepper | 0.06 | 0.40 | 72.23 |
| Sugar | 1.32 | 0.60 | 69.23 |
| Bread | 1.28 | 0.99 | 176.00 |
| Others | 12.45 | 20.58 | 1,418.50 |
| **Total** | **100** | **87.16** | **6,768.21** |

Source: Data analysis, 2007.

## RESULTS AND DISCUSSION

### Rural households' food consumption

In view of its importance to the nutritional wellbeing of individuals, food constitutes a larger proportion in the expenditure pattern of most rural households. Food consumption levels of rural households in Kwara State, Nigeria are as presented in Table 3.

Table 3 shows the percentage distribution of food consumption among the rural households in Kwara State. Food items which are consumed by less than 25% of the rural households in Kwara State were categorized as others. The food items are millet, soyabean, groundnut, cheese, bush meat, margarine, mango, pawpaw, cashew, beverage, honey, soft drink and alcohol. It can be seen that the rural households in the study area did not comply with dietary recommendation which specifies a minimum requirement for the supply of energy from carbohydrate, legumes, animal protein and fats/oils. For the purpose of this study, carbohydrate food includes cereals, fruits and vegetables. Rural households derived more of their energy from carbohydrates at the expense of other classes of food items. The study reveals that cereals together with root and tuber products contributed 67.88% of the households' weekly calorie intake. This was followed by fats and oils with 9.21%, animal products with 2.37% and lastly fruits and vegetables with 1.97% of the total weekly calorie intake among the rural households.

This study also shows that a rural household had an average weekly expenditure of ₦6,768.21 on food items. This was estimated at a sum of ₦119.96 daily per capita food expenditure. Hence, the next sub-section of this study was embarked upon with a view to developing least cost-food items that would include the right proportion of energy from the various food classes highlighted.

### Rural households' food insecurity status

Food security index with a food security line of 2470 Kcal daily per capita consumption was used to classify the rural households either as the food secure or food insecure. Rural households' food insecurity status is as presented in Table 4.

The study shows that the rural households with a mean daily per capita calorie consumption of 2290.60 Kcal are food insecure. Table 4 shows that about two-third of the rural households with an average daily per capita calorie

**Table 4.** Households' food insecurity situation.

| Food Security Status | Frequency | Percentage | Daily per capita energy consumption | | | |
|---|---|---|---|---|---|---|
| | | | Min | Max | Mean | Std. Dev |
| Insecure | 269 | 65.45% | 191.88 | 2461.19 | 1403.56 | 600.24 |
| Secure | 142 | 34.55% | 2479.93 | 14682.92 | 3970.98 | 1701.82 |
| Total | 411 | 100% | 191.88 | 14682.92 | 2290.60 | 165.48 |

Source: data analysis, 2007.

**Table 5.** Optimal food consumption plan for the average rural household.

| Food items | Optimum quantity (Kg) |
|---|---|
| Rice | 4.17 |
| Maize | 18.59 |
| Cowpea | 2.92 |
| Gari | 2.66 |
| Fish | 8.53 |
| Palmoil | 3.10 |

Value of the Plan: ₦ 4159.30.
Source: Data Analysis, 2007.

**Table 6.** Goals attainment.

| Goals | Target | Optimal plan Value | Under achievement | Over achievement | Degree of attainment |
|---|---|---|---|---|---|
| Rice Calorie | 14589.73 | 14589.73 | 0 | 0 | Achieved |
| *Gari* Calorie | 9056.32 | 9056.32 | 0 | 0 | Achieved |
| Other Carbohydrate Calorie | 66917.15 | 66917.15 | 0 | 0 | Achieved |
| Animal product calorie | 11253.41 | 11253.41 | 0 | 0 | Achieved |
| Legume Calorie | 9645.79 | 9645.79 | 0 | 0 | Achieved |
| Fat and oil Calorie | 27865.60 | 27865.60 | 0 | 0 | Achieved |

Source: Data Analysis, 2007.

**Table 7.** Analysis of constraint to optimal food plan.

| Constraint | Status | RHS | Slack or surplus | Shadow price |
|---|---|---|---|---|
| Available food | Loose | <= 66.603 | 26.638 | 0 |

Source: Data Analysis, 2007.

consumption of 1403.56 Kcal are food insecure. With a daily per capita calorie consumption of 3970.98 Kcal, about 35% of the rural households are food secure. This shows that food insecure households are 43.18% short of the food security line while the food secure households are above the food security line by 60.77%.

## Optimal food plan for rural households in Kwara State

Rural households require good health to be able to carry out agricultural production activities which are usually labour intensive. The optimal food plan ensures that resource-poor rural households meet their minimum food requirements at least possible cost. Apart from having the least cost to the rural households, the plan also satisfies the food preferences of the rural households. This is because the plan was based on the commonly consumed food items in the study area. Table 5 presents the optimal food plan.

The minimized value for the food plan in the study area was ₦ 4159.30 per week. With mean adjusted household sizes of eight male adult equivalent, this amount to ₦

73.73 per person per day. The food plan has cost implication that is below the poverty line set by the World Bank at One U.S. Dollar per person per day.

## Attainment of goals

The extent to which the food plan satisfies the goals of rural households' food consumption is as presented in Table 6.

As shown in Table 6, all the goals of the rural households' food consumption are achieved. As such, the plan satisfies the minimum food requirement necessary for the rural households' food security. This shows that the goals of meeting the minimum calorie requirement from carbohydrate, animal products, legumes and animal products are achieved without any deviation.

## Constraint to the attainment of optimal food consumption plan

Attainment of optimal food plans among rural households is affected by the level of available food. Table 7 presents the results of the analysis of the constraint to the attainment of optimal food plan among the rural households.

The average food available was in excess by the amount indicated in the slack/surplus column. This implies that the available food can be combined in such a way that an average rural household can meet the minimum food requirement with the possibility of such household having some returns from the sale of the surplus.

## Conclusion

There is high incidence of food insecurity among the majority of rural households whose calorie consumption comprise about 70% carbohydrates. Optimal food plan for the rural households in Kwara State is about 38% lower than the one U.S. Dollar World Bank poverty line per person per day. With a food expenditure of just about ₦ 73.73 per person per day, poor rural households may be food secure by merely readjusting their food combination toward rice, maize, cowpea, *gari*, fish and palm oil which constitute the least - cost food items for the rural households.

## Recommendations

Considering their limited resource base, rural households need to be educated on the nutritional implication of the various food items. This is necessary for the food insecure households to be able to make appropriate choices in matters of food consumption.

Farmers should be assisted to embark on dry season farming. This is to guarantee that the necessary food items are made available for the rural households at reasonable prices all year round.

There is the need for improvement in the processing of the local food produce. This is with particular reference to *gari* to minimise contamination and impurity. This is particularly necessary as the food item is an important component of the least cost food items in the rural areas of Kwara State.

Since agriculture constitutes the main stay of the rural households' livelihood, their food security may be seriously threatened in event of crop failure due to bad weather or pests and diseases. It may therefore be necessary for the efforts at improving food security in rural areas to put in place functional and accessible agricultural insurance scheme for farmers in the rural areas. Apart from helping the farmers to improve their agricultural production activities, it will enable the rural households to gain better economic access to their food need.

Rural households should be educated on the need to diversify their source of income from agriculture. This will ensure regular incomes for the households. Participation of poor rural households' in cooperative societies may be necessary for them to be able to acquire the necessary funds required. Besides, agencies of government such as the National Directorate of Employment should be made to go and give training to rural households on small scale enterprises.

## REFERENCES


Adegboye RO (2004). Land, Agriculture and Food Security in Nigeria. Faculty of Agriculture Lecture, University of Ilorin, Feb. 25th p. 27.

Adejobi AO (2004). Rural Poverty, Food Production and Demand in Kebbi state, Nigeria, Unpublished Ph.D. Thesis, Department of Agricultural Economics, University of Ibadan. Unpublished.

Adewumi MO (2002). A Multi- Objective Approach to Small Scale Farm Planning in Kwara State, Nigeria. Unpublished Ph.D. Thesis, Dept. of Agric. Econs. and Farm Management, University of Ilorin. Unpublished.

Akanji B (1993). "Some critical considerations in rural nutritional health status in Nigeria" Perspectives on Food Security in Nigeria. Olomola A. and S. Nwosu (eds) Nigerian Rural Sociological Association pp. 39-51.

Aromolaran AB (1992). Multiple Objectives And Resources Allocation Behaviour of Small Farmers in Ifedapo Area Of Oyo State, Nigeria, Unpublished Ph.D. Thesis, Department of Agricultural Economics Univeristy of Ibadan.

Barlet ET, Clawson WT (1978). Profit, Meat Production or Efficiency Use of Energy in Ranching, J. Anim. Sci. 46: 812-818.

Barnett D, Blake B, Mc Carl BA (1982). Goal Programming Via Multi-Dimensional Scaling Applied to Senegalese Subsistence Farms, Am. J. Agric. Econ. 64: 720-727.

Bazara MS, Bouzaer A (1981). A Linear Goal Programming Model for Developing Economies with an Illusion from the Agricultural Sector in Egypt, Science. 27: 396-413.

Bliss C, Stern N (1978). Productivity, Wages and Nutrition, J. Dev. Econs. North–Holland Publishing Company 5: 363–398.

De ville de Gouyet C, Seaman J, Geifer U (1978). The Management of Nutritional Emergencies in Large Population WHO Geneva pp. 88-89.

FAO (2004a). Food Insecurity and Vulnerability in Vietnam: Profiles of Four Vulnerable Groups Food Security and Agricultural Projects

Analysis Service (ESAF), ESA Working Paper, Agricultural and Development Economics Division The Food and Agriculture Organization of the United Nations. www.fao.org/es/esa. Retrieved on 05/12/2005 pp: 4-11.

FAO (2005a). FAO Warns World Cannot Afford Hunger", FAO Newsroom retrieved on December, 2005. www.fao.org Retrieved on 05/12/2005.

Feleke S, Kilmer RL, Gladwin (2003). Determinants of Food Security in Southern Ethiopia. A selected paper presented at the 2003 American Agricultural Economics Association meetings in Montreal. Canada. retrieved on 12th December, 2005 from http://ageconsearch.umn.edu/bit stream/22010/1/sp03fe02.pdf .

Fiske WA, D'Souza GE, Fletcher JJ, Philips TT, Bryan WB, Prigge EC (1994). "An Economic and Environmental Assessment of Alternative Forage- Resource Production Systems: A Goal-Programming Approach" Agricultural Systems 45 Elsevier Science Limited pp. 259-270.

Goldsmith GA, Revelle R, Beale CL, Brackett JW, Engel RW, Gartner WA, Johnson OC, Meyers T, Milner M, Shank RE (1967). The World Food Problem: Report of the Panel on the World Food Supply volume 11.the White House.

Honfoga BG, Van den Boom GJM (2003). Food-Consumption Patterns in Central West Africa, 1961 to 2000, and Challenges to Combating Malnutrition. Food and Nutrition Bulletin, The United Nations University 24(2): 167-182.

Joseph JK, Ajayi S (2002). Agricultural Products and Man. History and Philosophy of Science: General Studies Approach: Akanji MA, Jolayemi ET (eds.) pp. 118-141.

Kennedy G (2001). Global Trends in Dietary Energy Supply from 1961 to 1999, Paper prepared for the FAO/WHO/UNU, Expert Consultation on Energy in Human nutrition. retrieved on 12th December, 2005 from www.fao.org/DOCREP/005/Y3800M/y3800m07.htm.

Kwara State Agricultural Development Project (KWADP) (1996). Agronomic Survey Report, KWADP, Ilorin.

Makinde OK (2000). Determinants of Food Security in Bauchi Area of Northern Guinea Savanna of Nigeria. Unpublished Ph.D Thesis University of Ibadan, Nigeria, April. Unpublished.

Manyong MA, Degand J (1995). Sustainability of African Smallholder Farming Systems: Case Study of Highland Areas of Central Africa. J. Sustainable Agric. 6(4): 17- 41.

Marshall EP, Homans FR (2001). A Spatial Analysis of the Economic and Ecological Efficacy of Land Retirement. Paper prepared presented at the 2001 Annual Meeting of the American Agricultural Economics Association, Chicago IL, August 5-8. on 12th December, 2005 from http://agecon.lib.umn.edu/cgi-bin/pdf_view.pl?paperid=2501.

Marton GG, Sanchohez LA (1982). Economical land use planning and carrying Capacity evaluation in the Jalapa region, Veralcruz, Mexico. Agro- Ecosystems 8: 83-124.

Monarchi DE, Weber JF, Duckstein I (1976). An interactive multiple objective decision – making aid using non linear goal programming. Multiple Criteria Decision Making (Zeleny M. (ed.) New York, springer-verlag.

National Academies of Sciences (2002). Dietary Reference Intakes For Energy, Carbohydrate, Fiber, Fat, Fatty Acids, Cholesterol, Protein, and Amino Acids, Press Release Institute of Medicine National Academies of Sciences available., http://www.nap.edu.

National Population Commission 1991: Census `91 Final Results, Kwara State National Population Commission.

Njiti CF, Sharpe DM (1994). A Goal Programming Approach to the Management of Competition and Conflict among Land Uses in the Tropics: The Cameroon Example Ambio. 23(2): 112-119.

Ogunyinka EO, Norman DW (2003). Optimal Dietary Blending for Household Food Security in South Western Nigeria: Lessons from Published Data Selected Paper Prepared for Presentation at the Western Agricultural Economics Association Annual Meeting at the Denver Adam's Mark Hotel, Denver.Colorado July 11-15. retrieved on 15th December, 2005 from www.ageconomics.ksu.edu/DesktopModules/ViewDocument.aspx?... 4176.

Olayemi JK (1998). Food Security in Nigeria Research Report N0 2. Development Policy Centre, Ibadan p. 88

Olayide SO (1982). Food and Nutrition Crisis in Nigeria Ibadan University Press.

Onyido (1997). Poverty Alleviation: Agenda for Food Security, Integrated Agricultural Production in Nigeria: Strategies and Mechanisms for Food Security NARP and FMANRP., pp. 114-130.

Osundare F (1999). Analysis of the Demand for Certified Maize Seed in Ondo State: Implications for Food Security, Poverty Alleviation and Food Security in Nigeria Fabiyi Y.L. and E.O Idowu (eds.) Nigeria Association of Agricultural Economists pp. 238-242.

Romero C, Rehman T (1985). Goal Programming and Multiple Criteria Decision Making in Farm Planning: Some Extensions, J. Agric. Econs. 136: 171-185.

Sandiford F (1986). An Analysis of Multi Objective Decision Making for the Scottish on Shore Fishery, Am. J. Agric. Econs.

Taiwo S (2005). Rapid Assessment of The Impact of Liberalization and Foreign Private Investment in Agriculture for Food Security and Food Sovereignty in Nigeria: A Case Study of Kwara State Report of a research conducted as part of the "right to food as human right" project, Trade and Sustainable Development Series No. 2, Development Information Network and Heinrich Boll Foundation.

United State Department of Agriculture (USDA) (2007). Food Distribution Commodity Acceptability Processing (CAP) Report (for Institutions and Households). Food and Nutrition Service Retrieved on 17th May, 2007 from www.fns.usda.gov/fdd/forms/FNS-663.pdf

Wheeler, B.M. and J.R.M. Russel, (1977). Goal Programming and Agricultural Planning, Oper. Res. Q. 28(1): 21-32.

World Food Summit (1996). World Food Summit Plan of Action FAO, Rome. 13th – 17th November, 1996 retrieved from http://www.fao.org/docrep/003/w3613e/w3613e00.HTM

Yusuf TM (2000). Socio – Economic Analysis of Poverty Level among Rural Dwellers in Kwara State, Nigeria, Unpublished M. Phil. Thesis in the Department of Agric. Economics, University of Ibadan.

# Histopathology of *Raphia hookeri* leaf Infected with *Glomerella cingulata* causing seedling blight

**E. A. Oruade-Dimaro, I. B. Omamor and E. I. Eziashi**

Nigerian Institute for Oil Palm Research (NIFOR), Plant Pathology Division, P. M. B. 1030 Benin City, Edo State, Nigeria.

**The host-pathogen relationship between the *Raphia hookeri* seedling leaf and *Glomerella cingulata* showed that the host cells were macerated. This allowed penetration of the pathogen through the hypodermis to the palisade tissue which resulted in dark coloration and disruption of the transfusion tissues. Hypersensitivity reaction was observed. The xylem and phloem were left without any attachment or support. Their intact cells were darkened as a result of disease development which resulted in complete disintegration of the palisade tissues and disappearance of transfusion cells. The uninoculated seedling leaf remained healthy with glossy appearance.**

**Key words:** *Glomerella cingulata, Raphia hookeri,* seedling blight.

## INTRODUCTION

*Glomerella cingulata* was found to be the causal organism of seedling blight of *Raphia hookeri* (Oruade-Dimaro and Ekundayo 1992). *G. cingulata* has been reported as mulberry anthracnose fungus. The organism was isolated from leaf spot and blight symptoms on Kiwifruit (Ho, 2008). Histopathological study of leaf infected by fungus helps in elucidating how the fungus colonizes leaf surface, palisade and vascular tissues. It helps in showing the mode of entry of the pathogen (William, 1957). It is also used to study the nature of antagonism between the fungi and their hosts. Such study further enhances the formulation of most suitable means of controlling plant pathogens (Porto et al., 1988).

Literature review has shown that no work has been reported on the histopathology of *Raphia hookeri* leaf infected with *G. cingulata* in Nigeria. This study was to examine the host-pathogen relationship using histopathological method.

## MATERIALS AND METHODS

The *G. cingulata* (Herbarium IMI 283849) used in this study was isolated from naturally infected *R. hookeri* seedling leaf showing symptom of seedling blights. Re-infecting healthy seedling leaf, the re-isolated fungus was *G. cingulata* which showed symptoms of seedling blight.

### Sample preparations and microtome sections

Leaf samples were collected from healthy seedling leaf with no symptom which served as the control treatment (Plate 1) and healthy seedling leaf which was artificially infected with *G. cingulata* showing the symptoms of seedling blight (Plate 2). They were dehydrated, fixed, impregnated with wax, embedded, sectioned (6-8 µm thick) with a rotary microtome (Erma, Tokyo), and stained according to the methods of Sass (1958).

### Observations

The mounted sections in Canada balsam from microtome sections were carefully studied under a light microscope (ZEISS West Germany). Observations were made with regards to intact cells of healthy tissues, impact of the pathogen on host tissue and histopathological changes resulting from presence of the pathogen. Photomicrographs were made where necessary with an attached camera (Motic MCCamera) to the microscope connected to a computer.

## RESULTS

Microtome sections of healthy *R. hookeri* seedling leaf revealed that it was made up of transfusion tissue which is concerned with the translocation of water and food

*Corresponding author. E-mail: eziashius@yahoo.com.

**Plate 1.** Healthy *R. hookeri* seedling leaf with no symptom of disease.

**Plate 2.** *R. hookeri* seedling leaf artificially infected with *G. cingulata.*

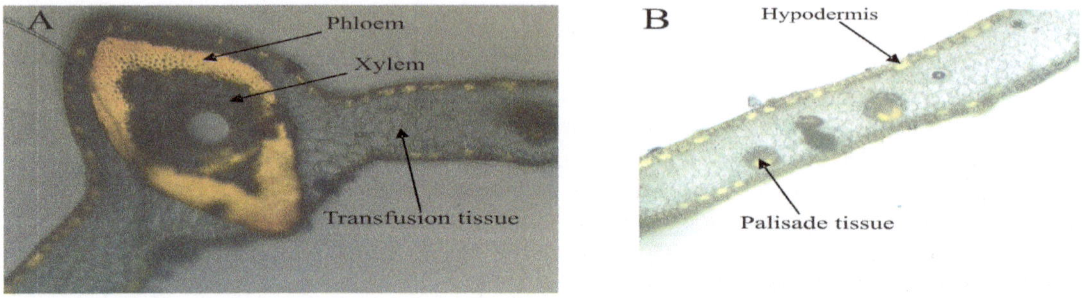


**Figure 1.** (×100). Microtome transverse section (TS) of healthy *R. hookeri* seedling leaf. A, mid-vein. B, blade of *R. hookeri* leaf (×100).

materials between the vascular bundles. The xylem is concerned with water conduction, storage and support while the phloem is a complex tissue which contains parenchyma cells. It is also concerned with conduction, storage of food and with support. The palisade tissue occurred on the lower side, arranged on the axis of the leaf, while the hypodermis occurred on the upper side (Figure 1b).

Sections of infected *R. hookeri* seedling leaf revealed that disease development started from the hypodermis with brown coloration which extended to the palisade tissue. The coloration is deepened, showing hypersensitivity reaction (Figure 2b and c). Complete disintergration of palisade and disappearance of transfusion tissues were seen, leaving the xylem and phloem without any attachment or support. The intact cells in the xylem and phloem were darkened as a result of disease development (Figure 2d) when compared with

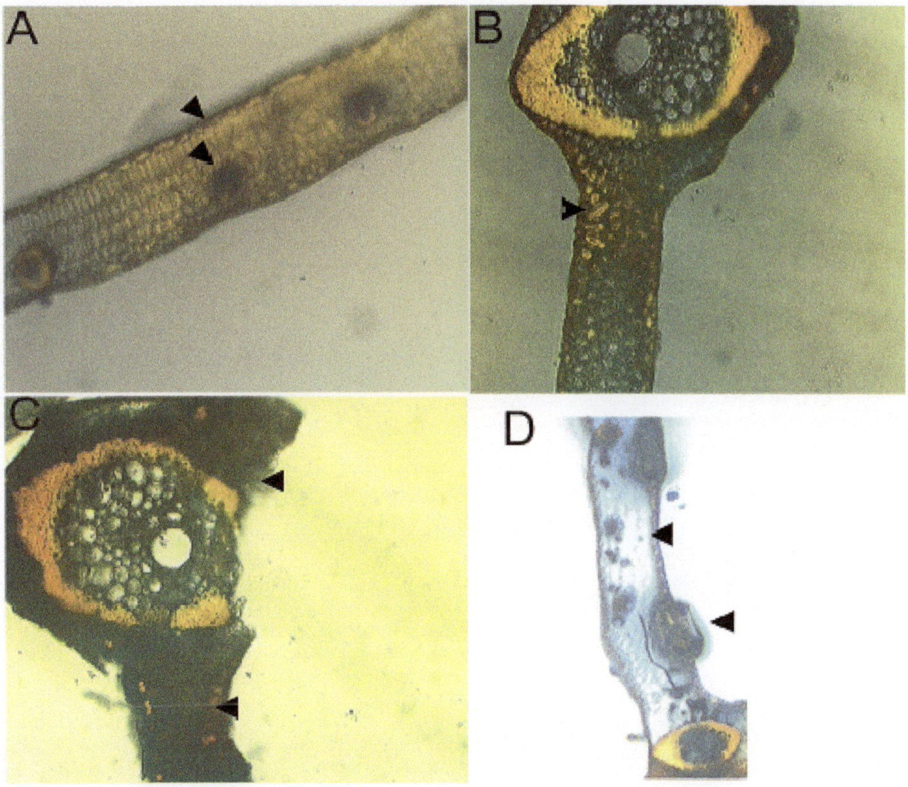


**Figure 2.** Microtome transverse sections of *Raphia hookeri* seedling leaf infected with *G. cingulata* (x100). (A. Arrow heads show initial infection points from the hypodermis to the palisade tissue; B. Arrow head shows progression of invasion and coloration of hypodermis, and part of the transfusion tissues; C. Arrow head shows deepened coloration. D. Arrow heads show disintegration of palisade tissue and disappearance of transfusion cells eight days after inoculation.

the control (Figure 1a).

## DISCUSSION

The host cell showed hypersensitive reaction in the presence of *G. cingulata.* A similar observation was made by Eziashi et al. (2007). Interaction between the host and the pathogen also showed that the cell was macerated. It was observed that disease development from the epidermis gave rise to maceration in and around the palisade tissue. The maceration of cells may have resulted from the action of hydrolytic enzymes produced by *G. cingulata.* Products of the enzymatic breakdown of cell components will most probably serve as a source of nutrients to the pathogen. Extracellular hydrolytic enzymes have previously been found to be produced by rot pathogens (Nwufo and Fajola, 1988; Adisa and Fajola, 1983).

In advanced stage of *G. cingulata* infection, the infected tissues turned black. The black coloration is still not fully understood, but Clark et al. (1981) reported the accumulation of furanoterpenoids in sweet potato following inoculation with *Botryodiplodia theobromae.* The mode of penetration of *G. cingulata* into cells was not investigated. It might have been mechanical, enzymatic or a combination of both. Enzymatic penetration will mean that cellulase may have been produced by this pathogen. Adisa and Fajola (1983), Turner and Ogundana (1983) and Arinze et al. (1976) have reported production of cell wall-degrading enzymes by *B. theobromae.* The total disappearance of the intact cells confirmed invasion of the host by the pathogen. However, the enzymes involved in this study were not determined. The result of this investigation could be exploited in the management of folia disease associated with *G. cingulata.*

### REFERENCES

Adisa VA, Fajola AO (1983). Cellulolytic enzymes associated with the fruit rots of *Citrus sinensis* caused by *Aspergillus aculeatus* and *Botryodiplodia theobromae.* Mycopathologia 82(1): 23-27.

Arinze AE, Naqvi SHZ, Ekundayo JA (1976). Production of extracellular cellulolytic and pectolytic enzymes by *Lasiodiplodia theobromae* on sweet potato (*Ipomoea batatus*) tubers. Int. Biodeterioration Bull., 12(1): 15-18.

Clark CA, Lawrence A, Martin FA (1981). Accumulation of furano terpenoids in sweet potato tissues following inoculation with

pathogens. Phytopathology 71: 708-711.
Eziashi EI, Omamor IB, Oruade-Dimaro EA (2007). Histopathology of *Ceratocystis paradoxa* causing Black Seed Rot on Oil Palm Sprouted Seeds. Advances in Food Sciences 29(4): 224-227.
Ho J (2008). http://www.riceblast.snu.ac.kr/file/58/57371 04-24 (2)-06.pdf.
Nwufo ML, Fajola AO (1988). Production of amylolytic enzymes in culture by *Botryodiplodia theobromae* and *Sclerotium rolfsii* associated with corn rots of *Colocasia esculenta*. Acta Microbiologica Hungaria 35: 371-378.
Oruade-Dimaro EA, Ekundayo CA (1992). Seedling blight of *Raphia hookeri* Man and Wendl. caused by *Glomerella cingulata* (Stonem). Sprauld and Shrunk in Nigeria. Princepes J. Int. Palm Soc. USA. 35(4): 41-44.
Porto MD, Grau CR, De Zoeten GA, Gaard G (1988). Histopathology of *Collectotrichum trifolii* on alfalfa. Phytopathology 78: 345-349.
Sass JE (1958). Botanical Microtechnique Jowa State University Press, Jowa p. 229.
Turner JG, Ogundana SK (1983). Production of yam cell-wall degrading enzymes by *Bothryodiplodia theobromae*. Transactions of British Mycological Society 80 (1): 163-166
William RJ (1957). The control of cowpea diseases in the IITA grain leagume improvement program. In: Tropical diseases of Legumes (ed. Bird, J. and Marancrosch, K.). Academic Press Inc. New York p. 172.

# Effect of NPK fertilization on yield and inflorescence blight of cashew (*Anacardium occidentale*)

**T. O. Adejumo**

Department of Microbiology, Adekunle Ajasin University, P. M. B. 01, Akungba-Akoko, Nigeria.
E-mail: toadejumo@yahoo.com.

**Application of three levels of nitrogen as Urea (0, 60 and 120 kg/ha), phosphorus as single super phosphate (SSP) (0, 72 and 144 kg/ha) and potassium as muriate of potash (MOP) (0, 24 and 48 kg/ha) fertilizers on inflorescence blight of cashew caused by *Lasiodiplodia theobromae* and yield of cashew were studied on acid sand soils of Ochaja, Nigeria in 2000 and 2001. Results showed that single application of SSP at 144 kg/ha significantly reduced infection and slightly increased yield during the second year. Combination of urea at 120 kg/ha with MOP at 48 kg/ha gave the highest yield of 3.53 kg/tree and lowest disease infection (90 inflorescence panicles infected) for the first year, while NPK fertilizer when applied as urea at 60 kg/ha, SSP at 144 Kg/ha and MOP at 24 kg/ha gave the highest yield for the second year and significantly reduced the disease infection (125 inflorescence panicles infected). Appropriate management of NPK fertilizers leading to a reduction in infection of inflorescence blight and increase in yield is an additional tool for minimizing the use of pesticides on cashew.**

**Key words:** *Anacardium occidentale*, cashew, fertilizer, inflorescence blight, yield.

## INTRODUCTION

Cashew, *Anacardium occidentale* L. is an important commodity cash crop which grows in most agro-ecological zone of Nigeria, right from the coast in the south to the sub-desert areas of the North (Topper et al., 2001). It has tremendous potential as a cash crop to generate foreign exchange, create employment, to curb desertification in the Northern states, while ornamental and alley trees are used to prevent soil erosion by protecting the watersheds and dams in the South.

Inflorescence blight caused by *Lasiodiplodia theobromae* is one of the major limiting factors in cashew production, causing about 40 - 45% crop loss annually (Ohler, 1979; Olunloyo, 1979). The symptoms include withering of petals and other parts of the flower, followed by a progressive dieback of small peduncles from the tips and downward to the main floral shoots (Adejumo, 2005). The disease spreads through insects which create wounds and predisposes inflorescence axes to infection. Among the effective ways of managing this disease are the use of resistant varieties, chemicals and botanicals (Adejumo and Otuonye, 2002; Adejumo, 2005). Cashew is grown as a casual crop by smallholder farmers and as a result its fertilizer requirements are overlooked. The trees are long standing and frequently grown in poor quality soils, which become depleted and results in decline productivity. Yields of trees grown in this way are very much lower than the potential that could be gained if fertilizer was applied (FAO, 2004). Azam-Ali and Judge (2001) reported that potential yield of 10 to 15 kg per tree could be obtained under optimum conditions.

The objective of this experiment was to assess the effect of NPK fertilizers, singly and in combinations on inflorescence blight infection and yield of cashew on acid sand soils of Ochaja, Nigeria.

## MATERIALS AND METHODS

The experiment was carried out in the South West 4 plots occupying 2 ha at Ochaja substation of the Cocoa Research Institute of Nigeria, Ibadan. The cashew cultivars were planted in 1982 and maintained at a plant density of 120 trees/ha. The fertilizers applied were urea (nitrogen source) applied at 0, 500 and 1000 g per tree (or 0, 60 and 12 kg/ha); single super phosphate (phosphorus source) at 0.600 and 1200 g per tree (0, 72, and 144 kg/ha) and muriate of potash (potassium source) at 0, 200 and 400 g per tree (0, 24 and 48 kg/ha) in a 3 x 3 factorial experimental with a total of 27 treatments.

Each treatment combination was replicated 5 times with one tree per treatment. Fertilizer was applied after weeding and cleaning of the base of the individual trees within a 1 - 2 m radius, the fertilizer was broadcast in a circular strip (1 - 1.5 m wide) and about 0.5 - 1.0 m away from the base of the tree. All the treatment combinations

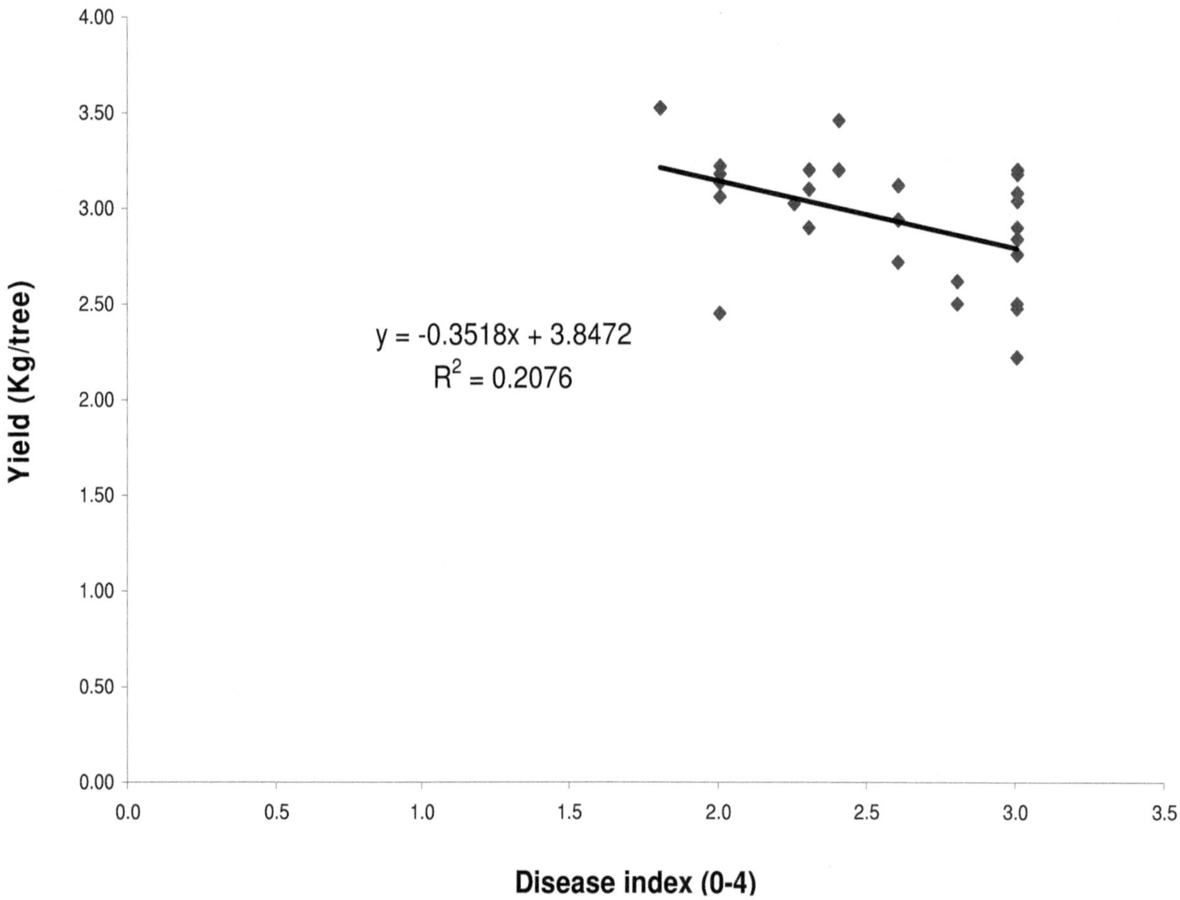


**Figure 1.** Correlation of inflorescence blight with yield for year 1.

were applied during the rainy seasons. Inflorescence infection was assessed on a disease rating scale of 0 to 4, where 0 = no floral infection, 1 = one to one hundred inflorescence panicles infected (1 - 100), 2 = between one hundred and one to one hundred and fifty panicles infected (101 - 150), 3 = between one hundred and fifty one and two hundred inflorescences infected (151 - 200), while 4 = over two hundred and one inflorescence panicles infected.

#### Statistical analysis

The two-year data on disease index and yield were arcsine transformed and analyses of variance (ANOVA) were performed using Statistix 8.1 analytical software. Tukey HSD all-pair wise comparesons test at 5% significance level was used to compare the means. Pearson correlation was used to evaluate the strength of the relationships between the disease index and cashew yield.

## RESULTS

The results in Table 1 show the floral infection intensity and yield in relation to fertilizer application for years 1 and 2. For the first year, the data shows that only single application of potassium as muriate of potash (MOP) at 24 kg/ha showed slight increase in yield over the control, 3.22 kg/tree as against 3.19 kg/tree for control and 100 inflorescence panicles infected over 150 inflorescence panicles for control. The interaction of PK, NK, NP and NPK also showed increase in yield over the control as 3.20, 3.53, 3.20, and 3.46 kg per tree, respectively. Single application of MOP at 24 kg/tree and urea at 60 Kg/ha also showed a decrease in disease infection of inflorescence blight over the control (190 inflorescence panicles infected). The interaction of PK, NP, NK and NPK showed appreciable decrease in disease infection, with the least floral index of 1.8 (90 inflorescence panicles infected) when Urea at 120 kg/ha was combined with MOP at 48 kg/ha. It was observed that the highest yield of 3.53 kg/ha was obtained when the disease infection was the lowest (90 inflorescence panicles infected). Despite that increase in yield and reduction in disease infection of inflorescence blight was observed as a result of the fertilizer application, the differences were not significant.

The correlation coefficient (r) = 0.45 indicates that as cashew yield increases, disease infection decreases. Figure 1 shows the graph of correlation of disease index with yield for year 1 with $R^2$ value of 0.21. This means that 21% of the variance in disease index can be explained by variation in yield and vice versa, and that

**Table 1.** Results of Years 1 and 2 of infection of inflorescence blight disease and yield of cashew.

| No. | Treatments | Year 1 | | Year 2 | |
|---|---|---|---|---|---|
| | | [1]Disease index (0 - 4) | Mean yield (kg/tree) | [1]Disease index (0 - 4) | Mean yield (kg/tree) |
| 1 | $N_0P_0K_0$ | 4.0a | 3.19 | 3.4 ab[3] | 2.00 ab |
| 2 | $N_0P_0K_1$ | 2.0a | 3.22 | 2.8 abcd | 2.00 ab |
| 3 | $N_0P_0K_2$ | 3.0a | 2.84 | 2.4 bcd | 2.20 ab |
| 4 | $N_0P_1K_0$ | 3.0a | 2.76 | 2.6 abcd | 2.50 ab |
| 5 | $N_0P_1K_1$ | 2.3a | 3.20 | 2.6 abcd | 2.50 ab |
| 6 | $N_0P_1K_2$ | 2.0a | 3.18 | 2.2 cd | 1.80 ab |
| 7 | $N_0P_2K_0$ | 3.0a | 3.04 | 2.0 d | 2.30 ab |
| 8 | $N_0P_2K_1$ | 3.0a | 3.08 | 2.4 bcd | 1.70 ab |
| 9 | $N_0P_2K_2$ | 2.8a | 2.50 | 3.6 a | 2.20 ab |
| 10 | $N_1P_0K_0$ | 2.0a | 2.45 | 2.5 abcd | 3.00 ab |
| 11 | $N_1P_0K_1$ | 3.0a | 2.48 | 3.0 abcd | 2.20 ab |
| 12 | $N_1P_0K_2$ | 3.0a | 3.18 | 2.6 abcd | 2.20 ab |
| 13 | $N_1P_1K_0$ | 3.0a | 2.90 | 3.0 abcd | 3.30 a |
| 14 | $N_1P_1K_1$ | 3.0a | 2.22 | 2.8 abcd | 2.30 ab |
| 15 | $N_1P_1K_2$ | 2.3a | 2.90 | 2.3 cd | 2.30 ab |
| 16 | $N_1P_2K_0$ | 2.0a | 3.06 | 2.6 abcd | 1.60 ab |
| 17 | $N_1P_2K_1$ | 2.3a | 3.10 | 2.3 cd | 3.60 a |
| 18 | $N_1P_2K_2$ | 2.0a | 3.13 | 3.3 abc | 2.10 ab |
| 19 | $N_2P_0K_0$ | 3.0a | 2.50 | 2.8 abcd | 2.60 ab |
| 20 | $N_2P_0K_1$ | 2.6a | 2.94 | 2.4 bcd | 2.20 ab |
| 21 | $N_2P_0K_2$ | 1.8a | 3.53 | 2.3 cd | 2.96 ab |
| 22 | $N_2P_1K_0$ | 2.4a | 3.20 | 3.6 a | 2.56 ab |
| 23 | $N_2P_1K_1$ | 2.6a | 2.72 | 2.8 abcd | 1.84 ab |
| 24 | $N_2P_1K_2$ | 2.8a | 2.62 | 2.0 d | 1.04 b |
| 25 | $N_2P_2K_0$ | 2.6a | 3.12 | 2.4 bcd | 1.22 b |
| 26 | $N_2P_2K_1$ | 2.4a | 3.46 | 2.2 cd | 2.44 ab |
| 27 | $N_2P_2K_2$ | 2.3a | 3.02 | 2.3 cd | 2.32 ab |
| CV (%) | | 19.3 | 26.7 | 27.1 | 59.0 |
| Error mean square | | 0.2526 | 0.6231 | 0.5016 | 1.7757 |

[1]Mean disease incidence (0 - 4) in five replications.
[2]Mean Yield (kg/tree) in five replications.
[3] The same letter(s) on each column are not significantly different at $P = 0.05$.
N levels in 3 × 3 factorial experiment: $N_0 = 0$ g, $N_1 = 500$ g, $N_2 = 1000$ g Urea/ tree (0, 60 and 120 Kg/ha).
P levels in 3 × 3 factorial experiment: $P_0 = 0$ g, $P_1 = 600$ g, $P_2 = 1200$ g SSP/ tree (0, 72 and 144 Kg/ha).

21% of the variance is shared between disease index of inflorescence blight and cashew yield. Results for the second year (Table 1) show that single and combined applications of fertilizer had significant effect on both the disease index and cashew yield. Single application of MOP at 48 kg/ha, SSP at both 72 kg/ha and 144 kg/ha, urea at both 60 and 120 kg/ha gave 2.20, 2.50, 2.30, 3.00 and 2.60 kg/tree, respectively.

This observed appreciable increase in yield over the control of 2.00 kg/tree was not significantly different at P = 0.05. However, a significantly different (reduced) floral infection of inflorescence blight (100 inflorescence panicles infected) was observed when SSP was applied singly at 144 kg/ha.

The interaction of PK (yield of 2.50 kg/tree and disease infection of 130 inflorescence panicles infected were obtained when SSP was applied at 72 kg/ha with MOP at 24 kg/ha), NK (urea at 120 kg/ha and MOP at 48 kg/ha), NP (urea at 60 kg/ha and SSP at 72 kg/ha) showed increase in yield and reduction in floral infection of inflorescence blight over the control. The highest yield of 3.60 kg/tree and reduction in floral infection of inflorescence blight was obtained when combined application (NPK) of Urea at 60 kg/ha, SSP at 144 kg/ha and MOP at 24 kg/ha were applied. At this fertilizer level, floral infection of inflorescence blight was significantly reduced to 115 inflorescence panicles infected at P = 0.05. Lowest floral infection of 100 inflorescence panicles infected was observed when SSP was applied singly at 144 kg/ha and combination of Urea at 120 kg/ha, SSP at 72 kg/ha and MOP at 48 kg/ha. However, the latter did not produce a corresponding increase in the yield of cashew.

## DISCUSSION

Applications of nutrients to crops are to avoid plant stress, which may allow crops to better withstand pathogen attack, and to manipulate nutrients to the advantage of the plant and disadvantage of the pathogen (Palti, 1981). It is known that an understanding of disease interactions with each specific nutrient, the effects on the plant, pathogen and the environment can be effectively modified to improve disease control, enhance production efficiency and increase crop quality (Walters and Bingham, 2007).

Soils of the cashew growing areas in Ochaja, Kogi state, Nigeria have low levels of exchangeable bases, nitrogen and phosphorus (Falade, 1978) indicating that possible response to fertilizer by cashew crop was possible. The report in this experiment that urea at 120 kg/ha when combined with MOP at 48 Kg/ha gave the highest yield of 3.53 kg/tree and lowest disease infection (90 inflorescence panicles infected) for the first year agree with the report of Katan (2009) that $NO_3$ nutrition stimulates K uptake and vice versa promoting the synthesis of organic N compounds. The author highlighted that blast severity in rice was low when there was a high K:N ratio in leaf tissue, whereas a low K: N ratio increased the disease.

NPK fertilizer when applied as urea at 60 kg/ha, SSP at 144 Kg/ha and MOP at 24 kg/ha gave the highest yield for the second year and reduced disease infection (125 inflorescence panicles infected) significantly. This might be attributed to a balanced NPK nutrition, especially SSP at high concentrations. Mineral nutrients may reduce the incidence of diseases in certain cases or increase them in others, depending on the particular mineral nutrient, the host plant, the pathogen and other factors. Appropriate management which takes into account fertilizer form and rates, and time and mode of application, has the potential to achieve high crop productivity while reducing the incidence of diseases, or at least avoiding their increase (Katan, 2009). The observation of increased yield and reduced disease infection of single application of urea at 60 and 120 kg/ha and SSP at 72 and 144 kg/ha observed for the second year was not surprising. Nitrogen is the most commonly used fertilizer and is essential for the production of many cellular components (Huber and Thompson, 2007). The rapid rate of nitrification in cultivated soils provides nitrate ($NO_3$) for plant uptake which is internally reduced to amino acids prior to utilization by cells. The two forms of nitrogen absorbed by the plants in either a reduced or an oxidized form can have a profound effect on diseases. The uptake and assimilation of nitrate leads to an increase in pH at the root/soil interface, the rhizosphere, whereas with ammonium ($NH_4$) nutrition, the rhizosphere is acidified. Phosphorus is an essential element of the building blocks of life, the ribonucleic acids (RNA), as well as being required for many additional biochemical and physiological processes including energy transfer, protein metabolism and other functions (Prabhu et al., 2007a). Addition of Phosphorus has been found to have predominantly beneficial effect on disease reduction. Phosphorus supplied in adequate amounts and in the appropriate form and methods of application has been reported to reduce disease incidence of some rusts and other foliar diseases, beneficial in counteracting high levels of nitrogen, suppressed *Fusarium* wilt in tomato at pH 7.0 and 7.5 and decreased severity of *Rhizoctonia* disease in soybean (Katan, 2009). Potassium application has been shown to reduce the incidence of both foliar and soil-borne diseases like *Verticillium* wilt; deficiency also pre-disposes cotton to *Fusarium* wilt (Prabbu et al., 2007b).

It was observed that there were decreased and increased disease infections on different fertilizer levels. This probably might be because of the inadequate methods of application of the fertilizers and limited knowledge of the nature of the interactions between the mineral nutrients and plant disease. Koike et al. (2000) reported that selection and application of fertilizers can influence disease development, while Huber and Thompson (2007) highlighted that contradictions in reports of the effects of mineral nutrients on diseases may result from failure to recognize the different effects of the various forms, rate of application, time of application and soil conditions. Although, excessive use of nitrogen fertilizers can result in leaf growth that is overly succulent and more susceptible to some diseases low yields were observed in the second year compared to the first year. This might be due to imbalanced nutrition, soil moisture and low fertilizer response due to early stoppage of rain during the second year. Tree crops are also known to show appreciable response to fertilizer application only after three years.

Fertilizer application has been used to control plant diseases of vegetables and crops. Woltz and Jones (1973) reported that tomato studies showed better *Fusarium* wilt control when nitrogen was used in soil with high pH. The nitrate nitrogen form also produced the lowest levels of *Fusarium* on chrysanthemum, king asters and carnations (Woltz and Ebgelhard, 1973).

This is the first available report of the use of fertilizers to reduce disease infection of cashew in Nigeria. Adequate crop nutrition makes plants to be more tolerant or resistant to disease, while the nutrient status of the soil, the use of particular fertilizers and amendments can have significant impact on the pathogen's environment (NCAT, 2004). Regular application of major plant nutrients (nitrogen, potassium and phosphorus) is beneficial for healthy trees and increased yields in cashew (Fernandopulle, 2000; FAO, 2004). Further research on the influence of fertilizers on plant diseases in nutrient deficient soils is therefore important; this is because this approach is more environmentally friendly than the use of costly and harmful pesticides to control the disease.

## ACKNOWLEDGEMENT

The author would like to thank Mr. Fred, the officer in charge of Ochaja substation of Cocoa Research Institute of Nigeria, Ibadan for technical assistance.

### REFERENCES

Adejumo TO (2005). Crop protection strategies for major diseases of cocoa, coffee and cashew in Nigeria. Afr. J. Biotechnol. 4(2): 143-150.

Adejumo TO, Otuonye AH (2002). The use of botanicals in the control of inflorescence blight disease of cashew, *Anacardium occidentale*. Niger. J. Sci., 36(1): 75-80.

Azam-Ali SH, Judge EC (2001). Small-scale cashew nut processing. FAO, Rome. p. 80 www.fao.org/inpho/content/documents/.../ac306e05.htm.

Falade JA (1978). Soils of cashew growing areas of Nigeria. East Afr. J. Agric., 2: 31-36.

FAO, Azam-Ali SH, Judge EC (2004). Small-scale cashew nut processing. (ed.) Publishing Management Service, Information Division, FAO, Viale delle Terme di Caracalla, 00100 Rome, Italy. FAO. http://www.fao.org/inpho/content/documents/vlibrary/ac306e/ac306e00.htm

Fernandopulle MN (2000). Fertilizer use in cashew. Cashew Bull., 3(1): 1-6.

Huber DM, Thompson IA (2007). Nitrogen and plant disease. In: Datnoff, Elmer and Huber (eds.). Mineral Nutrition and Plant Disease. APS Press, St. Paul, MN, pp. 31-44

Katan J (2009). Nutrient Management and Plant Disease. *e-ifc* 21: 6-8.

Koike ST, Gaskell M, Fouche C, Smith R, Mitchell J (2000). Plant disease management for organic crops. UC Cooperative; University of California, Division of Agriculture and Natural Resources, Publication (7252): 6.

National Center for Appropriate Technology, (NCAT) (2004). Sustainable Management of Soil-borne Plant Diseases. Preston Sullivan (ed.). Soil Systems Guide. Slot 131, Version 071604. http://www.attra.org/attra-pub/soilborne.html, 173: 16.

Ohler JG (1979). Cashew. Koninklijk Instituut voor de Tropen: Amsterdam, Netherlands p. 260.

Olunloyo AO (1979). The role of fruit flies in rot disease of cashew apples in the plantation. Ann. Rep, Cocoa Res. Inst. Niger. 1978/79: 101-103.

Palti J (1981). Cultural Practices and Plant Diseases. Springer-Verlag, Berlin.

Prabhu AS, Fageria ND, Berni RF, Rodrigues FA (2007a). Phosphorous and plant disease., In: Datnoff, Elmer and Huber (eds.). Mineral Nutrition and Plant Disease. APS Press, St. Paul, MN pp. 45-55.

Prabhu AS, Fageria ND, Huber DM, Rodrigues FA (2007b). Potassium and plant disease. In: Datnoff, Elmer and Huber (eds.). Mineral Nutrition and Plant Disease. APS Press, St. Paul, MN, pp. 57-78.

Topper CP, Caligari PDS, Camara M, Diaora S, Djaha A, Coulibay F, Asante AK, Boamah A, Ayodele EA, Adebola PO (2001). West African Regional Cashew Survey Report (Guinea, Guinea Bissau, Cote d'Ivore, Ghana and Nig.). Sustainable Tree Crop Programme (STCP) and Biohybrids Agrisystem Ltd. UK. 1: 10.

Woltz SS, Jones JP (1973). Tomato Fusarium wilt control by adjustments in soil fertility. Proceedings Florida State Hortic. Soc., 86: 157-159.

Woltz SS, Ebgelhard AW (1973). Fusarium wilt of chrysanthemum: effect of Nitrogen source and lime on disease development. Phytopathol., 63: 155-157.

# Design, construction and performance evaluation of solar maize dryer

**Ahmed Abed Gatea**

Department of Agricultural mechanization, College of Agriculture, University of Baghdad, Iraq.
E-mail: ahmedabd192000@yahoo.com.

**This paper presents the design, construction and performance evaluation of solar drying for maize, the solar drying system consists of V-groove collector of 2.04 m² area, drying chamber and blower. It was designed in such a way that solar radiation is not incident directly on the maize. K-type thermocouples were used for temperature measurement, while solar radiation was measured by solar meter m od.776. The thermal energy and heat losses from solar collector were calculated for each three tilt angles (30º, 45º, 60º). The results obtained during the test period denoted that the maximum gained energy occurred at 11 o'clock hour and then gradually declined since the maximum solar radiation occurred at this time. The performance of the solar drying system was highly dependent on the solar radiation, tilt angle and ambient temperature. The total loss factor of the collector increases with the increase in the intensity of solar radiation intensity. The theoretical thermal energy, the experimental actual heats gain increase by increasing radiation intensity.**

**Key words:** Solar energy, solar air collector, v-corrugated collector, solar drying system performance study.

## INTRODUCTION

In many parts of the world there is a growing awareness that renewable energy have an important role to play in extending technology to the farmer in developing countries to increase their productivity (Waewsak et al., 2006). Solar thermal technology is a technology that is rapidly gaining acceptance as an energy saving measure in agriculture application. It is preferred to other alternative sources of energy such as wind and shale, because it is abundant, inexhaustible and non-polluting, wind (Akinola, 1999; Akinola and Fapetu, 2006; Akinola et al., 2006).

Solar air heaters are simple devices to heat air by utilizing solar energy and it is employed in many applications requiring low to moderate temperature below 80°C, such as crop drying and space heating (Kurtbas and Turgut, 2006). They are defined as a process of moisture removal due to simultaneous heat and mass transfer (Ertekin and Yaldiz, 2004). According to Ikejiofor (1985) two types of water are present in food items; the chemically bound water and the physically held water.

In drying, it is only the physically held water that is removed. The most important reasons for the popularity of dried products are longer shelf-life, product diversity as well as substantial volume reduction. This could be expanded further with improvements in product quality and process applications. The application of dryers in developing countries can reduce post harvest losses and significantly contribute to the availability of food in these countries.

Estimations of these losses are generally cited to be of the order of 40% but they can, under very adverse conditions, be nearly as high as 80%. A significant percentage of these losses are related to improper and/or untimely drying of foodstuffs such as cereal grains, pulses, tubers, meat, fish, etc. (Bassey, 1989; Togrul and Pehlivan, 2004). Traditional drying, which is frequently done on the ground in the open air, is the most widespread method used in developing countries because it is the simplest and cheapest method of conserving foodstuffs.

Some disadvantages of open air drying are: exposure of the foodstuff to rain and dust; uncontrolled drying; exposure to direct sunlight which is undesirable for some foodstuffs; infestation by insects; effect by animals; etc (Madhlopa et al., 2002). Solar drying may be classified into direct, indirect and mixed-modes. In direct solar dryers the air heater contains the grains and solar energy passes through a transparent cover and is absorbed by the grains. Essentially, the heat required for drying is provided by radiation to the upper layers and subsequent conduction into the grain bed.

In indirect dryers, solar energy is collected in a separate solar collector (air heater) and the heated air then passes through the grain bed, while in the mixed-mode type of dryer, the heated air from a separate solar collector is passed through a grain bed and at the same time, the drying cabinet absorbs solar energy directly through the transparent walls or roof. Therefore, the objective of this study is to develop a mixed-mode solar dryer in which the grains are dried simultaneously by both direct radiation through the transparent walls and roof of the cabinet and by the heated air from the solar collector. The performance of the dryer was also evaluated.

## THEORY, MATERIALS AND METHODS

### Basic theory

The energy balance on the absorber is obtained by equating the total heat gained to the total heat lost by the heat absorber of the solar collector. Therefore,

$$IA_c = Q_u + Q_{cond} + Q_{conv} + Q_R + Q_p \quad (1)$$

Where

I = Rate of total radiation incident on the absorber's surface ($Wm^{-2}$).
$A_c$ = Collector area ($m^2$).
$Q_u$ = Rate of useful energy collected by the air (W).
$Q_{cond}$ = Rate of conduction losses from the absorber (W).
$Q_{conv}$ = Rate of convective losses from the absorber (W).
$Q_R$ = Rate of long wave re-radiation from the absorber (W).
$Q_p$ = Rate of reflection losses from the absorber (W).

The three heat loss terms Qcond, Qconv and QR are usually combined into one-term (QL), that is,

$$Q_L = Q_{cond} + Q_{conv} + Q_R \quad (2)$$

$$IA_c = \tau I_T A_c \quad (3)$$

The reflected energy from the absorber is given by the expression:

$$Q_p = \rho \tau I_T A_c \quad (4)$$

where

ρ = Reflection coefficient of the absorber.
τ = Transmittance of cover.

Substitution of Equations (2), (3) and (4) in Equation (1) yields

or

$$\tau I_T A_c = Q_u + Q_L + \rho \tau I_T A_c$$

or

$$Q_u = \tau I_T A_c (1 - \rho) - Q_I$$

For an absorber (1 – ρ) = α and hence

$$Q_u = (\alpha \tau) I_T A_c - Q_I \quad (5)$$

The total heat loss coefficient is given by the following Eq. (6) (Karim and Hawlader, 2006)

$$U_L = U_C + U_t \quad (6)$$

$$Ut = \left( \frac{N}{\frac{C}{Tp}\left(\frac{TP - Ta}{N + f}\right)^e} + \frac{1}{hw} \right)^{-1} + \frac{\sigma (Tp^2 + Ta^2)(Tp + Ta)}{\frac{1}{d} + \frac{2N + f - 1}{\varepsilon g} - N}$$

$$C = 520(1 - 0.000051\beta^2$$
$$d = \varepsilon p + 0.0425N(1 - \varepsilon p)$$
$$f = (1 + 0.089hw - 0.1166hw\, \varepsilon_p)(1 + 0.07866N)$$
$$hw = 5.7 + 3.8u$$
$$e = 0.252$$
$$U_b = \frac{k_{ins}}{L_{ins}}$$

QL composed of convection conduction and radiation parts. It is presented in the following form ;

$$Q_I = U_I A_c (T_c - T_a) \quad (7)$$

Where

$U_L$ =Overall heat transfer coefficient of the absorber ($Wm^{-2}K^{-1}$).
Tc = Temperature of the collector's absorber (K).
Ta = Ambient air temperature (K).
From Eqs. (5) and (7) the useful energy gained by the collector is expressed as:

$$Q_u = (\alpha \tau) I_T A_c - U_I A_c (T_c - T_a) \quad (8)$$

Where

Absorption = α
Therefore, the energy per unit area (qu) of the collector is:

$$q_u = (\alpha \tau) I_T - U_I (T_c - T_a). \quad (9)$$

If the heated air leaving the collector is at collector temperature, the heat gained by the air Qg is:

$$Q_g = m' cp (T_c - T_a) \quad (10)$$

$\dot{m}$ = Mass of air leaving the dryer per unit time (kg s– 1).
Cp = Specific heat capacity of air ($kJkg^{-1}K^{-1}$).

The collector heat removal factor, FR, is the quantity that relates the actual useful energy gained of a collector, Eq. (8), to the useful

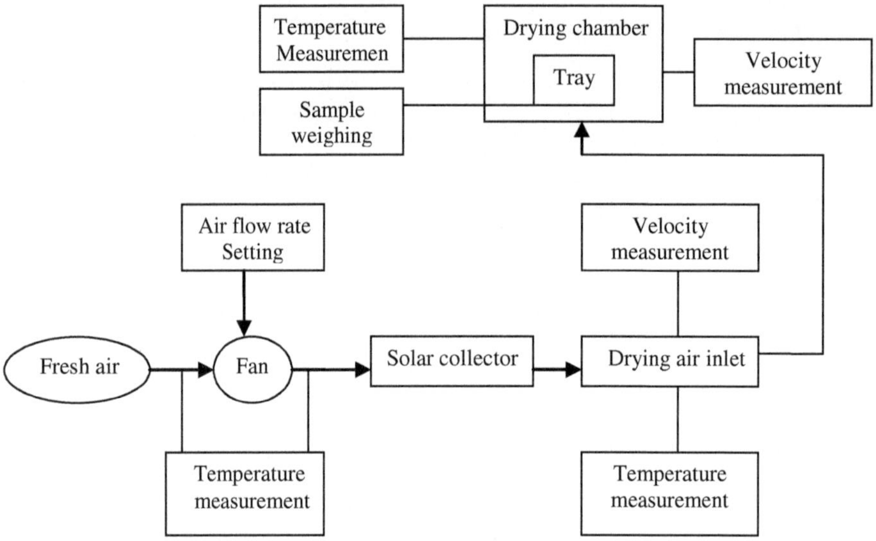


**Figure 1.** Flow diagram of drying process.

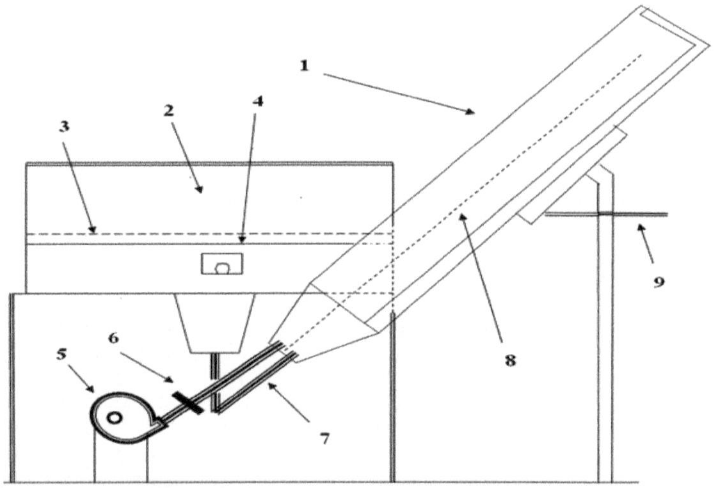


**Figure 2.** Section of the solar drying system: 1. Solar collector. 2. Drying chamber. 3. Drying tray. 4. Thermostat temperature. 5. Air blower. 6. Air valve. 7. Connecting pipes. 8. Absorption plates of two air passes. 9. Slide rule.

energy gained by the air, Eq. (10). Therefore

$$F_R = \frac{m' c_{pa} (T_c - T_a)}{A_c[(\alpha\tau) I_T - U_L (T_c - T_a)]} \tag{11}$$

$$Q_g = A_c F_R \left[ (\alpha\tau) I_T - U_L A_c (T_c - T_a) \right] \tag{12}$$

$Q_g$ = The useful energy gained.

The thermal efficiency of the collector is defined as (Itodo et al., 2002):

$$\eta_c = \frac{Q_g}{A_c I_T} \tag{13}$$

**The solar drying system**

The solar drying system, which includes collector, drying chamber and the blower, the use of V-grooved absorbers improves the heat transfer coefficient between the absorber plate and the airs as well as it increase the absorptivity to solar radiation (Goel et al., 1987; Chaudhory et al., 1988; Hafiz et al., 1999). It is made of galvanized steel with black paint containing (5%) black chromium powder to increase its absorbing capability (Figures 1 and 2).

The collector was insulated with rock wool of 10 mm thickness

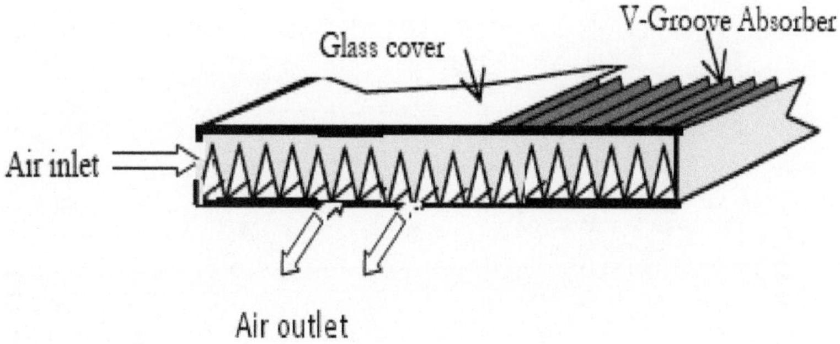


**Figure 3.** Schematic of the V-groove absorber collector.

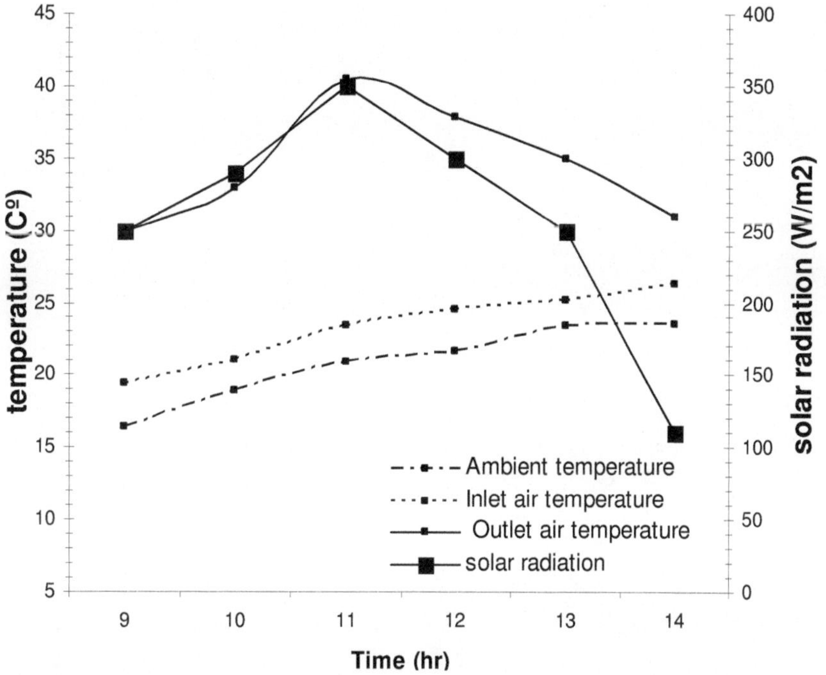


**Figure 4.** Time variation of air ambient temperature, air inside, outside of the collector and solar radiation for angle 30º (Date 21/11/2008).

from the bottom and 5 mm thickness from sides. The collector was contained within galvanized steel frame. The drying chamber used in this work was [1.06 x 0.66 x 0.56 m] [width, depth and height] respectively. All the system components are connected with each other by plastic pipe 63 mm diameter, as well as to transport the air between the collector and the drying chamber (Figure 3).

## EXPERIMENTAL SET-UP

The performance of the solar drying system was evaluated experimentally using three collector tilt angles for horizontal (30º, 45º, 60º). The following parameters are measured (a) radiation incident on the collector, (b) air temperatures at various locations in the collector and dryer and (c) relative humidity of air.

To measure the temperature of air at various locations of the collector and dryer, K-type thermo-couples were installed at various points along the length and with of the solar dryer system. All temperature data were registered at an interval of 15 min. Drying test was started at 9:00 h and stopped at 14:00 h. Preliminary calculations that to be dried at (38 kg) of maize , After field drying the moisture content of the maize 21% (wet basis) and should be reduced to 13% during drying.

## RESULTS AND DISCUSION

### Performance of the solar drying system

Solar drying performance was compared between ambient temperatures for the period of experimentation. The performance of the solar drying was highly dependent on the solar radiation and ambient temperature. Figure 4 shows each change of air ambient

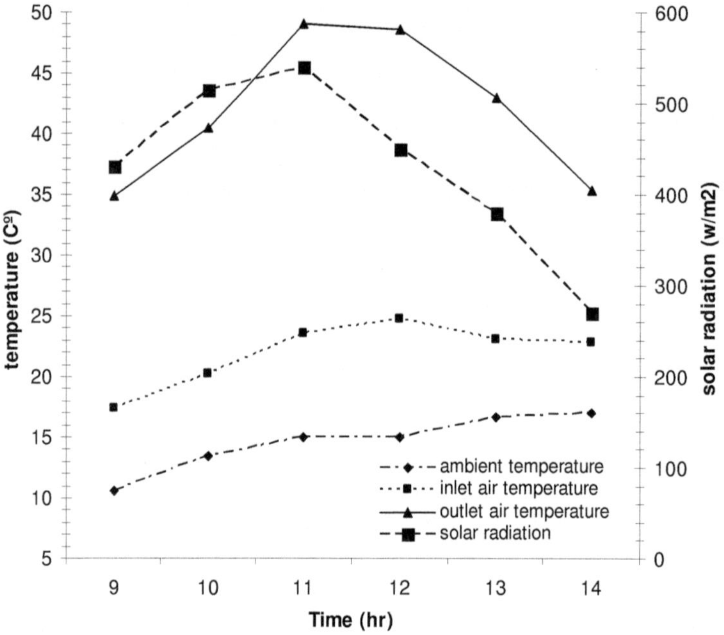


**Figure 5.** Time variation of air ambient temperature, air inside, outside of the collector solar and solar radiation for angle 45º (Date 22/11/2008).

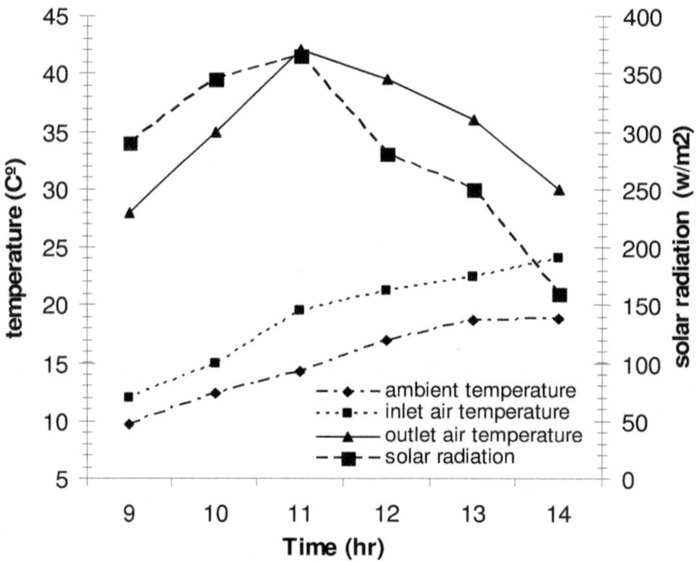


**Figure 6.** Air ambient temperature, air inside, outside of the collector solar and solar radiation for angle 60º (Date 23/11/2008).

temperature and the air inside and outside of the collector solar during daylight hours for the inclination angle of 30º. It was the highest temperature out of the collector (40.5°C) of the whole at the 11 o'clock hour at the highest intensity of radiation 350 w/m². 

Figure 5 shows the change of the temperature of the ambient air and the temperature of the air inlet and outlet of the collector solar during daylight hours for the inclination angle of 45°. It was the highest temperature out the collector 49°C of the whole at the 11 o'clock hour at the highest intensity of radiation 540 w/m². The highest temperature out of the collector at inclination angle of 60º (42ºC) at the highest intensity of radiation 365 w/m², of the whole at the 11 o'clock hour (Figure 6).

Figure 7 shows the relation between the theoretical thermal energy and the experimentally actual heat gain

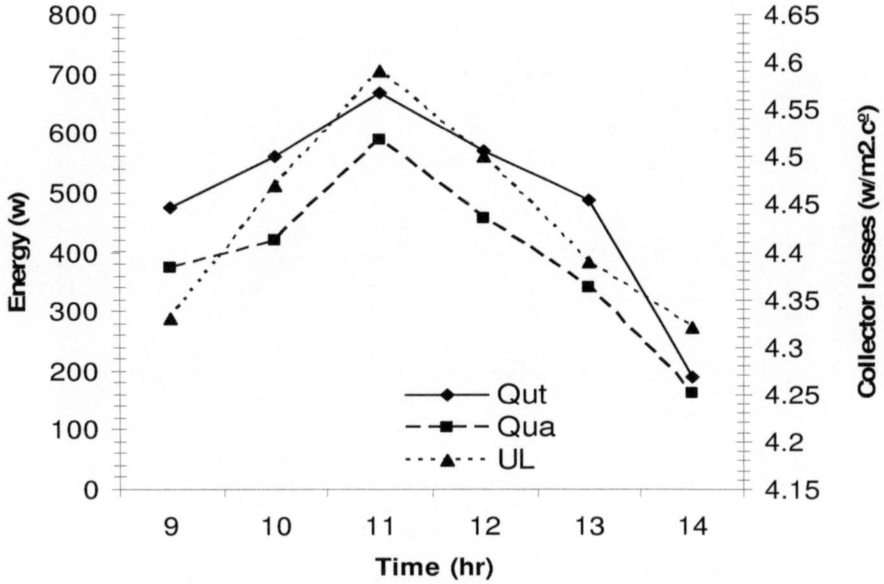


**Figure 7.** The theoretical thermal energy, the experimentally actual heat gain and the total loss factor of the collector versus the daylight hours for the inclination angle of 30º (Date 21/11/2008).

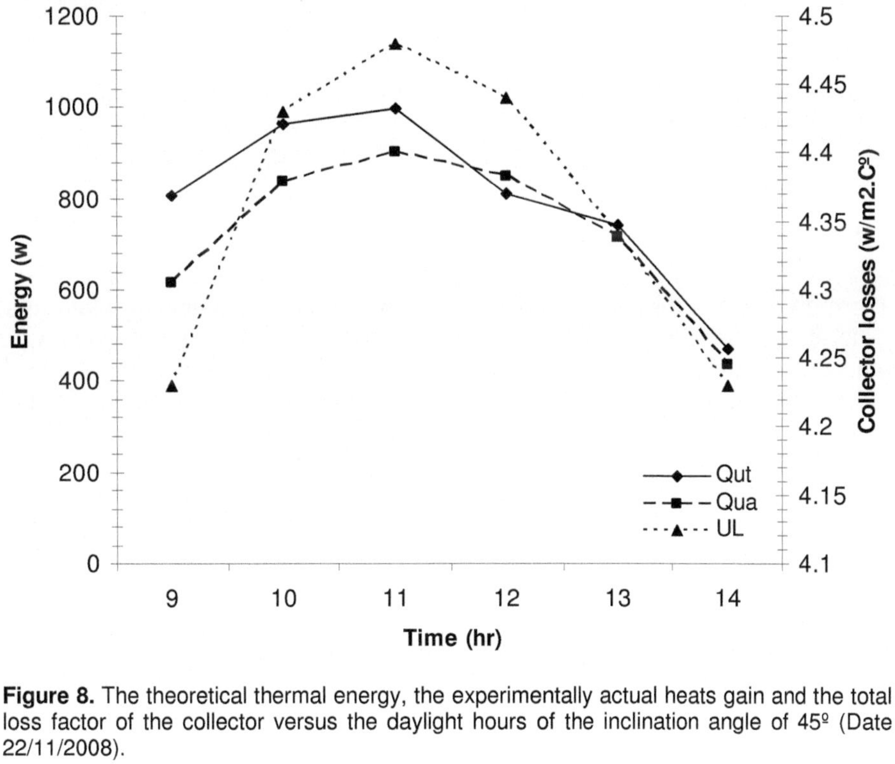


**Figure 8.** The theoretical thermal energy, the experimentally actual heats gain and the total loss factor of the collector versus the daylight hours of the inclination angle of 45º (Date 22/11/2008).

for the inclination angle 30º, the highest values 669.6 and 589.4 w for the theoretical thermal energy and the experimentally actual heat gain respectively at 11 am and then gradually declined. Theoretical energy and actual heat gain (which were calculated through FORTRAN 99 program computer) depend on the intensity of radiation. We also note that the total loss factor of the collector is increasing with the increase in solar radiation, the highest value (4.59 w/m².Cº) of these losses occurred at 11 am. From Figure 8, the highest values of theoretical thermal

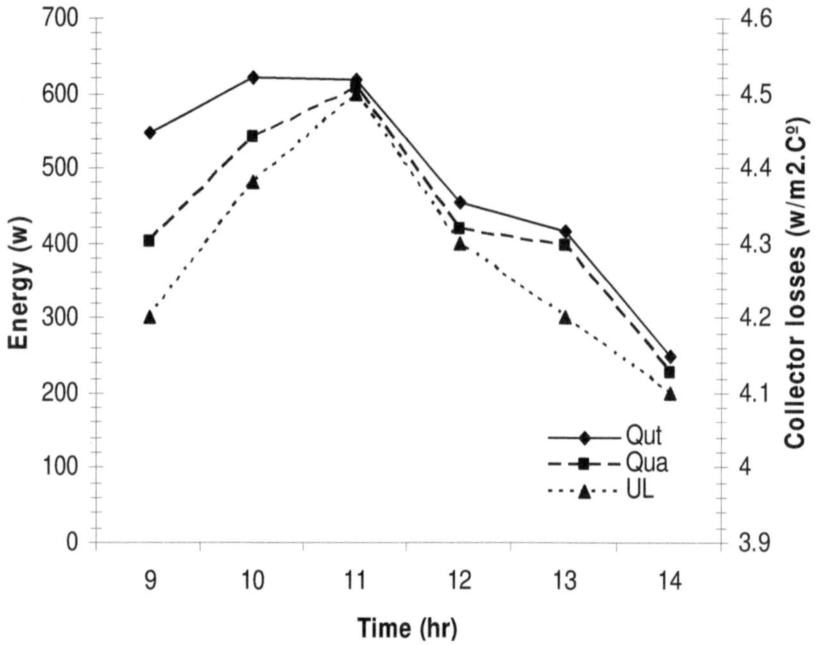


**Figure 9.** The theoretical thermal energy, the experimentally actual heats gain and the total loss factor of the collector versus the daylight hours of the inclination angle of 60º (Date 23/11/2008).

energy, experimentally actual heat gain and total loss factor of the collector at the inclination collector tilt -45º at the 11 am are (996.4, 900.7 w and 4.48 w/m².Cº) respectively (Figure 9).

## Conclusion

1. The theoretical thermal energy, the experimentally actual heats gain increase by increasing radiation intensity, the maximum values occurred at the 11 am and then gradually declined.
2. The energy gained obtained at the angle tilt 45º is higher than the corresponding values obtained at 60º, 30º tilt.
3. The total loss factor of the collector increases with the increase in the intensity of solar radiation intensity, as well as this factor depends on the convection coefficient for the air between the top glass cover and environment.
4. The performance of the solar drying system was highly dependent on the solar radiation and ambient temperature.

**Abbreviation: IT,** Radiation on tilted surface (W/m²); $\mathbf{A_c}$, collector area ($m^2$); $\mathbf{Q_u}$, rate of useful energy collected by the air (W); $\mathbf{Q_{cond}}$, rate of conduction losses from the absorber; $\mathbf{Q_{conv}}$, rate of convective losses from the absorber (W); $\mathbf{Q_R}$, rate of long wave re-radiation from the absorber (W); $\mathbf{Q_\rho}$, rate of reflection losses from the absorber (W); $\mathbf{U_L}$, overall heat transfer coefficient of the absorber ($Wm^{-2}K^{-1}$); $\mathbf{U_T}$, top loss coefficient, ($Wm^{-2}K^{-1}$); $\mathbf{U_B}$, bottom loss coefficient, ($Wm^{-2}K^{-1}$); $\mathbf{T_c}$, temperature of the collector's absorber (K); $\mathbf{T_a}$, ambient air temperature (K); $\dot{m}$, mass of air leaving the dryer per unit time ($kgs^{-1}$); $\mathbf{C_p}$, specific heat capacity of air ($kJkg^{-1}K^{-1}$); **mw,** mass of water evaporated from the food item (kg); **hfg,** latent heat of evaporation, kJ/kg H2O; **ma,** mass of drying air (kg); **hW,** wind induced heat transfer coefficient, W/m ² .K; **k,** thermal conductivity of plate material, W/m .K; **kINS,** thermal conductivity of rock wool insulation (= 0.04), W/m. K; **LINS,** thickness of insulation (=0.05), m; $\mathbf{T_{pr}}$, product temperature ºC; **T,** temperature ºC; **Ta,** ambient temperature, K; $\mathbf{T_p}$, plate temperature, K; **v,** wind speed, m/s; **N,** number of glazing (= 1); $\mathbf{T_1}$ **and** $\mathbf{T_2}$, initial and final temperatures of the drying air respectively (K); $\mathbf{m_i}$, initial mass of the food item (kg); $\mathbf{M_e}$, equilibrium moisture content (% dry basis); $\mathbf{M_i}$, initial moisture content (% dry basis); **α,** Absorption; **τ,** transmittance of cover; **ρ,** reflection coefficient of the absorber; **η,** thermal efficiency, %; **σ,** Stefan-Boltzmann constant, W/m²

## FORTRAN 99 program computer

```
! Computer program for calculation overall heat loss coefficient
! From solar air collector ,useful energy and efficiency

Open (1,file = 'as.txt')
open (2,file = 're.txt')
Do 10 I = 1, 6
```

```
Read (1,*)ti,ta,tp,ai,v,to
 tp = tp+273
 ti = ti+273
 ta = ta+273
 to = to+273
 hw = 5.7+(3.8*v)
 f = ((9./hw)-(30./((hw)**2)))*(ta/316.9)*1.091
 w = .00000005667
 u1 = 1./(434.61/tp)*((tp-ta)/(1.+f))**0.252
 u2 = 1./(u1+(1./hw))
 u3 = w*(tp**2+ta**2)*(tp+ta)
 u4 = 1.24+(1.13*f)
 u = u +(u3/u4)
 ui = u+0.4
 write (*,'(1x,5f10.4)')u,ui
 write (2,'(1x,5f10.4)')u,ui
 ru = 348.3/tp
 b = 1.02
 dh = 0.1125
 y = 0.0000179
 pr = 0.71
 Cp = 1006
 ag = 0.01485
 uii = ui/3600
 ac = 2.04
 re = (ru*b*dh)/y
 nu = (0.0192*(re**(0.75))*pr)/(1.+(1.22*(re**(-.125))*(pr-
2.)))
 hc = (nu*cp*y)/(pr*dh)
 fd = hc/(hc+ui*0.5)
 fr = ((ag*cp)/uii)*(1.- exp((-fd*uii)/(cp*ag)))
 qi = ai*0.817-(ui*(ti-ta))
 q = 2.04*fr*qi*1.75
 eff = (q/(ai*2*ac))
 ruu = 348.3/((to+ti)/2.)
 w1 = 0.025*ruu
 qu = w1*cp*(to-ti)
 effu = qu/(ai*2*ac)
 tm = to-ti
 write (*,'(1x,5f10.4)')ru,re,hc,fd,fr,qi,q,eff,qu,effu,tm
 write (2,'(1x,5f10.4)')ru,re,hc,fd,fr,qi,q,eff,qu,effu,tm
 10 read*
 !9 format (1x,2f9.4,1i5,6f9.4)
 stop
 end.
```

**REFERENCES**


Audsley E, Wheeler J (1978). The annual cost of machinery calculated using actual cash flows. J. Agric. Eng. Res. 23: 189-201

Akinola AO (1999). Development and Performance Evaluation of a Mixed-Mode Solar Food Dryer. M. Eng. Thesis, Federal University of Technology, Akure, Nigeria.

Akinola AO,Fapetu OP (2006). Exergetic Analysis of a Mixed-Mode Solar Dryer. J. Eng. Appl. Sci. 1: 205-210.

Akinola, OA, Akinyemi AA, Bolaji BO (2006). Evaluation of traditional and solar fish drying systems towards enhancing fish storage and preservation in Nigeria. J. Fish. Int. Pakistan 1(3-4): 44-49.

Bassey MW (1989). Development and use of solar drying technologies, Niger. J. Solar Energy 89: 133-164.

Choudhury C, Anderson SL, Rekstad J (1988). A Solar Air Heater for Low Temperature Applications, Solar Energy 40: 335–344.

Ertekin C, Yaldiz O (2004). Drying of eggplant and selection of a suitable thin layer drying model, J. Food Engr. 63: 349-359.

Fisk MJ, Anderson, HC (1982). Introduction to solar technology, Addison-Wesley,Company Inc. 23: 2269

Goel VK, Ram Chandra, Raychaudhuri BC (1987). Experimental Investigations on Single-Absorber Solar Air Heaters. Energy Convers. Manage. 27: 343-349.

Hafiz MR, Othman MY, Salleh MM, Sopian K (1999). Design and Indoor Testing of the V-Groove Back Pass Solar Collector. Renewable Energy 16: 2118-2122.

Itodo IN, Obetta SE, Satimehin AA (2002). Evaluation of a solar crop dryer for rural applications in Nigeria. Botswana J. Technol. 11(2): 58-62.

Ikejiofor ID (1985). Passive Solar Cabinet Dryer For Drying Agricultural Products. In: O. Awe (Editor), African Union of Physics. Proc. Workshop Phys. Tech. Solar Energy Convers. Univ. of Ibadan, Nigeria pp. 157-165.

Kurtbas I, Turgut E (2006). Experimental investigation of solar air heater with free and fixed fins: efficiency and energy loss. Int. J. Sci. Technol. 1: 75-82.

Karim MA , Hawlader MN (2006). Performance evaluation of a v-groove solar air collector for drying applications. Appl. Thermal Eng. 26: 121-130.

Madhlopa A, Jones SA, Kalenga-Saka JD (2002). A solar air heater with composite absorber systems for food dehydration. Renewable Energy 27: 27–37.

Togrul IT, Pehlivan D (2004). Modelling of thin layer drying kinetics of some fruits under open-air sun drying process. J. Food Engr. 65: 413-425.

Waewsak J, Chindaruksa S, Punlek C (2006). A mathematical modeling study of hot air drying for some agricultural products. Thammasat. Int. J. Sci. Technol. 11: 14-20.

Youcef-Ali S, Messaoudi H, Desmons JY, Abene A, Le Ray M (2001). Determination of the average coefficient of internal moisture transfer during the drying of a thin bed of potato slices. J. Food Engr. 48: 95-101.

# Partial purification of leaf lectin from *Manihot esculenta* and screening its fungicidal activity

**Maria Cristina Silva, Angelita Duarte Corrêa*, Custódio Donizete dos Santos, Flávia Cristina Almeida Marcos and Celeste Maria Patto de Abreu**

Departamento de Química, Universidade Federal de Lavras, 37200-000, Lavras, MG, Brazil.

**Lectins have various biotechnological applications including their use in agriculture as antimicrobials and pesticides. Lectin from manioc leaf powder has been extracted, partially purified and evaluated in terms of its haemagglutination activity and fungicidal properties against *Fusarium oxysporum*. Ground, dried and the entire leaves of *Manihot esculenta* (Cacao cultivar) were extracted with distilled water and the protein extract was concentrated by precipitation in 80% saturated ammonium sulphate solution. The concentrated protein solution was partially purified on a Sepharose column and the unbound protein fraction was submitted to native and denaturing polyacrylamide gel electrophoresis. Lectin from manioc leaf powder exhibited a haemagglutination (HU) activity of 0.5 HU/100 µl and was stable over a wide temperatures range (up to 70°C). Four protein bands (of 13, 58, 61, 64 kDa) were identified in the unbound protein fraction obtained by column chromatography, and the presence of a lectin was demonstrated by the determination of an HA value of 0.5 HU/100 µl. The concentrated protein solution was not, however, active against *F. oxysporum* at concentrations up to 100 µg/ml.**

**Key words:** *Manihot esculenta,* Euphorbiaceae, manioc, fungicidal activity, haemagglutination activity, lectin.

## INTRODUCTION

Lectins are proteins that bind reversibly to carbohydrates without inducing chemical modifications in the covalent structure of the glycosyl ligands. Such proteins exhibit high levels of specificity and affinity towards the sugar moieties of the glycoconjugates, but do not possess enzymatic activity (Peumans and Van Damme, 1995; Gabius et al., 2002; Loris, 2002).

The interaction between a lectin and a sugar moiety occurs at the carbohydrate-recognition domain and involves Van der Waals forces and the formation of hydrogen bonds between sugar hydroxyls and amino acid residues at the active site of the lectin (Weis and Drickamer, 1996). The cross-linkages formed between adjacent cells through binding between lectins and glycoconjugates causes cell agglutination (Peumans and Van Damme, 1995). When erythrocytes are involved in this type of interaction, the phenomenon of haemagglutination occurs (Alonso et al., 2001).

Ingestion of lectins by animals may cause anti-nutritional effects, degeneration of cell membranes and inhibition of digestive enzymes (Vasconcelos and Oliveira, 2004), and the resulting interference with the absorption of nutrients may give rise to serious physiological consequences. On the other hand, the strong sugar specificity of lectins has led to their exploitation in many biological and medicinal areas, and they have found application as fungicidal, bactericidal and insecticidal agents in agriculture (Peumans et al., 2000; Grahamstown and Van Staden, 2002), in the purification and characterisation of polysaccharides and in the glycoconjugates industry (Lima et al., 1997). Furthermore, it is used in the identification of blood groups (Lis and Sharon, 1998), as stimulators of lymphocyte mitogenesis and immunological functions (Sharon and Lis, 2004), in the agglutination of cancer cells and in studies of oncogenesis (Sharon and Lis,

*Corresponding author. E-mail: crisiria@yahoo.com.br.

**Abbreviations: MLP,** Lectin from manioc leaf powder; **HA,** haemagglutination activity.

2004), and in the investigations concerning the structures of proteins and carbohydrates in cells (Silva and Silva, 2000).

The use of synthetic chemicals in the control of agricultural pests is not only costly, but may also be environmentally unacceptable in terms of pollution and the induction of pathogen resistance. Application of natural insecticides and fungicides could provide a cheap and environmental friendly alternative (Macedo et al., 2000). Within this context, the leaves of manioc (*M. esculenta* Crantz) represent a rich source of lectins that could be used in the protection of crops against diseases. Currently, manioc leaves have no economical value and is normally treated as agricultural waste.

The aim of the present study was, therefore, to extract and partially purify the lectins from manioc leaf powder (MLP) in order to evaluate their haemagglutination activity (HA) and to test their fungicidal activity on *F. oxysporum*, the causal agent of a wilt disease that attacks a variety of commercial crops. The possibility of using manioc leaf lectin as a fungicidal agent could transform an abundant waste material into a valuable commodity.

## MATERIALS AND METHODS

### Preparation of MLP

Mature aerial parts of manioc (Cacao cultivar) were collected from 12 month old plants grown in Fazenda Rio Grande, Lavras, MG, Brasil. The whole leaves and petioles were washed with tap water and dehydrated in a fan-assisted oven at 30 to 35°C for 24 h, following which the petioles were removed and the leaves were dried further for 24 h. Dried leaves were ground to a powder (MLP) using a Willy type mill (Tecnal, Piracicaba, Brasil).

### Extraction and concentration of protein

A 25 g portion of MLP was extracted with distilled water (1:20; w/v) under constant stirring at room temperature for 15 min. The suspension was filtered through muslin cloth and centrifuged at 8000 x *g* for 15 min at 7°C. The supernatant was removed and the residue submitted to two successive extractions under the conditions as described. The crude MLP extracts were combined, cooled in an ice bath and ammonium sulphate was added under constant stirring, and pH control until 80% saturation was achieved (Cooper, 1942; Pereira et al., 2008). The mixture was maintained in the refrigerator for 24 h and subsequently centrifuged at 8000 x *g* for 15 min at 7°C. The residue was resuspended in 0.15 M NaCl solution and dialysed against the same solution. All procedures were carried out in triplicate, and the concentrated protein solution was submitted to further purification.

### Assessment of the protein concentration, haemagglutination activity and thermostability

The protein concentration, HA and thermostability of the crude MLP extract and of the concentrated protein solution were estimated according to standard procedures. All assays were performed in triplicate. Total protein content was determined using the method of Bradford (1976) with bovine serum albumin (BSA) as standard. The HA assay was performed according to Calderón de la Barca et al. (1985). Briefly, aliquots (100 µL) of buffered saline solution were added to each well of a 96-well microtiter plate, and a 100 µl of the sample was transferred to the first well. Serial dilutions (base 2) were prepared by transferring a 100 µl aliquot of a more concentrated sample to an adjacent well containing diluent alone, it was mixed and dilution procedure continued. An aliquot (100 µl) of 2% suspension of erythrocytes prepared from human blood (type A, Rh +) was added to each well and the microtiter plate was maintained at room temperature. HA was determined visually at 60 and 90 min, and the results were expressed in haemagglutination units (HU), defined as the reciprocal of the maximum dilution presenting visible agglutination of red blood cells (that is, if the 1/4 titre represented the maximum dilution exhibiting agglutination and the volume of sample assayed was 100 µl, then the HA value would be 4 HU/100 µl). Negative control samples consisted of buffered saline solution and erythrocyte suspension only.

The thermostability of MLP lectin was assessed by submitting aliquots (1 ml) of the concentrated protein solution to temperatures in the range 40 to 90°C for 30 min, cooled in an ice bath and the HA and protein concentration of the samples was determined.

### Partial purification of MLP lectin

Chromatographic procedures were conducted in a refrigerator at 4°C. A glass column (40 x 2 cm) was packed with Sepharose CL-4B resin (GE Healthcare, Chalfont St. Giles, UK) according to the manufacturer's instructions and equilibrated with 0.15 M NaCl until the $A_{280}$ of the effluent was < 0.02. Subsequently, a 30 ml sample of the concentrated protein solution (containing 5.84 mg of protein) was loaded onto the column and eluted with 0.15 M NaCl at a constant flow rate of 1 ml/min with the aid of a peristaltic pump (Amersham Pharmacia Biotech, Piscataway, EUA). Unbound proteins fractions (2.5 ml) were collected, monitored at 280 nm, and those showing absorption at this wavelength were combined. The protein content was determined and, following lyophilisation, an aliquot was submitted to polyacrylamide gel electrophoreses (PAGE) and sodium dodecyl sulphate (SDS)-PAGE. The column was finally eluted with 0.15 M NaCl containing 0.1 M galactose in order to elute any proteins that remained bound to the Sepharose.

### Gel electrophoresis

Native PAGE was performed according to the method of Laemmli (1970) and Robyt and White (1987) with a stacking gel of 5% polyacrylamide in 0.6173 M Tris/HCl buffer (pH 6.8) and a separation gel of 12.5% polyacrylamide in 0.3777 M Tris/HCl buffer (pH 8.9). The lyophilised unbound protein fraction was prepared in 0.6173 M Tris/HCl buffer (pH 6.8) containing 20% glycerol. Electrophoresis was carried out at 60 volts for 5 h using an Amersham Pharmacia Biotech model EPS 1001 vertical electrophoresis system. Protein bands were visualised by staining the gel with Coomassie Blue R–250.

SDS-PAGE was performed according to Laemmli (1970). The stacking gel consisted of 6% polyacrylamide in 0.5 M Tris/HCl buffer (pH 6.8) containing 0.4% SDS, concentrated *N,N,N',N'*-Tetramethylethylenediamine (TEMED) and 10% ammonium persulphate, while the separation gel comprised 12.5% polyacrylamide in 1.5 M Tris/HCl buffer (pH 8.8) containing 0.4% SDS, concentrated TEMED and 10% ammonium persulphate. The lyophilised unbound protein fraction and molecular markers (BSA, ovalbumin, glyceraldehyde-3-phosphate dehydrogenase, carbonic anhydrase, trypsinogen, soy-bean trypsin inhibitor and lactoalbumin) were prepared in 62.5 M Tris/HCl buffer (pH 6.8) containing 10% glycerol, 2% SDS and 5% 2-mercaptoethanol.

**Table 1.** Total protein content and haemagglutination activity of the manioc leaf powder (MLP) samples.

| Sample | Total protein (mg) | Haemagglutination activity (HU /mg protein) |
|---|---|---|
| Crude MLP extract [a] | 41.58 ± 5.42 | 2164.0 ± 386.0 |
| Concentrated protein solution [b] | 24.05 ± 3.21 | 2856.0 ± 430.0 |

[a] Obtained by extraction of 25 g of MLP with distilled water. [b] Obtained by precipitation of the crude MLP extract with 80% saturated ammonium sulphate solution. Values shown represent the mean ± standard deviation.

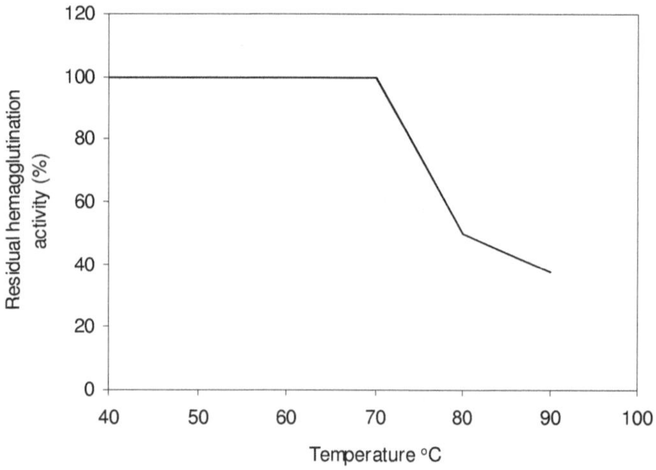


**Figure 1.** Effect of temperature on the haemagglutination activity of the concentrated protein solution obtained from manioc leaf powder. Incubations were carried at 40, 60, 70, 80 and 90ºC for 30 min.

Moreover, the running conditions and visualisation of bands had been described.

#### Fungicidal activity

Assays were conducted in the form of a random design comprising 5 treatments with 4 repetitions each. Lyophilised protein material was redissolved in distilled water at concentrations of 1, 10, 50 and 100 µg/ml and was added to molten (45 - 50°C) potato dextrose agar (PDA) medium. Aliquots (25 ml) of the medium were poured into Petri dishes (9 cm diameter) and left to solidify. Negative control plates were prepared with PDA medium alone. Cultures of *F. oxysporum* were obtained from the culture bank of the Departmento de Fitopatogia, Universidade Federal de Lavras. Disks (9 mm) of 6 day-old fungal mycelia, previously cultivated in PDA medium, were placed in the centre of Petri dishes containing the analyte and the dishes sealed with plastic film. Incubation was carried out under biological oxygen demand at 25°C. After incubation for 7 days, the diameters of the growing mycelia were measured (cm) and compared with the control culture, and the results were submitted to analysis of variance.

## RESULTS AND DISCUSSION

### Extraction and concentration of MLP lectin

Initially, purification of lectins followed the scheme used for proteins in general without exploiting their special characteristics. The methods used included precipitation by salts, acids and organic solvents. Progress in this area was achieved by the introduction of preparative chromatographic methods with ion exchangers, gel filtration (size exclusion chromatography) media and, above all, affinity adsorbents (Rüdiger and Gabius, 2001).

In this work, the partial purification of the lectin, we used ammonium sulphate as precipitating agent.

The amounts of protein present in the crude MLP extract and in the concentrated protein solution derived therefrom are shown in Table 1. Approximately 58% of the proteins present in the crude MLP extract could be recovered following precipitation in 80% ammonium sulphate and dialysis against 0.15 M NaCl solution. The recovery efficiency obtained in the present study was much greater than that reported previously (11%; Pereira et al., 2008). The haemagglutination activity (expressed in HU/mg protein) of the concentrated protein solution was approximately 30% higher than that of the crude MLP extract (Table 1).

### Thermostability of MLP lectin

The HA of the concentrated protein solution was stable at temperatures up to 70°C (Figure 1), but was diminished by 50 and 62.5%, respectively, at 80 and 90°C. Stability over a wide temperature range is typical of plant lectins and can be attributed to their role in plant defence mechanisms against environmental stress (Peumans and Van Damme, 1995).

### Partial purification of MLP lectin

Column chromatography of the concentrated protein solution over Sepharose CL-4B resin eluted with 0.15 M NaCl gave rise to a single 280 nm peak corresponding to the unbound protein fraction (fractions 24 to 51; Figure 2). The total protein content of these combined fractions was 1.75 mg, giving a recovery efficiency of 29.91%, and the HA was 0.5 HU/100 µl.

Normally, elution of proteins bound to a stationary phase can be achieved by introducing a soluble ligand that competes with the immobilised ligand to which the target protein is attached (Collins et al., 1997). In the case of lectins, the linkage with a soluble ligand (e.g. galactose) could occur either through the carbohydrate

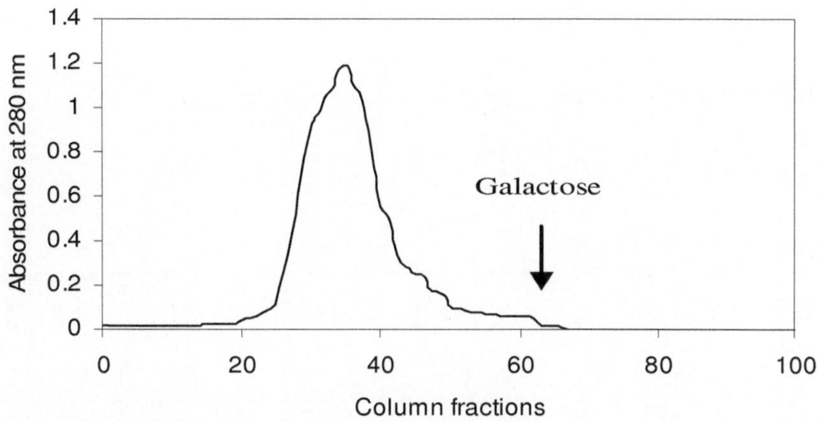


**Figure 2.** Chromatographic profile of the concentrated protein solution derived from manioc leaf powder. Column chromatography was carried out using a Sepharose CL-4B glass column (20 x 2 cm i.d.; approximate volume 63 ml). Non-bound proteins were eluted with 0.15 M NaCl at a flow rate of 1 ml/min, whilst matrix-bound proteins were eluted with 0.15 M NaCl and 0.1 M galactose (initiation of elution with galactose arrowed).

receptor site (specific linkage) or through ionic, hydrophobic and hydrogen interactions (non-specific linkage). In the present study, the Sepharose column was eluted with 0.15 M NaCl containing 0.1 M galactose in an attempt to elute the bound protein fraction, but none of the eluted fractions showed absorption at 280 nm as shown in Figure 2. There are two possible explanations for the interaction between the MLP lectin and the stationary matrix and the absence of lectin in the eluted bound fraction. The first hypothesis is that the carbohydrate receptor site of the MLP lectin is not specific for the galactose ligand such that 0.15 M NaCl was sufficient to break the weak non-specific interactions, thus displacing the bound lectin from the stationary phase. The second proposition is based on the fact that the agarose gel of the stationary phase is highly cross-linked forming pores of different sizes such that the proteins were separated according to molecular weight.

Column chromatography of the concentrated protein solution over Sepharose CL-4B resin eluted with 0.15 M NaCl was found only at the peak of lectin activity, but it was found heterogeneous on native electrophoresis. It is possible since plants contain families of lectin genes that may have altered patterns of expression in a tissue or in different tissues of the same plant. Common lectin gene products may also differ during posttranslational modification (Etzler, 1985). However, incomplete separation of proteins may have occurred.

## Gel electrophoresis

When the unbound protein fraction obtained from the Sepharose column was submitted to native PAGE (Figure 3A), four bands were stained with Coomassie Blue indicating the presence of four separate proteins. Under denaturing conditions, the SDS-PAGE showed separate bands at 64, 61, 58 and 13 kDa (Figure 3B) confirming the presence of four distinct proteins. It is not possible at this stage to determine, which of these bands correspond to MLP lectin since plant lectins present molecular weights that vary over a wide range, as exemplified by those isolated from *Urtica dióica* (8.5 kDa; Broekaert et al., 1989) and *Phaseolus lunatus* (265 kDa; Moreira et al., 1991).

## Fungicidal activity

With the insecticidal action of plant lectins, the spectrum of external functions is certainly not completely covered yet. Binding to cell wall constituents in fungi can interfere with their growth. Fungal cell walls contain chitin, the β1 → 4 linked polymer of GlcNAc. It is, therefore, likely that a fungicidal action is exerted by GlcNAc-binding lectins (Sharon and Goldstein, 1998).

MLP lectin showed no inhibitory activity against *F. oxysporum* as demonstrated by the growth of mycelia when cultured on medium containing variable concentrations (0 to 100 µg/ml) of protein solution (Figure 4). It has been shown previously that some plant lectins do exhibit fungicidal activity against *F. oxysporum*, e.g. lectins isolated from *Talisia esculenta* (Freire et al., 2002) and *Pouteria torta* (Boleti et al., 2007) were both active at a concentration of 280 µg/ml.

## Conclusions

This study indicated that Leaf lectin from Manihot

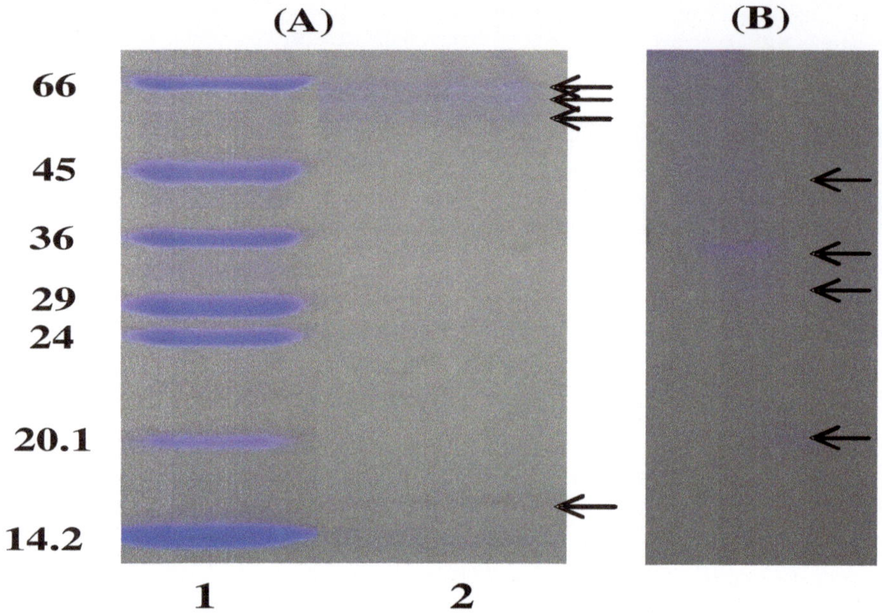


**Figure 3.** Gel electrophoresis of the unbound protein fraction obtained via column chromatography over Sepharose CL-4B of a manioc leaf powder solution. (A) Native PAGE: Bands corresponding to the unbound protein pool; (B) SDS- PAGE : Lane 1 - Bands corresponding to the molecular markers BSA (66 kDa), ovalbumin (45 kDa), glyceraldehyde-3-phosphate dehydrogenase (36 kDa), carbonic anhydrase (29 kDa), trypsinogen (24 kDa), soy-bean trypsin inhibitor (20.1 kDa) and lactoalbumin - 14.2 kDa; Lane 2 - Bands corresponding to the unbound protein pool.

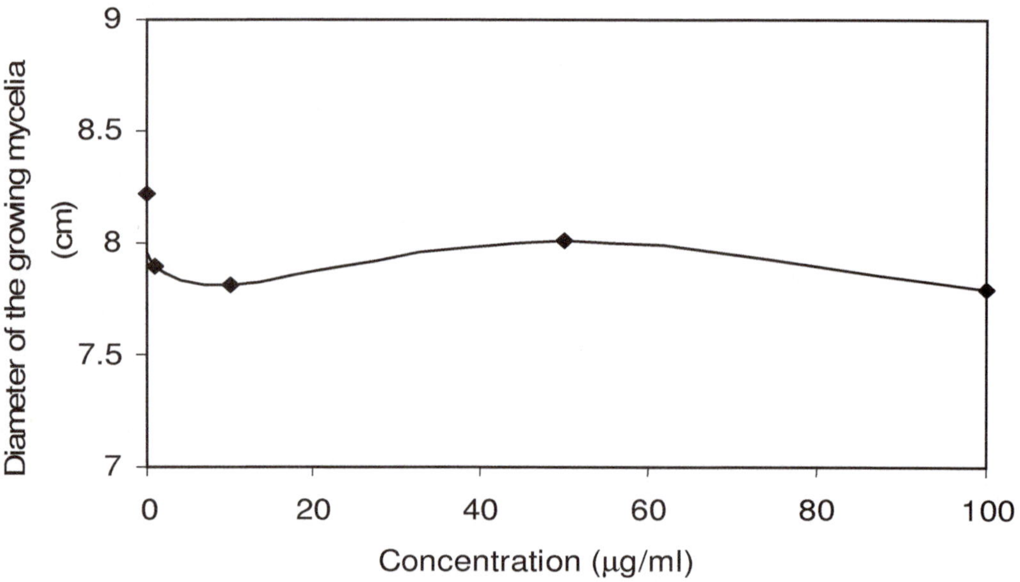


**Figure 4.** Effect of the concentrated protein solution derived from manioc leaf powder on the growth of mycelia of *F. oxysporum* after 7 days incubation at 25°C under biological oxygen demand.

esculenta has four protein bands (13, 58, 61, 64 kDa), which were identified in the unbound protein fraction. This fraction exhibited a haemagglutination activity of 0.5 HU/100 µl indicating the presence of lectin, although it was not possible to associate the activity with a specific band. MLP-lectin was stable over a wide temperature range (up to 70°C), but was not active against *F. oxysporum* in concentrations up to 100 μg/ml.

## ACKNOWLEDGMENTS

The authors wish to thank Conselho Nacional de Desenvolvimento Científico e Tecnológico and Coordenação de Aperfeiçoamento de Pessoal de Nível Superior for their scholarships awarded to M. C. Silva and Flávia C. A. Marcos, respectively.

## REFERENCES

Alonso R, Rubio LA, Muniz M, Marzo F (2001). The effect of extrusion cooking on mineral bioavailability in pea and kidney bean seeds meal. Animal Food Sci. Technol., 94: 1-13.

Boleti APA, Freire MGM, Coelho MB, Baldasso PA, Silva W, Gomes VM, Novello JC, Marangoni S, Macedo MLR (2007). Insecticidal and antifungal activity of a protein from *Pouteria torta* seeds with lectin-like properties. J. Agric. Food Chem., 55: 2653-2658.

Bradford MM (1976). A rapid and sensitive method for the quantitation of microgram quantites of protein utilizing the principle of protein-dye binding. Anal. Biochem., 72: 248-254.

Broekaert WF, Van Parijs J, Leyns F, Joos H, Peumans WJ (1989). A chitin-binding lectin from stinging nettle rhizomes with antifungal properties. Sci., 245: 1100-1102.

Calderón de la Barca AM, Ochoa JL, Valencia ME (1985). Effect of the extraction of a hemagglutinin on the nutritive value of *Amaranthus leocarpus* seeds. J. Food Sci. 50: 1700-1702.

Collins CH, Braga GL, Bonato PS (1997). Introdução a Métodos Cromatográficos. UNICAMP, Campinas.

Cooper TG (1942). The Tools of Biochemistry. Wiley-Interscience, New York.

Etzler ME (1985). Plant lectins: molecular and biological aspects. Annu. Rev. Plant. Physiol., 36: 209-234.

Freire MGM, Gomes VM, Corsini RE, Machado OLT, De Simone SG, Novello JC, Marangoni S, Macedo MLR (2002). Isolation and partial characterization of a novel lectin from *Talisia esculenta* seeds that interferes with fungal growth. Plant Physiol. Biochem., 40: 61-68.

Gabius HJ, Andre S, Kaltner H, Siebert HC (2002). The sugar code: functional lectinomics. Biochim. Biophys. Acta, 1572: 165-177.

Grahamstown M, Van Staden J (2002). Lectin-like proteins from South African plant species used in traditional medicine. S. Afr. J. Bot., 68: 36-40.

Laemmli UK (1970). Cleavage of structural proteins during the assembly of the bacteriophage T4. Nature, 227: 680-685.

Lima VLM, Correia MTS, Cechinel YMN, Sampaio CAM, Owen JS, Coelho LCBB (1997). Immobilized *Cratylia mollis* lectin as a potential matrix to isolate plasma glycoproteins, including lecithin-cholesterol acyltranferase. Carbohydrate Polym., 33: 27-32.

Lis H, Sharon N (1998). Lectins: carbohydrates specific proteins that mediate cellular recognition. Chem. Revs., 98: 637-674.

Loris R (2002). Principles of structures of animal and plant lectins. Biochim. Biophys. Acta, 1572: 198-208.

Macedo MLR, Coelho MB, Freire MGM, Machado OLT, Marangoni S, Novello JC (2000). Effect of a toxin protein isolated from *Zea mays* seeds on the development and survival of the cowpea weevil *Callosobruchus maculatus*. Protein Pept. Lett., 7: 225-231.

Moreira RA, Cavada BS, Oliveira JTA, Ainouz IL (1991). Plant Lectins. In: Proceedings of the Brazilian Congress on Proteins, 1st ed. UNICAMP, Campinas, pp. 77-96.

Pereira CA, Corrêa AD, Santos CD, Abreu CMP, Sousa RV, Magalhães MM (2008). Hemaglutinina de folhas de mandioca (*Manihot esculenta* Crantz): purificação parcial e toxicidade. Ciência e Agrotecnologia, 32: 900-907.

Peumans WJ, Van Damme EJM (1995). Lectins as plant defense proteins. Plant Physiol., 109: 347-352.

Peumans WJ, Zhang W, Barre A, Astoul CA, Balinti-Kurti PJ, Rovira P, Rougé P, May GD, Van Leuven F, Truffa-Bachi P, Van Damme EJM (2000). Fruit-specific lectins from banana and plantain. Planta, 211: 546-554.

Robyt JF, White BJ (1987). Biochemical Techniques: Theory and Practice. Brooks/Cole, Monterey.

Rüdiger H, Gabius Hans-J (2001). Plant lectins: Occurrence, biochemistry, functions and applications. Glycoconjugate J. 18: 589-613.

Sharon N, Goldstein IJ (1998). Lectins: More than insecticides. Sci. 282, 1049.

Sharon N, Lis H (2004). History of lectins: from hemagglutinins to biological recognition molecules. Glycobiol., 14: 53R-62R.

Silva MR, Silva MAAP (2000). Fatores antinutricionais: inibidores de proteases e lectinas. Revista de Nutrição, 13: 3-9.

Vasconcelos IM, Oliveira JTA (2004). Antinutritional properties of plant lectins. Toxicon, 44: 385-403.

Weis WI, Drickamer K (1996). Structural basis of lectin-carboydrate recognition. Annu. Rev. Biochem., 65: 441-473.

Sharon N, Goldstein IJ (1998). Lectins: More than insecticides. Sci., 282: 1049.

# 6

# Production and characterization of flour produced from ripe "apem" plantain (*Musa sapientum* L. var. *paradisiacal;* French horn) grown in Ghana

**H. D. Zakpaa, E. E. Mak-Mensah and J. Adubofour**

Department of Biochemistry and Biotechnology, Faculty of Biosciences, College of Science, Kwame Nkrumah University of Science and Technology, Private Mail Bag, Kumasi, Ghana.

**The importation and processing of wheat grain into flour by economies like Ghana does constitute an enormous economic drain. To remedy this situation, FAO of the United Nations launched its Compost Flour Programme, which aimed to formulate flour mixtures consisting mainly of indigenous raw materials and raw wheat to a lesser extent. The production of plantain baked products, instant breakfast meals and baby food formulation have been proven to be technically feasible with powder from both ripe and unripe fruit. Further, the production of plantain powder and processing of the flour into baked goods could reduce the cost of the plantain baked products and would also reduce dependence on imported wheat and increase self-sufficiency in food. In this investigation, proximate and mineral analyses and some functional properties of ripe *Musa sapientum* L. powder were analyzed. The powder contained 3.14% moisture, 2.68% protein, 2.68% ash, 0.334% fat, 91.16% carbohydrate and 1603.09 KJ of energy. Bulk density was 0.76 gmL$^{-1}$, water-binding capacity of 71.0 g/100g, solubility of 18.87% and swelling power of 5.237 gg$^{-1}$. Mineral content per 100 g was analyzed to be 1.125 g sodium, 0.297 mg phosphorus, 2.900 g iron, 0.419 g calcium and 435.200 g potassium.**

**Key words:** Plantain, *Musa sapientum*, French horn, food product development, flour, proximate analysis, functional properties, Ghana.

## INTRODUCTION

This investigation aimed at processing a local cultivar of plantain into a stable flour as a way of extending the shelf life of ripe plantain fruits, to add value to plantain for both the local markets and for export, thereby ensuring food security. Low cost processing methods, such as solar drying etc., suitable for use in local Ghanaian homes, were employed to obtain a product that was then subjected to various analyses to determine the quality of the resulting product.

In Ghana, two main cultivars of plantain, the "apem" (a triploid French plantain called French horn) and "apantu" (gaint horned) are grown. Plantain belongs to the genus *Musa* in the family *Musaceae*. Plantain (*Musa* sp. AAB group) is a giant perennial herb. It is a natural inter-specific cross between the two wild species *Musa auminata* colla which contributes genome A and *Musa balbisiana* colla, which contributes genome B (Simmonds and Shepherd, 1955). Morphological description has proven very useful for the identification of the large diversity of plantain cultivars that exist in the tropics (Tezenas du Montcel, 1987; Swennen, 1990a; Jarret and Gawel, 1995). In addition to the use of morphological description in identifying specific plantain cultivars, various DNA-based marker techniques have also been employed. These techniques can supply additional information not available from the examination of morphological characteristics alone (Jarret and Gawel, 1995; Shaibu et al., 2003).

Plantain requires about two and a half to four months after shooting before the fruit becomes ready for harvesting or a total of about eight to twelve months after planting (Simmonds, 1948; Swennen, 1990a). At maturity, a constant weight is maintained for two to four days, and then the weight starts to decrease with changes in the

*Corresponding author. E-mail: hzakpaa@hotmail.com.

peel colour from green to yellow and then to black. The maturity of the fruit may be determined by the weight of the pulp to peel ratio, brittleness of floral ends and disappearance of angularity of the fingers.

The chemical composition of plantain varies with the variety, maturity, degree of ripeness and where it is grown (soil type). The water content in the green plant is about 61% and increases on ripening to about 68%. The increase in water is presumably due to the break down of carbohydrates during respiration. Green plantain contains starch which is in the range of 21 to 26%. The starch in the unripe plantain is mainly amylose and amylopectin and this is replaced by sucrose, fructose, and glucose during the ripening stage due to the hydrolysis of the starch (Marriott et al., 1981). The carbohydrate content reduces to between 5 and 10% when ripe. The sugar content is between 0.9 to 2.0% in the green fruit but becomes more prominent in the ripe state. The titrable acidity of plantain is about twice that of sweet potato (Aurand et al., 1987).

Plantains therefore have a high carbohydrate content (31 g/100 g) and low fat content (0.4 g/100 g). They are good sources of vitamins and minerals (Adeniji, et al., 2006), particularly iron (24 mg/kg), potassium (9.5 mg/kg), calcium (715 mg/kg), vitamin A, ascorbic acid, thiamin, riboflavin and niacin. The sodium content (351 mg/kg) is low in dietary terms hence recommended for low sodium diets (Izonfuo and Omuaru, 1988; Stover and Simmonds, 1987; Welford et al., 1988). The amino acid components include β-alanine, aminobutyric acid, glutamine, asparagine, histidine, serine, arginine and leucine. The ascorbic acid is high compared to that of banana. As a starchy staple food, plantain supply about 1 g protein/100 g edible portion (USDA, 2009). As a healthy adult requires about 0.75 g protein $kg^{-1}$ $day^{-1}$ (Burton and Willis, 1976), plantain alone cannot meet adult protein needs.

The fat content of plantains and bananas is very low, less than 0.5%, and so fats do not contribute much to the energy content. Although the total lipid content remains essentially unchanged during ripening, the composition of fatty acids, especially within the phospholipids fraction has been observed to change, with a decrease in their saturation (Ogazi, 1996).

The energy value of a food derives from the sum of its carbohydrates, fat and protein content. In the case of plantain, the carbohydrate fraction is by far the most important. The sugars and starches that make up this fraction are present in varying concentrations according to the state of the ripeness of the fruit. The two main components of this starch are amylose and amylopectin, present in a ratio of about 1:5. Sugars comprise only about 1.3% of total dry matter in unripe plantains, but rises to around 17% in the ripe fruit (Ogazi, 1985). Plantains are considered palatable at lower water content than maize, thus boiled and mashed plantain may prove to be a higher energy staple than maize porridge (Chandler, 1995). For example the energy value/100 g plantain is as follows: raw fresh fruit, 112 kcal; boiled fruit, 122 kcal; ripe fried fruit, 267 kcal (Chandler, 1995).

Plantains and bananas are a good source of vitamin A (carotene), vitamin B complex (thiamin, niacin riboflavin and $B_6$) and vitamin C (ascorbic acid). Processing and cooking will affect the vitamin content. In comparison with other starchy staples; vitamin C content is similar to those of sweet potato, cassava and potato. Plantains provide a better source of vitamin A than most other staples (Aurand, 1987; Kirk and Sawyer, 1991; USDA, 2009). Although plantains do not provide a particularly good source of several important minerals in human nutrition, such as calcium, iron and iodine, they are notably high in potassium and low in sodium (Marriott et al., 1983; Stover and Simmonds, 1987; USDA, 2009).

Non starch polysaccharides (collectively known as fibers) include crude fiber, cellulose, pectic substances, hemicelluloses and other polysaccharides. Unripe plantain pulp has a total of 3.5% dry matter as cellulose and hemicellulose and therefore constitutes a good source of dietary fiber (Kirk and Sawyer, 1991).

Plantain is grown in 52 countries with world production of 33 million metric tonnes (FAO, 2005). Locally, plantain constitutes about 13% of the county's agricultural gross domestic product of Ghana (SRID-MOFA, 2006), and ranks third in volume of production among starchy staples in Ghana. In the agricultural sector, plantain is ranked fourth in Ghana (FAO, 2005). Total annual national production is 2.00 million tonnes (SRID-MOFA, 2006) with per capita consumption of 101.8 kg (SRID-MOFA, 2006; FAO, 2005; Lescot, 2000). Plantain for local consumption undoubtfully, plays a role in food and income security and has the potential to contribute to national food security and reduce rural poverty. This crucial role is still largely ignored by policy makers and therefore special public awareness effort is required to sensitize policy makers in both producing and donor countries. Despite the importance of plantain, major constraints threaten its cultivation in terms of pest and disease infestations, soil fertility, planting materials, post-harvest losses, marketing constraints, particularly poor road system and lack of infrastructure and storage facilities and much of the fruits harvested go waste.

In Ghana during the glut season from late August to late April, before the "Easter storms" set in, markets are filled with vehicles laden with loads of the fruits from the remote production areas.

The production of food powders is on the increase as a result of gradual urbanization and globalization. For example, in Ghana the production and export of powder from local tuber crops and cereals is gradually gaining market status. Apart from extending the storage life of these staples (due to the low moisture of the dried product), powder products are easy to handle in terms of transport and space. In Ghana a company such as Rosafrik Co. Ltd., produces and exports powdered starchy stables such as "fufu", "tuozaafi", "kokonte" (cassava meal), "banku"

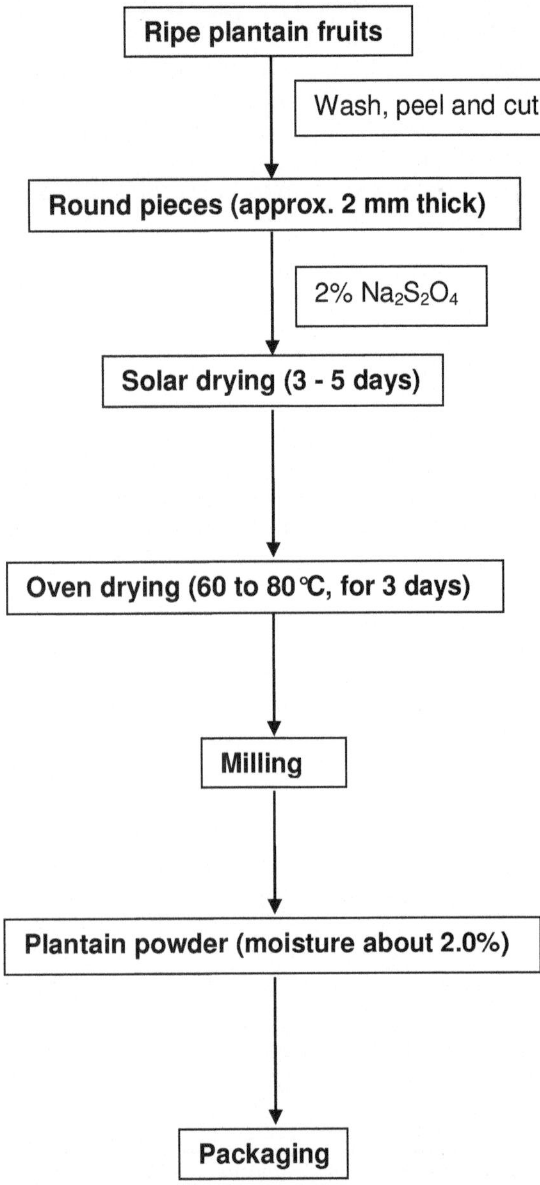


**Figure 1. Flow** chart of plantain flour preparation

etc. to countries such as the USA and UK.

In the manufacture of plantain powder, fully ripe plantain pulp is converted into a paste by passing through a chopper followed by a colloid mill. 1.0 to 2.0 sodium metabisulphite solution is added to improve the colour of the final product. Spray- or drum-drying may by used, with drum-drying being favoured as all the solids are recovered. A typical spray dryer can produce 70 kg powder per hour to give yields of 8 to 11% of the fresh fruit, while drum-drying gives a final yield of about 13% of the fresh fruit. In the latter method, the moisture content is reduced to 8 to 12% and further decreased to 2% by drying or cabinet dryer at 60°C (Ogazi, 1996).

There are different ways of preserving plantain fruits. However, the type of method employed should make economic sense. An enquiry from a food processing company in Kumasi, Ghana revealed that, plantain processed into "fufu" flour can store for up to a maximum of two years.

Although much work or research has been done in the areas of disease and pests (MUSACO research team: Musa network for West and Central Africa) that is, integrated strategies to reverse plantain losses in Ghana (IITA, 1997) through Sigatoka-resistant hybrid production, not much has been done in the area of post harvest losses in terms of long term keeping quality of the matured fruits.

## MATERIALS AND METHODS

Mature plantain fruits (maturity age of 3.5 months) were obtained from the CSIR (Centre for Scientific Research and Industrial Research), Crop Research Institute, Fumesua-Kumasi.

### Methodology of ripe plantain flour production

1. Washed plantain fruits were ripened by covering in jute sack for four days (ripening was induced by addition of ripe orange peels to fruits in the sack).
2. Washed ripe plantain fruits were sorted from damaged or rotten fruits and then weighed.
3. Fruits were hand peeled and chopped into pieces (approx. 2 mm thick) using a stainless steel knife.
4. 2.0% of sodium metabisulfite ($Na_2S_2O_4$) was prepared by dissolving 2 g of the salt in approximately 100 ml of distilled water.
5. 25 ml of the prepared 2.0% $Na_2S_2O_4$ was added to each 500 g weighed chopped pulp.
6. The round chopped pieces were initially subjected to solar drying (3 - 5 days).
7. The sulphited pulp was then dried in a Gallenkamp oven dryer maintained between 60 to 80°C for 72 h to obtain dry chips which were milled with a ball mill to get the dry powder.
8. Proximate and mineral analyses, as well as functional properties determinations were carried out on the dried powder.

The procedure for dry powder preparation is shown in Figure 1. The yield of powder was calculated from the weight measurement of the powder obtained after drying ($P_w$)/g, the total weight of fresh whole plantain fingers taken (PT)/g. The weight of the pulp and peels were also measured. The yield was calculated as:

$$\frac{(P_w)/g \times 100\%}{(P_T)/g}$$

### Proximate analysis: Moisture determination

About 2.0 g of powder was weighed and transferred into a previously weighed crucible. The crucible was then placed into the drying oven at 105°C for 5 h. After this, they were removed and placed in a desiccator to cool. The cooled crucibles were reweighed. This was done in triplicate. The loss in weight after drying was then calculated as the percentage moisture (AOAC, 1990).

### Proximate analysis: Ash determination

About 20 g of sample was weighed into a previously weighed crucible and placed in the muffle furnace (600°C) for 2 h. The crucibles were cooled and reweighed. The loss in weight was then calculated as percentage as or mineral content of the sample (AOAC, 1990).

**Proximate analysis: Crude fat determination**

The dry sample from moisture determination was transferred to a 22 x 80 mm paper thimble. A small ball of cotton wool or glass wool was put into the thimble to prevent loss of sample. Anti-bumping granules were added to a previously dried 250 ml round bottom flask and weighed. 150 ml of petroleum spirit was added to the flask and the apparatus was assembled. A quick fit condenser was connected to the Soxlet extractor and refluxed for six hours on low heat. The flask was removed and evaporated on a steam bath. The flask with the fat was heated for 30 min in an oven at 103°C. The flask and its contents were cooled to room temperature in a desiccator after which it was weighed and percentage fat calculated (AOAC, 1990).

**Proximate analysis: Crude fiber determination**

A defatted sample of about 2.0 g was transferred into a 750 ml Erlenmeyer flask and approximately 0.5 g of asbestos was added. 200 ml of boiling 1.255% $H_2SO_4$ was added and the flask was immediately placed on a hot plate and a condenser was connected. The sample came to a boil within a minute. The flask was removed at the end of 40 min and the contents immediately filtered through cheese cloth and washed with large volume of boiling water. The washing was continued until the filtrates were no longer acidic. The charge and asbestos were scrapped with a spatula back into the flask with 1.25% NaOH solution using a wash bottle calibrated to deliver 200 ml. The flask again was connected to the condenser and boiling was undertaken for exactly 30 min. The content of the flask was filtered through fine linen with large volumes of boiling water as before. The residue was transferred to a crucible and washed with 15 ml alcohol. The crucible and contents were dried for 1 h at 100°C, cooled and then weighed. The crucible was ignited in the muffle furnace for 30 min, cooled and reweighed. The loss in weight was then determined (AOAC, 1990).

Digestion: 2.0 g of sample was weighed and transferred into a digestion flask with 0.5 g selenium- based catalyst crystal and a few anti-bumping agents. 25.0 ml of concentrated $H_2SO_4$ was added and the flask was shaken to wet the entire sample.

The flask was placed on a digestion burner and heated slowly until bubbling ceased and the resulting solution was clear. The solution was cooled to room temperature. The digested sample was transferred into a 100 ml volumetric flask and made up to the mark (AOAC, 1990).

**Protein determination: Distillation**

25 ml of 2% boric acid was pipetted into a 250 ml conical flask and 2 drops of mixed indicator was added. The conical flask with its content was placed under a condenser such that the tip of the condenser was completely immersed in the solution. 10 ml of the digested sample was transferred through the stop cork of the funnel on the steam jacket into the chamber. 15 ml of 40% NaOH was added to the decomposition flask. The funnel stop cork was then closed. The stop cork on the steam trap outlet was closed which forced steam through the decomposition chamber and in the process drove the librated ammonia into the collection flask. The boric acid changed to yellow as soon as it came into contact with the ammonia liberated and the distillation continued for 5 min. At the end of 5 min, the receiving flask was lowered such that the condenser tip was just above the liquid and washed with a little distilled water. Distillation was continued for another 30 min after which the burner was removed from the steam generator (flask). The apparatus was flashed as was done in the digestion step above.

**Protein determination: Titration**

The distillate was titrated with 0.1 M HCl solution. The same procedure was followed for the blank except that the sample was omitted. The titre values of the duplicate samples were recorded and the percentage protein calculated (%Protein = % Nitrogen x 6.25).

**Functional properties: Water-binding capacity (WBC) determination**

The water-absorption capacity of the flours was evaluated by placing 2 g samples in a centrifuge tube. 40.0 ml of distilled water was added and the resultant slurry was shaken for one hour before centrifugation at 2,200 rpm for 15 min. The supernatant was decanted and the amount of water in grams gained by a 100 g sample was determined (Yamazaki, 1953; Medcalf and Gilles, 1965).

**Bulk density determination**

The bulk density of the flour was determined by placing10 g of sample in a 50 ml graduated cylinder with gentle uniform tapping during filling. The cylinder was filled to the mark and the weight of the flour was measured. The bulk density was calculated as mass by volume in grams per milliliter ($gmL^{-1}$). The average of two determinations is reported (Medcalf and Gilles, 1965).

**Solubility and swelling power determination**

1.0 g of powder was placed into a pre-weighed centrifuge tube. 40.0 ml of distilled water was then added and stirred. The mixture was placed on a water bath thermostatically controlled at 85°C with continuous stirring for 30 min. It was cooled to room temperature and then centrifuged at 2,200 rpm for 15 min.

The supernatant was poured into a pre-weighed crucible and then placed in the oven to evaporate. The solid residue in the crucible was weighed again and the difference in weight calculated as percentage solubility. The paste in the tube was then weighed and the swelling power determined by the following equation:

$$\% \text{ Solubility} = \frac{\text{Weight of soluble starch} \times 100\%}{\text{Weight of original sample}} \quad (1)$$

$$\% \text{ Swelling power} = \frac{\text{Weight of sedimented paste}}{\text{Weight of original sample} \times (100 - \text{percentage solids})} \times 100\% \quad (2)$$

(Leach et al., 1959)

**Mineral analysis**

**Phosphorus (P)**

Analytical grade $KH_2PO_4$ was used to prepare serial standards. To 20.0 ml each of the serial standards was added 1.0 ml CDR (50 ml of 2.5 M $H_2SO_4$; 15 ml of 4.0% ammonium molybdate; and 10.0 ml of ascorbic acid) colour-developing reagent. This was repeated for the sample. A blank was also prepared using distilled water. The tubes were incubated at room temperature for 1 to 1½ h and the absorbance read from the spectrophotometer (LKB Biochrom Ultrospec II, Model 4050, Cambridge, England) at 770 nm.

### Calcium ($Ca^{2+}$)

Analytical grade calcium chloride was used to prepare serial dilutions. 3.5 M 2-amino-2-methyl-1-propanol was also prepared. In this case the colour-developing reagent (CDR) was made from 15 ml of 3.5 M 2-amino-2-methyl-1-propanol; 40 ml ethanediol as buffer; 0.002 g 0f *O*-cresol phthalein complexone; 0.1 g of 8-hydroxylquinolin and made up to 500 ml). 5 ml of each serial standard was added to 3.0 ml CDR and their absorbance taken. This was repeated for the sample in duplicate. The samples were incubated at room temperature for 20 to 30 min and their absorbance read at 570 nm.

### Iron ($Fe^{2+}$)

An analytical grade iron (II) compound was used to prepare serial standards. 10% of ascorbic acid was also prepared. The function of the ascorbic acid was to reduce any iron (III) in the sample to iron (II).To 1.0 ml of each sample in a test tube was added 1 ml of 10% ascorbic acid. This was followed by the addition of 0.5 ml of 0.5% phenanthroline. This was repeated for the sample in duplicate. The tubes were incubated at room temperature for 30 min and their absorbance read at 520 nm.

### Sodium and potassium (Na+ and K+)

These minerals were determined by flame photometry. Flame photometry works on the principle that when an atom of a metal is energized, its energy level changes, that is, it goes to the excited state. Energy of a specific wavelength and intensity is released when the atom returns to its ground state. The intensity of a characteristic wavelength produced by the atom in the flame is proportional to the number of atom excited in the flame which is proportional to the concentration of the metal that is the mineral. A sodium filter was used to filter out the intensity of the light given off by the sodium in the mixture. A detector detects the intensity of the light which then sends signals to a galvanometer. The same principle was used for potassium. Sodium gives off a yellow colour and potassium gives off a red-violet colour. In determining the mineral content (P, Na, Fe, K, and Ca) of a food or feed, the residue left after ashing is dissolved in a known volume of concentrated HCl, usually 2.0 ml. This is filtered and made up to 100 ml with distilled water. The solution is then used for mineral analysis. All determinations were done in triplicates.

## RESULTS AND DISCUSSION

Analysis of proximate composition provides information on the basic chemical composition of foods/ feeds. The compositions are moisture, ash, crude fat, protein, crude fiber, and carbohydrate (Table 1). These components are fundamental to the assessment of the nutritive quality of the food being analysed. Proximate analysis is usually sufficient to establish the general category of foodstuff to which a particular food sample belongs and the similarity of a particular food sample to materials previously reported in literature. Careful sampling is required to obtain accurate and reproducible values or results. The chemical and physical properties of foodstuffs exhibit a certain inherent variability among different samples; variability within a given sample, however, can be minimized by proper sampling. Analyses are usually performed on small discrete samples rather than the entire amount of food stuff (a so-called perfect sample). Various techniques (grinding, mixing, etc.) are used to insure that such small samples are representative of the entire material and provide a true measure of its overall content.

The moisture content for the plantain flour was determined to be 3.14% (± 0.22%) as compared to a value 5.43% quoted in literature (USDA, 2009) Moisture content of food or processed product gives an indication of its shelf life and nutritive values. Low moisture content is a requirement for long storage life. The moisture content of a fresh fruit is related to its dry matter content and therefore to the yield obtained from a particular crop or cultivar. The yield obtained using the “apem” cultivar was 25.86%

Crude fat determines the free fatty lipids (neutral fats-triglycerides) of a product. This property can be used as the basis in determining processing temperatures as well as auto-oxidation which can lead to rancidity (can affect flavour of food). Crude fat is always determined on defatted samples. A value of 0.336% (± 0.054) was obtained for crude fat. Work carried out by Fagbemi (1999) in Nigeria on unripe plantain fruit (powder) gave a value of 3.00%. However, the range in literature was given to be between 0.2 to 0.5% (Jaffe et al., 1963; Ketiku, 1973; Ogazi, 1996; USDA, 2009). The low content of fat will enhance the storage life of the flour due to the lowered chance of rancid flavour development (Table 1).

The protein content of foodstuffs is commonly done by analysis of total nitrogen according to the method of Kjeldahl. Total nitrogen is the sum of that derived from amino acids, which generally represent the vast majority, and that from non-protein nitrogen (NPN) sources, generally minor in quantity, existing in the foodstuff. Total nitrogen derived from the analysis is converted into protein by multiplying by a factor (6.25), that takes into account the nitrogen content of a known or average amino acid composition. The total nitrogen (protein) content of the flour was determined to be 2.68%. Ketiku (1973) in his work determined the protein level in plantain to be between 3.0 to 3.5%, as-is, or 1.3 to 1.8%, on dry-matter basis. The level of protein in unripe plantain powder was determined by Fagbemi (1999), to be 3.6%. Protein content in plantain flour is low compared to other widely eaten stable roots, tubers and fruits (Aurand, 1987; USDA, 2009). Although the protein content is lower, the nutritive quality of food is not only scored by its protein content. The underlying factor is the overall energy value it can supply to the consumer. The carbo-hydrate and energy contents (91.162% and 1603.091 kJ,

**Table 1.** Proximate composition, mineral composition and functional properties of plantain flour.

| **Proximate composition value standard deviation** | | |
|---|---|---|
| Moisture (%) | 3.14 | ± 0.220 |
| Ash (%) | 2.68 | ± 0.050 |
| Protein [N x 6.25 (%)] | 2.682 | ± 0.050 |
| Crude fat (%) | 0.336 | ± 0.054 |
| Fiber (%) | 0.979 | ± 0.167 |
| Carbohydrate (%) | 91.162 | ± 2.010 |
| Energy (kJ) | 1603.09 | - |
| Yield (%) | 25.86 | - |
| **Mineral composition** | | |
| Sodium (mg/100 g) | 1.125 | ± 0.030 |
| Phosphorus (mg/100 g) | 0.297 | ± 0.005 |
| Iron (mg/100 g) | 2.900 | ± 0.050 |
| Calcium (mg/100 g) | 0.419 | ± 0.011 |
| Potassium (mg/100 g) | 435.200 | ± 2.200 |
| **Functional properties** | | |
| Solubility (%) | 18.870 | ± 0.004 |
| Swelling power (gg-1) | 5.237 | ± 0.030 |
| Water-binding capacity (g/100 g) | 71.003 | ± 0.004 |
| Bulk density ( $gml^{-1}$) | 0.755 | ± 0.020 |

respectively) are comparable with those of the common stables (Irvine, 1982; Sampson, 1986; Aurand, 1987; Kirk and Sawyer, 1991).

Crude fiber measures the cellulose, hemicellulose and lignin content of food. Lignin comprises polymers of phenolic acids and hemicellulose is made up of hetero-polymers of polysaccharides. Crude fiber is reported as the loss in weight on ignition of dry residue remaining after digesting material with 1.25% $H_2SO_4$ and 1.25% NaOH under specified conditions (This removes proteins and carbohydrates). This was determined to be 0.98%, a value very close to 0.9% stated in the works of Izonfou and Omuaru (1988) and 1.1% by Ketiku (1973).

The ash content was determined to be 2.68% ± 0.055 (Table 1), comparable to 1.66 to 2.00% for unripe plantain analysed by other researchers (Ketiku, 1973; Irvine, 1982; Sampson, 1986; Aurand, 1987; Kirk and Saywyer, 1991). The ash of foodstuff is the inorganic content residue remaining after the organic matter has been burnt away. The ash content can provide an estimate of the quality of the product, since high levels may indicate contamination. The slight increase in the value could be attributed to the differences in soil or growth conditions (Wilson, 1987). Of the 104 natural elements, only few serve as nutrient in living system and only few are essential (Burton and Willis, 1976). In determining the mineral content of a foodstuff or feed, first the total inorganic content is determined, that is, either by dry ashing at 600℃ or wet ashing. From a nutritional stand point the ash content is not important as contaminations do occur. Analysis of minerals therefore had to be carried out to ascertain the actual mineral value of the food sample. The minerals usually determined are Ca, Fe, I, Na, Zn and P. Phosphorus (P) is widely present in foods and other sources so deficiency is rare (Burton and Willis. 1976).

The mineral content of the powder was generally low. Sodium, potassium, iron and phosphorus levels were determined as 1.125 mg/100 g, 435.20 mg/100 g, 2.9 mg/100 g and 0.297 mg/g, respectively (Table 1); compared with 4, 499 and 0.6 mg/100 g, for sodium, potassium and iron respectively stated in other works (USDA, 2009). The iron and potassium levels were higher in the flour produced from the "apem" cultivar of plantain than for most of the others cultivars used in other projects, and this could be attributed to differences in soil conditions (soil type and mineral content) as well as different conditions of experimental analysis (Wilson, 1987; Swennen, 1990b).

Ripe plantain flour is rich in minerals like potassium, sodium and phosphorus and this could be formulated into instant flours for convalescence and in the formulation of baby foods as these categories of humans require high levels of minerals for growth and repair.

Plantain is low in sodium (Chandler, 1995), contains very little fat and no cholesterol, therefore it is useful in managing patients with high blood pressure and heart disease (Dzomeku et al., 2007). Plantain flour is free from substances that give rise to uric acid and therefore is ideal for patients with gout or arthritis (Ahenkora et al., 1996). Due to the low sodium and protein content, plantain is used in special diets for kidney disease sufferers.

For functional property analysis (Table 1), bulk density of 0.755 ± 0.02 $gml^{-1}$ was indicated for the flour produced, comparable to that obtained by other researchers (Fegbemi, 1999). This value is also comparable to that obtained for sweet potato powder (0.7453$gml^{-1}$) which is used as a thickener or base in foods like yoghurt (USDA, 2009). Based on this similarity, plantain flour could find use as food thickener in the food industries to give body and mouth-feel to food products.

The water-binding capacity (WBC) of the flour was determined to be 71.003 ± 0.004 g/100 g; a value lower than that obtained by Fegbemi (1999) for unripe plantain (250 - 338%). Unripe fruit has high amylose/ amylopectin content implying high -OH groups to form H-bonding and hence ability to bind more water. The ripe fruit on the other hand, has all the starches broken down to form sugars and as such less water binding ability. This property can however be exploited in the formulation of baby foods since less moisture is absorbed, thus making the food less bulky and hence more palatable and easily digested. Work carried out by Fegbemi (1999), on the effect of blanching on the functional properties of green plantain powder gave higher values for the functional properties. Ripening is said to lower the values obtained for all functional properties with the exception of bulk density and gelation.

Percentage solubility and swelling power were determined to be 18.890 ± 0.004% and 5.237 ± 0.03 $g/g^{-1}$ (Table 1), respectively. Swelling power indicates how much water a product can absorb to swell in the presence of heat. This confirmed by the low water binding capacity of the power (71.003 g/100 g). The "apem" plantain flour in foods or as an instant flour would not absorb much water hence would be less bulky, that is, more palatable. The water-binding capacity and swelling power values obtained suggest that the powder might absorb moisture upon storage, especially in the humid forest belt in Ghana were plantain is cultivated, it would therefore be advisable to store the plantain power in moisture-proof bags to extend the storage life of the product. It has been shown that plantain flour has a good potential for use as a functional agent in bakery products on account of its high water absorption capacity, but evaluation of the functionality of flour in test baking has not been demonstrated (Akubor, 1998). Mepba et al. (2007) investigated the feasibility of partially replacingwheat flour with plantain flour in bread and biscuit making and determined that the addition of plantain flour to wheat flour decreased the resistance of dough to extension, extensibility of dough and mechanical work of dough deformation and recommended that technically, organoleptically acceptable breads and biscuits were formulated from wheat-plantain composite flours using up to 80:20 and 60:40 ratios of wheat: plantain flour as maximum acceptable levels of substitution for breads and biscuits, respectively. Olaoye et al. (2006), investigated the use of soy flour and plantain flour substitution in wheat flour, from 0 to 15% each, for the production of bread and stated that there were no significant differences ($p < 0.05$) between the whole wheat bread and the plantain supplemented breads up to 10% plantain flour substitution in all the sensory attributes tested; crust, taste, aroma, shape, internal texture, appearance and general acceptability.

Hence the plantain supplemented breads had comparable sensory and nutritional qualities to the whole wheat bread, while the soy supplemented breads had higher proteins contents than the latter. However the whole wheat bread had highest hedonic mean scores in all the sensory attributes tested.

## Conclusion

Plantains are undoubtfully of cultural, food and socio-economic importance in the Ghanaian society. Due to the low moisture content of the plantain powder produced, it is expected to have a longer shelf-life compared to ripe plantain; hence the product could ensure food security all year round as a result of the long storage life and the plentiful supply of ripe plantain available on the local Ghanaian markets.

Based on the functional characteristics and proximate composition of the plantain flour product, it can be incorporated as a food thickener in foods for people of all ages.

### REFERENCES

Adeniji TA, Sanni LO, Barimalaa IS, Hart AD (2006). Determination of micronutrients and colour variability among new plantain and banana hybrids flours. World J. Chem. 1(1): 23-27.

Ahenkora K, Kyei MA, Marfo EK, Banful B (1996). Nutritional composition of False Horn Apantu plantain during ripening and processing. Afr. Crop Sci. J. 4(2): 243-247.

Akubor PI (1998). Functional properties of cowpea-plantain flour blends. Proceedings: 22nd annual NIFST Conference, University of Agriculture, Abeokuta.

AOAC (1990). Official Methods of Analysis of the Association of Official Analytical Chemists, 15th ed., AOAC, Arlington, Virginia, USA.

Aurand WL (1987). Food composition and analysis of food, Von Nostrand Reinhold, New York, pp.135-138.

Aurand WL, Wood AE, Wells RM (1987). Food composition and analysis, 4th edition, Van Nostrand Reinhold, 115 fifth avenue, New York, pp. 19-34.

Burton TB, Willis RF (1976). Human nutrition, 4th edition, McGraw Hill Book Company, New York, pp. 566-588.

Chandler S (1995). The nutritional value of bananas. In: Gowen S (ed.). Bananas and plantain. Chapman and Hall, London. pp. 468-480.

Daily Graphic Newspaper, Tuesday, April 3, 2001, 16.

Fagbemi TN (1999). Effect of blanching and ripening on the functional properties of plantain (*Musa aab*) flour. Plant Foods Human Nutr. 54: 261-269.

FAO (Food and Agricultural Organization) (2005). Food and agriculturalindicators. ESSA Oct. 2005. FAO Rome. http://www.fao.org/es/ess/ top/country.html.

Dzomeku BM, Armo-Annor F, Adjei-Gyan K, Darkey SK (2007). Consumer preference for three selected *Musa* hybrids in Ghana. Am. J. Food Technol. 2(7): 684-688.

International institute of tropical agriculture (IITA) (1997). Annu. Rep. pp. 34-35.

Irvine FR (1982). West African crops, 4th edition, Oxford University Press, pp. 205-206.

Izonfuo W-AL, Omuaru VOT (1988). Effect of ripening on the chemical composition of plantain peels and pulps (*Musa paradisiaca*). J. Sci. Food Agric. 45: 333-336.

Jaffe WG, Chavez JF, Keigman BDe (1963). Concerning the nutrient value of bananas and plantain. Arch. Venez Nutr. 13: 9-23.

Jarret RL, Gawel N (1995). Molecular markers, genetic diversity and systematics in *Musa*. In: Gowen S (ed.). Bananas and Plantains. Chapman and Hall, London, UK. pp. 66-83.

Ketiku AO (1973). Chemical composition of unripe (green) and ripe plantains (*Musa paradisiaca*). J. Sci. Agric. 24: 703-707.

Kirk RS, Sawyer R (1991). Pearson's composition and analysis of food. 9th ed., Longman Singapore publishers, Singapore. pp. 330-333.

Leach HW, McCowen DL, Schoch TJ (1959). Swelling and solubility patterns of various starches. Structure of Starch Granules, 1(36): 534-544.

Lescot T (2000). The importance of plantain and cooking bananas in Africa: Outlets for the subtropical zones. InfoMusa, 9: 25-28.

Marriott J, Robinson M, Karikari SK (1983). Starch and sugar transformations during ripening of plantains and bananas. J. Sci. Food Agric. 32: 1021-1026.

Medcalf DG, Gilles KA (1965). Wheat starches. Comparison of physicochemical properties. Cereal Chem. 42: 558-568.

Mepba HD, Eboh L, Uwaojigwa SU (2007). Chemical composition, functional and baking properties of wheat-plantain composite flours. Afr. J. Food Agric. Nutr. Dev. 7(1): 1-22.

Ogazi PO (1985). The biscuit making potentials of plantain flours. Nig. Food J. 3: 168-171.

Ogazi PO (1996). Plantain: production, processing and utilisation. Paman and Associates publishers, Okigwe, Nigeria. pp. 1-29.

Olaoye OA, Onilude AA, Idowu OA (2006). Quality characteristics of bread produced from composite flours of wheat, plantain and soybeans. Afr. J. Biotechnol. 5(11): 1102-1106.

Sampson JA (1986). Tropical fruits. 2nd edition, co-published in the United States with Wiley and Sons Inc., 605 third avenue, New York, p.143.

Shaibu AA, Dimkpa C, Orburia MN (2003). Bio-Diversity in Diploid Musa acuminata (AA Group) and their Hybrids using AFLP markers. Afr. J. Agric. Res. 1: 67-72.

Simmonds NW (1948). Genetical and cytological studies in Musa X and stomatal size in plants vigour in relation to polyploidy. J. Genet. 49: 57-68.

Simmonds NW, Sherperd K (1955). The taxonomy and origin of the cultivated bananas. J. Linn. Soc. Lond. Bot. 55: 302-312.

SRID-MOFA (2006). Statistics, Research and Information Directorate, Ministry of Food and Agriculture, Ghana.

Stover RH, NW Simmonds (1987). Bananas. Tropical agricultural series, third edition. John Wiley and Sons, Inc., N.Y. 10158: p. 468.

Swennen R (1990a). Limits of mophotaxonomy of plantain in Africa and elsewhere. Genet. Res. Crop Evol. 18: 172-210.

Swennen R (1990b). Plantain cultivation under West African conditions: A reference manual. International Institute for Tropical Agriculture, Ibadan, Nigeria, Amarin Printing Group Co. Ltd., Thailand. p. 24.

Tezenas du Montcel H (1987). Plantain bananas. Macmillan Publisher Ltd, London, UK. pp. 4-6.

USDA (2009). National Nutrient Database for Standard Reference, http://www.nal.usda.gov/fnic/foodcomp/plantain/.

Welford A, Izonfuo L, Victor F (1988). Effect of ripening on the chemical composition of plantain peels and pulps (*Musa paradisiaca*). J. Sci. Food Agric. 45: 333-336.

Wilson GF (1987). Status of banana and plantain breeding strategies. In: Persley GJ and De Langhe E (eds.). Proceeding of an international workshop held at Cairns, Australia, 13-17 Oct. 1986.

Yamazaki WT (1953). An alkaline water retention capacity test for the evaluation of cookies in baking potentialities of soft winter wheat flours. Cereal Chem. 30: 242-246.

# Assessment of large population of cassava accessions for resistant to cassava bacterial blight infection in the screen house environment

A. A. Ogunjobi[1, 2*], O. E. Fagade[1] A. G. O. Dixon[2] and R. Bandyopadhyay[2]

[1]Department of Botany and Microbiology, University of Ibadan, Nigeria.
[2]International Institute of Tropical Agriculture (IITA), P. M. B. 5320, Ibadan, Nigeria.

**Cassava bacterial blight (CBB) is caused by *Xanthomonas axonopodis* pv manihotis *(Xam)* a major cassava disease in all cassava growing area of the world. Resistance to the disease is found in *Manihot esculenta* and, in addition, has been introduced from a wild relative, *M. glaziovii*. We introduced a novel method of screening large population of cassava accessions in a screen-house environment using planting tray with a planting tree that can accommodate 68 different cassava stems cuttings in one planting tray. A total of 1,090 cassava accessions were screened with this method in the screen-house. The experimental design was augmented with randomized complete block design (ARCBD). Out of 490 land races cassava accessions that were assessed for susceptibility to CBB, 14.3 % were resistant to the disease. Land races populations were more susceptible to CBB infection having 30.1% highly susceptible and 12.3% susceptible cultivars. Six hundred cassava accessions of the improved and the new improved were screened in the greenhouse. Within the improved cultivars, 11.1% showed very strong resistance and 30.1% were resistant to the bacterial infections while 4.3% of these improved cassava accessions were observed as being highly susceptible to the disease and 36.6% were tolerant. Our results demonstrated that resistance to CBB is broadly distributed in cassava germplasm and for quick evaluation, the planting tray system should be employed.**

**Key words:** CBB, *Xanthomonas axonopodis* pv manihotis, *Manihot esculenta,* resistance, susceptibility.

## INTRODUCTION

*Xanthomonas axonopodis* pv manihotis causes bacterial blight, the most important bacterial disease of cassava in all cassava growing areas of the world (Lozano, 1986). Compared to the long history of cassava cultivation, the deployment of resistance cassava varieties to *X. axonopodis* pv manihotis in commercial cassava cultivars is relatively recent. The introduction of these resistance cassava cultivars correlated with a change in the pathogenic diversity of *X. axonopodis* pv manihotis populations, that is, new strains of the pathogen emerge and overcome deployed resistance (Dixon et al., 2002; Ogunjobi, 2005). These observations have stimulated much curiosity concerning the contribution of host genotype and other factors to the genetic diversity of the pathogen.

Molecular markers have been used in conjunction with virulence typing to evaluate the diversity and structure of *X. axonopodis* pv manihotis populations within and between countries in Africa (Assigbetsé et al., 1998; Ngeve, 1999; Dixon et al., 2002; Ogunjobi, 2005). In general, regionally defined pathogen populations in Africa were not found to be distinct (Wydra et al., 1998; Ogunjobi et al., 2001). Although populations within a region generally were similar, in some cases genetically similar strains were detected in different regions, suggesting the migration of strains between countries, (Assigbetsé et al., 1998; Wydra et al., 1998). Within Nigeria, significant differentiation in virulent populations was observed between different agro-ecosystems; that is, virulence of *X. axonopodis* pv manihotis in the rain forest and Southern Guinea Savannah agro-ecological systems in which improved cassava cultivars and land races varieties

*Corresponding author. E-mail: aaogunjobi@hotmail.com.

**Plate 1.** Screen house experimental layout showing cassava accessions planted in a plating tray. Each plating tray holds 68 cassava cultivars. The experimental design is augmented with randomised complete block design (ARCBD).

were grown together and were different from those in the Northern Sudan Savannah, where land races varieties dominated (Ogunjobi et al., 2001; Dixon et al., 2002).

To overcome the threat of CBB in cassava cultivation, there should be a strategy for screening large population of cassava accessions in other to determine resistant varieties for deployment to high disease pressure areas. Here we proposed a new method of screening large population of cassava in the green house to assist in quick identification of CBB resistant cultivars.

## MATERIALS AND METHODS

The method of using cassava planting tree to screen cassava for CBB resistance was introduced and over 1000 cassava accessions held in International Institute of Tropical Agriculture (IITA) Ibadan were assessed for resistance to cassava bacterial blight with this method. This method is design to hold 68 cassava cultivars on a single tray of planting tree. The planting tree is made up of a long white polyethylene tube perforated at the lower end to allow water to drip out and the other upper side with an open end for in filling of the tube as shown in plate 1. The long white polyethylene tube (hereafter referred to as planting tree) in this case was filled with coco-peat which served as soil that supports the roots of the plants and the open end was tighten with a cupper wire. The polyethylene tubes were hung with an iron rod on an aluminum tray in such away that the perforated end was down. The design of the planting tray allow the entire set up to be covered with a covering net and also allow a holes pipe to be inserted at the top of the tube for irrigation as shown in plate 1.

In this experiment, all the different cassava genotypes stem cuttings were well labeled with ribbon to differentiate them and they were planted on the tree of the planting tray. Ten planting trays were arranged with 600 different genotypes to be screened and 80 checks representing 8 checks per block in the greenhouse as shown in plate 1, and the experiment was done twice to avoid over crowding and repeated again for reproducibility. Augmented with completely randomized block design was used as the experimental design. The cassava cultivars were planted on coco-peat and were irrigated once in a week for the first three weeks and later they were irrigated every other day with 5 liters of water. The water drips into the planting tree through the 2 mm rubber holes connected from the water source to the tube. The excess water drained off from the perforated end of the tube (Plate 1). Isolates 131B, 119A and 115A of *X. axonopodis* pv manihotis that has been identified to be consistent in virulence across the six cassava genotypes (Ogunjobi, 2005) were used for screening experiment. The bacterial inocula were developed by washing 20 plates of 48 h old YGCA cultures with 500 ml sterile distilled water for each isolate. This was homogenized on a magnetic stirrer to obtain a uniformly dispersed bacterial suspension. Exactly 1000 ml of this bacterial cells suspension was made up to 20 litres with sterile distilled water to give $10^6$ - $10^8$ cfu/ml. About 50 ml of Tween 80 was then added and shaken together to facilitate adhesion of the inoculum on the under surface (abaxial surface) of the leaves.

**Plate 2.** Spraying cassava germplasm after fourth weeks of growth on the abxial surface of the leaves in the screen house with bacterial suspension using motorized sprayer (Made in Germany, Kleinmotore GMBH, Model: Solo TYP 422).

The cassava germplasms were sprayed, inoculated with the suspension of bacterial isolates after fourth weeks of growth on the abaxial surface of the leaves with motorized sprayer (Model: Solo TYP 422) shown in Plate 2. This was carried out very early in the morning before sunrise to facilitate condition for infection. The cassava plants were covered on the plating tray with the net designed to cover the tray as shown in Plate 1 for 24 h before and 48 h after inoculation to enhance maximum humidity. Also the floor of the screen house was kept wet with running water to increase the humidity and to reduce the greenhouse temperature. The symptoms manifestations were monitored and scored periodically for two months based on the responses of the cassava clones to the disease symptom development; a score on disease rating scale of 1 to 5 (where 1 is no symptom development and 5 is severe damage by the disease). The cassava accession were grouped as: (i) Resistant cultivars if they scored less than 2 on the scale; (ii) Mild resistant clones if score was between 2 and 2.5 on the disease scoring scale; (iii) Tolerant cultivars when score ranged from > 2.5 and 3.5 while (iv) Susceptible clones scored > 3.5 and 4. (v) Any cultivar that scored value that is greater than 4 was regarded as highly susceptible.

## RESULTS

All the cassava germplasm populations of the International Institute of Tropical Agriculture Ibadan (which included landraces or local accessions, improved germplasm and new improved germplasm populations) screened for the inherent ability to withstand CBB attack in the screen-house condition, revealed different levels of susceptibility. A total 1,090 cassava cultivars were inoculated in the screen house using this method of planting as shown in Plate 1 and 2.

A total of 490 local cassava accessions were assessed for susceptibility to CBB. Out of these, 14.3% were resistant to the disease as shown Figure 1. The resistant cultivars were those that manifested symptoms less than or equal to 2.5 while the very resistant ones had disease symptom less than 2 on the rating scale. Majority of the land races population were susceptible to CBB infection, 30.1% were highly susceptible while 12.3% were susceptible to the disease. The susceptible cassava clones were those that developed disease symptom greater than 3.5 while the highly susceptible had symptom greater than 4 on the scale. The tolerant groups were those that developed the symptoms but resisted the spread of the disease beyond the initial blight stage that was scored 3 on the scoring scale, consequently 24.3% were tolerant to the infection among the local cassava accessions. Six hundred cassava accessions of the improved and the new improved were screened in the greenhouse. Of these, 11.1% showed very strong resistance and 30.1%

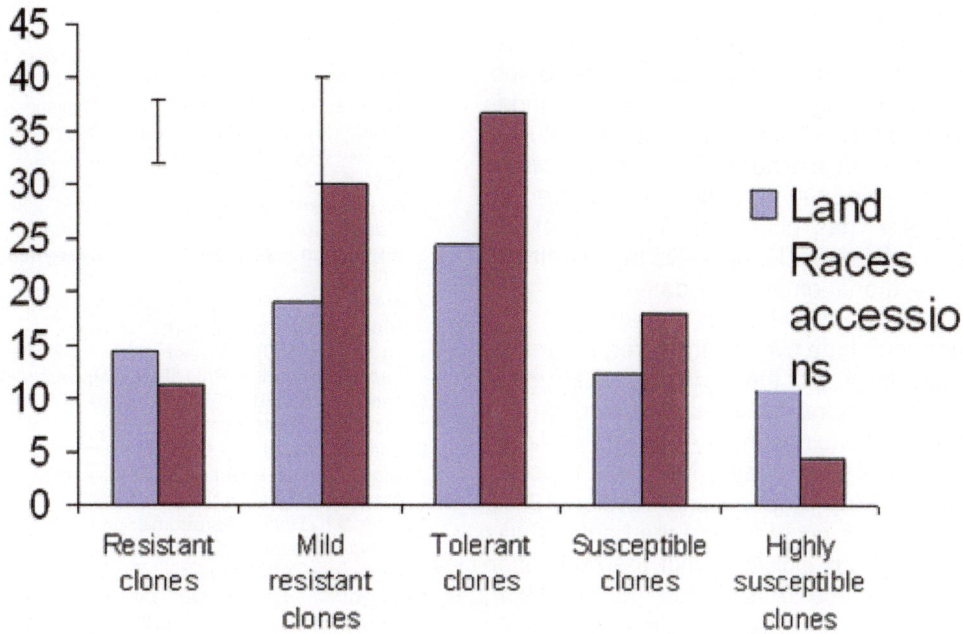


**Figure 1.** Susceptibility level of improved and local cassava accessions to cassava bacterial blight in International Institute of Tropical Agriculture (IITA) cassava germplasm collection

were resistant to the bacterial infections while 4.3% of these cassava accessions were observed as being highly susceptible to the disease while 36.6% were tolerant. The method showed very efficient way of screening large population of cassava cultivars within a very short period in a control environment.

## DISCUSSION

The cassava populations were grouped into three categories based on their susceptibility level to this disease caused by Cassava Bacterial Blight (CBB). The resistant cassava varieties were the fewest (4.3%) and the most numerous were the tolerant varieties (36.6%) among the improved genotypes. The tolerant clones cannot be relied on for either planting in high disease pressure areas or in breeding for disease resistant cassava since they might revert to being susceptible in another environment. Some of the clones that exhibited strong resistance to CBB infection could be developed for mass deployment to areas where there is high level of disease pressure, although caution must be exercised in such deployment since it has been established that some cassava varieties show regional adaptability (Ngeve, 1999) and most cassava adaptability is narrow (Dixon et al., 1994; Dixon and Nukenine, 1997; Dixon et al., 2002). Such resistant cassava cultivars should be examined in the areas where the deployment is intended before mass production could be embarked upon.

Close observations showed that some of these cassava cultivars such as TME 9 and TME 3 that were tolerant in the screen-house were susceptible to the disease from the field record, such cultivars could be deceptive if it is deployed to area with high disease pressure. According to Dixon et al. (2002), Ibadan and Ubiaja were identified as locations with high levels of disease pressure. These two locations in Nigeria are suitable for screening cassava genotypes for resistance to CBB because genotypes that show resistance to CBB at either of these locations could as well be able to resist the disease at other locations where the disease pressure is not as high. The screening carried out in this study was done in Ibadan, one of the locations identified as having high disease pressure with one of the most aggressive isolates (strain 131B) originating from Ibadan. This was in line with the report of Johnson (1984) who proposed that the most powerful test for durability of resistance in plants was for the plant to be grown in an environment favouring the disease. Another advantage of screening all the cassava accessions was to suggest possible cassava cultivars that could be useful in molecular mapping for resistant markers and marker-assisted selection, characterizing resistance to CBB in the cassava population was also to assess the likely effects of selection for this character on the overall genetic diversity.

Host plant resistance has been used extensively for disease control in many crop species. The deployment of resistant varieties is also the major method of CBB control in cassava. Resistance has been found in *M. esculenta* and in its wild relative *M. glaziovii* (Hahn, 1978). Resistance in cassava is said to be polygenic and additively inherited (Lozada, 1990), but the genes involved in the resistance response have not yet been identified (Verdier et al., 1997). Many types of resistance, however, are not long-lasting as a result of rapid changes in pathogen populations (Vera Cruz et al., 2000). Changes in the population structure of *X. axonopodis* pv manihotis has been reported in Colombia where the predominant haplotype has changed (Restrepo et al., 2004). Durability of resistance (R) genes is thus central to sustainable disease management. To date, it has been possible to assess the durability of an *R* gene only after its deployment for a long time over a large area (Johnson, 1984). The capacity to predict the durability of *R* genes would be highly desirable for making sound investments in breeding and genetic engineering. Although, variation in the aggressiveness of *Xam* strains has been reported (Maraite and Meyer, 1975; Restrepo and Verdier, 1997), and in recent study, Africa population of the bacteria was reported to be heterogynous indicating that the population *Xam* is not static in Africa (Assigbetsé, et al., 1998; Ogunjobi et al., 2006, 2007). Also, no clear interactions between cassava cultivars and *Xam* strains have yet been established. The most limiting factor in such studies has been the lack of a suitable way of choosing the cassava varieties that can be used for establishing an appropriate set of host differentials. Since there are very large populations of cassava accessions available in the world cassava collection held at CIAT (Sánchez et al., 1999).

However, with this method of screening large population of cassava cultivars at once in a control environment, it is now possible to establish clear interaction between cassava cultivars and *Xam* strains. This method of screening also is of immense advantage because of the environmental safety regulation that prohibits the release of pathogenic organisms in the environment. This had since being a very serious restriction to massive screening of cassava cultivars in this country for deployment to other neighbouring African countries. Although, the method is labour intensive, it is a task that must be done if we are going to survive the onslaught of Cassava bacterial blight and ensure food security in Africa.

## REFERENCES

Assigbetsé K, Verdier V, Wydra K, Rudolph K, Geiger JP (1998). Genetic variation of the cassava bacterial blight pathogen, *Xanthomonas campestris* pv. *manihotis*, originating from different ecoregions in Africa, pp. 223-229. In IX International Conference on Plant Pathogenic Bacteria, Madras, India.

Dixon AGO, Asiedu R, Hahn SK (1994). Genotype stability and adaptation: Analytical methods and implications for cassava breeding for low input agriculture. In Ofori F and Hahn SK. (eds). Tropical Root Crops in a Developing Economy Proceedings of the 9th Symposium of the International Society forr Tropical Root Crops Accra Ghana. pp. 130-137.

Dixon AGO, Ngeve JM, Nukenine EN (2002). Genotype X environment effects on severity of cassava bacterial blight disease caused by *Xanthomonas axonopodis* pv *manihotis*. Eur. J. Plant Pathol. 108: 763-770.

Dixon AGO, Nukenine EN (1997). Statistical analyses of cassava yield trials with the additive main effects and multiplication interaction (AMMI) Model Afr. J. Root Tuber Crops, 3: 46-50.

Hahn SK (1978). Breeding of cassava for resistance to cassava mosaic disease and bacterial blight in Africa. *In* Diseases of tropical food crops. Proceedings of the International Symposium of the UCL, Louvain-La-Neuve, Belgium. Edited by Maraite H and Meyer YA. pp. 211-219.

Johnson R (1984). A critical analysis of durable resistance. Annu. Rev. Phytopathol. 22: 309-330.

Lozada T (1990). Cruzamientos dialéticos em mandioca (*Manihot esculenta* Crantz). PhD Thesis, Universidade de São Paulo, Brazil.

Lozano JC (1986). Cassava Bacterial Blight: A manageable disease. Plant Dis. 70: 1089-1093.

Maraite H, Meyer J (1975). *Xanthomonas manihotis* (Arthaud- Berthet) Starr, causal agent of bacterial wilt, blight and leaf spots of cassava in Zaire. PANS, 21: 27-37.

Ngeve JM (1999). Agronomic performances, yield stability and field disease reaction of cassava genotypes in the sub-humid forest of Cameroon. Afr. Crop Sci. J. 7: 129-142.

Ogunjobi AA, Fagade OE, Dixon AGO (2006). Molecular variation in population structure of *Xanthomonas axonopodis* pv *manihotis* in the South Eastern Nigeria. Afr. J. Biotechnol. 5(20): 1868-1872.

Ogunjobi AA (2005). Molecular fingerprinting of *Xanthomonas axonopodis* pv *manihotis* Bondar 1915 isolated in Nigeria. PhD thesis University of Ibadan, Nigeria, p. 160.

Ogunjobi AA, Dixon AGO, Fagade OE (2001). Prevalence of cassava bacterial blight in Nigeria. Afr. Crop Sci. Conf. Proc. 5: 479-482.

Ogunjobi AA, Dixon AGO, Fagade OE (2007). Molecular Genetic study of Cassava Bacterial Blight Casual agent in Nigeria Using Random Amplified Polymorphic DNA. EJEAF Chem. 6(9): 2364-2376.

Restrepo S, Verdier V (1997). Geographical differentiation of the population of *Xanthomonas axonopodis* pv. manihotis in Colombia. Appl. Environ. Microbiol. 63: 4427-4434.

Restrepo S, Velez CM, Duque MC, Verdier V (2004). Genetic structure and population dynamics of *Xanthomonas axonopodis* pv. *manihotis* in Colombia from 1995 to 1999. Appl. Environ. Microbiol. 70: 255-261.

Sánchez G, Restrepo S, Duque MC, Fregene M, Bonierbale M, Verdier V (1999). AFLP assessment of genetic variability in cassava accessions (*Manihot esculenta*) resistant and susceptible to the cassava bacterial blight (CBB). Genome, 42: 163-172.

Vera Cruz CM, Bai J, Ona I, Leung H, Nelson RJ, Mew TW, Leach JE (2000). Predicting durability of a disease resistance gene based on an assessment of the fitness loss and epidemiological consequences of avirulence gene mutation. Proc. Natl. Acad. Sci. USA, 97(25): 13500-13505.

Verdier V, Restrepo S, Boher B, Nicole M, Geiger JP, Alvarez E, Bonierbale M (1997). Cassava bacterial blight: Recent achievement in understanding the disease. Afr. J. Root Tuber Crops, 2: 64-68.

Wydra K, Fessehaie A, Fanou A, Sikirou R, Janse, Verdier V, Rudolph K (1998). Variability of isolates of *Xanthomonas campestris* pv. *manihotis* (*XCM*), incitant of cassava (*Manihot esculenta* Crantz) bacterial blight, from different geographic origins in pathological, physiological, biochemical and serological characteriztics. In: Balasubramanian R, Lalithakumari D, Mahadevan A, eds. Proceedings of the 9th International Congress of Plant Pathogenic Bacteria 1996. University of Madras, Madras, India, pp. 317-323.

# 8

# Studying the interactive effect of potassium application and individual field crops on root penetration under drought condition


**Sayed Alireza Valadabadi[1] and Hossein Aliabadi Farahani[2*]**

[1]Islamic Azad University, Shahr-e-Qods Branch, Iran.
[2]Islamic Azad University, Takestan Branch, Iran.



**In this study, interactive effects of different potassium application and water restrict on: root penetration at Iran. The experimental unit had designed by achieved treatment in factorial on the basis completely randomized block design with three replications. Certain factors including potassium and non-potassium applications (0 and 200 kg/ha), certain field crops (maize, sorghum and millet) and water supply were studied. In this study crops water supply was determined by indicated irrigation conditions by keeping leaf relative water content (RWC) > 95% (non-drought stress condition or irrigation conditions) and drought condition by RWC = 60 - 70%. We noticed, however K fertilizer significantly increased the root penetration of plants. Although the non-drought stress treatment significantly increased root penetration and highest root penetration of maize was achieved under irrigation conditions (non-drought), but the highest roots penetration of sorghum and millet were indicated under the drought condition. Whereas K application persist less damaging of drought stress result and it enabled plant to significantly grew its root and penetrating along soil depth under the drought condition. Our finding may give applicable advice to farmers and agricultural researchers for management and concern on fertilizer strategy and carefully estimate soil potassium supply within arid or semi arid regions as most challengeable issues of environmental safety.**



**Key words:** Potassium fertilizer, drought stress, root penetration, maize, sorghum, millet.


## INTRODUCTION

A sufficient supply of soil water is essential for crop growth, replacement of transpiration loss, and transportation of crop nutrients to roots. Drought occurs when there is a deficit in soil water supply to the crop. Severe drought limits crop water use and reduces yield. Earl and Davis (2003) summarized three main mechanisms through which corn yield is reduced by soil water deficit: (i) reduced canopy absorption of incident photosynthetically active radiation, (ii) decreased radiation use efficiency, and (iii) reduced harvest index. The ability to tolerate drought and have acceptable yields is limited among cultivars within a species (Serraj and Sinclair, 2002; Purcell et al., 2003). The objectives of a study were to; (i) quantify the relationship between crop water use and dry matter (DM) yield for corn (*Zea mays* L.), foxtail millet (*Setaria italica* L. Beauv.), and winter triticale (*X Triticosecale* Wittmack); and (ii) determine the range and distribution of expected DM yields for these three crops in the central Great Plains based on historical precipitation records. The three crops were grown in a drought stress condition no-till corn–millet–triticale sequence. Dry matter production was linearly correlated with water use for all three crops, with regression slopes ranging from 24.2 (corn) to 33.0 kg $ha^{-1}$ $mm^{-1}$ (millet). Precipitation use efficiency for the millet–triticale–corn forage system was 8.7 kg $ha^{-1}$ $mm^{-1}$, suggesting this as an efficient forage system for the region (Nielsen et al., 2006). Bandaru et al. (2006) studied the growing grain sorghum plants in clumps would result in fewer tillers and less vegetative growth so that more soil water would be available during the grain-filling period. The results suggest that planting grain sorghum in clumps rather than spaced uniformly conserves soil water use until later in the season and may enhance grain yield in semiarid dryland environments.


*Corresponding author. E-mail: farahani_aliabadi@yahoo.com.

**Table 1.** The results of soil analysis.

| Soil texture | Sand (%) | Silt (%) | Clay (%) | K (mg/kg) | P (mg/kg) | N (mg/kg) | Na (Ds/m) | EC (1: 2.5) | pH (In soil) | Depth of sampling (cm) |
|---|---|---|---|---|---|---|---|---|---|---|
| Sandy loam | 49 | 29 | 22 | 145.2 | 4.2 | 36.7 | 0.04 | 0.18 | 8.0 | 0-15 |
| | 56 | 25 | 19 | 121.3 | 3.7 | 25.2 | 0.03 | 0.15 | 7.8 | 15-30 |

Drought stress reduced dry matters of balm by reduction in the leaf area, plant height and lateral stem number (Aliabadi et al., 2009). Also, drought stress decreased biological yield, grain yield and root length of barley (Khalvati et al., 2005).

Potassium is important in the growth of crops and an important ion in the physiology of plant water relations. Management practices have a direct effect on P, K, S and Ca availa-bility and utilization by crops. Manure, as opposed to inorganic fertilizers, supplies nutrients over time through mineralization. Also, the addition of organic matter with manure or with the use of an efficient crop rotation will affect soil properties such as cation exchange capacity and pH, and therefore root and nutrient interactions (Hickman, 2002). In addition, the presence or absence of certain elements can affect the general soil quality. For example, K is a soil aggregating agent which is known to have a positive effect on soil physical properties and subsequently crop yields (Hamza and Anderson, 2003). Recommendations of Wortmann et al. (2009) for P, K, and S were evaluated using results from 34 irrigated corn (*Z. mays* L.) trials conducted in diverse situations across Nebraska. The results indicate a need to revise the current recommendation for P, to maintain the current K and S recommendations, and to use soil organic matter and pH in addition to soil test nutrient values in estimating applied nutrient requirements for irrigated high yield corn production. In present study, there has been a growing interest in brown midrib (BMR) sorghum (*Sorghum bicolor* (L.) Moench.) x Sudan grass (*Sorghum sudanense* Piper) hybrids (SxS) as a replacement for silage corn (*Z. mays* L.) in the North-eastern USA.

The objective was to evaluate the impact of K fertilizer management (0, 112 or 224 kg $K_2O$ $ha^{-1}$ $cut^{-1}$) under optimum N management (112-168 kg N $ha^{-1}$ $cut^{-1}$) on yield, quality and K concentrations of BMR SxS over a 2-year period. Potassium application increased forage K concentration up to 13 mg K $kg^{-1}$ dry matter (in the first cut in 2003). Without K addition in the 2-year period, K concentrations in the forage decreased from 23 g $kg^{-1}$ for the first cutting in 2002 to 15 g $kg^{-1}$ for the second cut in 2003. Low K forage was obtained for all second-cut forage unless 224 kg $K_2O$ $ha^{-1}$ $cut^{-1}$ had been added. These results suggest low K BMR SxS forage can be harvested from initially high K soils without loss in dry matter yield as long as no additional K is added (Ketterings et al., 2005). Rosolem et al. (2005) evaluated K leaching from millet straw grown under potassium rates (0, 100, 200, and 300 mg $dm^{-3}$), and submitted to five levels of simulated rain (5, 10, 20, 40 and 80 mm). Plants were grown in soil filled pots in a greenhouse. Potassium deficiency speeds up millet dehydration after herbicide application and increases lightly rain water retention in the straw. The amount of K leached right after plant desiccation is correlated with the residue nutrient content and can be as high as 64 kg $ha^{-1}$ considering a mulch of 8 t $ha^{-1}$. Although well-nourished millet plants release considerable amounts of K with the first rains, a large percentage of the nutrient is still retained in the straw. Therefore, this study was undertaken to examine the interactive effects of potassium fertilizer and water in the soil on root penetration of maize, sorghum and millet.

## MATERIALS AND METHODS

This study was conducted on experimental field of the Ishmael Abad station in Qazvin at Iran (36°15' N, 49°55' W; 1300 m above sea level) from 10 June to 20 October 2005, with sandy loam soil (Table 1). The mean annual temperature (27°C) and rainfall in the study area is distributed with an annual mean of 309 mm. The experimental design was a factorial experiment in a completely randomized block design with three replications. The treatments included potassium ($K_1$) and non-application ($K_2$) (0 and 200 kg/ha), certain field crops: maize ($C_1$), sorghum ($C_2$) and millet ($C_3$)) and irrigation levels with estimate leaf relative water content (RWC) by > 95% (non-drought stress condition ($S_1$)) and irrigation under RWC = 60-70% (drought stress condition ($S_2$)). Maize (*Z. mays* L. Cul. Single-cross 704), sorghum (*S. bicolor* L. cul. Speed Feed) and millet (*Panicum italicum* L. cul. Notre Feed) were used in this experiment. Initially, plant nutrient feed of nitrogen supply was added by applying 300 kg/ha urea twice; 150 kg/ha urea at the cultivation time and further 150 kg/ha urea at beginning of stem elongation stage as well as 200 kg/ha phosphorus (ammonium phosphate) and potassium ($K_2O$) through cultivation time once. At the plants growth period and between both irrigation, we selected 20 young leaves from each plot for determination of RWC by following formula (Aliabadi Farahani et al, 2008).

$$\text{RWC} = \frac{\text{Leaves fresh weight} - \text{Leaves dry weight}}{\text{Leaves turgid weight} - \text{Leaves dry weight}} \times 100$$

Also, depth of root penetration was determined by sampling of deep soil. Thus, 24 h after irrigation (field capacity condition) depth of root penetration was determined by Auger (Height = 100 cm and Radius = 15 cm). Sampling was performed twice to determination of the depth of root penetration > 100 cm (Newman, 1995). Finally, obtained data were subjected to analysis of variance (ANOVA) using Statistical Analysis System (SAS institute Cary, USA, 1988) and means considered significant at $P < 0.05$.

**Table 2.** Analysis of variance.

| S.O.V | df | Mean squares penetration depth of root |
|---|---|---|
| Replication | 2 | 311.41 |
| Drought stress | 1 | * 598.05 |
| Crops | 2 | * 701.406 |
| K fertilizer | 1 | ** 712.316 |
| Stress × Crops | 2 | * 632.011 |
| Crops × K fertilizer | 2 | 281.04 |
| Stress × K fertilizer | 1 | 203.008 |
| Stress × Crops × K fertilizer | 2 | * 531.106 |
| Error | 22 | 37.084 |
| CV (%) | | 7.11 |

Note** and *: Significant at 1 and 5% levels, respectively.

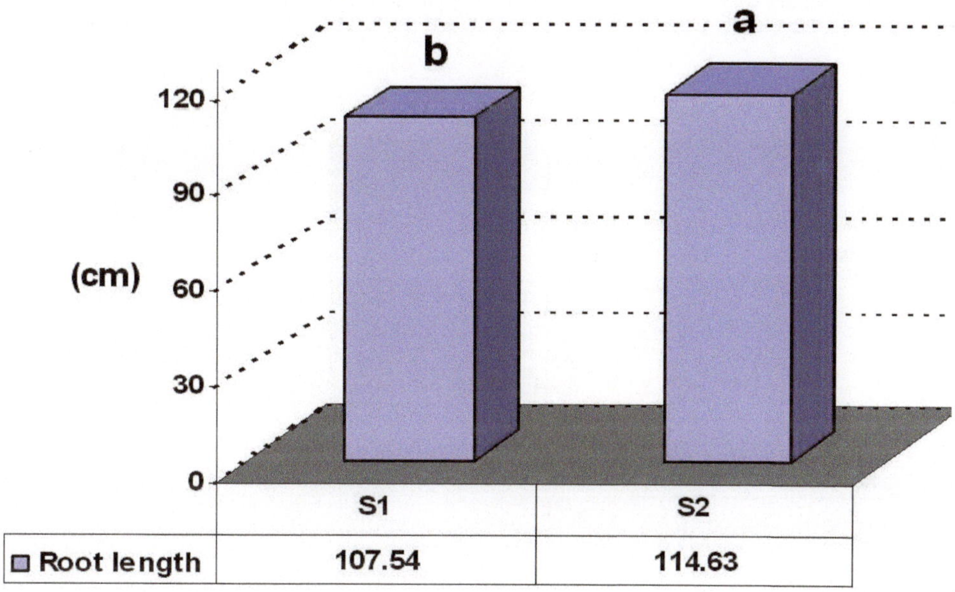


**Figure 1.** Effect of drought stress on root penetration.

## RESULTS AND DISCUSSION

Final results of plants features showed that drought stress had significant effect on root penetration ($P < 0.05$) and we noted a highest depth of root penetration (root length) under the drought stress (Table 2 and Figure 1). Those findings are in agreement with the observations of Katerji et al. (2004) and Ajayi and Olufayo (2004). As it was shown in the results, reducing water supply in soil achieved a situation for plant to pursue root growth though soil depth. This shows that in order to resist drought stress, the plant employed different strategies throughout individual survival straggle by drought conditions. In addition, K application had a significant effect on root penetration ($P > 0.01$), therefore, our findings indicated higher root penetration under K application (Table 2 and Figure 2). Those results were similar with the findings of Wortmann et al. (2009); Yin and Vyn (2002) and Parsons et al. (2007). Potassium is important for a plant's ability to withstand extreme drought stress. Some field crops showed ability of water relations adjustment which refers to water use efficiency. Soil nutrients like as potassium ions affect water transport in whole plant, maintain cell pressure and regulate the opening and closing of stomata (small openings found on the leaf responsible for cooling and taking in carbon dioxide for photosynthesis). Regarding to this fact, our data indicated that crops plant had a significant effect on root penetration ($P < 0.05$) and was preformed higher root penetration in millet (Table 2 and Figure 3).

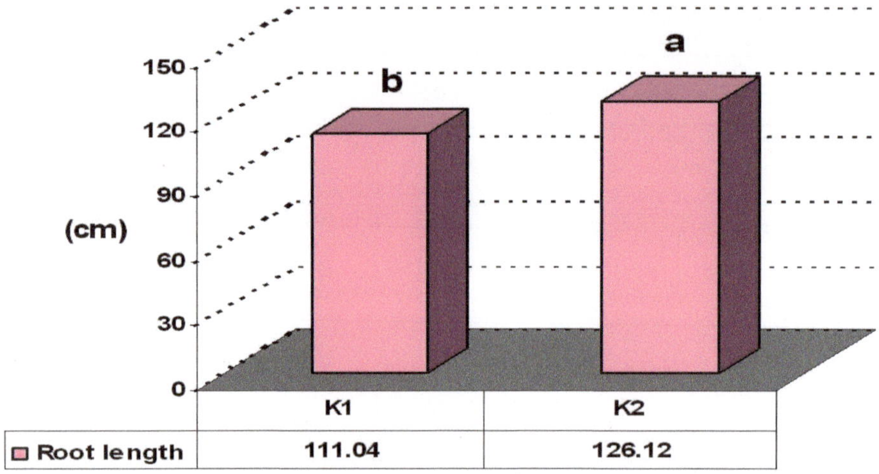


**Figure 2.** Effect of K application on root penetration.

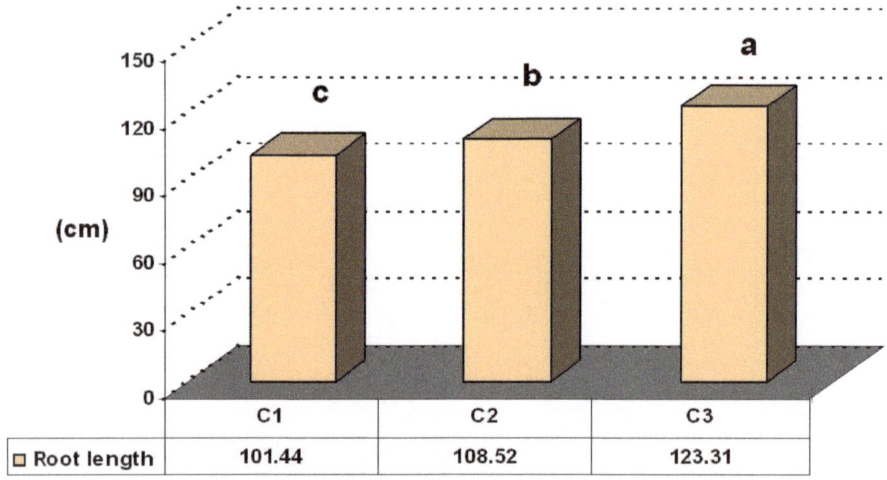


**Figure 3.** Root penetration in crops plant.

Significant different between crops plants treated with K application under water restriction was highlighted to compare with non-application K under drought condition and treatment had significant effect on depth of root penetration ($P < 0.05$) and however, higher root penetration was indicated millet with K application and none stress condition. Surprisingly, K application contributed plants to approach longer root length as well through the condition of drought stress (Table 2 and Figure 4). Our results of treatments interaction were similar to the results of Rosolem et al. (2005) and Ashraf et al. (2003).

### Conclusion

Our study showed that K application reduces the damaging effects of drought stress by increasing depth of root penetration. Control of drought stress has been paying attentions due to the most important environmental factor in arid and semi –arid regions. It may useful to consider on screen certain local crops and fertilizer strategies to gain higher predictability under scope of limited available water recourse in these regions. Practically, findings may suggest farmers and agricultural researchers to consider carefully on limiting or control the huge among of soil potassium application in suffered soils by water restriction as current challenge of scientist in global changes.

### ACKNOWLEDGEMENTS

We thank Dr. Jahanfar Daneshian and Dr. Amir Hossein Shiranirad for their great support, encouragement and help with the experiments.

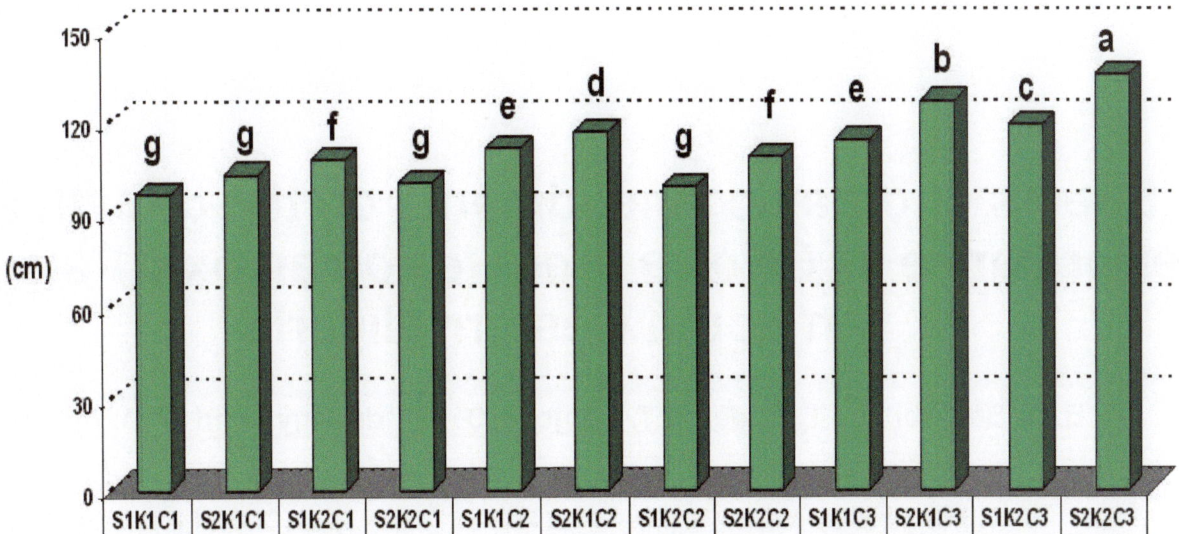


| | S1K1C1 | S2K1C1 | S1K2C1 | S2K2C1 | S1K1C2 | S2K1C2 | S1K2C2 | S2K2C2 | S1K1C3 | S2K1C3 | S1K2C3 | S2K2C3 |
|---|---|---|---|---|---|---|---|---|---|---|---|---|
| Root length | 96.25 | 102.5 | 107.25 | 99.75 | 111.25 | 116.25 | 98.25 | 108.25 | 113.75 | 126.25 | 118.5 | 134.75 |

**Figure 4.** Effect of interaction on root penetration.

## REFERENCES


Ajayi AE, Olufayo AA (2004). Evaluation of two temperature stress indices to estimate grain sorghum yield and evapotranspiration. Agron. J. 96: 1282-1287.

Aliabadi Farahani H, Lebaschi MH, Shiranirad AH, Valadabadi SAR, Daneshian J (2008). Effects of arbuscular mycorrhizal fungi, different levels of phosphorus and drought stress on water use efficiency, relative water content and proline accumulation rate of Coriander (*Coriandrum sativum* L.). J. Med. Plants Res. 2(6): 125-131.

Aliabadi Farahani H, Valadabadi SAR, Daneshian J, Khalvati MA (2009). Evaluation changing of essential oil of balm (*Melissa officinalis* L.) under water deficit stress conditions. J. Med. Plants Res. 3(5): 329-333.

Ashraf M, Arfan M, Ahmad A (2003). Evaluation of the usefulness of senescing agent potassium iodide for assessing inter-cultivar variation for drought tolerance in pearl millet [*Pennisetum glaucum* (L.) R.Br.]. Aust. J. Exp. Agric. 43(11): 1337-1343.

Bandaru V, Stewart BA, Baumhardt RL, Ambati S, Robinson CA, Schlegel A (2006). Growing dryland grain sorghum in clumps to reduce vegetative growth and increase yield. Agron. J. 98: 1109-1120.

Earl H, Davis RF (2003). Effect of drought stress on leaf and whole canopy radiation use efficiency and yield of maize. Agron. J. 95: 688–696.

Hamza MA, Anderson WK (2003). Responses of soil properties and grain yields to deep ripping and gypsum application in a compacted loamy sand soil contrasted with a sandy clay loam soil in Western Australia. Aust. J. Agric. Res. 54: 273-282.

Hickman MV (2002). Long term tillage and crop rotation effects on soil chemical and mineral properties. J. Plant Nutr. 25: 1457-1470.

Katerji N, Hoorn JWV, Hamdy A, Mastrorilli M (2004). Comparison of corn yield response to plant water stress caused by salinity and by drought. Agric. Water Manage. 65(2): 95-101.

Ketterings QM, Godwing G, Cherney JH, Kilcer TF (2005). Potassium management for brown midrib sorghum x sudangrass as replacement for corn silage in the North-eastern USA. J. Agron. Crop Sci. 191(1): 41-46.

Khalvati MA, Mzafar A, Schmidhalter U (2005). Quantification of water uptake by arbuscular mycorrhizal hyphae and its significance for leaf growth, water relations, and gas exchange of barley subjected to drought stress. Plant Biol. Stuttgart 7(6): 706-712.

Nielsen DC, Vigil MF, Benjamin GJ (2006). Forage yield response to water use for dryland corn, millet, and triticale in the Central Great Plains. Agron J. 98: 992-998.

Parsons KJ, Zheljazkov VD, MacLeod J, Caldwell CD (2007). Soil and tissue phosphorus, potassium, calcium and sulfur as affected by dairy manure application in a no-till corn, wheat and soybean rotation. Agron. J. 99: 1306-1316.

Purcell LC, Keisling TC, Sneller CH (2002). Soybean yield and water extraction in response to deep tillage and high soil aluminum. Commun. Soil Sci. Plant Anal. 33: 3723–3735. http://agron.scijournals.org/cgi/external_ref?access_num=11841674&link_type=MED.

Rosolem CA, Calonego JC, Foloni JSS (2005). Potassium leaching from millet straw as affected by rainfall and potassium rates. Commun. Soil Sci. Plant Analy. 36(7-8): 1063-1074.

SAS Institute (1998). SAS user's guide: Statistics. SAS Inst., Cary, NC.

Serraj R, Sinclair TR (2002). Osmolyte accumulation: Can it really help increase crop yield under drought conditions? Plant Cell Environ. 25: 333-341.

Wortmann CA, Dobermann AR, Ferguson RB, Hergert GW, Shapiro CA, Tarkalson DD, Walters DT (2009). High-yielding corn response to applied phosphorus, potassium, and sulfur in Nebraska. Agron. J. 101: 546-555.

Yin X, Vyn TJ (2002). Soybean responses to potassium placement and tillage alternativesfollowing no-till. Agron. J. 94: 1367-1374.

# Effects of organic amendments on the growth and performance of *Celosia argentea* on an oxic Paleustalf in South western Nigeria

**B. A. Senjobi*, O. T. Ande, C. T. Senjobi, O. A. Odusanya and O. D. Ajilore**

College of Agricultural Sciences, Olabisi Onabanjo University, Ogun State, Nigeria.

**The problems associated with availability of space for cultivation due to high demands on land for other uses has made virtually all farmers to embark on continuous cropping at the expense of soil fertility management. This study was therefore conducted to assess the effect of organic amendments on the yield and yield components of *Celosia argentea*. A field experiment was carried out to examine the effects of four different levels 0, 100, 150, and 200 g of different organic amendments (maize stovers, legume and cassava) on growth and yield of *C. argentea* and soil fertility. The experiment was laid out in a randomized complete block design (RCBD) with three replications. Data collected were analyzed and significant means was compared using analysis of variance (ANOVA) and Duncan multiple range test (DMRT) at 5%. Application of organic amendments had significant effect ($p < 0.05$) on growth and short yield of *Celosia*. The leguminous residues gave the highest plant growth and cumulative yield while the least was obtained with control. Cassava residues particularly, gave its best yield (133.36 g/ha) mainly at the second rate (150 g) of treatment while legumes gave the highest yield at the highest level of treatment (345.6 g/ha). There is no significant increase between the second (150 g/ha) and third (200 g/ha) rates of maize stovers treatment although, the third rate still gave the highest, that is, yield the same as legume. It was concluded that organic amendments significantly increased the soil fertility and yield of *Celosia* planted on the soil.**

**Key words:** Organic amendments, yield and yield components, soil fertility, *Celosia* spp.

## INTRODUCTION

Soil nutrients depletion is one of the most serious problems currently affecting agricultural productivity in developing countries of the tropics, including Nigeria. The intensification of cultivation with little or no fertility management has been one of the critical factors contributing to this decline in natural ecological conservatory balances in the landscape. This in turn poses great difficultly for productivity increase to meet the food and fibre needs of a rapidly growing population, thus, endangering food security (Senjobi, 2007). As population increases geometrically and more intensive demand are made on the earth's natural resources, including land, available land for cultivation will reduce. The problems of urbanization, incessant demand for food crops, land tenure system, indifferent or outdated government policies and industrialization make farmlands scarce and inaccessible to farmers. Consequently, continuous cultivation is prevalent. This in turn reduces fallow period and other cultural practices that can help to replenish the soil nutrients (Akinsehinwa, 2007; Adedipe, 1991). The depletion of soil nutrients due to continuous cropping reduces the soil organic matter, cause significant acidification and yield reduction (Batiano and Makwunye, 1991; Anonymous, 1995; Ajilore, 2008). There is therefore need for adequate fertilization or manuring in order to sustain soil productivity for optimum growth and yield of planted crops to ensure food security for the ever increasing growing population. The impor-tance of mineral and organic fertilizers in improving soil productivity has been documented (Cooke, 1982; Setia

*Corresponding author. E-mail: bolasenjobi@gmail.com.

and Sharma, 2004; Odu and Mba, 1991; Ketiku et al., 1985; Palada et al., 1992; Adenawoola et al., 2005).

Although, there is a wealth of information on the dietary importance of *Celosia argentea*, a leafy vegetable which is extensively grown in West Africa (Tindall, 1983; Akanbi, 2000a), little is known about its nutritional requirements generally. This is because in West Africa's farming system, leafy vegetables are regarded as back-yard crops. However, this idea is changing, with the development of private and government vegetable gardens on a large scale. This, coupled with the scarcity and high cost of obtaining mineral fertilizer, calls for search for alternatives (Akanbi et al., 2006), and thus necessitate this research. Specifically, this work focuses on the assessment of effect of organic amendment on the growth and performance of *C. argentea* and soil properties.

## MATERIALS AND METHODS

### Description of the study area

The study was carried out at the Teaching and Research Farm of Olabisi Onabanjo University, College of Agricultural Sciences, Ayetoro, Ogun State. The site is located on latitude 7˚12'N and longitude 3˚0'E. The site is situated in a derived guinea savannah part of South Western Nigeria. The geology of these is mainly lateritic with low activity clay mineral soils that are more of an Alfisol. The land is mainly used for farming, especially for the cultivation of cassava, maize and vegetables.

### Field work

The experimental site was manually cleared, ploughed and harrowed before cultivation. Organic amendments: legumes, maize stovers, and cassava were incorporated into the soils two weeks before planting to allow decomposition and mineralization to take place effectively. The soil samples were then taken and analyzed for physico-chemical constituents. Soil pH was determined in both water and 0.01 M potassium chloride solution (1:1) using glass electrode pH meter (Mclean, 1965). Total nitrogen was determined by the macro-kjeldahl digestion method of Jackson (1962) available P was extracted using Bray-1 extract followed by molybdenum blue colorimetry. Exchangeable cations were extracted with 1M NH4OAC (pH 7.0), potassium, calcium and sodium were determined using flame photometer and exchangeable Mg by atomic absorption spectrophotometer (Sparks, 1996). Exchangeable acidity was determined by the KCl extraction method (Mclean, 1965), organic carbon was determined using dichromate wet oxidation (Walkley and Black, 1934). Organic matter was got by multiplying the percent organic carbon by 1.72. Cation exchange capacity (CEC) was calculated from the sum of all exchangeable cations. Available micronutrients were determined by atomic absorption spectrophotometer (AAS) method after leaching on NH4Cl (Senjobi, 2007). Particle size analysis was by the Bouyoucos (1951) hydrometer method using calgon as dispersing agent.

Seeds of TLV8 variety of *Celosia* spp were used as test crop. The experimental design was a randomized complete block design (RCBD) with three replicates. Agronomic attributes such as plant height, leaf area etc. were measured. The data collected were subjected to analysis of variance (ANOVA) and means were separated using the Duncan multiple range test as at 5%.

## RESULTS AND DISCUSSION

The pH of the soil is slightly acidic ranging from 6.1 - 6.9 for cassava and legume treated soils. The pH increases slightly with increase in concentration or level of materials applied. Maize treated soils have a relatively constant pH value except in the second level of treatment that is, 150 g (Table 1). Legume treated soil gave the highest total nitrogen value. This may probably be due to its nitrogen fixing capacity. This was followed by cassava, with the least value in maize. Organic carbon level in the soil also follows the same trend, with the leguminous treatment having the highest value and maize treatment soil had the least. However, this is lower than the value obtained in the control soil. There was an appreciable increase in the level of available phosphorus over control in all the treatments. Available P in cassava treated soils increases with the concentration of amendment. The highest values were obtained at the third rate in maize. Other nutrients e.g. Ca, Mg, Fe, K etc. followed the same trend either with N or available P (Table 1).

Enhancement of various attributes with significant increase in growth (at $p < 0.05$) varying proportionally with increase in the level of organic amendment applied. Virtually all the measurable parameters such as root weight, whole shoot for market, leaf stalk, leaf girth etc. increase significantly compared to control (Tables 2 - 20). They all increase at an increasing rate, except in cassava and maize treatments where the whole leaf stalk weight gave appreciable lower value at first level. In case of legume, the second level gave a value relatively lower than that of the control (Table 2).

The results showed that the third level of legume treatment (PL C3) and the second level of cassava (Pc C3) gave higher values of root weights that are significantly different at $p < 0.05$ than other treatments (Table 5). However, the PL C3 and Pc C3, respectively, gave the values that are significantly different ($p < 0.05$) from other treatments with respect to which shoot and plant heights (Tables 6 and 7). All the treatments are not significantly different from one another ($p < 0.05$) with respect to root biomass of the planted crops except the PL C3 which gave appreciably significant value (Table 8). The similar trend occurred in number of leaves and whole leaf stalk weights where PC C2 gave a significant higher value (Tables 9 and 11). The following trend was observed in leaf girth PL C1 < PL C3 < Po Co < Pc C2 with organic amendments (Table 10).

The use of organic amendments produced significant effect on the performance of *Celosia* when compared with the control plants. All the growth parameters taken were positively improved by the organic amendments Adequate nutrients availability had been indicated to improve crop growth and yield parameters. For instance, it has been reported that when N supply is not limiting, dry matter production, partitioning and assimilation as well as production of certain organic compounds (e.g. Proteins) would not be hampered. But a shortage in any

**Table 1.** Physico-chemical properties of the soil used.

| No | Treatment Sample | pH $H_2O$ | pH KCl1:1 | Total N | Organic C | Avail. P | Ca | Mg | K | Na | Al | Exchangeable acidity | ECEC | Fe | Mn | Cu | Zn | % Clay | % Silt | % Sand |
|---|---|---|---|---|---|---|---|---|---|---|---|---|---|---|---|---|---|---|---|---|
| | | | | g kg-1 | | mg $kg^{-1}$ | | | | C mol kg-1 | | | | | mg $kg^{-1}$ | | | | | |
| 1 | $P_o C_o$ | 6.40 | 5.90 | 3.73 | 36.01 | 138.06 | 12.22 | 6.71 | 1.53 | 0.94 | Nil | 0.15 | 21.55 | 66.90 | 98.74 | 8.92 | 27.32 | 16.80 | 23.40 | 59.80 |
| 2 | $P_m C_1$ | 6.70 | 5.80 | 3.52 | 33.95 | 141.02 | 10.06 | 5.54 | 1.40 | 0.94 | Nil | 0.10 | 18.05 | 71.77 | 98.92 | 11.72 | 36.17 | 20.80 | 16.40 | 62.80 |
| 3 | $P_m C_2$ | 6.90 | 5.58 | 3.36 | 32.44 | 135.17 | 8.98 | 4.95 | 1.25 | 0.86 | Nil | 0.13 | 16.11 | 57.93 | 98.14 | 8.52 | 22.33 | 18.80 | 17.40 | 63.80 |
| 4 | $P_m C_3$ | 6.70 | 5.60 | 3.67 | 35.38 | 150.02 | 10.50 | 5.33 | 1.41 | 0.92 | Nil | 0.17 | 18.33 | 64.30 | 97.93 | 9.02 | 23.65 | 18.90 | 20.40 | 60.50 |
| 5 | $P_c C_1$ | 6.30 | 5.60 | 3.81 | 36.72 | 136.11 | 15.99 | 6.78 | 1.48 | 0.98 | Nil | 0.20 | 25.43 | 66.95 | 96.57 | 11.53 | 29.77 | 16.80 | 18.40 | 64.80 |
| 6 | $P_c C_2$ | 6.30 | 5.70 | 3.85 | 37.12 | 150.07 | 9.43 | 5.07 | 1.19 | 0.81 | Nil | 0.20 | 16.70 | 56.03 | 86.17 | 9.21 | 20.93 | 18.80 | 17.40 | 61.80 |
| 7 | $P_c C_3$ | 6.20 | 5.60 | 3.55 | 34.25 | 161.15 | 15.15 | 6.32 | 1.38 | 0.88 | Nil | 0.20 | 23.93 | 55.87 | 91.74 | 9.23 | 17.45 | 16.80 | 15.40 | 67.80 |
| 8 | $P_L C_1$ | 6.40 | 5.70 | 3.61 | 34.85 | 135.53 | 15.05 | 6.03 | 1.56 | 1.02 | Nil | 0.15 | 23.75 | 66.15 | 104.63 | 10.78 | 31.18 | 16.10 | 18.40 | 65.50 |
| 9 | $P_L C_2$ | 6.10 | 5.30 | 4.09 | 39.42 | 132.53 | 10.06 | 5.26 | 1.25 | 0.82 | Nil | 0.17 | 17.56 | 54.73 | 91.67 | 8.85 | 19.93 | 18.10 | 17.40 | 64.50 |
| 10 | $P_L C_3$ | 6.00 | 5.30 | 7.76 | 37.42 | 143.16 | 9.39 | 5.42 | 1.42 | 0.91 | Nil | 0.12 | 17.30 | 57.19 | 94.04 | 9.22 | 23.20 | 18.10 | 16.40 | 65.50 |

Type of plant residue: $P_o$ - Control (No plant residue); $P_m$ - Maize residue; $P_c$ - Cassava residue; $P_L$ - Legume (Mucuna) residue. Concentration of plant residue: $C_o$ - Control (No inclusion); $C_1$ - 100 g inclusion level; $C_2$ - 150 g inclusion level; $C_3$ - 200 g inclusion level.

**Table 2.** Mean weight (g) of yield/agronomic parameters.

| No | Morphological properties | Control | Legume | | | Maize | | | Cassava | | |
|---|---|---|---|---|---|---|---|---|---|---|---|
| | | | $C_1$ | $C_2$ | $C_3$ | $C_1$ | $C_2$ | $C_3$ | $C_1$ | $C_2$ | $C_3$ |
| 1 | Root WT (g) | 6.80 | 12.60 | 8.20 | 8.20 | 10.50 | 11.00 | 17.00 | 6.10 | 47.90 | 15.70 |
| 2 | WSM (g) | 39.80 | 90.20 | 58.30 | 58.30 | 85.70 | 62.20 | 86.40 | 56.10 | 62.80 | 133.60 |
| 3 | WLS (g) | 38.30 | 41.80 | 30.80 | 30.80 | 29.90 | 30.20 | 51.20 | 23.50 | 82.40 | 47.70 |
| 4 | DRYRT | 6.10 | 3.90 | 2.80 | 2.80 | 6.70 | 6.60 | 6.80 | 2.40 | 17.10 | 8.60 |
| 5 | BMASS (g) | 8.30 | 8.70 | 5.40 | 5.40 | 6.10 | 10.40 | 13.20 | 6.90 | 14.30 | 13.20 |
| 6 | NL | 24.80 | 22.30 | 24.80 | 24.80 | 13.40 | 15.20 | 32.20 | 18.70 | 17.80 | 17.80 |
| 7 | PH (cm) | 20.20 | 18.30 | 19.40 | 19.40 | 13.30 | 15.50 | 19.40 | 16.70 | 18.50 | 17.60 |
| 8 | G (cm) | 2.70 | 2.80 | 2.50 | 2.50 | 1.90 | 2.00 | 2.50 | 2.00 | 2.90 | 2.40 |
| 9 | NB | 8.00 | 5.00 | 5.50 | 5.50 | 6.00 | 7.00 | 5.90 | 5.20 | 5.30 | 4.60 |
| 10 | LA (mm) | 14.80 | 28.70 | 19.60 | 19.60 | 22.70 | 12.90 | 20.40 | 11.30 | 33.30 | 20.20 |

RTWT - Root weight; WSM - Whole shoot for market; WLS - Whole leaf stalk weight; NL - Number of leaves; DRYTRT - Dry root weight; BMASS - Root biomass; G - Leaf girth; NB - Number of branching; LA - Leaf area (cm); PH - Plant height.

of the required nutrients may have pronounced effect on some or all these physiological processes (Akanbi et al., 2000 cited from Ajilore, 2008). This explains the better performance and attributes of the crops treated with leguminous residues (mucuna), especially the third level of application, which have high nitrogen content and is able to fix nitrogen into the soil through the help of micro-organisms in their root nodules, over other crops treated with non-leguminous treatments and the control.

Also, phosphorus, when optimally available, helps in cellular processes, being a component of RNA and DNA (Ajilore, 2008). It aids cell division and fast growth and can be found mainly in large quantities in young plants (Schlegel, 1986 cited from Ajilore, 2008). This explains the higher values of some morphological properties e.g. leaf girth, leaf area etc. of crops treated with cassava residues which have a higher level of available P compared to the control and other treatments. The

**Table 3.** Mean weight (g) of morphological properties.

| Obs | TRT | _Type | _Freq | NL | PH | G | NB | LA |
|---|---|---|---|---|---|---|---|---|
| 1 | $P_o C_o$ | 0 | 3 | 26.4 | 20.2 | 2.7 | 8.6 | 30.0 |
| 2 | $P_c C_2$ | 0 | 3 | 17.8 | 18.5 | 3.0 | 5.3 | 33.3 |
| 3 | $P_c C_3$ | 0 | 3 | 17.8 | 17.6 | 2.4 | 5.7 | 18.4 |
| 4 | $P_L C_o$ | 0 | 3 | 22.0 | 22.4 | 2.5 | 6.2 | 21.6 |
| 5 | $P_L C_1$ | 0 | 3 | 22.3 | 18.3 | 2.8 | 5.0 | 28.7 |
| 6 | $P_L C_2$ | 0 | 3 | 24.8 | 19.4 | 2.5 | 5.5 | 19.6 |
| 7 | $P_L C_3$ | 0 | 3 | 33.2 | 23.1 | 2.8 | 6.7 | 26.6 |
| 8 | $P_m C_1$ | 0 | 3 | 13.4 | 13.3 | 1.9 | 6.0 | 22.7 |
| 9 | $P_m C_2$ | 0 | 3 | 15.2 | 15.5 | 2.0 | 7.0 | 12.9 |
| 10 | $P_m C_3$ | 0 | 3 | 32.2 | 19.4 | 2.5 | 7.5 | 20.4 |

Type of plant residue: $P_o$ - Control (no plant residue); $P_m$ - Maize residue; $P_c$ - Cassava residue; $P_L$ - Legume (mucuna) residue. Concentration of plant residue: $C_o$ - Control (No inclusion); $C_1$ - 100 g inclusion level; $C_2$ - 150 g inclusion level; $C_3$ - 200 g inclusion level. NL - Number of leaves; G - Leaf girth; NB - Number of branching; LA - Leaf area (cm); PH - Plant height (cm).

**Table 4.** Weight (g) of morphological properties with standard deviation and co-efficient of variation.

| Obs | TRT | RTWT | WSM | WLS | DRYRT | BMASS | NL | PH | G | NB | LA |
|---|---|---|---|---|---|---|---|---|---|---|---|
| 1 | $P_o C_o$ | 6.80 | 39.80 | 38.30 | 6.10 | 8.30 | 24.80 | 20.20 | 2.70 | 8.00 | 14.80 |
| 2 | $P_c C_2$ | 6.10 | 56.10 | 23.50 | 2.40 | 6.90 | 18.70 | 16.70 | 2.00 | 5.20 | 11.30 |
| 3 | $P_c C_3$ | 47.90 | 362.80 | 182.40 | 17.10 | 14.30 | 17.80 | 18.50 | 2.90 | 5.30 | 33.30 |
| 4 | $P_L C_o$ | 15.70 | 133.80 | 47.70 | 8.60 | 13.20 | 17.80 | 17.60 | 2.40 | 4.60 | 20.20 |
| 5 | $P_L C_1$ | 12.60 | 90.20 | 41.80 | 3.60 | 8.70 | 22.30 | 18.30 | 2.80 | 5.00 | 28.70 |
| 6 | $P_L C_2$ | 8.20 | 58.30 | 30.80 | 2.80 | 5.40 | 24.80 | 19.40 | 2.50 | 5.50 | 19.70 |
| 7 | $P_L C_3$ | 49.30 | 345.60 | 128.50 | 16.20 | 33.10 | 33.20 | 23.10 | 2.80 | 5.70 | 26.60 |
| 8 | $P_m C_1$ | 10.50 | 85.70 | 29.90 | 6.70 | 6.10 | 13.40 | 13.30 | 1.90 | 6.00 | 22.70 |
| 9 | $P_m C_2$ | 11.00 | 62.20 | 30.20 | 6.60 | 10.40 | 15.50 | 15.50 | 2.00 | 7.00 | 12.90 |
| 10 | $P_m C_3$ | 17.00 | 86.40 | 151.20 | 6.80 | 13.20 | 19.40 | 19.40 | 2.50 | 5.90 | 20.40 |
| | MEAN | 18.50 | 132.10 | 70.40 | 7.70 | 11.90 | 22.00 | 18.20 | 2.50 | 5.80 | 21.00 |
| | STDEV | 16.00 | 119.90 | 59.50 | 5.10 | 8.10 | 6.80 | 2.70 | 0.40 | 1.00 | 7.00 |
| | SE | 3.40 | 24.80 | 12.50 | 1.10 | 1.80 | 1.50 | 0.70 | 0.10 | 0.50 | 1.60 |
| | CV | 82.90 | 86.00 | 83.80 | 62.90 | 66.50 | 27.40 | 14.60 | 13.90 | 33.50 | 30.60 |

TRT – Treatment; RTWT - Root weight; WSM - Whole shoot for market; WLS - Whole leaf stalk weight; DRYTRT - Dry root weight; BMASS – Root biomass; NL - Number of Leaves; G - Leaf girth; NB - Number of branching; LA - Leaf area (cm); PH - Plant height.

**Table 5.** ANOVA test of significant difference in root weight.

| Source | Df | Sum of square | Mean square | F-Value | Pr > F |
|---|---|---|---|---|---|
| Treatment | 17 | 8737.87526 | 513.99266 | 2.34 | 0.0205 |
| Residual | 30 | 6600.45094 | 220.01503 | | |
| Total | 47 | 15338.32620 | | | |

**Table 6.** Result after ANOVA (Post hoc) test conducted using Duncan multiple range test procedure.

| Mean | TRT |
|---|---|
| $49.33_a$ | $P_L\ C_3$ |
| $47.97_a$ | $P_c\ C_2$ |
| $17.00_b$ | $P_m\ C_3$ |
| $15.73_b$ | $P_c\ C_3$ |
| $12.57_b$ | $P_L\ C_1$ |
| $11.03_b$ | $P_m\ C_2$ |
| $10.47_b$ | $P_m\ C_1$ |
| $8.17_b$ | $P_L\ C_2$ |
| $6.75_b$ | $P_o\ C_0$ |
| $6.07_b$ | $P_c\ C_1$ |

Means with same letter in same column are not significantly different at 5%.

**Table 7.** ANOVA test of significant difference in whole shoot for market.

| Source | Df | Sum of square | Mean square | F-value | Pr > F |
|---|---|---|---|---|---|
| Treatment | 17 | 448687.4298 | 26393.3782 | 1.63 | 0.1173 |
| Residual | 30 | 485143.5150 | 16171.4505 | | |
| Total | 47 | 933830.9448 | | | |

**Table 8.** Result after ANOVA (Post hoc) test conducted using Duncan multiple range test procedure.

| Mean | TRT |
|---|---|
| $362.8_a$ | Pc $C_2$ |
| $345.6_a$ | $P_L\ C_3$ |
| $133.6_{ab}$ | $P_c\ C_3$ |
| $90.2_b$ | $P_L\ C_1$ |
| $86.4_b$ | $P_m\ C_3$ |
| $85.7_b$ | $P_m\ C_1$ |
| $62.2_b$ | $P_m\ C_2$ |
| $58.3_b$ | $P_L\ C_2$ |
| $56.1_b$ | $P_c\ C_1$ |
| $39.8_b$ | $P_o\ C_0$ |

Means with same letter in same column are not significantly different at 5%.

**Table 9.** ANOVA test of significant difference in plant height.

| Source | Df | Sum of square | Mean square | F-value | Pr > F |
|---|---|---|---|---|---|
| Treatment | 17 | 1888.711696 | 111.100688 | 15.60 | <.0001 |
| Residual | 30 | 213.597296 | 7.119910 | | |
| Total | 47 | 2102.308992 | | | |

significant increase in the % clay and of all the treatments compared to the control indicates that, apart from supplying nutrients to the soil and plant, the organic amendments applied also improve the soil texture and

**Table 10.** Result after ANOVA (Post hoc) test conducted using Duncan multiple range test procedure.

| Mean | TRT |
|---|---|
| $33.223_{a}$ | $P_L C_3$ |
| $32.223_{ab}$ | $P_m C_3$ |
| $24.777_{abc}$ | $P_o C_0$ |
| $24.777_{abc}$ | $P_L C_2$ |
| $22.330_{abc}$ | $P_L C_1$ |
| $18.670_{bc}$ | $P_c C_1$ |
| $17.780_{c}$ | $P_c C_2$ |
| $17.777_{c}$ | $P_c C_3$ |
| $15.223_{c}$ | $P_m C_2$ |
| $13.443_{c}$ | $P_m C_1$ |

Means with same letter in same column are not significantly different at 5%.

**Table 11.** ANOVA test of significant difference in root biomass.

| Source | Df | Sum of square | Mean square | F-value | Pr > F |
|---|---|---|---|---|---|
| Treatment | 17 | 2580.012783 | 151.765458 | 2.38 | 0.0186 |
| Residual | 30 | 1916.334517 | 63.877817 | | |
| Total | 47 | 4496.347300 | | | |

**Table 12.** Result after ANOVA (Post hoc) test conducted using Duncan multiple range test procedure.

| Mean | TRT |
|---|---|
| $33.090_{a}$ | $P_L C_3$ |
| $14.337_{b}$ | $P_c C_2$ |
| $13.213_{b}$ | $P_m C_3$ |
| $13.173_{b}$ | $P_c C_3$ |
| $10.413_{b}$ | $P_m C_2$ |
| $8.717_{b}$ | $P_L C_1$ |
| $8.260_{b}$ | $P_o C_0$ |
| $6.930_{b}$ | $P_c C_1$ |
| $6.117_{b}$ | $P_m C_1$ |
| $5.357_{b}$ | $P_L C_2$ |

Means with same letter in same column are not significantly different at 5%.

**Table 13.** ANOVA test of significant difference in number of leaves.

| Source | Df | Sum of square | Mean square | F-value | Pr > F |
|---|---|---|---|---|---|
| Treatment | 17 | 10202.30954 | 600.13586 | 12.11 | < 0.0001 |
| Residual | 30 | 1487.15720 | 49.57191 | | |
| Total | 47 | 11689.46673 | | | |

stability (Lee, 1988; Anonymous, 1995; Akanbi 2006). Finally, the similarity in the growth and yield parameters and soil physico-chemical properties, of crops treated with 150 and 200 g of organic amendments might be an

**Table 14.** Result after ANOVA (Post hoc) test conducted using Duncan multiple range test procedure.

| Mean | TRT |
|---|---|
| $33.223_{a}$ | $P_L C_3$ |
| $32.223_{ab}$ | $P_m C_3$ |
| $24.777_{abc}$ | $P_o C_0$ |
| $24.777_{abc}$ | $P_L C_2$ |
| $22.330_{abc}$ | $P_L C_1$ |
| $18.670_{bc}$ | $P_c C_1$ |
| $17.780_{c}$ | $P_c C_2$ |
| $17.777_{c}$ | $P_c C_3$ |
| $15.223_{c}$ | $P_m C_2$ |
| $13.443_{c}$ | $P_m C_1$ |

Means with same letter in same column are not significantly different at 5%.

**Table 15.** ANOVA test of significant difference in leaf girth.

| Source | Df | Sum of square | Mean square | F-value | Pr > F |
|---|---|---|---|---|---|
| Treatment | 17 | 11.59551875 | 0.68208934 | 5.95 | < 0.0001 |
| Residual | 30 | 3.44011250 | 0.11467042 | | |
| Total | 47 | 15.03563125 | | | |

**Table 16.** Result after ANOVA (Post hoc) test conducted using Duncan multiple range test procedure.

| Mean | TRT |
|---|---|
| $2.9533_{a}$ | $P_c C_2$ |
| $2.8333_{ab}$ | $P_L C_3$ |
| $2.7567_{ab}$ | $P_L C_1$ |
| $2.7333_{ab}$ | $P_o C_0$ |
| $2.5367_{abcd}$ | $P_m C_3$ |
| $2.4667_{abcd}$ | $P_L C_2$ |
| $2.3767_{abcd}$ | $P_c C_3$ |
| $2.0233_{abcd}$ | $P_m C_2$ |
| $2.0000_{abcd}$ | $P_c C_1$ |
| $1.9333_{abcd}$ | $P_m C_1$ |

Means with same letter in same column are not significantly different at 5%.

**Table 17.** ANOVA test of significant difference in whole leaf stalk weight.

| Source | Df | Sum of square | Mean square | F-Value | Pr > F |
|---|---|---|---|---|---|
| Treatment | 17 | 125110.0217 | 7359.4130 | 1.33 | 0.2382 |
| Residual | 30 | 165470.7820 | 5515.6927 | | |
| Total | 47 | 290580.8037 | | | |

**Table 18.** Result after ANOVA (Post hoc) test conducted using Duncan multiple range test procedure.

| Mean | TRT |
|---|---|
| $182.37_a$ | $P_c\ C_2$ |
| $151.17_{ab}$ | $P_m\ C_3$ |
| $128.53_{ab}$ | $P_L\ C_3$ |
| $47.73_{ab}$ | $P_c\ C_3$ |
| $41.83_{ab}$ | $P_L\ C_1$ |
| $38.25_b$ | $P_o\ C_0$ |
| $30.77_b$ | $P_L\ C_2$ |
| $30.17_b$ | $P_m\ C_2$ |
| $29.97_b$ | $P_m\ C_1$ |
| $23.50_b$ | $P_c\ C_1$ |

Means with same letter in same column are not significantly different at 5%.

**Table 19.** ANOVA test of significant difference in dry root weight.

| Source | Df | Sum of square | Mean square | F-value | Pr > F |
|---|---|---|---|---|---|
| Treatment | 17 | 836.963450 | 49.233144 | 1.84 | 0.0700 |
| Residual | 30 | 802.292617 | 26.743087 | | |
| Total | 47 | 1639.256067 | | | |

**Table 20.** Result after ANOVA (Post hoc) test conducted using Duncan multiple range test procedure.

| Mean | TRT |
|---|---|
| $17.067_a$ | $P_c\ C_2$ |
| $16.240_{ab}$ | $P_l\ C_3$ |
| $8.560_{abc}$ | $Pc\ C_3$ |
| $6.823_{bc}$ | $P_m\ C_3$ |
| $6.683_{bc}$ | $P_m\ C_1$ |
| $6.597_{bc}$ | $P_m\ C_2$ |
| $6.090_c$ | $P_o\ C_0$ |
| $3.850_c$ | $P_L\ C_1$ |
| $2.810_c$ | $P_L\ C_2$ |
| $2.437_c$ | $P_c\ C_1$ |

Means with same letter in same column are not significantly different at 5%.

indication that the optimum requirement is in between the two levels of treatment (Akanbi 2006; Adenawoola, 2005).

## REFERENCES


Adedipe NO (1991). Environemntal and Technical Support Factors for Large-Scale Farming in Nigeria. National Seminar on Large-Scale Farming. Federal Ministry of Agriculture and Natural Resources, Lagos.

Adenawoola AR, Akanbi WB, Akinfasoye JO (2005). Influence of Poultry Manure on Growth, Yield and Quality of 'Oniyaya' Cultivar of Jew's Mallow (Cochorus olitorius). Int. J. Appl. Agric. Apic. Res., 2(1): 93-101.

Ajilore OD (2008). Effects of Continuous Cultivation and Organic Amendments on the Growth and Performance of *C. argentea* in an Oxic Paleustalf in South-Western Nigeria., Unpublished B. Agric. Thesis. Soil Science and Farm Mechanization, Department, Olabisi Onabanjo University, Ago-Iwoye.

Akanbi WB, Baiyewu RA, Tairu FM (2000). Effect of Organic Based Fertilizer and Spacing on Growth and Yield of *Celosia* (*C .Argentea* L.). J. Agric. For. Fish., 1: 5-10.

Akanbi WB, Oyediran GO, Olaniran AO, Adeyeye SA, Akande MO,

Adediran JA (2006). Effects of Organic and Inorganic Fertilizers and Their Combination on Growth, Nutrient Uptake, and Shoot Yield of *Celosia* (*Celosia argentea* L.) Sci. Focus., 2(1): 84-90.

Akinsehinwa OE (2007). Interaction Between Farm Fragmentation and Farm Productivity in Yewa North Local Government of Ogun State. Unpublished B. Agric. Thesis. Agrilcultural Economics Department, Olabisi Onabanjo University, Ago-Iwoye.

Anonymous (1995). Effects of Continuous Cultivation on Ferrosols in Subtropical Southeast Queensland I. Site Characterization, Crop Yields and Socio-Chemical Status. Australian J. Agric. Res., 46(1): 237-253.

Batiano A, Mokunye AU (1991). Roles of Manure and Crop Residue in Alleviating Soil Fertility Constraints to Crop Production: With Special Reference to the Sahelian and Sudanian Zones of West Africa. Nutrient Cycling in the Agrosystem, 29(1).

Bouyoucos GH (1951). A Recalibration of the Hydrometer for making Mechanical Analysis of Soil. Agronomy J., 43: 434-438.

Cooke GW (1982). Fertilizing for Maximum Yield. English Language Book Society ELBS/Collins Publishers (3rd Edition).

Jackson ML (1962). Soil Chemical Analysis. Prentice Hall, New York.

Ketiku AO, Kogbe JOS, Omololu A (1985). Studies on the Manorial Requirements of Nigerian Local Leafy Vegetables I. Effect of Poultry Manure on the Nutrient Composition of the Leaves of Amaranth and Bush Okra. Niger. Agric. J., 19(20): 136-144.

Lee KK (1988). Significance of Biological N Fixation and Organic Manures in Soil Fertility and Fertilizer Management in Semi Arid Tropical India. Christainson (eds). Proceedings of a Colloquium held at ICRISAT Centre, Patanchem, India, Oct. 10-11, pp. 89-108.

Mclean EO (1965). Aluminum: In Methods of Soil Analysis (ed. C. A. Black). Agronomy No. 9 Part 2, Am. Soc. Agronomy, Madison, Wisconsin, pp. 978-998.

Odu CTT, Mba CC (1991). Microbial Consideration for Maximizing Nutrient Availability through Organic Fertilization. Proceedings of National Organic Fertilizer Seminar, Durbar Hotel, Kaduna, March 26-27: 69-80.

Palada MC, Kang BT, Classen SL (1992). Effect of Alley Cropping Leucaena Leucocephala and Fertilizer Application on Yields of Vegetable Crops. Agroforestry Systems, 19(2): 139-147.

Senjobi BA (2007). Comparative Assessment of the Effect of Land Use and Land Type on Soil Degradation and Agricultral Productivity in Ogun State, Nigeria. Unpublished Ph.D. Thesis. Department of Agronomy, University of Ibadan.

Setia RK, Sharma KN (2004). Effect of Continuous Cropping and Long Term Differential Fertilization on Profile Stratification of DPTA-extractable Micronutrients. Food, Agric. Environ., 2(1): 260-264.

Sparks DL (1996). Methods of Soil Analysis. Part 3. Chemical Methods. SSSA and ASA. Madison, W1., pp. 551-574.

Tindall HD (1983). Vegetables in the Tropics. Macmillan Educational Limited Houndmills, Hampshire, p. 5333.

Walkey A, Black IA (1934). Determination of Organic Matter in Soil. Soil. Sci., 37: 549-556.

# Effect of water deficit at grain repining stage on rice grain quality

**M. Fofana[1*], M. Cherif[2], B. Kone[2], K. Futakuchi[1] and A. Audebert[3]**

[1]Africa Rice Center (AfricaRice); 01 B.P. 2031 Cotonou, Benin.
[2]Université de Cocody-Abidjan, 22 BP 582 Abidjan.
[3]CIRAD, Département cultures annuelles, B.P. 5085, 34032, Montpellier.

**Rice production is usually reduced by water stress that can evenly occur during rice cycle in West Africa under bimodal rainfall pattern. In order to determine the effects of water stress on rice grain quality, experiments were conducted on upland site (on ferralsol) at the main AfricaRice research center at M'be, 30 km North of Bouaké, Côte d'Ivoire. The rice varieties CG14 (*Oryza glaberrima*), WAB56-104 (*Oryza sativa*), and NERICA1 (cross WAB56-104 x CG14) were sown at 25 × 25 cm spacing during the dry season cropping period of 2000, 2001 and 2002. Irrigation line (Boon irrigation) was used to supply water until flowering stage. Water was then supply manually from the milky stage of each variety to its full ripening stage. Physical (husking yield, milling recovery, and head rice ratio), chemical (amylose and proteins contents) and cooking parameters (cooking time, volume expansion, rice flour gelatinization temperature, consistency and viscosity) of the harvested grains were determined in the laboratory. The results showed a significant difference ($p < 0.05$) between all the parameters in comparison with the checks samples and stressed crop. In general, NERICA 1 showed better physical and cooking quality traits than its parents. Rice samples from plots subject to lower water availability during repining stage showed higher protein content for all varieties studied. Increase in the average protein content of stressed samples were 31, 11.8 and 13.3% times, respectively for NERICA 1, CG14 and WAB56 -104, where (using the protein content of check plots as 100%) NERICA 1 showed higher husking yield, total mean milling recovery and head rice ratio for samples collected on stressed plots than the g*laberrima* and the *sativa* samples recorded on similar plots. Finding showed that cooking properties that meet West African rice consumers' preferences for cooked rice were more improved for NERICA 1 than its parents in comparison with samples collected from stressed plots. It is concluded that moisture stress at ripening stage should be further investigated as potential indirect means of improving rice grain quality.**

**Key words:** Interspecific, grain quality, water stress, *Oryza glaberrima*, *Oryza sativa*.

## INTRODUCTION

Although rice is consumed worldwide, there is no universal rice quality attribute (Veronic et al., 2007). Nevertheless, rice appearance and cooked rice texture are the characters considered as main quality attributes by consumers (Okabe, 1979; Rousset et al., 1999). Thus, measuring and understanding factors that influence appearance and texture properties are a great challenge for industries and breeders in meeting consumer preferences. About 13% of the world's 147 million ha of rice is cultivated as rainfed rice under upland conditions (Crosson, 1995) where moisture stress affects rice growth and reduces grain yield and quality (Carlos et al., 2008). Water is a major constituent of plant tissue as reagent for chemical reactions and solvent for translocation of metabolites and minerals as well as an essential component for cell enlargement through increasing turgor pressure (Carlos et al., 2008). The occurrence of soil llmoisture stress affects many of the physiological processes such as photosynthesis and transpiration resulting in reduced growth and poor grain filling (Samonte et al., 2001). In West Africa, rice annual consumption increase at the rate of 6.5% (WARDA, 2008), whereas, there is increase in water stress occurrence for

*Corresponding author. m.fofana@cgiar.org.

**Table 1**. Physical and chemical properties of the soil at 0 - 15 cm depth.

| Parameters | Contents |
|---|---|
| pH water | 5.4 |
| Clay (%) | 31 |
| Silt (%) | 18 |
| Sand (%) | 51 |
| Pf 2.0 (%) | 29 |
| Pf 4.2 (%) | 6 |
| Carbon (%) | 0.84 |
| Nitrogen (%) | 0.03 |
| Organic matter (%) | 1.44 |
| Ca (meq/100 g) | 1.2 |
| Mg (meq/100 g) | 0.82 |
| K (meq/100 g) | 0.21 |
| Na (meq/100) | 0.1 |
| Exchange Cation Capacity (meq/100 g) | 6.16 |

rainfed crop as rice (Koné et al., 2008, Koné et al., 2009) essentially induced by notable climate changes (CNRS, 2000). Therefore, soil moisture stress appears to be a major factor for the gap in producing enough quantity of rice as well as grain quality (Tomlins et al., 2005). However, drought effect on crop is depending on soil nutrient content (Gandah et al., 2003) as well as climate and cultivar (Cooper et al., 1987). Except characterization study of tolerance of the New rice of Africa (NERICA) under mid-season drought (Kone et al., 2008), there is missing knowledge on grain quality on acid soil of West Africa. These cultivars (NERICA) are generated by AfricaRice crossing *sativa* from Asia and *glaberrima* from Africa. The main purpose of this study therefore, was to examine the effect of late season soil moisture stress on the most popular upland NERICA rice variety (NERICA 1) grain quality. Results should provide Knowledge for planning interventions as supplementary irrigation.

## MATERIALS AND METHODS

### Site description

In the upland field at Africa Rice Center's headquarter at M'bé, Bouaké, Côte d'Ivoire (7°52N, 5° 6W and 300 m of altitude). The soil was a ferralsol type (Table 1).

### Experimentation

NERICA 1 was sown during three years from 2000 - 2002 similarly with WAB 56 -104 and CG14. NERICA 1 is one of the most adopted of the 18 released NERICA in West Africa. Four grains of rice was sown per hill, spaced by 25 × 25 cm and tinned latter (21 days) to two plants per hill. The total area of each plot was 15 m$^2$. A split plot design was used. The main factor was the irrigation with two levels (well watered and stressed). Variety was the sub-factor with three levels. The check was irrigated at the rate of 30 mm per square meter each three days from seeding to the rice harvest. The second main plot was stressed from milky stage to full ripening. Each irrigated plot was designed as one factor in randomized complete design with three replications. The space between the two blocks was fifteen meters to avoid water migration from well watered plots to stress plots.

An irrigation line (boom irrigation system) was used until the reproductive stage. Since the beginning of the ripening stage (milky stage) of varieties differing by few days, water was supplied manually throughout reproductive stage to ripening. Rice growth patterns registered by national and international research centers in West Africa were followed. Basal fertilizer application was 200 kg/ha of N, PO5, K2O. Sixty days after (60 DAS) seedlings emergence, only nitrogen fertilizer was applied as top-dress at the rate of 40 kg/ha.

### Weather data collection

Daily weather parameters (Precipitation, air temperature and moisture) were collected from AfricaRice's weather station located near the trials.

### Soil moisture measurement

Soil moisture measurement stated when the last variety reaches its milky stage and last fifteen days. Volumetric moisture content (%) in soil was weekly measured by Time domain reflectometer (Delta-T Devices Ltd, UK) in all the plots throughout the stress period. Due to large variation of measurements within each plot as the result of substantial heterogeneity within soils, 10 measurements were taken per plot and the average values considered. The moisture-meter was calibrated with the experimental soil samples and gravimetric methods to generate a calibration curve that was used for transforming the moisture-meter reading to true soil moisture status values (Hamlyn, 2007).

### Yield determination

At grain maturity, rice was harvested in 6 m$^2$ of each micro-plot leaving two lines in the border. Samples of rice grain collected from three plots of same treatment were mixed together. After threshing, the grains were sun-dry, sieved and weighted after the measurement of the moisture content. The grain yield was determined for corresponding weight of standard moisture of 14%. Grain moisture was monitored by a Kett Riceter (KETT C600).

### Milling characteristics

Husking yield (HY), milling yield (MY) and head rice ratio (HRR), the ratio of, respectively, brown rice to paddy, milled rice to brown rice and head rice to milled rice on a weight basis- were determined. For each interspecific line, the 250 g of winnowed paddy was hulled with a Satake testing husker (THU 35H). Brown rice was weighed and HY was determined. The 150 g of brown rice was milled with a milling machine for laboratory use (Yamamoto Test Rice Whitener VP-31 T). After milling, rice bran was removed with 1.7 mm sieve. A cleaned sample of milled rice was weighed and MY was determined. From 20 g of cleaned milled rice, all head rice were taken and weighed. HRR was then calculated. Milled rice grains with a length greater than three-quarters of complete grains were considered as head rice. Colored and damaged grains were also removed from the category of head rice.

### Determination of amylose content

The cleaned milled rice was ground and powdered by an Udy Cyclone

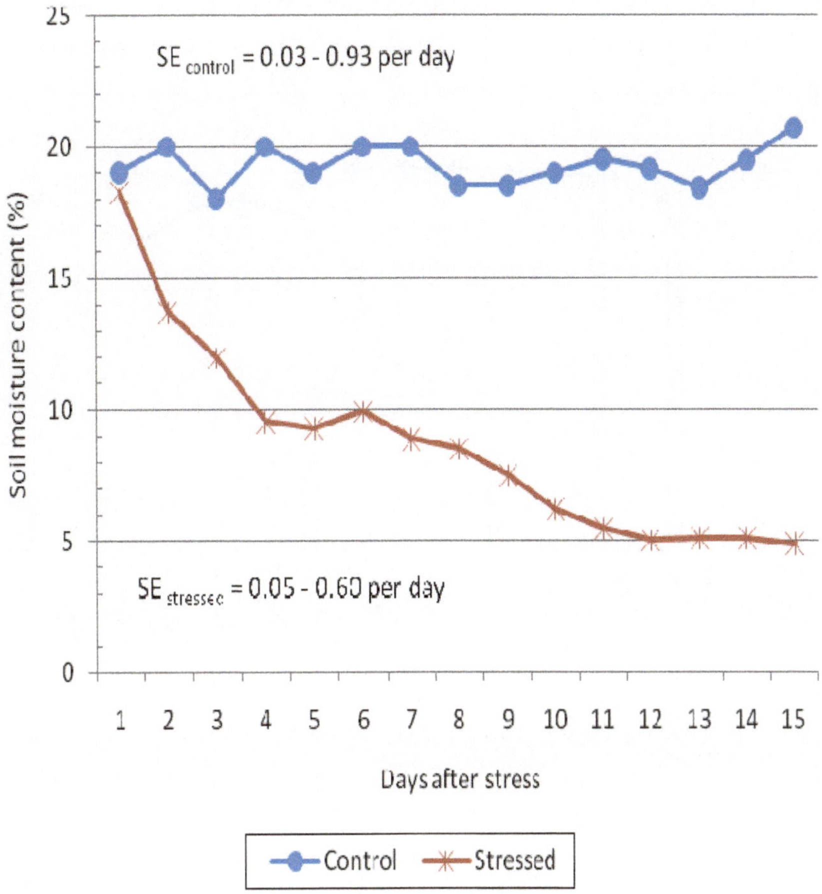


**Figure 1**. Trend of average daily soil moisture content of the three years of experimentation and their standard error (SE) in stressed and control plots during fifteen days after water stress.

Mill with a 1.0 mm mesh screen. Amylose content was determined by auto-analyzer based on the iodine-colorimetric method (Juliano, 1971).

### Determination of protein content

The powdered samples prepared for the determination of amylose contents were used. Total nitrogen content was determined by colorimetric based on Berthelot reaction (Walinga et al., 1989). Protein content was calculated by multiplying total nitrogen content by 5.95, the constant to convert nitrogen content to protein content in rice-and expressed on a dry matter basis of milled rice.

### Determination of viscosity characteristics

The interpretation of the viscograph curve was done according to the indication of the AfricaRice Center grain quality laboratory manual. Powdered samples mentioned were passed through a mesh (NO. 40). 50 g of powdered sample of the varieties at 13.3% moisture contents was mixed in 450 ml of distilled water. The following viscosity parameters were determined with a Brabender viscogram by using the rapid method (Watanabe and Futakuchi, 2000): Peak Viscosity (PV), Minimum Viscosity (MIN), Breakdown Viscosity (BD) and Final Viscosity (FV).

### Statistical analysis

Analysis of variance (ANOVA) and regression analysis were performed using Statistical Analysis system (SAS 11.0) and Sigma Plot 11.0.

## RESULTS AND DISCUSSION

During the drought simulation period, the precipitation during the three years experiments was far much few to cancel the water deficit induced. The average temperatures were higher enough (Figure 2) to allow soil drying during the drought period jugged by the fast decrease of soil moisture content during the stressed period (Figure 1). NERICA 1 showed better yields in both control and drought treatments (Table 2) confirming its ability to produce more than its two parents as discussed by Jones et al. (1997). The low availability of water during this work had less effect on the yields (Table 2). This might be due to its late occurrence during the rice plants life cycles. Similar results were found by Boonjung and Fukai (1996). Husking yield gave an idea of the total amount of edible grains after

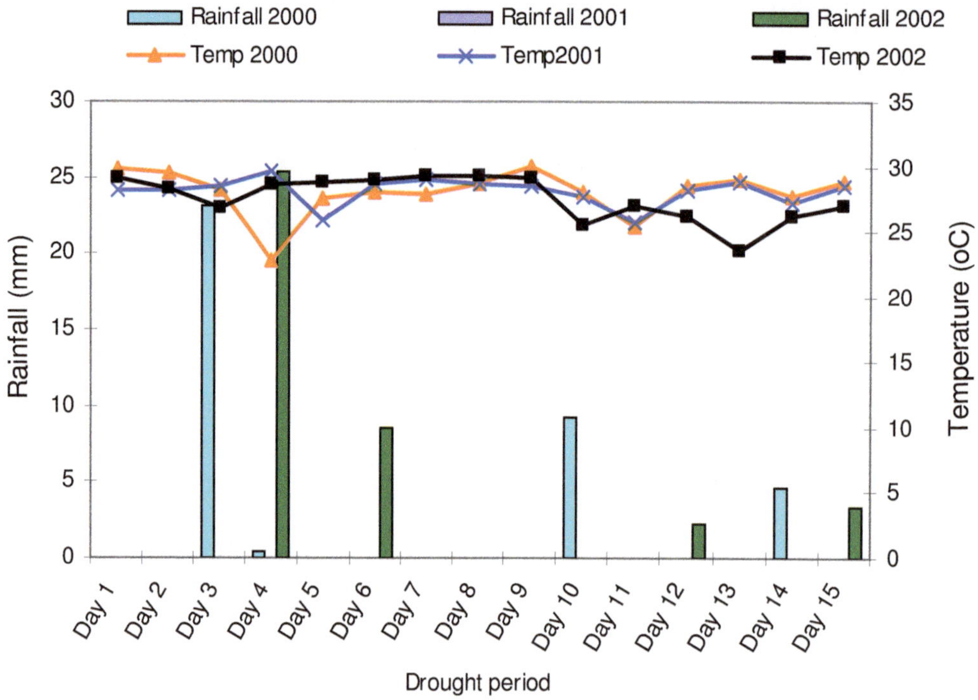


**Figure 2**. Trend of average daily temperature and precipitation of the three years of experimentation during the drought period.

**Table 2**. Probability (P) of source of variations) and their significance.

| Source | Traits | | |
|---|---|---|---|
| | Yield | Husking yield | Total milling yield |
| Treatment | 0.2490 ns | 0.7846 ns | 0.0688* |
| Variety | < 0.0001 *** | 0.3996 ns | < 0.0001*** |
| Year | 0.9871 ns | 0.2645 ns | 0.1920 ns |
| Treatment*Variety | 0.1560 ns | 0.8884 ns | 0.3340 ns |
| Treatment*Year | 0.7419 ns | 0.7467 ns | 0.9569 ns |
| Year*Variety | 0.7970 ns | 0.7046 ns | 0.6671 ns |

Significance: influence * = 10%; **= 5%; *** = 1%, ns = non significant.

husking. Results showed that this trait was very slightly affected by the occurrence of drought at ripening stage (Table 2). Total milling recovery (TMR) is a measure of milling quality and hence, economic value. A proportion of 50% of head rice or less is undesirable since it means that 50% of the rice is discarded as husk and bran after milling. Although stressed samples had a higher mean value of TMR compared to that of control samples for all the cultivars during the three years trials, these differences were not significant at 5% (Table 3). Base on this fact, HY and TMR could not all be considered as the best indicators of grain quality by using water management as in the present study. The percentage of HRR is one of the most important criteria of selection of rice (milled rice) in West Africa (Sakurai et al., 2006). In this study, the increase of the head rice ratio (HRR) is 8.0 - 13.7% for NERICA 1 with stressed samples (Table 4). Therefore, the occurrence of drought during grain ripening stage could be considered as a useful factor that might help to reduce broken grain in milled rice with high level of head whole. Several factors are generally recognized as probable cause of breakage of rice during milling. Delay in harvesting and threshing (Rhind, 1962) and too rapid drying (Angladette, 1963) enable cracking which always increase the number of broken grains. Autrey et al. (1955) showed that rice breakage was related to milling conditions. From this study it could be concludes that unknown factors such as water management determine during ripening stage can reduce grain breakage as discussed by Adu-Kwarteng et al. (2003). Study conducted by Sakurai et al. (2006) in urban area in Ghana showed that rice varieties with high proportion of head rice were sold with higher price in the market.

**Table 3**. Yield (t/ha) of studied verities during the three year trial.

| Variety | 2000 | | 2001 | | 2002 | |
|---|---|---|---|---|---|---|
| | Control | Stressed | Control | Stressed | Control | Stressed |
| CG 14 | 1.2 c | 1.1 c | 1.3 c | 1.0 c | 1.5 b | 1.0 c |
| NERICA 1 | 4.1 a | 4.2 a | 3.8 a | 4.3 a | 4.0 a | 3.9 a |
| WAB 56-104 | 2.5 b | 2.1 b | 2.6 b | 2.3 b | 2.5 b | 2.3 b |
| LSD(0.05) | 0.7 | 0.9 | 0.8 | 1.1 | 1.1 | 0.6 |
| CV (%) | 14.7 | 17 | 16.8 | 21 | 21 | 13 |

**Table 5**. Protein content (%) of studied verities during the three year trial.

| Variety | Protein content in check plots samples | | | Protein content in stressed plot samples | | |
|---|---|---|---|---|---|---|
| | 2000 | 2001 | 2002 | 2000 | 2001 | 2002 |
| CG 14 | 7.1 a | 7.7 a | 7.2 a | 6.2 c | 6.3 c | 6.8 c |
| NERICA 1 | 7.8 a | 8.0 a | 7.5 a | 10.5 a | 10.1 a | 10.2 a |
| WAB 56-104 | 5.5 b | 5.9 b | 6.0 b | 9.0 b | 8.4 b | 7.8 b |
| LSD (0.05) | 1.3 | 0.8 | 0.8 | 0.7 | 0.8 | 0.8 |
| CV (%) | 31.0 | 19.2 | 19.5 | 13.6 | 15.5 | 15.6 |

Therefore, the occurrence of water deficit during grain ripening stage can be considered as a factor that increases grain quality as preferred by local consumer because of low rate of broken grain (Sakurai et al., 2006). Protein and Amylose are factors that mainly influence the eating quality of rice (Adu-Kwarteng et al., 2003, Futakuchi et al., 2008). Negative correlation was found between Amylose content and protein content for stressed samples (Figure 3). These results were similar to those of Koutroubas et al. (2004) on 318 European *sativa* varieties but they were not consistent with the findings of Futakuchi et al. (2008) who found a positive correlation of 0.075 between protein and amylose contents with interspecifics. NERICA 1 is known for its higher protein content than WAB 56 -104 and CG14 in adequate growing conditions (Watanabe et al., 2006). The increase of protein content in NERICA 1 (Table 5) will greatly improve the nutritional status of West African people (Futakuchi et al., 2008) but the protein content should be taken with care since it negatively correlates to rice taste (Ishima et al., 1974). Care should also be taken to decrease the Amylose content (Table 6) since it is well known that less amylose lead to sticky rice after cooking, which West African Rice consumers dislike (Sakurai et al., 2006).

The various parameters of Brabender viscograph results, provide useful information on the texture of cooked rice, although, the test is performed on powdered sample as discussed in the AfricaRice grain quality laboratory manual (1999). Results showed that in general, stressed samples had lower gelatinization temperature and lower peak viscosity (Table 7). This means that NERICA 1 stressed samples tend to gelatinize quicker than its parents (WARDA, 1999). It is well known that there is a positive correlation between gelatinization temperature and cooking time (Veronic et al., 2007). Therefore, the occurrence of the moisture stressed at the ripening stage could shorten NERICA 1's grain cooking time than that of the two other varieties. Cooking time reduction is very important in terms of energy and time saving that can lead to the reduction of cooked rice price on West Africans' rural markets, where rice is usually sold when cooked. The value of final viscosity (Table 7) indicates that NERICA 1 stressed samples tenderness should be intermediate between those of its two parents. The increase of the peak viscosity time and value of the 'breakdown' observed in stressed samples showed that moisture stress during ripening stage induces higher grain volume expansion upon cooking for all studied varieties (Table 8). In most West African countries, especially in areas with large family size, consumers' first criteria of choice of a given variety is its capacity of expanding after cooking (Sakurai et al., 2006). From this point of view NERICA 1 could meet people demand. From this study, water deficit occurrence at ripening stage could be considered as an important positive factor since it could enable West Africans' rice consumers to afford the type of rice grains they prefer.

## Conclusion

The study clearly showed physico-chemical variation in rice grain when moisture stress occurs during grain maturity stage. There were significant differences among both the stressed and the check samples. Late drought that occurs during ripening stage appears to increase the main characteristics defining rice grain quality including total milling rate, head rice ratio, and protein content. Since the criteria of choice of a given variety depend on each

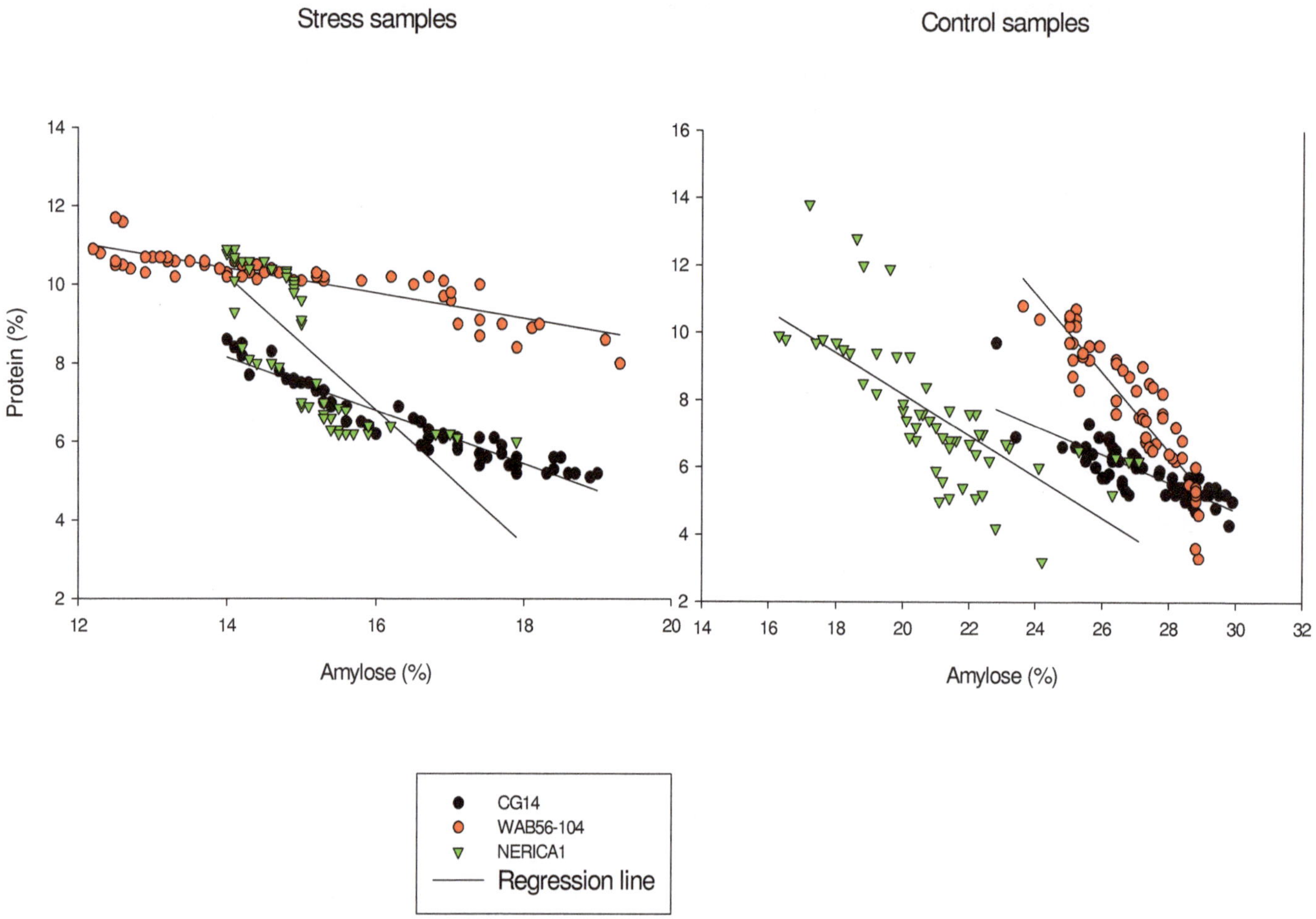


**Figure 3**. Interaction between amylose and protein content of rice grain.

**Table 6**. Amylose content (%) of studied verities during the three year trial.

| Variety | Amylose content in check plots samples | | | Amylose content in stressed plot samples | | |
|---|---|---|---|---|---|---|
| | 2000 | 2001 | 2002 | 2000 | 2001 | 2002 |
| CG 14 | 26.9 a | 27.1 a | 27.6 a | 17.0 a | 16.3 a | 16.0 a |
| NERICA 1 | 27.2 a | 26.5 a | 26.9 a | 14.71 b | 14.7 b | 15.1 ab |
| WAB 56-104 | 20.9 b | 21.4 b | 21.0 b | 14.9 b | 15.3 b | 14.8 b |
| LSD (0.05) | 0.8 | 1.5 | 1.0 | 0.9 | 0.8 | 0.9 |
| CV (%) | 6.5 | 9.4 | 6.5 | 9.6 | 8.5 | 10.2 |

**Table 7**. Gelatinization temperature (0C) and final viscosity (BU*) and Peak viscosity of the flour of the studied verities.

| Variety | Gelatinization temperature of stressed plots samples | | | Final viscosity down of stressed samples | | |
|---|---|---|---|---|---|---|
| | 2000 | 2001 | 2002 | 2000 | 2001 | 2002 |
| CG 14 | 67.4 a | 65.9 a | 62.7 b | 1511 a | 1415 a | 1360 a |
| NERICA 1 | 66.6 a | 66.8 | 64.2 ba | 1010 b | 1014 b | 1018 b |
| WAB 56-104 | 67.3 | 67.2 a | 66.2 a | 1001 b | 1003 b | 1009 b |
| LSD(0.05) | 4.2 | 3.2 | 2.3 | 56.5 | 24.4 | 58.4 |
| CV (%) | 6.8 | 5.3 | 4.0 | 5.2 | 2.3 | 5.6 |

BU: Brabender unit

**Table 8**. Break down (BU) and peak viscosity (BU) of the flour of the studied verities.

| Variety | Break down of stressed plots samples | | | Peak viscosity down of stressed samples | | |
|---|---|---|---|---|---|---|
| | 2000 | 2001 | 2002 | 2000 | 2001 | 2002 |
| CG 14 | 171 c | 170 c | 173 c | 701 b | 706 b | 706 b |
| NERICA 1 | 233 b | 228 b | 234 b | 703 b | 706 b | 696 b |
| WAB 56-104 | 399 a | 398 a | 403 a | 770 a | 769 | 768 a |
| LSD(0.05) | 11.9 | 9.2 | 11.7 | 38 | 31 | 23.2 |
| CV (%) | 4.8 | 3.7 | 4.7 | 5.7 | 4.7 | 3.5 |

BU: Brabender unit.

consumer, it might not be well advised to conclude that the occurrence of water deficit during ripening stage necessarily or not enhance rice grain quality. However, the reproducibility of this trial is possible only in the case where weather conditions remain similar throughout consecutive years. The progressive higher carbondioxide concentration due to climate change might be a limiting factor since the biosynthesis of protein and amylose which are the key elements of grains component is significantly influenced by atmospheric conditions especially after heading stage. More investigations should be done on water stress at rice plant reproductive stage to better understand the physiological phenomenon that happened when this constrain occurs.

## ACKNOWLEDGMENTS

We thank Ibrahim Teslim and late Ade Sobambo and other technicians in grains laboratory during the study years for data collection

### REFERENCES

Adu-Kwarteng E, Ellis WO, Oduro I, Manful JT (2003). Rice grain quality: a comparison of local varieties with new varieties under study in Ghana. Food Control, 14: 507-514.

WARDA (1999). Laboratory Manual for Rice Grain quality Analysis. Bouaké (Côte d'Ivoire), p. 171.

WAARDA (2008). Rice trends in Sub-Saharan Africa 2007. Cotonou, Benin: Africa rice center (WARDA). p. 84.

Angladette A (1963). Rice drying. Principles and techniques. Agr. Eng. Br. F.A.O. (informal work). Bull. p. 23.

Autrey HS, Gricorieff WW, Altschul AM, Hogan JT (1955). Effect of milling conditions on breakage of rice grain. J. Agr. Food Chem. 3: 596-599.

Boonjung H. Fukai S (1996). Effects of soil Water deficit at different growth stages on rice growth and yield under upland conditions. Phenol. Biomass Prod. Yield, 48: 47-55.

Carlos ACC, Orivaldo A, Rogério PS, Gustavo PM (2008). Grain quality of upland rice cultivars in response to cropping systems in the Brazilian tropical savanna. Sci. Agric. (Piracicaba, Braz.), 65(5): 468-473

CNRS (2000). La recherche française sur le climat : le cycle de l'eau-la Mousson d'Afrique de l'Ouest. Lettre nº11 de PIGB-PMRC, CNRS, Paris. En ligne: http://www.cnrs.fr/dossiers/dosclim/rechfran/4them/cyclel'eau/moussonAfOuest.htm. [21 Novembre 2007].

Crosson P (1995). National Resource and environmental Consequences of rice Production. In: IRRI Fragile Lives In fragile Ecosystems. Proceeding of the International Rice Research Conference, 13-17 Feb. 1995., P.O. Box 933, Manila, Philippines, pp. 83-100.

Cooper PJM, Gregory PJ, Keating JDH, Brown SC (1987). Effect of fertilizer, variety and location on barley production under rainfed conditions in northern Syria. 2. Soil water dynamics and crop water use. Field crops Res. 16: 67-84.

Futakuchi K, Watanabe H, Jones MP (2008). Relationship of grain Protein Content to Other Grain Quality Traits in Interspecific Oryza sativa L. x Oryza glaberrima Steud. Progenies Agric. J. 3(1): 50-57.

Gandah M, Bouma J, Brown J, Hiernaux P, Van-Duivenbooden N (2003). Strategies to optimize allocation of limited nutrients to sandy soils of the Sahel: a case study from Niger, West Africa. Agric. Ecosyst. Environ. 94: 311-319.

Hamlyn GJ (2007). Monitoring Plant and soil Water status: Established and Novel Methods Revised and their Relevance to studies of Drought Tolerance. J. Exp. Bot. 58(2): 119-130.

Ishima T, Taira H, Mikoshiba K (1974). Effect of nitrogenous fertilizer application and protein content in milled rice on organoleptic quality of cooked rice. Rep. Nat. Food Res. Inst. 29: 9-15.

Jones MP, Mande S, Aluko K (1997). Diversity and potential of Oryza glaberrima Steud in upland rice breeding. Breed Sci. 47: 395-398.

Juliano BO (1971). A simplified assay for milled rice amylose. Cereal Sci. Today, 12: 334-360.

Koné B, Ettien JB, Amadji G, Diatta S (2008). Caractérisation de la tolérance de NERICA à la sécheresse de mi-saison en riziculture pluviale. Afr. Crop. Sci. 6(2): 133-145.

Koné B, Diatta S, Saıdou A, Akintayo I, Cisse B (2009). Réponses des variétés interspécifiques du riz de plateau aux applications de phosphate en zone de fore^t au Nigeria. Can. J. Soil Sci. pp. 555-564.

Koutroubas SD, Mazzini B, Ntanos DA (2004). Grain quality variation and relationship with morpho-physiological traits in rice (*Oryza sativa* L.) genetic. Resour. Eur. Field Crops Res. 86: 115-130.

Okabe M (1979). Texture measurement of cooked rice and its relationship to the eating quality. J. Texture Stud. 10: 131-152.

Rhind D (1962). The breakage of rice on milling: review. Trop. Agr. Trin. 39: 19-28.

Rousset S, Pons B, Martin JM (1999). Identifying objective characteristics that product clusters produced by sensory attributes in cooked rice. J. Texture Stud. 30: 50-532.

Sakurai T, Furuya K, Futakuchi K (2006). Effects of industrial amassment on the improvement of efficiency and quality. A case study for rice milers in Ghana. In Market and economic development. Sonobe T and Sawada Y (Eds), Tokyo Keizai Shinpou Sya, Tokyo, pp. 151-179.

Samonte S, Wilson LT, McClung AM, Tarpley L (2001). Seasonal dynamics of non-structural carbohydrate in 15 diverse rice genotypes. Crop Sci. 41: 902-909.

Tomlins KI, Manful JT, Larwer P, Hammond L (2005). Urban consumer preference and sensory evaluation of locally produced and imported rice in West Africa. Food Qual. Preference, 16: 79-89.

Veronic V, Brigitte P, Judith B, Stephan H, Xavier R, Christian M (2007). Cooking behavior of rice in Relation to Kernel Physicochemical Properties. J. Agric. Food Chem. 55: 336-346.

Watanabe H, Futakuchi K (2000). Rapid method of Brabender Viscograph and its application to analysis of large number of samples. Nippon Shpkuhin Kagaku Kogaku Kaishi, 47: 926-931.

Watanabe H, Futakuchi K, Jones MP, Sobambo A (2006). Grain protein content of interspecific progenies developed from the cross of African rice (Oryza glaberrima Steud.) and Asian rice O.*sativa* L.). Plant Prod.

Sci. 9: 287-293.
Walinga I, Van VW, Houba VJG, Van der LJJ (1989). Plant Analysis Procedures (Soil and Plant Analysis, Part 7). Syllabus. Wageningen Agricultural University.

# Nutritional and mycoflora changes during storage of plantain chips and the health implications

**E. D. Fagbohun*, O. K. Abegunde and O. M. David**

Department of Microbiology, University of Ado-Ekiti, P. M. B. 5363, Ado-Ekiti, Ekiti State, Nigeria.

**Mycoflora of sundried plantain chips during storage and the effects of these fungi on the nutrient composition of chips were determined. Eight fungi: *Aspergillus niger, A. flavus, A. fumigatus, Mucor* sp., *Neurospora* sp., *Fusarium* sp., *Rhizopus* sp., and *Penicillum* sp were found associated with the sundried plantain chips in storage while the freshly prepared chip harbored *A. niger, A. flavus, Mucor sp.*and *Fusarium sp.* The various fungi were isolated using washing, direct and dilution methods. The fungal count was found to increase as the time of storage of increased. Moreover, the carbohydrate, crude protein, crude fat and crude fibre of the freshly prepared plantain chips were significantly higher than those of the stored ones. The moisture content, ash and Phosphorus content of the stored sundried plantain chips were significantly higher than the freshly prepared plantain chips. The increase in moisture content and the ash content with storage time may be due to the degradative activities of the fungi present on (and or in) the samples. All the minerals (Na, K, Ca, Mg, Zn, Fe, Mn, Co, Cu and P) assessed were found to be in high concentrations, in the freshly prepared plantain chips than in the stored samples except Phosphorus.**

**Key words:** Fungi, plantain chips, stored product, minerals, proximate analysis.

## INTRODUCTION

Plantain is one of the most important crops of the tropical plants. It belongs to the family *Musaceae* and the genus *Musa.* It has a life span of about 15 years (Philips, 1982). The fruit average 125 kg (Xiao et al., 1998). Plantain is very rich in minerals and vitamins. Mature unripe plantain can be made to chips and sundried to reduce the moisture content to a barest minimal. This will discourage the growth of spoilage organisms (Fayemi, 1999) and increases the shelf-life of the stored product. The chips that are produced in Nigeria may be blanched for consumption after preparation. This may be used as animal feedlots and human consumption (Amusa, 2001).

Fungi dominate the microflora of stored products due to their ability to grow at low water activity (Deible and Swamson, 2001). The growth of these mycoflora deplete the mineral composition and the food value of the chips (Egbebi et al., 2007). In the dry products where the equilibrium relative humidity is less than 18% only (or water activity, $a_w$), fungi are classified into two groups xerophilic fungi can grow. Based on this relative humidity according to Christensen (1957) namely field fungi which attack developing and matured seeds in the field, and storage fungi which are predominantly species of *Aspergillus* and *Penicillum* which attack the stored plant products. It is important to know the quality of mycoflora of sundried plantain chips and cut-off point of the sundried product.

Therefore, the objective of this work was to determine the mycoflora of sundried plantain chips during storage and to know the effects of these fungi on the nutrient composition of chips.

## MATERIALS AND METHODS

### Collection of samples

Samples of healthy unripe plantain were collected from a market in Ado-Ekiti, Ekiti-State, Nigeria. The unripe plantains were made into chips by peeling and cutting them into slices, and dry in the sun for one week. The samples were stored for four months in an insect free labelled container and kept in the laboratory. The samples were examined for the changes in the mycoflora and nutrient

*Corresponding author. E-mail: fagbohundayo@yahoo.com.

composition after each month of storage.

### Isolation of fungi from the stored sun dried plantain chips

**Direct plating method:** From the stored sun dried plantain chips, ten slices were examined randomly for external mouldness. They were surfaced sterilized with ethanol and later washed with sterile distilled water. Using a sterile dissecting forceps, the surface of the stored sun dried plantain chips were scrapped and were aseptically plated on Potato Dextrose Agar (PDA) plate and incubated at room temperature for 5-7days as described by Amusa (2001). The fungi cultures were further subcultured until pure colonies were obtained by successive hyphae tip transfer (Egbebi et al., 2007). The cultures were examined under the microscope for fruiting bodies, hyphae to determine the common fungi present.

**Dilution plate methods:** This method was used to determine the type of fungi present in the stored sun dried plantain chips. About one gram of the sample was sterilized with ethanol and grinded with 10ml of sterile distilled water. This was shaken thoroughly and 1ml of suspension was pipetted into a sterile test tube containing 9ml of distilled water. This was thoroughly mixed together. The sample was serially diluted and 1ml each of aliquots of $10^{-4}$ and $10^{-5}$ were added to molten PDA plates. The plates were swirled gently to obtain thorough mixing and were allowed to solidify and incubated at room temperature for 5 - 7 days. The fungal colonies were counted every 24 h. Successive hyphae tip were transferred until pure cultures of each of fungus was obtained.

**Washing method:** This was carried out by weighing 1 g of the plantain chips into 10 ml of sterile distilled water in a beaker. This was shaken thoroughly and drops of suspension of contaminated water were introduced into petri dishes containing Potato Dextrose Agar. This was evenly spread on the agar plate with aid of a sterile glass spreader. The plates were incubated at room temperature for 5 - 7 days and were observe for visible fungi growth.

**Identification of mycoflora:** The associated fungi were identified by their cultural and morphological features (Alexopoulous et al., 1996). The isolates were examined under bright daylight for the colour of the culture and further examination were carried out.

**Needle mount preparation method:** Fragments of the sporing surface of the intial culture was taken midway or between the centre and the edge of the colony. This was teased out in drop of alcohol on a sterilized glass slide using a botany needle. The fragments were stained by adding a drop of lactophenol blue. A cover slip was applied and the preparation was examined under X10 and X40 objective len of the microscope (Tuite, 1961; Crowley et al., 1969 and Egbebi et al., 2007).

**Slide culture technique:** From a plate approximately 2 mm deep, 1 $cm^2$ PDA was cut and placed on a sterile glass slide. Fungus was innoculated into the four vertical sides using a sterile needle. A sterile coverslip so that it over lapped the medium on all sides. The preparation was placed on a suitable support in a petri dish containing blotting paper soaked in 20% glycerol in water. The preparation was kept moist at 28°C until adequate growth was observed. After removing the medium with scalpel, the fungus adhering to both coverslip and slide was examined (Crowley et al., 1969). A drop of alcohol was added followed by a drop of lactophenol blue and the preparation was covered and examined under the low power objective of microscope.

### Proximate analysis

The proximate analysis of the samples for moisture, ash, fibre and fat were done by the method of AOAC (2005). The nitrogen was determined by micro-Kjeldahl method as described by Pearson (1976) the percentage Nitrogen was converted to crude protein by multiplying 6.25. Carbohydrate was determined by difference. All determinations were performed in triplicates.

### Mineral analysis

The mineral was analyzed dry ashing the samples at 550°C to constant weight and dissolving the ash in volumetric flask using distilled water, deionized water with a few drop of concentrated HCl. Sodium and Potassium were determined by using a flame photometer (Model 405 Corning, UK), using NaCl and KCl to prepare the standards. Phosphorus was determined colometrically using Spectronic 20 (Gallenkap, UK) as described by Pearson (1976) with $KH_2PO_4$ as standard. All other metals were determined by atomic absorption spectrophotometer (Pekin-Elmar Model 403, Norwalk CT, USA). All determinations were done in triplicates. All chemicals used were analytical grade (BDH, London). Earlier, the detection limit of the metals has been determined according to Techtron (1975). The optimum analytical range was 0.1 - 0.5 absorbance unit with a coeffient of variation of 0.87 - 2.20%. All the proximate values were reported as percentage while the minerals were reported as milligram/100 g.

## RESULTS AND DISCUSSION

A total of eight of fungi were isolated from stored sun dried plantain chips based on their cultural and morphological characteristics. The fungi include *Aspergillus niger, A. flavus, A. fumigatus, Mucor* sp., *Neurospora* sp., *Fusarium* sp., *Rhizopus* sp., and *Penicillum sp.* The microorganisms isolated from sun dried plantain chips using different methods are represented in Table 1 and the summary of fungi isolated from stored sun dried plantain chips using various methods is as shown in Table 2. Results of the mineral and proximate analyses are shown on Tables 3 and 4, respectively.

The results indicated that *Aspergillus niger, A. flavus, A. fumigatus, Mucor* spp., *Neurospora* spp., *Fusarium* spp., *Rhizopus* spp., and *Penicillum* spp. were found with stored sundried plantain chips. Most of these (Species of *Aspergillus, Mucor, Fusarium, Rhizopus,* and *Penicillum*) fungi are known to be surface contaminant of most agricultural products that induces decay also most of the fungi isolated by washing methods are those capable of growing inside the chips (Amusa et al., 2002). In this study there was an increase in quality of fungi in dry agricultural products as against the report of Ogundana et al. (1970) that noticed a decrease in fungi quantity in stored products. The fungi were likely to originate mainly from contamination from air which was not detected until the 8th week of storage. It have been reported in immunosuppressed hosts such as in AIDS patients, non-AIDS patients with hematological malignancies and those receiving antifungal antibiotics which could alter the microbiota of human (Selik et al., 1997). *Penicillium*

**Table 1.** Fungi isolated from stored sun dried plantain chips using different isolating methods.

| Fungal spp | Weeks of storage | | | | | | | | | | | | | | |
|---|---|---|---|---|---|---|---|---|---|---|---|---|---|---|---|
| | 0 | | | 4 | | | 8 | | | 12 | | | 16 | | |
| | A | B | C | A | B | C | A | B | C | A | B | C | A | B | C |
| *A. niger* | . | + | + | . | + | + | . | + | + | . | + | + | . | + | + |
| *A. flavus* | - | - | + | . | + | + | . | + | + | . | + | + | . | + | + |
| *A. fumigatus* | - | | - | . | - | - | . | + | + | . | + | + | . | + | + |
| *Mucor* spp. | - | + | - | . | + | + | . | + | + | . | + | + | . | + | + |
| *Neurospora* spp. | - | - | - | . | + | + | . | + | + | . | + | + | . | + | + |
| *Fusarium* spp | . | + | + | . | + | + | . | + | + | . | + | + | . | + | + |
| *Rhizopus* spp. | - | - | - | - | - | - | - | - | - | . | - | + | . | + | + |
| *Penicillium* spp. | - | - | - | - | - | - | . | + | + | . | + | + | . | + | + |

A = Direct plating method; B = Dilution plate method; C = Washing method; + = present (isolated). -= absent (not isolated).

**Table 2.** The summary of fungi isolated from stored sun dried plantain chips using various methods.

| Fungal spp | Weeks of storage | | | | |
|---|---|---|---|---|---|
| | 0 | 4 | 8 | 12 | 16 |
| *A. niger* | + | + | + | + | + |
| *A. flavus* | + | + | + | + | + |
| *A. fumigates* | - | + | + | + | + |
| *Mucor* spp. | + | + | + | + | + |
| *Neurospora* spp. | - | + | + | + | + |
| *Fusarium* spp | + | + | + | + | + |
| *Rhizopus* spp. | - | - | - | + | + |
| *Penicillium* spp. | - | - | + | + | + |

+ = present (isolated). - = absent (not isolated).

infections results in keratitis, endophtalmitis, otomycosis, necrotizing esophagitis, pneumonia, endocarditis, peritonitis, urinary tract infections, mucocutaneous, genitourinary, gastrointestinal, pulmonary and disseminated infections as the clinical features (Lueg et al., 1996; Mitchell et al., 1996; Kontogiorgi et al., 2007).

Three different species (*Aspergillus niger, Aspergillus flavus* and *Aspergillus fumigatus*) isolated and identified from sundried plantain chips. *Aspergillus* spp are common mould living in soil, hay etc. and the second most commonly recovered fungus in opportunistic mycoses. Transplantation, extensive use of immunosuppressive drugs which include corticosteroids predisposes human to *Aspergillus* infections (Douglas, 2007). The clinical features of *Fusarium* infections include keratitis, endophthalmitis, otitis media, onychomycosis, cutaneous infections pulmonary infections, endocarditis and fungemia (Lueg et al., 1996). Proper heating of food, elimination of infected and suspected food by *Fusarium* spp are the major preventive measures (Odds et al., 1998).

*Rhizopus* spp. was the least frequently encountered fungi in this study. The isolation was made at the 12th weeks of storage. Kontogiorgi et al. (2007) reported *Rhizopus* to causes rhinocerebral mucormycosis, mucocutaneous, genitourinary, gastrointestinal, pulmonary and disseminated infections. It is also responsible for the damage of blood vessels and nerves. Vascular invasion by *Rhizopus* causes necrosis of the infected tissue. Treatment of *Rhizopus* infections remains difficult due to its property to invade vascular tissues, infarction of the infected tissue is common and mortality rates are very high *Rhizopus* infections can be prevented by avoiding contact with contaminated object as well as maintaining a proper hygiene (Welsh and Kaplan, 1998).

Result of the proximate analysis (Table 4) revealed that the freshly prepared plantain chips had Crude protein(CP) content of 4.42%, Carbohydrate (CHO) of 87.45%, crude fat (CF) of 0.78% ash content of 2.0%, moisture content of 4.67% and fibre content of 0.64%. However after 4 months of storage the %CP, CF, CHO and fibre content decreased to 3.30, 0.52, 77.14 and 0.06% respectively.

The moisture content and ash content increased to 15.44 and 3.55% respectively, this agreed with the work of Taylor, (1983). The nutrient composition and nutritive

**Table 3.** Mineral analyses of sun dried plantain chips during storage (mg/100 g).

| Minerals | Period of sample storage (weeks) | | | | |
|---|---|---|---|---|---|
| | 0 (freshly prepared) | 4 | 8 | 12 | 16 |
| Na | 0.38 | 0.94 | 0.8 | 0.65 | 0.58 |
| K | 8.52 | 11.91 | 11.45 | 10.32 | 9.19 |
| Ca | 4.98 | 10.50 | ND | 7.63 | 5.40 |
| Mg | 10.92 | 17.06 | 16.01 | 14.51 | 11.37 |
| Zn | 6.42 | 12.28 | 19.23 | 9.48 | 8.34 |
| Fe | 0.56 | 0.47 | 0.37 | 0.28 | 0.19 |
| Mn | 0.19 | ND | 0.01 | ND | ND |
| Co | 0.28 | 0.09 | ND | ND | ND |
| Cu | ND | ND | ND | ND | ND |
| P | 49.53 | 34.6 | 38 | 40.0 | 45.11 |
| Na/K | 0.045 | 0.078 | -- | 0.063 | 0.063 |
| Ca/P | 0.303 | 0.222 | 0.191 | 0.120 | 1.101 |

Data are the means of three determinations. ND = Not detected.

**Table 4.** Changes in proximate composition (in mg/100mg) of sun dried plantain chips during storage.

| (Week of storage) | Parameters | | | | | |
|---|---|---|---|---|---|---|
| | Ash | Moisture content | Crude protein | Fat | Fibre | Carbohydrate |
| 0 (Freshly prepared) | 2.06 | 4.67 | 4.42 | 0.78 | 0.64 | 87.45 |
| 4 | 2.34 | 5.28 | 4.36 | 0.68 | 0.15 | 87.21 |
| 8 | 2.84 | 14.8 | 3.81 | 0.62 | 0.12 | 77.77 |
| 12 | 3.39 | 15.0 | 3.58 | 0.56 | 0.09 | 77.38 |
| 16 | 3.55 | 15.44 | 3.30 | 0.52 | 0.06 | 77.14 |

value of the leaf protein concentrate from two solanaceous vegetables (Taylor, 1983). The increase in ash and moisture contents may be due to the degrading activity of the fungi as reported by (Egbebi et al., 2007).

The mineral analyses of the plantain chips during storage (in mg/100g) of the sample (Table 3) revealed the following minerals Na (0.94), K(11.91), Ca (10.50), Mg (17.06), Zn (12.28), Fe(0.56), Mn (0.19), Co (0.28) and P (34.6) in the freshly prepared samples. Among the major minerals evaluated for, the trend of concentration in the sun dried plantain chip was Mg>Zn>K>Ca. The trend was maintained except for the 8th week of storage. Magnessium recorded the highest value during the period of storage. It helps in phosphorus metabolism (Ahn, 1970). There was low levels of Ca/P that were below 0.5. This might not allow strong bone development because absorption of Calcium under this situation would be low (Nieman et al. 1992). To avoid hypertension from food sources, the ratio of Na/K should be about 0.60 (Wise, 1983). Our current report showed that the levels of Na/K were lower than 0.60. The sample at any stage of storage would not promote the development of hypertension if consumed by human beings.

In conclusion, handling and processing agricultural products, apart from good hygiene caution must be taken to reduce contamination by pathogens. The isolated fungi can degrade plantain chips as substrate, and pose a threat to the consumers by either infecting them or elaborating metabolites that can affect organs of the body.

This study provided evidence for pathogenic fungi to enter, survive and grow within locally processed food (plantain chips). There is a need to provide rational basis for designing intervention technologies that are needed to assure the microbiological safety of such products so they can be safe for consumption and meet international standards.

## REFERENCES

Ahn PM (1970). West African Soils. Oxford University Press, Oxford pp. 20-100.

Alexopoulous CJ, Mims CW, Blackwell M (1996). Introductory Mycology. 4th edition. John Wiley & Sons, Inc., New York pp. 127-171.

Amusa NA (2001). Fungi associated with yam chips in storage and the effect on the chips nutrients composition. Moor J. Agric. Res. 2: 35-39.

Amusa NA, Kehinde IA, Ashaye OA (2002). Biodeterioration of bread fruit (*Artocarpus communis*) in storage and its effects on the nutrient composition. Afr. J. Biotechnol. 1(2): 57-60.

AOAC (2005). Official Method of Analysis. 14th Edition, Association of Official Analytical Chemist, Washington D. C.

Christensen CM (1957). Deterioration of stored grains by fungi. Annu. Rev. Phytopath. 3: 69-84.

Crowley N, Bradley JM, Darrell JH (1969). Practical Bacteriology. Butterworth and Co. Ltd. London pp. 164-168.

Deible KE, Swanson KMJ (2001). Cereal and cereal products. In F. P. O. Downes and K. Ito (eds). Compendium of Methods for the Microbiological Examination of Foods. Blackwell Pub. Co, London pp. 98-102.

Douglas F (2007). Mycotic Infections. Del Med. J. 69: 1-4.

Egbebi AO, Anibijuwon II, Fagbohun ED (2007). Fungi associated with dry cocoa beans during storage in Ekiti State, Nigeria. Pak. J. Nutr. In press.

Fayemi PO (1999). Nigerian vegetables. Heinemann Educational Books (Nigeria) Plc. pp. 15-20.

Kontogiorgi M, Floros I, Koroneos A, Vamvonka C, Paniara O, Roussos C, Routi C (2007). Fatal post-truamatic zygomycosis in an immunocompetent young patient. J. Med. Microbiol. 56: 1243-1245.

Lueg EA, Ballagh RH, Forte J (1996). Analysis of the recent cluster of invasive fungal sinusitis at the Toronto Hospital for sick children. J. Otolaryngol. 25: 366-370.

Mitchell SJ, Gray J, Morgan MEI, Hocking MD, Durbin GM (1996) Nosocomial infection with *Rhizopus microsporus* in preterm infants: association with wooden tongue depressors. Lancet 348: 441-443.

Nieman DC, Butterworth DE, Nieman CN (1992). Nutrition. Wm. C. Brown Publishers, Dubuque, USA pp. 237-312.

Odds FC, Van Gerven F, Epsinel-ingroff A, Bartlett MS Ghannoum MA, Lancaster MV, Pfaller MA, Rex JH, Rinaldi MG, Walsh TJ (1998). Evaluation of possible correlations between antifungal susceptibilities of filamentous fungi *in-vitro* and antifungal treatment outcomes in animal infection models. Antimicrob. Agents Chemother. 42: 282-288.

Ogundana SK, Naqui SH, Ekundayo JA (1970). Fungi associated with soft rot of yam (*Discorea* spp) in Nigeria. Trans. Br. Mycol. Soc. 54: 445-451.

Pearson DH (1976). Chemical Analysis of Foods. Churchhill London. pp. 335-336.

Philips TA (1982). An Agricultural Notebook. Longman, Nigeria p.125.

Selik RM, Karon JM, Ward JW (1997). Effect of the human immunodeficiency virus epidemic of mortality from opportunistic infections in the United States in 1983. Infect. Dis. 176: 632-636.

Taylor OO (1983). The nutrient composition and nutritive value of the leaf protein concentrate from two selanaceous vegetables. Acta Hortic. 123: 10-13.

Techtron V (1975). Basic Atomic Absorption Spectroscopy: A Modern Introduction, Domican Press, Victoria, Australia pp. 104-106.

Tuite J (1961). Fungi isolated form unstored corn seed in Indiana in 1956-1988. Plants Dis. Rep. 45: 212-215.

Welsh TS, Kaplan J (1998). The role of postmortem examination in medical education. Mayo Clin. Proc. 73: 802-805.

Wise A (1983). Dietary factors determining and moisture the biological activities of phytate. Nutrition foods. Hum. Nutr. Appl. Nutr. 40A: 49-59.

Xiao R, Beck O, Hjemdahn P (1998). The accurate measurement of serotoxin in white blood. Scandavia J. Clin. Lab. Invest. 58(6): 23-26.

# 12

# Optimizing gari quality attributes for different groups of consumers with response surface methodology

**Udofia, P. G[1]*, Udoudo, P. J[1], Eyen, N. O[1] and Udoekong, N. S.[2]**

[1]Department of Hotel and Catering Management, Akwa Ibom State Polytechnic, Ikot Ekpene, Nigeria.
[2]Department of Science Technology, Akwa Ibom State Polytechnic, Ikot Ekpene, Nigeria.

**Different groups of consumers prefer different sensory attributes of gari. Therefore it is difficult to produce one stock of the product for all sectors of the market. Fermentation time (h), frying temperature (°C) and resident time (min.) and their interaction effect on quality attributes of gari; swelling capacity, texture, colour, taste and acceptability of gari was investigated. Optimization analysis showed that the optimum quality for all the parameters; 3.51, 3.83, 4.14, 4.70 and 4.70 of swelling capacity, texture, colour, taste and acceptability, respectively were obtained with the fermentation temperature of 12 h, frying temperature of 95°C and resident time of 34.64 h. Other runs in the design were also accepted by some members of the taste panel. The design provides various experimental runs for the production of gari with varying quality attributes to suit the traditional varying market preference.**

**Key words:** Optimization, response surface methodology, CCD.

## INTRODUCTION

Cassava (*Manihot esculenta* Crantz) is a major food crop in Nigeria (Kim, 2009). It supplies about 70% of the daily calorie of over 50 million people (Oluwole et al., 2004) and about 500 million people in the world (Abu et al., 2006). Some cassava varieties show high resistance to drought and mosaic disease, high yield, even in agroecological conditions where other starch bearing crops do not thrive. The prevailing climate change, the threatening global warming and the expected negative impact on the yield of less hardy crops highlights further advantages of cassava as the crop of the century. Traditionally, cassava was cultivated by farmers at subsistence levels as the 'poor man's food'. Currently, semi-commercial and commercial farms are available due to increasing awareness and proof of functional versatility of cassava flour especially in the food manufacturing sector (Ernesto et al., 1999; Cardoso et al., 2000; FAOSTAT, 2002). Cassava is a staple crop in Nigeria, some other West African countries and Brazil. It is produced into gari, lafun, tapioca kokote and achke. Gari is an important by-product of cassava, an important item in the menu of most Nigerians. It is particularly popular because of its ready-to-eat (Abu et al., 2006). The importance of gari is also spreading to many countries of West Africa and some of the international markets. Gari is a good source of energy and fiber. Other nutrients are also present in marginally nutritional significance (Ikegwu et al., 2009). The product is consumed dry, smoked with milk, sugar and salt, or in hard paste (eba*)* with soups. Acceptability of gari by all sections of the consumer population is still suffering a set back. This trend is due to its varied sensory attributes and failure to meet the international quality standard (FAO/WHO, 1991; Damadjali et al., 1993; Owuamanam and Achinewu, 2002). Physical, chemical and biochemical processes are employed in the production of gari. Processing factors and ingredients are applied singly or interactively to modify the native cassava starch (Achinewu et al., 1998). The extent of the modification determines the levels of physicochemical and sensory attributes of the final product and consequently, its acceptability. For instance, Oduro et al. (2000), Okafor and Uzuegbu (1987) and Collard and Levi (1987) acknowledged that sensory attributes of gari is determined by level of fermentation, frying temperature, quantity of palm oil added, post-harvest storage of

*Corresponding author: E-mail: paddofff@yahoo.com.

**Table 1.** Levels of dependent and independent variables.

| Variable | Unit | Low | Medium | High |
|---|---|---|---|---|
| **Independent variable** | | **Real (code)** | **Real (code)** | **Real (code)** |
| X1 - Duration of fermentation | H | 12(-1) | 24(0) | 48(+1) |
| X2 - Frying temp. | ℃ | 70(-1) | 80(0) | 100(+1) |
| X3 - Frying time | min | 45(-1) | 56(0) | 70(+) |
| **Dependent variable** | | | | |
| Y1 - Swelling capacity | | | | |
| Y2 –Texture | | | | |
| Y4 – Colour | | | | |
| Y3 –Taste | | | | |
| Y4 – Acceptability | | | | |

cassava tubers before processing, method of grating, and rate of dewatering of cassava mash during fermentation. Other factors include age of cassava plant at harvest, cassava variety, soil quality of location of farm and storage condition of gari before consumption. In Nigeria, gari quality varies along traditional/cultural lines. The east prefers it creamy of different shades coloured by red palm oil; sweet, imparted by short period of fermentation. The west prefers creamy to slightly-golden colour, varied degrees of sourness imparted by longer period of fermentation. Due to these variations, it is difficult to source the same quality of gari more than once, hence the consumers are forced to consume what is available. For optimum quality attributes, the processing variables must be optimized singly or in combination in order to produce gari of consistent quality and consumer acceptability. The multivariate approach applied in the central composite design of the response surface methodology is the most widely used response surface designs. It reduces the number of experiments, cost and time, improves statistical interpretations and indicates whether parameters interact (Huguo, 2002). The objective of this work was to optimize fermentation time, frying temperature and frying time to produce gari with the optimum values of swelling capacity, texture, colour, taste, and consumer acceptability.

## MATERIALS AND METHODOLOGY

About 100 kg of 12 month-old ATM123 cassava variety was obtained from Akwa Ibom State Agriculture Development Project, (AKADEP), Ikot Ekpene, Nigeria. The frying pot was locally fabricated with vulcanized metal with provision to monitor temperature; gas cooker was also locally fabricated. Gari processing was done according to the method adopted by (Ngaba and Lee, 1999; Achinewu et al., 1998) with slight modifications to suit the experimental design and method. The roots were peeled, washed and grated 6 h post-harvest with a locally fabricated grating machine and bagged to dewater the mash to moisture content of about 25%. Fermentation was accomplished according to the design in Table 1.

The dewatered fermented cassava mash was sifted through a local sifter of about 5 mm pore size. Bits of the sifted cassava mash were introduced into hot frying pan with a surface temperature of about 120℃. This was done gradually until mash of about 600 g was introduced; 10 ml of red palm oil was added to the cooking mash cascading gently and steadily with a wooden T stirrer to prevent formation of starch lumps.

### Determination of statistical design

The small $3^3$ central composite design of the response surface methodology was adopted to study the simultaneous effect of fermentation time, frying temperature and frying time on gari sensory quality. The design is cheap, only a fraction of the experimental runs is required for a full three-level factorial analysis, without compromising accuracy, reproducibility. It is robustness and reliable. It is based on a matrix of multifactor method which measures linear, interaction, quadratic and cubic effects and encompasses the entire multidimensional experimental region (Anderson and Whitcomb, 2007). It is an important tool for a more rapid food process and product development (Huguo, 2002).

$$Y_i = b_0 + b_1X_1 + b_2X_2 + b_3X_3 + b_{12}X_1 X_2 + b_{13} X_1 X_3 + b_{23} X_2 X_3 + b_{11} X_1^2 + b_{22}X_2^2 + b_{33}X_3^2 \quad (1)$$

where
$Y_1$ = measures the predicted response as a result of each factor levels combination; $b_o$ = the intercept, $b_1$ - $b_{33}$ = regression coefficient determined experimentally, $X_1$, $X_2$ and $X_3$ = independent variables (Box and Behnken, 1960).

The dependent and independent variables for the experimental runs are given in Tables 1 and 2 respectively. The response profile for the 14 experimental runs, analysis of variance and response surface plots optimization analysis were generated with the trial version of Design-Expert (Ver. 7.1.1, Statease, Minneapolis, Mn).

### Determination of swelling capacity

Swelling capacity for the samples was determined according to the method adopted by Sanni et al. (2001). A 50 ml measuring cylinder was filled with samples up to the 10 ml mark. Distilled water was added gently up to the 50 ml mark. The mouth of the measuring cylinder was secured tightly covered with a clean plastic cork, the

**Table 2.** Analysis of variance of models describing effects of fermentation time, frying temperature and resident time on basic qualities of gari.

| Variable | Swelling capacity | Texture | Colour | Taste | Acceptability |
|---|---|---|---|---|---|
| Model | 0.3332 | 0.0229 | | 0.0229 | 0.0027 |
| $X_1$ | 0.8487 | 0.0012 | 0.2908 | 0.0229 | 0.0008 |
| $X_2$ | 0.5264 | 0.7332 | 0.6523 | 0.7332 | 0.6682 |
| $X_3$ | 0.1943 | 0.4351 | 0.5230 | 0.4351 | 0.0723 |
| $X_1X_2$ | 0.2145 | 0.1722 | 1.0000 | - | - |
| $X_1X_3$ | 0.3761 | 0.3226 | 0.7838 | - | - |
| $X_2X_3$ | 0.4473 | 0.0829 | 0.5271 | - | - |
| $R^2$ | 0.4847 | 0.8193 | 0.2814 | 0.8103 | 0.7420 |
| Mean | 3.30 | 3.95 | 3.99 | 3.95 | 3.95 |

**Table 3.** The responses profile for the 15 experimental runs.

| Run | Std | $X_1$ (H) | $X_2$ (H) | $X_3$ (H) | S. C. | Texture | Colour | Taste | Accept |
|---|---|---|---|---|---|---|---|---|---|
| 1 | 5 | 12.00 | 95.00 | 30.00 | 3.44 | 3.30 | 4.30 | 4.70 | 4.00 |
| 2 | 13 | 30.00 | 95.00 | 40.00 | 3.20 | 4.00 | 4.00 | 3.30 | 3.30 |
| 3 | 12 | 30.00 | 120.00 | 50.00 | 3.37 | 3.67 | 4.70 | 4.70 | 4.00 |
| 4 | 1 | 12.00 | 70.00 | 40.00 | 3.45 | 3.30 | 4.70 | 4.00 | 4.00 |
| 5 | 11 | 30.00 | 70.00 | 50.00 | 3.41 | 4.70 | 4.00 | 4.00 | 4.30 |
| 6 | 14 | 30.00 | 95.00 | 40.00 | 3.40 | 4.30 | 4.30 | 4.00 | 4.00 |
| 7 | 7 | 12.00 | 95.00 | 50.00 | 3.21 | 3.30 | 3.30 | 4.30 | 4.70 |
| 6 | 9 | 30.00 | 70.00 | 30.00 | 3.22 | 4.00 | 4.00 | 3.00 | 3.00 |
| 9 | 10 | 30.00 | 120.00 | 30.00 | 3.41 | 4.30 | 4.00 | 2.60 | 3.30 |
| 10 | 3 | 120.0 | 120.00 | 40.00 | 3.22 | 3.00 | 4.00 | 4.30 | 4.70 |
| 11 | 4 | 48.00 | 120.00 | 40.00 | 3.27 | 4.70 | 3.30 | 2.60 | 3.30 |
| 12 | 8 | 48.00 | 95.00 | 50.00 | 3.01 | 4.70 | 3.30 | 3.30 | 3.30 |
| 13 | 2 | 48.00 | 70.00 | 40.00 | 3.11 | 4.00 | 4.00 | 2.30 | 3.00 |
| 14 | 6 | 48.00 | 95.00 | 30.00 | 3.51 | 4.00 | 4.00 | 2.60 | 3.00 |

$X_1$ = Fermentation time, $X_2$ = Frying temperature, $X_3$ = Resident time, S. C = Selling capacity.

content was mixed by inversion at 2 min intervals, the cylinder was left to stand for 3 min before the final volume of the gari in the cylinder was read. The swelling capacity was determined by dividing the volume of sample in water by initial volume of sample.

$$\text{Swelling capacity} = \frac{\text{Final volume of sample in water}}{\text{Initial volume of sample in water}}$$

### Sensory evaluation

A descriptive sensory evaluation of the samples in terms of texture, colour, taste, and overall acceptability was carried out on a 5-point hedonic scale, where 1 = poor, 2 = fair, 3 = satisfactory, 4 = good, and 5 = excellent (Venet and Nyla, 2001). A 30-member semi-trained panel from National Youth Service Corp (NYSC) members comprising 15 females and 15 males from the northern, southern and eastern and western regions of Nigeria who are regular gari consumers were used for the sensory evaluation. Samples were coded and randomly presented in clean ceramic plates all at the same time and were assessed in dried and reconstituted forms, cold and hot water was supplied for personal reconstitution of samples. Data collected from the sensory evaluation were subjected to ANOVA, regression analysis and response surface plots.

## RESULTS AND DISCUSSION

Table 2 is a summary of analysis of variance and multiple regression analysis showing p-values of models for the parameter and of the processing variables at linear and interactive levels, their corresponding coefficient of determinations and means obtained by fitting the data to the 2nd order response model.

Bold face characters in the models 2 to 6 are not significant in statistics, but there are retained since sensory differences in the products may be hidden by the apparently insignificant terms.

The table shows that the developed models for swelling capacity, texture, texture colour appeared to be inadequate ($p>0.05$, $R^2$ = 0.4646). However, the developed models for taste and acceptability were significantly appropriate ($p<0.05$). Table 3 shows that the

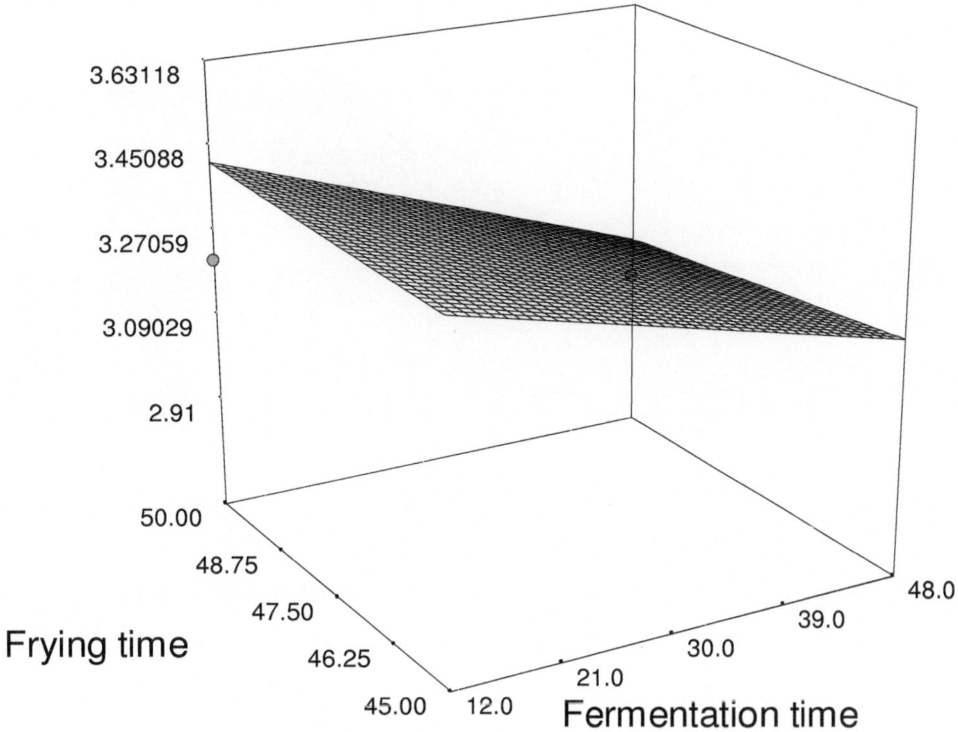


**Figure 1.** Response surface plot of frying time and fermentation time on swelling capacity.

swelling capacity of the samples ranged from 3.01 to 3.51.

Run 6, corresponding to 48 h, 98°C and 30 min of fermentation time, frying temperature and time respectively showed the highest value (3.50) while Run 8 with 48 h, 95°C and 50 min of fermentation time, frying temperature and time respectively showed the least value (3.01) with a low range of (0.50). This observation was in agreement with those reported by Achinewu et al. (1998) and the recommended value of 300 and 500% for good quality gari (Achinewu et al., 1990). Swelling capacity, the ability of gari particles to absorb water and swell, a very important sensory and mechanical attribute which determines the engineering property including the mouldability of eba and the flow of gari for 'smoking'. According to Sanni et al. (2005), swelling capacity depends on the free amylose and associative forces within the starch granules and moisture content (Achinewu et al., 1998) availability of hydrophilic substituent groups ($OH^-$, $O^=$) may also be a factor and time of contact with water (Igathinathane et al., 2007). Supporting the trend, Figure 1 shows that swelling capacity (S.C) decreases with fermentation period and a decrease in frying time.

$$\text{Swelling capacity} = 2.002X_1 + 0.0085X_2 + 0.0099X_1X_2 + 0.0415X_1X_3 \quad (2)$$

$$\text{Texture} = 2.3767X_1 - 0.04856X_2 + 0.0432X_3 + 0.0007X_1X_2 - 0.0014X_1X_3 + 0.0006X_2X_3 \quad (3)$$

$$\text{Colour} = 8.7534 - 0.0588X_1 - 0.0363X_2 - 0.0793X_3 + 0.000325X_1X_2 \quad (4)$$

$$\text{Taste} = 6.63 - 0.0.0780X_1 + 0.\ 0713X_2 + 0.124X_3 + 0.0011X_1X_2 - 0.0011X_1X_3 + 0.0009\ X_2X_3 \quad (5)$$

$$\text{Acceptability} = 1.9465 - 0.0136X_2 - 0.0251X_2 + 0.1078X_3 + 0.00091X_1X_2 - 0.0021X_1X_3 \quad (6)$$

Texture is an important attribute which influences level of acceptability of eba. Texture is a complex sensory parameter which is difficult to describe in isolation of swelling capacity, bulk density (not shown here), moisture content (water activity). According to Ikegwu et al. (2009). Variety of the tuber is a major influence on the kinetics of cassava mash during fermentation.

The result of sensory evaluation of texture is presented in Table 2. The table shows that contribution of linear and interaction of the factors on the parameter were not significant ($p>0.05$) except frying temperature, ($p<0.05$), however, the model was appropriate ($R^2 = 0.9222$).

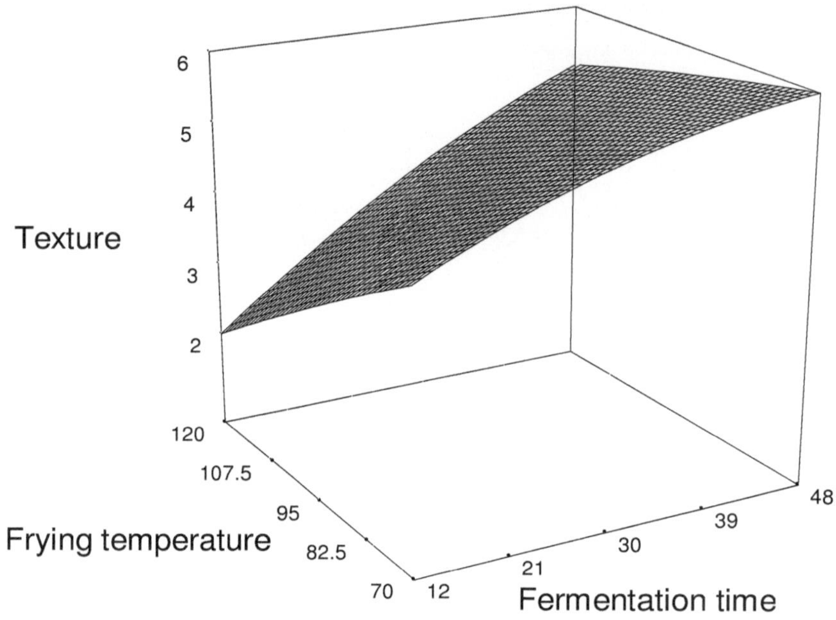


**Figure 2.** Response surface plot of frying temperature.

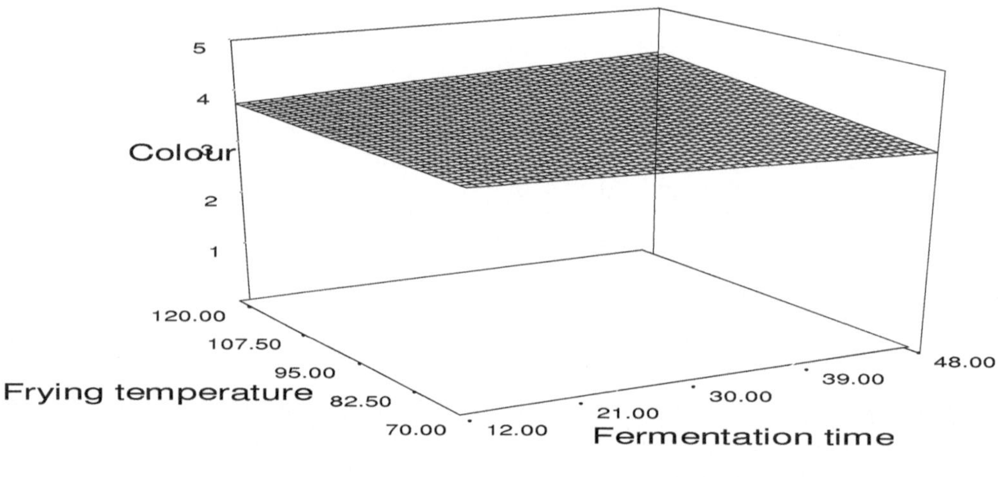


**Figure 3.** Response surface plot of frying temperature and fermentation time on colour and fermentation time on texture.

According to Table 3, texture sensory score ranged between 3.30 and 4.70. The highest was observed at 48 h, 95°C and 50 min of fermentation time, frying temperature and time respectively while the lowest was observed at 120 h, 20°C and 40 min of fermentation time, frying temperature and time respectively. Figure 2 shows that value of the parameter is increased with increasing fermentation time, and decreasing frying temperature. This observation may be due to the depletion of amylose of the cassava mash during fermentation, slow cooking of the mash with low temperature, this trend is in agreement with was reported by Cadoso et al. (1987) and Eresto et al. (2002), the slight variation may be due to varietals, procedural and idiotic differences of the experimenters.

Colour of gari is normally white or cream. Traditionally, different shades of redness is obtained by addition of palm oil. Although an important sensory parameter in the choice of gari, in our study, the model of processing variables on colour of gari was not significant (p>0.05,) and appeared to be reliable ($R^2$ = 0.7490), showing that colour may not be a major determining factor of gari acceptability compared to texture and taste. Figure 3 shows a near flat response surface to support the observation.

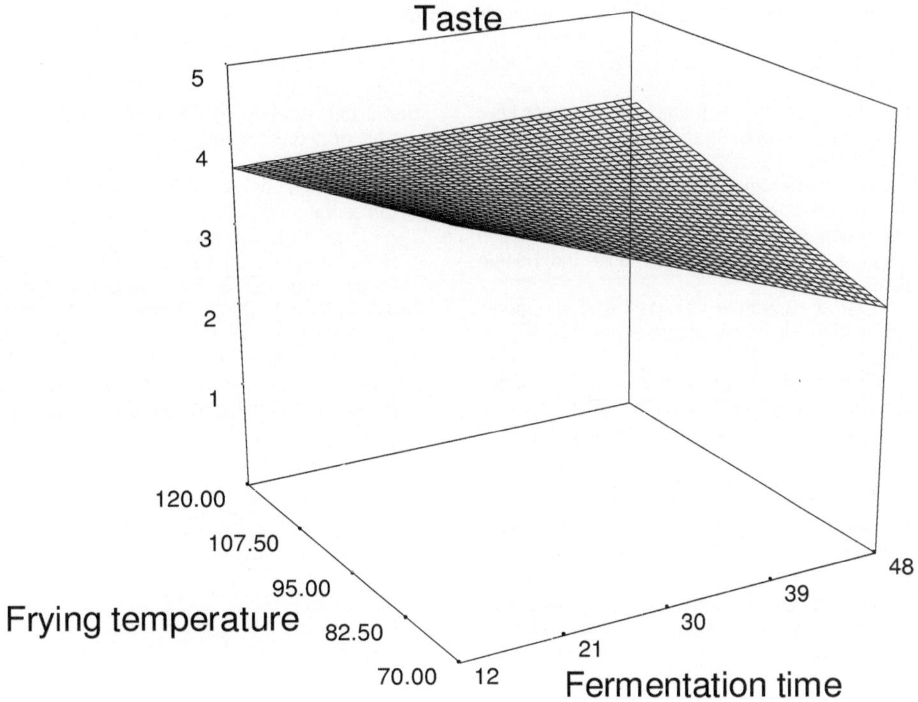


**Figure 4.** Response surface plot of frying temperature and fermentation time on taste.

Effect of resident time of cassava mash during frying is minimally significant ($p>0.05$). Taste sensor score ranged between 2.30 to 4.70, the highest scores were observed in products with fermentation period of between 12 and 30 h with interaction of frying temperature of between 70 to 95°C, the lowest were observed for samples from fermentation period of more than 30 h and frying temperature of more than 95°C (Table 3).

Figure 4 further elucidates the observation. This observation is not in agreement with that reported by Achinewu et al. (1998), the high acceptability of gari of 60 h of fermentation in his study may have been be due to the composition of the taste panel. Due to the high pH value of the product a few consumers would accept and make limited use of the product. Taste of gari emanates cassava fermentation, production, accumulation and breakdown of metabolites (aldehydes, ketones and organic acids) is a factor of fermentation period which in turn is proportional to the taste quality of the final product. Besides, prolornged fermentation of cassava mash introduces an unhygenic problem into gari production. In our study, it was observed that some panelists preferred the sour to the sweet sweet sample, this pattern of preference shows that one run of gari cannot satisfy the organoleptic requirement of all consumers.

Our experience shows that acceptability of a product reflects its preference and ready-to-buy attitude of consumers in the presence of close substitutes. The level of acceptability depends on the general product sensory properties, familiarity, attractiveness and traditional values. Non-availability of the desired quality product forces consumers to buy and consume what is available, with limited satisfaction. In this study the optimum quality for all the parameters of 3.51, 3.83, 4.14, 4.70 and 4.70 of swelling capacity, texture, colour, taste and acceptability, respectively were obtained with the fermentation temperature of 12 hours, frying temperature of 95°C and resident time of 34.64 h.

## Conclusion

The work revealed that the organoleptic attributes and preference of gari is relative to the levels of processing factors used and the psychology of the consumer (Wayne and Macinnis, 1997). Therefore, in the production of gari, adequate attention should be paid to the effect of traditional preference of the target consumer population.

The models produced in this study are not final to the production of consistently high quality of gari, effects of other variables on the quality and acceptability of the product should be studied.

## REFERENCES

Achinewu SC (1990). Toxic components of food and their mode of removal. Public lecture Rivers State University of Science and Technology, Port Harcourt, pp. 69-70.

Achinewu SC, Barber LI, Ijoma IO (1998). Physicochemical properties and garification (gari yield) of selected cassava cultivars in Rivers State, Nigeria. Plant Foods Hum. Nutri., 52: 133-140.

Anderson M, Whitcomb P (2002). RSM simplified: Optimizing processes method for design of experiments. Productivity Press, New York.

Box GEP, Behnken DW (1960). Some new 3-level design for the study of quadratic variables. Technometrics, 2: 455-475.

Cardoso AP, Cliff J, Bradbury JH (2000). Cyanogens in cassava flour and roots and urinary thiocyanate concentration in Mozambique. J. Food Consumption Anal., 13: 1-12.

Collard P, Levi S (1959). A two stage fermentation of cassava. Nature, London, (183): 620-621.

Damadjali DS, Widowate S, Rachin A (1993). Cassava flour production and consumers acceptance at village level in Indonesia. Indonesian Agric. Res. Dev. Lurnal, 15: 16-25.

Ernesto M, Cardoso AP, Cliff J, Bradbury JH (1999). Cyanogenic potential of cassava flour; field trial in Mozambique of a simple kit. Int. Food Sci. Nutr., 49: 93-99.

FAO/WHO (1991). Joint food standards programme Rome: Codex Alimentarius Commission XII, Supplement 4, FAO, p. 145.

FAO STAT (2002). The world production of cassava roots in 2001. http://www.fao.org.

Huguo GN (2002). Food product design: A computer-aided statistical approach. CRC Press, pp. 56-63.

Igathinathane C, Pdessimo LO (2007). Moisture sorption thermodynamic properties of corn stover fractions. Am. Soc. Agric. Biol. Eng., 50(6): 2151-2160.

Ikegwu OJ, Nwobasi VN, Odoh MO Lledinma NU (2009). Evaluation of pasting and some functional properties of staqarch isolated from some improved cassava varieties in Nigeria. Afr. J. Biotechnol., 8(10): 2310-2315.

Jenet S, Nyla T (2001). Eating qualities of muffins prepared with 10% and 20% soy flour. J. Nutr. Recipe Menu Dev., 3(2): 25-33.

Kim HS (2009). Cassava: Gabon seeks Nigeria's help; http://cassavamews/ blogspot.com/2009/02/cassava-gabon-seeks-nigerias-help.html.

Ngaba RR, Lee JS (1999). Fermentation of cassava (*Manihot esculenta* Crantz). Am. J. Food Sci., 44: 1570-1571.

Oduro I, Ellis WO, Dziedzoave NT, Nimaakoyeboah K (2000). Quality of gari from selected processing zones in Ghana. Food Control, 11: 297-303.

Okafor N, Uzuegbu JO (1987). Studies on the contributions of microorganisms to the organoleptic properties of gari: a fermented food derived from cassava. (*Manihot esculenta* Crantz). Nig. J. Food Agric., 2(13): 39-41.

Sanni LO, Adebaoale AA, Awoyale W, Fetuga GO (2008). Quality of gari (roasted cassava mash) in Lagos State, Nigeria. Nig. Food J., 26(2): 125-130.

Sanni LO, Ikuomola DP, Sanni SA (2001). Effect of length of fermentation and varieties on the qualities of sweetpotato garri. Proc. 8th Triennaial Symposium of the International Society for Tropical Root Crop- Africa Branch (ISTRC-AB), Ed. M. O. Akoroda, IITA, Ibadan, Nigeria, 12-16 November, 2001, pp. 208-211.

Janet S, Nyla T (2001). Eating qualities of muffins prepared with 10% and 20% soy flour. J. Nutr. Recipe Menu Dev., 3(2): 25-33.

Wayne D, Hoyer JM, Deborah JM (1997). Regional, ethnic, and religious influences on consumer behaviour. In: Consumer behaviour. Houghton Mifflin Company, USA, 1: 296-336.

# Effect of different land use types and their implications on land degradation and productivity in Ogun State, Nigeria

**B. A. Senjobi[1]* and A. O. Ogunkunle[2]**

[1]Department of Soil Science and Land Management, University of Agriculture, Abeokuta, Nigeria.
[2]Department of Agronomy, University of Ibadan, Ibadan, Nigeria.

**This study was designed to assess the extent to which land use types influence soil nutrients and productivity in Ogun State. Soils under four land uses were sampled for physical, chemical and biological analyses. The land use types studied were arable cropping (land use 1); oil palm (land use 2), secondary forest (land use 3) and building sites (land use 4). The results showed that soil nutrients vary from one land use type to another with respect to morphological properties such as colour, texture, structure and consistency. This soils have very high sand content (>90%), this decreases with depth thereby enhancing porosity, uptill a point of high clay content where permeability is hampered. Exchangeable bases, and acidity and silt: Clay ratios were generally low. Ca and Mg dominate the exchange complex, organic matter content, total nitrogen and effective cation exchange capacity were high in the surface soils compared to sub-surface. The reverse is the trend for available P. Extractable Zn and Cu relative to the critical levels established down the profile were very high and erratic. Soil nutrients depletion followed this order: Secondary forest < Oil palm < Arable cropping < Building sites. Soil depletion and degradation were very prominent under building sites and arable cropping. It was observed that the land use types employed on the study sites were not very compatible with the characteristics of the soil. However, since soil nutrients depletion were very high in all land uses, therefore, there must be a careful choice of appropriate use of land in order to reduce soil nutrients depletion and enhances soil productivity. In addition to the above, plausible land use approach, multiple cropping, organic mulching, contour ridge and cultivation of cover crops are recommended in order to minimize soil nutrients depletion which mostly accounted for major degradation in these plots.**

**Key words:** Soil nutrients depletion, land uses, nutrients dynamic, land degradation, productivity.

## INTRODUCTION

In all parts of Nigeria, there is noticeable evidence of land degradation. This varies from place to place in terms of the types, duration, severity, and socio-economic impact (Aruleba, 2004; Senjobi, 2007). A survey in 1990 by the United Nations suggested that a quarter of the world's total crop land is affected by degradation severely enough to restrict its productivity, 15.6% of this compromised agricultural land is strongly degraded land whose original biotic functions including nutrient cycling are largely destroyed. This has resulted in production loss of about 17% on degraded lands. Accelerated land degradation processes have been attributed not only to rapid human population growth, but also to concomitant intensification of land use. In the last 50 years alone, 20% of the world's agricultural land has been irreversibly damaged, and if the process of destruction continues at this pace, agriculture could lose 15 to 30% of its productivity (FAO, 1984). During 1945 to 1990, human induced land degradation in Africa exceeded 494 millions hectares, and only 22% of the total land area constitutes the producing biomass United Nations environmental programme (UNEP) and international soil reference and

*Corresponding author. E-mail: bolasenjobi@yahoo.com, bolasenjobi@gmail.com.

information centre (ISRIC), 1990 (cited from global assessment of soil degradation (GLASOD), 1998. According to Dudal (1982) and Lal (1990), annual estimate of land lost to land degradation was expected to rise from 5 to 7 million hectares in 1981 to 10 million hectares in year 2000. In an assessment of soil under different covers in Nigeria, it was observed that the average soil loss under bare fallow management was 5.5 kg/m$^2$ with average rainfall of 177 mm a month. Comparative loss in the same plot in the preceeding year which was 3.5 kg/m$^2$ corresponding to 55 and 50 tons/ha soil loss respectively (Senjobi, 2007). In view of the land degradation processes in South Western Nigeria due to human interference on the natural ecosystems, it then becomes pertinent to study the effect of this degradation on nutrients dynamic and their implications on soil productivity.

## MATERIALS AND METHODS

### Description of the studied area

The study was carried out at the Teaching and Research Farm of Olabisi Onabanjo University, Ago-Iwoye, Ogun State, Nigeria in November/December, 2004. The area is located between latitude 6°55´ and 7°00´N and between longitudes 3°45´ and 4°05´E. The area falls within the rain forest region of South Western Nigeria with two main alternating wet and dry seasons which can be sub-divided into the long wet season from mid March to July, the short dry season in the month of August, the short wet season from September to early November and the long dry season from early November to mid March during which the local harmattan period commence between December to February. The area has an annual rainfall of about 1150 mm. The mean relative humidity of the area is generally high (about 80%) with the peak between May and October. The diurnal temperature ranges from highest in November to May and the lowest from June to October, the annual mean temperature is 27°C (Senjobi, 2007).

The dominant vegetable type of the area is secondary forest, tree species found in the study area include *Pentaclethra macrophylla, Bosqueia angolensis, Piptadeniastrum africanum, Antiaris toxicaria var africana, Cola spp., Guarea spp. and Mimusop spp* among others. There is also the widespread of *Elaeis guineensis* and the dominance of *Chlomolaena odorata*; a perennial weed in fallow lands.

Most of the soils in the survey area are developed from undifferentiated igneous and metamorphic, pre-cambrian basement complex rocks such as granite, biotite, gneiss, biotite schist, quartz schist and quartzite (Senjobi, 2007). The major land use types in the study area are arable crop production, cash crop production and non-agricultural uses such as residential, commercial and roads construction.

### Field work

Four land use types namely: Arable cropping (land use 1), Oil palm (land use 2), Secondary forest (land use 3), and Building sites (land use 4) were studied. At each of the chosen land use type, an area of 50 hectares was identified with the aid of tape measurement. Bulk samples consisting of ten core samples from two depths (0 to15 cm and 15 to 30 cm) were collected randomly for physical, chemical and biological analyses. Profile pits (2 x 1 x 2 m) were dug at the three predominant different land types or slopes segment encountered at each of the chosen land use type. These were crest, middle slope, and valley bottom. The general site descriptions, such as climate, vegetable, land use, gradient of slope, drainage type, soil surface form, type and degree of erosion, field texture, micro-relief and depths to ground water table were recorded.

The profile pits were also described morphologically after FAO (2006), sampled and then followed by placing them in labelled bags and then transported to the laboratory for air-drying. After being air-dried for 72 h and the samples were crushed to pass through sieved using a 2 mm screens.

Soil samples were analyzed for the following parameters: Soil pH was determined in both water and 0.01 M potassium chloride solution (1:1) using glass electrode pH meter (Mclean, 1965). Total nitrogen was determined by the macro-kjeldahl digestion method of Jackson (1962). Bray-1 P was determined by molybdenum blue colorimetry (Bray and Kurtz, 1945) exchangeable cations were extracted with 1 M $NH_4OAC$ (pH 7.0), K and Na were determined using flame photometer and exchangeable Mg and Ca by atomic absorption spectrophotometer (Sparks, 1996). Exchangeable acidity was determined by the KCl extraction method (Mclean, 1965) and organic carbon was after dichromate wet oxidation method (Walkey and Black, 1934). The organic matter content was got by multiplying a factor of percent organic carbon by 1.72. Cation Exchange Capacity (CEC) was calculated from the sum of all exchangeable cations. Available micronutrients were extracted with 1N $NH_4Cl$ solutions and determined by Atomic Absorption Spectrophotmeter (AAS) (Water and Sammer, 1948) cited from Aruleba (2004). Particle size analysis was by the Bouyoucos (1951) method. Saturated hydraulic conductivity was determined using a constant head method, bulk density by core method, soil porosity was estimated from the bulk density data at an assumed particle density of 2650 kgm$^{-3}$. Water retention at 15 bar was determined in order to calculate available water holding capacities of the soil profile horizons (Mbagwu, 1985).

### Statistical analysis

#### *Multiple linear step-wise regression analysis (forward elimination method)*

The significance of this analysis is to estimate the relative contribution of each factor (land use types, land type and suitability) to land degradation for each site. To achieve this objective, a step-down model which adds one variable to the regression equation at a time was used (SAS Inst. 1990). As each variable is entered, the model incorporates a check on the variance test. The process of adding variables in turn continued until the contribution of the most recently entered variable is not significant at ($P < 0.05$) by the partial F-value. Any variable which provides a non-significant contribution is removed from the model.

## RESULTS AND DISCUSSION

### Results

### Morphological properties

The details of the morphological properties of the soils of land use types are shown in Table 1. The basic differences between pedons at different land use types locations were in colour, texture, structure and

**Table 1.** Morphological properties of the pedons and soils under the different land uses.

| Profile no. | Parent material | Land use | Horizon designation | Depth (cm) | Colour | | Boundary | Textural class | Structure | Consistency | Quartz | Concretions | Roots | Drainage | Mottles |
|---|---|---|---|---|---|---|---|---|---|---|---|---|---|---|---|
| | | | | | Moist | Dry | | | | | | | | | |
| **Slope (10%)** | | | | | | | | | | | | | | | |
| | | Fallow/cassava | Ap | 0-46 | 10YR 6/4 | 10YR 5/6 | cl | SCL | Sg – sbk | ss, sp, vfr, lo | - | c | vf, f | pd | - |
| $P_1$ (VB) | Basement complex | | Bc | 46-76 | 10YR 5/6 | – | gs | SCL | P | ns, np, lo | f | vm | vf, vf | pd | 5YR 5/4 |
| | | | Bt | 76-126 | 10YR 5/6 | – | cl | SCL | Co | s, p, h | f | f | vf, vf | pd | 5YR 4/2 |
| **Slope (5%)** | | | | | | | | | | | | | | | |
| | | Cassava/plantain | Apc | 0-18 | 7.5YR 3/2 | – | ab, s | SC | G | ns, np, lo | f | vm | vf, f | wd | - |
| $P_2$ (MS) | " | | Btc | 18-57 | 5YR 5/4 | – | cl, w | SC | Ab | s, p, fr | - | vm | vf, vf | wd | - |
| | | | Bt | 57-118 | 5YR 5/6 | – | ir | SCL | Sg | st, sp, fr | - | c | vf, vf | wd | - |
| **Slope (1%)** | | | | | | | | | | | | | | | |
| | | Cassava/maize | Apc | 0-14 | 5YR 6/8 | – | di, s | SCL | Ab | s, sp, f | f | vm | vf, vf | wd | - |
| $P_3$ (C) | " | | Bt | 14-23 | 5YR 5/8 | – | di, s | SCL | Ab | s, sp, f | f | vm | vf, vf | wd | - |
| | | | Btc | 23-53 | 7.5YR 6/6 | – | ab, s | SCL | G | vs, p, f | - | vm | vf, vf | wd | - |
| | | | Bt | 53-112 | 2.5YR 4/8 | – | cl | CL | M | vs, vp, vh | - | f | vf, vf | wd | - |
| **Slope (6%)** | | | | | | | | | | | | | | | |
| $P_4$ (VB) | " | Oil palm | Ap | 0-30 | 5YR 4/3 | | ab, s | SL | Sg | ns, ns, lo, vf | - | vm | c – m | wd | - |
| | | | Btc | 30-110 | 5YR 6/6 | 5YR5/8 | s | SCL | g/m | s, p, fr | - | vm | vf, vf | | |
| **Slope (3%)** | | | | | | | | | | | | | | | |
| $P_5$ (MS) | Basement complex | Oil palm | Ap | 0-34 | 5YR 2/2 | 5YR4/2 | ab, s | S | Sg | ns, np, vfr | | f | m, me-co | wd | - |
| | | | AB | 34-69 | 5YR 4/4 | 5YR6/3 | cl, ir | LS | sg | ns, np | - | c | c, me | wd | - |
| | | | B | 69-113 | 7.5YR 5/6 | 7.5YR5/8 | g, ir | LS | g | ns, np | - | vm | f,m-c0 | wd | - |
| | | | BC | 113-137 | 7.5YR 5/6 | 10YR6/6 | di | LS | g-cr | ns, np | f | vm | vf,vf | wd | 5YR 3/2 |

+ Boundary: ab = abrupt, cl = clear, g = gradual, di = diffuse, s = smooth, w = wavy, ir = irregular,[1]Textural class: LS = loamy sand, s = sand, SL = Sandy loam, SC = Sandy clay, SCL = Sandy clay loam,[2]Structure: sbk = Subangular blocky, ab = angular blocky, sg = single grain, m = massive, co = coarse, p = prismatic, cr = crumb, g = granular, [3]Consistency: ns = non-sticky, np = non-pastic, lo = loose, s = sticky, p – plastic, f = firm, h = hard VS = very sticky, vp – very plastic, vh = very hard, fr = friable, vfr = very friable, sp = slightly plastic, ss = slightly sticky,[4]Quartz: f = few, - = Absent,[5]Concretions: f = few, c = common, m = high, vm = higher,[6]Roots: vf = very few, f = few, c = common, m = many (concentration) vf = very fine, f = fine, me = medium, co = coarse (size),[7]Drainage: wd = well drained, pd = poorly drained,[8]VB=Valley-Bottom, MS = Middle Slow and C = Crest.

**Table 1.** Morphological properties of the pedons and soils under the different land uses (continued).

| Profile no. | Parent material | Land use | Horizon designation | Depth (cm) | Colour | | Boundary | Textural class | Structure | Consistency | Quartz | Concretions | Roots | Drainage | Mottles |
|---|---|---|---|---|---|---|---|---|---|---|---|---|---|---|---|
| | | | | | Moist | Dry | | | | | | | | | |
| | | Oil palm | Ap | 0-34 | 5YR 2/2 | 5YR4/2 | ab, s | S | Sg | ns, np, vfr | | f | m, me-co | wd | - |
| $P_5$ (MS) | Basement complex | | AB | 34-69 | 5YR 4/4 | 5YR6/3 | cl, ir | LS | sg | ns, np | - | c | c, me | wd | - |
| | | | B | 69-113 | 7.5YR 5/6 | 7.5YR5/8 | g, ir | LS | g | ns, np | - | vm | f, m-c0 | wd | - |
| | | | BC | 113-137 | 7.5YR 5/6 | 10YR6/6 | di | LS | g-cr | ns, np | f | vm | vf,vf | wd | 5YR 3/2 |
| **Slope (0.5%)** | | | | | | | | | | | | | | | |
| | | Oil palm | Ap | 0-33 | 7.5YR 3/2 | 7.5YR4/4 | cl, s | S | sbk –gb | ns, np | f | c | m, f-co | wd | - |
| $P_6$ (C) | " | | BC | 33-65 | 7.5YR 5/6 | 7.5YR6/4 | cl, w | SC | ab | ns, np, lo | f | vm | f, f–me | wd | - |
| | | | Bt | 65-112 | 10YR 5/8 | 7.5YR5/8 | di, s | SC | m | vs, p, fr | - | vm | vf, vf | wd | - |
| **Slope (4%)** | | | | | | | | | | | | | | | |
| $P_7$ (VB) | " | Secondary forest | Apc | 0-31 | 7.5YR 5/8 | - | ab, s | SL | g | vs, fr | f | f | f, vf | wd | - |
| | | | Btc | 31-108 | 2.5YR 4/8 | – | di, ir | SL | sg | Vs | - | f | vf, vf | wd | - |
| **Slope (2%)** | | | | | | | | | | | | | | | |
| | | Secondary forest | Ap | 0-30 | 7.5YR 5/8 | – | ab, s | LS | sg | ns, lo | - | f | f, f-me | wd | - |
| $P_8$ (MS) | " | | B | 30-62 | 7.5YR 5/6 | – | g, w | LS | sg | ns, lo | - | f | vf, vf | wd | - |
| | | | Btc | 62-112 | 7.5YR 7/8 | – | di | SL | g | vs, vf | f | vm | c,me-co | wd | - |
| **Slope (1%)** | | | | | | | | | | | | | | | |
| $P_9$ | | Secondary forest | Ap | 0-50 | 7.5YR 6/8 | – | cl, w | LS | sbk | vs, f | f | f | f, vf | wd | - |
| (C) | " | | Bt | 50-86 | 7.5YR 5/8 | – | di, | SCL | m | vs, vp, vf | - | f | vf, vf | wd | - |
| | | | Btc | 86-128 | 7.5YR 5/6 | – | g, w | LS | sg | ns, lo | - | vm | vf, vf | wd | - |
| **Slope (6%)** | | | | | | | | | | | | | | | |
| | | Building site | Ap | 0-30 | 5YR 5/4 | - | ab, s | SL | sg | ns, np, lo | - | m | m, vf-f | pd | - |
| $P_{10}$ (VB) | Sand stones | | B | 30-52 | 7.5YR 5/4 | – | cl, w | SL | sg | ns, np, lo | - | f | co, vf-f | pd | 5YR 3/2 |
| | | | Btc | 52-82 | 7.5YR 6/6 | – | w | LS | m | s, p, f | f | f | vf –vf | pd | - |
| **Slope (2%)** | | | | | | | | | | | | | | | |
| $P_{11}$ (MS) | Basement complex | Building site | Ap | 0-45 | 7.5YR 5/4 | – | cl, s | LS | sg | ns, np, lo | - | f | m, vf-co | wd | - |
| | | | $Bc_1$ | 45-72 | 7.5YR 5/6 | – | g, w | LS | g | ns, np, lo | f | vm | f, vf -f | wd | - |
| | | | $Bc_2$ | 72-104 | 7.5YR 5/6 | – | ir | SL | m | ns, np, lo | f | f | vf –vf | wd | - |

**Table 1.** Morphological properties of the pedons and soils under the different land uses (continued).

| | | | | | | | | | | | | | | | |
|---|---|---|---|---|---|---|---|---|---|---|---|---|---|---|---|
| **Slope (4%)** | | | | | | | | | | | | | | | |
| $P_{12}$ (C) | Basement complex | Building site | Ap | 0-26 | 7.5YR 3/2 | – | ab, s | SL | sg | ns, np, lo | - | f | m, f-co | wd | - |
| | | | AB | 26-57 | 7.5YR 5/2 | – | di, s | LS | sg | ns, np, lo | - | f | co, f-co | wd | - |
| | | | Bt | 57-102 | 7.5YR 6/8 | – | s | SL | m | s, p, f | - | f | vf –vf | wd | - |

+ Boundary: ab = abrupt, cl = clear, g = gradual, di = diffuse, s = smooth, w = wavy, ir = irregular,[1] Textural class: LS = loamy sand, s = sand, SL = sandy loam, SC = Sandy clay, SCL = sandy clay loam, [2]structure: sbk = subangular blocky, ab = angular blocky, sg = single grain, m = massive, co = coarse, p = prismatic, cr = crumb, g = granular,[3]consistency: ns = non-sticky, np = non-pastic, lo = loose, s = sticky, p – plastic, f = firm, h = hard VS = very sticky, vp – very plastic, vh = very hard, fr = friable, vfr = very friable, sp = slightly plastic, ss = slightly sticky,[4]Quartz: f = few, - = Absent, [5]concretions: f = few, c = common, m = high, vm = higher, [6]Roots: vf = very few, f = few, c = common, m = many (concentration) vf = very fine, f = fine, me = medium, co = coarse (size),[7] Drainage: wd = well drained, pd = poorly drained,[8]VB = Valley-Bottom, MS = Middle Slow and C = crest.

**Table 2.** Physical properties of the soil under cassava.

| Profile no. | Horizon designation | Depth (cm) | Gravel | Sand (g/kg) | Silt (g/kg) | Clay | Silt: Clay ratio | Porosity (%) | Textural Class | WHC (%) | BD ($g/cm^3$) | HC (cm/h) | Permeability (cm/h) |
|---|---|---|---|---|---|---|---|---|---|---|---|---|---|
| | Ap | 0-46 | 226 | 508 | 202 | 290 | 0.70 | 50.6 | SCL | 24.1 | 1.31 | 1.68 | 13.6 |
| $P_1$ (VB) | Bc | 46-76 | 775 | 642 | 148 | 210 | 0.70 | 47.4 | SCL | 28.3 | 1.40 | 1.53 | 12.5 |
| | Bt | 76-126 | 102 | 522 | 118 | 360 | 0.33 | 44.2 | SC | 28.5 | 1.41 | 1.55 | 12.2 |
| | Apc | 0-18 | 687 | 472 | 158 | 370 | 0.43 | 51.9 | SC | 34.8 | 1.28 | 2.82 | 23.8 |
| $P_2$ (MS) | Btc | 18-57 | 663 | 508 | 142 | 350 | 0.41 | 55.8 | SC | 31.6 | 1.17 | 0.54 | 3.24 |
| | Bt | 57-118 | 201 | 468 | 202 | 330 | 0.61 | 43.4 | SCL | 35.0 | 1.50 | 1.08 | 4.43 |
| | Apc | 0-14 | 808 | 492 | 220 | 288 | 0.76 | 41.7 | SCL | 33.3 | 1.55 | 1.32 | 12.1 |
| $P_3$ | Bt | 14-23 | 593 | 672 | 120 | 208 | 0.58 | 45.9 | SCL | 38.9 | 1.44 | 0.99 | 7.4 |
| (C) | Btc | 23-53 | 487 | 532 | 188 | 280 | 0.67 | 43.5 | SCL | 38.8 | 1.43 | 0.10 | 6.8 |
| | Bt | 53-112 | 135 | 418 | 202 | 380 | 0.53 | 42.6 | CL | 47.5 | 1.42 | 0.10 | 6.7 |

P = ploughed, c = concretion, t = illuvial accumulation of clay, LS = loamy sand, s = sand, SL = sandy loam, SC = sandy clay, SCL = sandy clay loam, VB = valley-bottom, MS = middle slow and C = crest, WHC = water holding capacity (%), BD = bulk density ($g/cm^3$), HC = hydraulic conductivity (cm/hr).

consistency. The soils of the land use types were characterized by brown (5YR 5/4 – 10YR 6/4), sandy loam / loamy sand, topsoil over reddish brown (2.5YR 4/8 – 7.5YR 6/8) sandy clay loam to sandy clay subsoil. The soils are deep and well drained. The structure is single grain at the top to sub-angular blocky and massive at the sub-horizons.

## Physical properties

The main physical properties of the soils are presented in Tables 2 to 5. The particle size distribution data showed that the pedons have very high sand content (> 90%) and this decreases with depth, except in some cases, particularly at the sub-soil horizon (>50 in depths) where they have very high percentage of it. Sand dominates at all the depth of the profiles. The clay content in the soils ranged from 4.8 to 37% in the

**Table 3.** Physical properties of the soil under oil palm.

| Profile no. | Horizon designation | Depth (cm) | Gravel | Sand (g/kg) | Silt (g/kg) | Clay | Silt: Clay ratio | Porosity (%) | Textural class | WHC (%) | BD ($g/cm^3$) | HC (cm/hr) | Permeability (cm/hr) |
|---|---|---|---|---|---|---|---|---|---|---|---|---|---|
| $P_4$ (VB) | Ap | 0-30 | 893 | 612 | 320 | 68 | 4.71 | 53.2 | SL | 43.8 | 1.24 | 0.57 | 5.01 |
| | Btc | 30-110 | 705 | 558 | 142 | 300 | 0.47 | 49.9 | SCL | 55.4 | 1.07 | 0.40 | 3.12 |
| $P_5$ (MS) | Ap | 0-34 | 174 | 912 | 40 | 48 | 0.83 | 30.6 | S | 18.6 | 1.84 | 0.84 | 6.12 |
| | AB | 34-69 | 253 | 872 | 80 | 48 | 1.67 | 37.4 | LS | 23.5 | 1.66 | 0.75 | 5.16 |
| | B | 69-113 | 631 | 872 | 40 | 88 | 0.45 | 41.0 | LS | 29.8 | 1.57 | 0.78 | 5.43 |
| | Bc | 113-137 | 612 | 812 | 140 | 48 | 2.92 | 42.8 | LS | 23.5 | 1.52 | 0.48 | 2.85 |
| $P_6$ (C) | Ap | 0-33 | 251 | 912 | 40 | 48 | 0.83 | 42.5 | S | 28.7 | 1.53 | 0.54 | 3.39 |
| | Bc | 33-65 | 627 | 508 | 142 | 350 | 0.41 | 35.1 | SC | 25.7 | 1.72 | 1.13 | 9.90 |
| | Bt | 65-112 | 899 | 532 | 120 | 348 | 0.34 | 53.4 | SC | 38.7 | 1.24 | 0.85 | 7.90 |

**Table 4.** Physical properties of the soil under secondary forest.

| Profile no. | Horizon designation | Depth (cm) | Gravel | Sand (g/kg) | Silt (g/kg) | Clay | Silt: Clay ratio | Porosity (%) | Textural class | WHC (%) | BD ($g/cm^3$) | HC (cm/hr) | Permeability (cm/hr) |
|---|---|---|---|---|---|---|---|---|---|---|---|---|---|
| $P_7$ (VB) | Apc | 0-31 | 133 | 712 | 160 | 128 | 1.25 | 42.5 | SL | 23.8 | 1.53 | 1.05 | 7.89 |
| | Btc | 31-108 | 99 | 752 | 120 | 128 | 0.94 | 43.8 | SL | 23.1 | 1.49 | 2.22 | 18.48 |
| $P_8$ (MS) | Ap | 0-30 | 60 | 812 | 140 | 48 | 2.92 | 54.2 | LS | 35.5 | 1.22 | 1.74 | 14.01 |
| | B | 30-62 | 157 | 812 | 120 | 68 | 1.76 | 38.9 | LS | 26.1 | 1.62 | 2.40 | 20.13 |
| | Btc | 62-112 | 444 | 672 | 140 | 188 | 0.74 | 31.2 | SL | 21.9 | 1.83 | 1.11 | 8.43 |
| $P_9$ (C) | Ap | 0-50 | 143 | 872 | 80 | 48 | 1.67 | 44.0 | LS | 53.2 | 1.49 | 0.69 | 1.77 |
| | Bt | 50-86 | 95 | 672 | 120 | 208 | 0.58 | 40.5 | SCL | 43.2 | 1.57 | 0.57 | 3.81 |
| | Btc | 86-128 | 495 | 872 | 80 | 48 | 1.67 | 45.5 | LS | 34.0 | 1.45 | 0.30 | 1.38 |

surface layer and 4.8 to 38% in the subsurface and the clay content increased down the profile. The silt content in the soil was generally low and there was no definite sequence in its distribution within the profile. The silt: clay ratios were generally low

The water holding capacity of the soils increases with increase in clay content of the soils. The porosity of the land use soils was directly proportional to percentage of sand and gravel concentration, whereas the permeability was inversely proportional to the clay content. The bulk density and hydraulic conductivity were generally low with no definite sequence in their distribution within the profile.

## Chemical properties

The data on the chemical properties of the soils are given in Tables 6 to 9. The pH values of the

**Table 5.** Physical properties of the soil under building site.

| Profile no. | Horizon designation | Depth (cm) | Gravel | Sand g/kg) | Silt (g/kg) | Clay | Silt: Clay ratio | Porosity (%) | Textural class | WHC (%) | BD ($g/cm^3$) | HC (cm/h) | Permeability (cm/h ) |
|---|---|---|---|---|---|---|---|---|---|---|---|---|---|
| | Ap | 0-30 | 229 | 812 | 140 | 88 | 1.59 | 39.6 | SL | 29.1 | 1.61 | 0.60 | 3.54 |
| $P_{10}$ (VB) | B | 30-52 | 101 | 712 | 200 | 88 | 2.27 | 37.6 | SL | 24.0 | 1.66 | 1.05 | 7.89 |
| | Btc | 52-82 | 124 | 772 | 80 | 108 | 0.74 | 36.4 | LS | 30.2 | 1.58 | 1.02 | 6.84 |
| | Ap | 0-45 | 78 | 852 | 80 | 68 | 1.18 | 48.3 | LS | 42.2 | 1.37 | 2.28 | 19.02 |
| $P_{11}$ (MS) | $Bc_1$ | 45-72 | 504 | 812 | 100 | 88 | 1.14 | 47.4 | LS | 19.9 | 1.58 | 0.84 | 6.12 |
| | $Bc_2$ | 72-104 | 152 | 712 | 120 | 168 | 0.71 | 43.2 | SL | 52.1 | 1.51 | 1.29 | 10.08 |
| | Ap | 0-26 | 61 | 772 | 140 | 88 | 1.59 | 47.8 | SL | 37.0 | 1.32 | 3.39 | 28.56 |
| $P_{12}$ (C) | AB | 26-57 | 30 | 792 | 140 | 68 | 2.06 | 46.8 | LS | 46.0 | 1.41 | 2.67 | 22.3 |
| | Bt | 57-102 | 103 | 672 | 180 | 148 | 1.22 | 37.4 | SL | 28.0 | 1.66 | 0.51 | 3.27 |

P = Ploughed, c = concretion, t = illuvial accumulation of clay, LS = loamy sand, s = sand, SL = sandy loam, SC = sandy clay, SCL = Sandy Clay loam, VB=Valley-Bottom, MS=Middle Slow and C = crest, WHC = water holding capacity (%), BD = bulk density ($g/cm^3$), HC = hydraulic conductivity (cm/hr).

**Table 6.** Chemical properties of soils under cassava/maize/plantain/banana.

| Profile no. | Horizon designation | Depth (cm) | pH $H_2O$ | pH KCl | cmol.$kg^{-1}$ Exchange bases | | | | | | B. Sat (%) | C (%) | Total N (%) | P | Zn | Cu |
|---|---|---|---|---|---|---|---|---|---|---|---|---|---|---|---|---|
| | | | | | Na | K | Ca | Mg | Ex AC | ECEC (soil) | | | | mg.$kg^{-1}$ | | |
| | Ap | 0-46 | 7.20 | 5.10 | 0.34 | 0.21 | 1.04 | 2.06 | 0.05 | 3.70 | 98.65 | 0.32 | 0.02 | 4.71 | 5.70 | 5.60 |
| $P_1$ (VB) | Bc | 46-76 | 7.15 | 4.75 | 0.48 | 0.30 | 1.18 | 1.60 | 0.05 | 3.64 | 98.63 | 0.38 | 0.04 | 1.88 | 5.40 | 5.30 |
| | Bt | 76-126 | 7.05 | 5.65 | 0.29 | 0.19 | 0.86 | 1.73 | 0.05 | 3.12 | 98.40 | 0.29 | 0.02 | 5.18 | 6.90 | 5.10 |
| | Apc | 0-18 | 7.00 | 6.15 | 0.39 | 0.43 | 1.15 | 1.73 | 0.06 | 3.76 | 98.40 | 1.54 | 0.13 | 2.82 | 6.20 | 5.60 |
| $P_2$ (MS) | Btc | 18-57 | 7.25 | 6.10 | 0.30 | 0.36 | 0.78 | 1.87 | 0.05 | 3.36 | 98.51 | 0.29 | 0.03 | 13.65 | 5.80 | 3.70 |
| | Bt | 57-118 | 6.05 | 5.10 | 0.25 | 0.24 | 1.14 | 1.21 | 0.11 | 2.95 | 96.27 | 1.22 | 0.13 | 10.35 | 4.60 | 5.20 |
| | Apc | 0-14 | 6.35 | 5.20 | 0.36 | 0.41 | 1.10 | 2.00 | 0.09 | 3.96 | 97.73 | 1.66 | 0.16 | 1.88 | 5.00 | 5.60 |
| $P_3$ (C) | Bt | 14-23 | 5.95 | 5.20 | 0.27 | 0.21 | 1.16 | 1.19 | 0.11 | 2.94 | 96.26 | 0.40 | 0.03 | 10.35 | 7.00 | 6.10 |
| | Btc | 23-53 | 6.15 | 5.05 | 0.24 | 0.15 | 1.23 | 1.50 | 0.10 | 3.22 | 96.89 | 1.10 | 0.10 | 6.12 | 6.10 | 5.60 |
| | Bt | 53-112 | 6.05 | 4.95 | 0.23 | 0.14 | 1.18 | 1.65 | 0.11 | 3.31 | 96.68 | 1.13 | 0.11 | 7.06 | 6.40 | 4.80 |

P = ploughed, c = concretion, t = illuvial accumulation of clay, VB=Valley-Bottom, MS=Middle Slope and C=Crest.

**Table 7.** Chemical properties of soils under oil palm.

| Profile no. | Horizon designation | Depth (cm) | pH $H_2O$ | pH KCl | cmol.kg$^{-1}$ Exchange bases Na | K | Ca | Mg | Ex AC | ECEC (soil) | B. Sat (%) | C (%) | Total N (%) | P (mg.kg$^{-1}$) | Zn | Cu |
|---|---|---|---|---|---|---|---|---|---|---|---|---|---|---|---|---|
|  | Ap | 0-30 | 7.10 | 6.15 | 0.25 | 0.21 | 1.05 | 1.56 | 0.05 | 3.12 | 98.40 | 1.57 | 0.16 | 11.76 | 5.30 | 5.90 |
| $P_4$ (VB) | Btc | 30-110 | 6.65 | 5.05 | 0.47 | 0.29 | 1.15 | 1.83 | 0.08 | 3.82 | 97.91 | 0.43 | 0.05 | 3.29 | 5.00 | 5.20 |
|  | Ap | 0-34 | 6.70 | 5.65 | 0.20 | 0.12 | 1.14 | 1.38 | 0.08 | 2.92 | 97.26 | 1.01 | 0.10 | 14.12 | 6.20 | 5.70 |
| $P_5$ (MS) | AB | 34-69 | 6.60 | 5.75 | 0.21 | 0.10 | 1.25 | 1.44 | 0.08 | 3.08 | 97.40 | 1.31 | 0.13 | 7.06 | 5.80 | 6.50 |
|  | B | 69-113 | 6.70 | 5.90 | 0.28 | 0.14 | 0.90 | 1.71 | 0.08 | 3.11 | 97.43 | 0.16 | 0.01 | 10.35 | 4.80 | 3.90 |
|  | Bc | 113-137 | 6.95 | 5.90 | 0.24 | 0.14 | 0.89 | 1.89 | 0.06 | 3.22 | 98.14 | 0.38 | 0.04 | 10.00 | 4.90 | 4.50 |
|  | Ap | 0-33 | 7.50 | 6.20 | 0.31 | 0.30 | 1.04 | 1.75 | 0.04 | 3.44 | 98.84 | 0.99 | 0.10 | 5.18 | 6.00 | 4.90 |
| $P_6$ (C) | Bc | 33-65 | 6.90 | 6.00 | 0.26 | 0.13 | 0.96 | 1.79 | 0.06 | 3.20 | 98.13 | 0.48 | 0.05 | 8.47 | 6.20 | 4.70 |
|  | Bt | 65-112 | 6.80 | 5.90 | 0.26 | 0.15 | 0.99 | 1.19 | 0.07 | 2.66 | 97.37 | 0.50 | 0.06 | 6.59 | 6.00 | 6.50 |

**Table 8.** Chemical properties of soils under secondary forest.

| Profile no. | Horizon designation | Depth (cm) | pH $H_2O$ | pH KCl | cmol.kg$^{-1}$ Exchange bases Na | K | Ca | Mg | Ex AC | ECEC (soil) | B. Sat (%) | C (%) | Total N (%) | P (mg.kg$^{-1}$) | Zn | Cu |
|---|---|---|---|---|---|---|---|---|---|---|---|---|---|---|---|---|
| P7 | Apc | 0-31 | 6.70 | 5.45 | 0.28 | 0.18 | 1.08 | 1.48 | 0.08 | 3.10 | 97.42 | 1.33 | 0.13 | 8.94 | 6.20 | 5.60 |
| (VB) | Btc | 31-108 | 6.40 | 5.65 | 0.23 | 0.13 | 1.03 | 1.38 | 0.09 | 2.86 | 96.85 | 0.67 | 0.07 | 5.18 | 6.00 | 5.40 |
|  | Ap | 0-30 | 6.90 | 6.45 | 0.32 | 0.50 | 1.08 | 1.93 | 0.06 | 3.89 | 98.46 | 0.27 | 0.03 | 3.76 | 5.70 | 5.30 |
| P8 (MS) | B | 30-62 | 6.80 | 5.80 | 0.23 | 0.15 | 1.06 | 1.98 | 0.07 | 3.49 | 97.99 | 0.32 | 0.03 | 1.14 | 4.80 | 4.70 |
|  | Btc | 62-112 | 6.65 | 5.75 | 0.33 | 0.28 | 0.90 | 1.87 | 0.08 | 3.46 | 97.69 | 0.21 | 0.02 | 0.94 | 4.80 | 4.50 |
| P9 | Ap | 0-50 | 6.80 | 5.60 | 0.28 | 0.24 | 1.01 | 1.93 | 0.07 | 3.40 | 97.50 | 0.54 | 0.05 | 8.47 | 5.10 | 4.10 |
| (C) | Bt | 50-86 | 6.75 | 5.80 | 0.26 | 0.29 | 0.91 | 1.93 | 0.08 | 3.47 | 97.69 | 0.66 | 0.06 | 13.17 | 5.30 | 4.40 |
|  | Btc | 86-128 | 6.30 | 5.30 | 0.36 | 0.21 | 0.99 | 1.73 | 0.09 | 3.38 | 97.34 | 0.61 | 0.06 | 1.14 | 5.40 | 4.40 |

land use soils ranged from 6.35 to 7.5 and KCl (5.10 and 6.30) in the surface layer. These values had no definite sequence in their distribution down the profile. The exchangeable bases (K, Ca, Mg and Na) in all the pedons were generally low and the exchange complex of all pedons were dominated by Ca and Mg. The Exchangeable Acidity (EA) was also low (0.04 – 1.10) cmol.kg$^{-1}$ and fluctuated irregularly with depth in most of

**Table 9.** Chemical properties of soils under building site.

| Profile no. | Horizon designation | Depth (cm) | pH $H_2O$ | pH KCl | Exchange bases (cmol.kg$^{-1}$) | | | | Ex AC | ECEC (soil) | B. Sat % | C (%) | Total N (%) | P (mg.kg$^{-1}$) | Zn (mg.kg$^{-1}$) | Cu (mg.kg$^{-1}$) |
|---|---|---|---|---|---|---|---|---|---|---|---|---|---|---|---|---|
| | | | | | Na | K | Ca | Mg | | | | | | | | |
| P10 (VB) | Ap | 0-30 | 6.85 | 5.70 | 0.29 | 0.18 | 1.09 | 1.89 | 0.07 | 3.52 | 98.01 | 0.64 | 0.06 | 0.94 | 4.10 | 5.10 |
| | B | 30-52 | 6.65 | 5.40 | 0.26 | 0.15 | 1.11 | 1.46 | 0.08 | 3.06 | 97.39 | 0.26 | 0.03 | 0.94 | 4.70 | 4.80 |
| | Btc | 52-82 | 7.10 | 5.50 | 0.32 | 0.26 | 1.11 | 1.95 | 0.05 | 3.69 | 98.64 | 0.21 | 0.02 | 16.47 | 5.60 | 5.30 |
| P11 (MS) | Ap | 0-45 | 6.85 | 5.50 | 0.25 | 0.24 | 0.80 | 1.67 | 0.07 | 3.03 | 97.69 | 0.32 | 0.02 | 2.35 | 4.90 | 5.30 |
| | $Bc_1$ | 45-72 | 7.10 | 5.85 | 0.35 | 0.26 | 1.15 | 1.80 | 0.05 | 3.61 | 98.61 | 0.41 | 0.05 | 15.06 | 4.50 | 5.10 |
| | $Bc_2$ | 72-104 | 7.00 | 5.85 | 0.30 | 0.32 | 0.88 | 1.56 | 0.06 | 3.12 | 98.08 | 0.32 | 0.03 | 2.82 | 4.60 | 5.20 |
| P12 (C) | Ap | 0-26 | 6.90 | 5.60 | 0.30 | 0.20 | 1.25 | 1.58 | 0.06 | 3.39 | 98.23 | 0.72 | 0.07 | 5.65 | 5.70 | 5.90 |
| | AB | 26-57 | 6.90 | 5.15 | 0.29 | 0.10 | 1.19 | 1.87 | 0.06 | 3.51 | 98.29 | 0.18 | 0.02 | 1.88 | 6.00 | 5.00 |
| | Bt | 57-102 | 6.50 | 5.05 | 0.31 | 0.10 | 1.13 | 1.48 | 0.08 | 3.10 | 97.42 | 0.12 | 0.01 | 7.87 | 4.80 | 5.40 |

P = Ploughed, c = concretion, t = illuvial accumulation of clay, VB = Valley-Bottom, MS = middle slope and C = crest.

the pedons. The Effective Cation Exchange Capacity (ECEC) was high and followed the same trend with exchangeable acidity with the highest values being recorded in most pedons at the surface horizon, the values of ECEC ranged from 2.92 – 3.96 cmol.kg$^{-1}$ on the surface and 2.66 – 3.82 cmol.kg$^{-1}$ at sub-surface. The base saturation by ECEC values were very high both in the surface and lower horizons in all the profiles of the land use types examined. While the distribution down the profiles decreases in some, in the others they were very erratic.

The organic carbon fluctuated irregularly with depth in nearly all the pedons, indicating deposition material particularly from soils of sand stones origin. The value of organic carbon ranged from 0.27 to 1.66% in the surface layers, with the highest concentration on the surface soils in most pedons. The total N was relatively low but high in most of the top soils where the range was between 0.02 and 0.16% and it fluctuated irregularly with depth following the same trend as organic carbon. Available P (Bray P1) in all the pedons was generally high and fluctuated irregularly with depth with high concentration at the sub-surface horizons in most pedons. This may be due to application of phosphate fertilizer. Extractable Zn and Cu relative to the critical levels established down the profile were very high and erratic.

The result of multiple linear stepwise regressions of land use type, land type and suitability with land degradation (physical, chemical and biological) is shown in Table 9 for all the land use types. The significance of the contribution of each factor was determined at 5% significance level ($P < 0.05$).

At the land use 1, suitability (oil palm) contributed significantly (24.22%) of degradation due to potassium content. There was no contribution and regression model at land use 2 because of the uniformity of independent variables of degradation indicators examined. At land use 3, land type contributed significantly 29.3% and suitability (cassava/maize) contributed significantly 35.1% due to nitrogen degradation. For degradation (organic matter content), suitability (in terms of cassava/maize) contributed 39.7. For degradation due to permeability in land use 4, land type significantly contributed 22.2% (Table 10).

## DISCUSSION

The morphological, physical and chemical properties of the soils studied varied from one land type to another in all the four land use types especially as regard colour, texture, structure, consistency, exchangeable bases, organic carbon, total nitrogen etc. The major differences observed were in structure, texture, colour, drainage and soil consistence. It was observed

**Table 10.** Stepwise multiple linear regression analysis of land use types, land types, land suitability and land degradation.

| Degradation indicator | Independent variable | Land use 1 | | Land use 2 | | Land use 3 | | Land use 4 | |
|---|---|---|---|---|---|---|---|---|---|
| | | Contribution | Cumulative | Contribution | Cumulative | Contribution | Cumulative | Contribution | Cumulative |
| BD | Suitability (Oil palm) (s) | - | - | - | - | - | - | - | - |
| Permeability | Land type | - | - | - | - | - | - | 22.2 | 22.2 |
| | Suitability (Oil palm) (s) | - | - | - | - | - | - | - | - |
| N | Land type | - | - | - | - | 29.3 | 29.3 | - | - |
| | Suitability (Cassava/Maize) (s) | - | - | - | - | 85.1 | 64.4 | - | - |
| K | Suitability (Cassava/Maize) (s) | 24.2 | 24.2 | - | - | - | - | 23.6 | 45.8 |
| Organic matter content | Suitability (Cassava/Maize) (s) | - | - | - | - | 39.7 | 39.7 | - | - |

Model for land degradation: Land use 1, Potassium, $Y = 0.38 + 0.49S_1$, $-0.08LUT - 0.16LT - 0.13S_2$; Land use 3, Nitrogen, $Y = 4.50 - 0.50S_1 - 0.25S_2 - 0.03LT$, humus content, $Y = 4.65 - 0.40S_1 - 0.14S_2 - 0.17LT$; Land use 4, Permeability, $Y = 2.93 - 0.15S_1 - 0.28LT$, Potassium, $Y = 0.53 + 0.30S_1 + 0.60LT$. Note: LUT – Land use type, LT – Land type, $S_1$ – Cassava/Maize, $S_2$ – Oil palm.

from the study that the pedons at the crest and middle slope were well drained, while the pedons at the valley bottom were poorly drained. This is due to the differences in clay contents and regional water table, with valley bottom section having higher clay content at the surface than the other land types which invariably reduces infiltration rate of the water (Ahn, 1970; McKeague, Day and Shields, 1970; Philipson and Drosdoff. 1972; Banin and Amiel, 1969; Smyth and Montgomery, 1962; Senjobi, 2007). The colour difference showed that pedons on the crest and middle slopes were more reddish to brownish but greyish at the valley bottom. This variation in soil colour was due to the obvious sequences of drainage (Senjobi, 2007).

Besides, the influence of drainage on soil colour variation, it has been reported by Majlis (1967) that soil colour has been expressed too as a function of iron and organic matter contents, pH and type of clay mineral in the soil.

The fluctuating water table in the lower slope and valley bottom soils also resulted in the occurrence of hydromorphic mottles in these areas.

The gravel content was very high ranging from 6.0 to 95.9% in nearly all the pedons except in pedons number 12 at building site location. The pedons are composed of concretions, ferruginous nodules, quartz gravel and stones, which concentrate as stone lines or stone layers (Sharpe, 1938; Nye, 1954). The erosion exposes a sheet of gravel mantle, which is subsequently buried by the sedimentation of fine textured materials. Stonelines interfere considerably with manual soil tillage and thereby limits the aggregate size of farmland that a farmer can put under cultivation. According to Babalola and Lal (1987), this type of physical feature is permanent and difficult to change.

The high percentage of soil texture (41.8 – 91.2%) in all the land use types which is sandy is a good indication of the observable high infiltration rate and low water holding capacity of the soils, thereby resulting into moisture stress as reported by Fagbami and Udoh (1982) andSenjobi (2007). In addition to the above, this scenario encourages rapid leaching of nutrients from the soils beyond the rooting zones of the planted crops – a situation that threatens increase in food productivity and food security.

The poor water holding capacity of the soils is as a result of coarse texture nature of the studied soils. This in turn enhances erodibility of the soil on exposure either through cultivation or construction of buildings and roads. As a result of this, topsoil is washed away leading to loss of nutrients and organic matter content of the soil and possibly soil compaction. These erosion problems can be checked to the barest minimal level through appropriate land use practices which are not only environmental friendly and acceptable by land users, but which also ensure the maintenance and continuous vegetable cover over the soil surface.

The organic carbon fluctuates irregularly with depth for most of the pedons and this is an indication of continuous deposition of organic material. The low organic carbon values in some pedons may be partly due to the high temperature and high relative humidity, which favour rapid mineralization. Organic matter has been reported to have positive influence on the CEC, base saturation, structure, pH, buffering capacity, soil colour, and water holding capacity (Majlis, 1967; Ahn, 1970; Senjobi, 2007; Senjobi et al., 2010a). The organic matter content of the surface horizons in all the land use types is acceptable. This may be due to the fact that most of the organic residues in both cultivated and virgin soils are incorporated or deposited on the surface. The organic matter content of soils under secondary forest and oil palm cultivation are higher than those of building site and arable cultivation. This may be because the soils under those land use systems were always covered and they had not been subjected to intense cultivation and use as in arable and building land use types (Senjobi and Ogunkunle, 2010b). The incorporation of organic residues into the soils through tillage practices at the arable land use type have contributed to the relatively higher level of organic matter in this site than that of building site.

The total nitrogen content in all the soils of land use types was generally low (0.01 – 0.19%) compared with the critical value of 0.15% (Agboola and Carey, 1973). The intense cultivation of the soils normally increases the rate of mineralization of the organic matter (Senjobi, 2007; Senjobi and Ogunkunle, 2010b), thus negatively affect the level of total N content in the soil. The available P was generally moderate in most of the pedons and high in the remaining pedons compared with the critical level of 10 – 16 ppm (Adeoye and Agboola, 1985).

The exchangeable cations Ca, Mg, K and Na were generally low in all the pedons. This may be attributed to intense cultivation, leaching of nutrients and weathering consequently leading to the inherent low fertility status of the soil.

According to FAO (1978), the decrease in organic matter status of the soils when the biomass or the crop residues are insufficient to replace the humus could be attributed to the rapid biological degradation of the soils. For most low activity clays, of the tropical soils, the organic matter is the major exchange site for the basic nutrients cations in the soils. Organic matter had an influence on the cation exchange capacity and clay minerals of the tropical soils (McKague et al., 1970). Onasanya (1992) however, observed that organic matter had a more controlling effect on ECEC than clay minerals, especially in tropical soils. Therefore steps should be taken to increase the organic matter content of the soils, so as to increase the ECEC values of the soils and other soil chemical properties.

This can be achieved through the use of organic manure as a supplement to inorganic fertilizers. The use of mulching materials which will drastically reduce the effect of erosion at the same time enrich the soil is valuable. The use of organic refuse to conserve, maintain favourable soil temperature and supply nutrients through biological activities of earthworms are advisable (Senjobi, 2001, 2007; Senjobi et al., 2010a; Senjobi and Ogunkunle, 2010b).

The result of the multiple linear stepwise regression analysis above showed that factors contributing to different forms of degradation differ from one land use type to another. This analysis indicates that virtually most of the degradation forms are contributed to by suitability and land type nearly all the land use types. This could be attributed to the exploitative nature of the crops which are mainly cassava/maize and oil palm as well as exposure of land by the arable crops.

These results further showed that these crops were not planted to where they are best suited for. Again inappropriate allocations of agriculturally good lands to non-agricultural purposes constitute an immense factor to physical, chemical and biological degradation of the land.

## Conclusion

It was observed from the study that soil nutrients vary and changes from one land use type to another. Soil nutrients depletion and degradation processes at all land use types were high, very high and prominent under building site and arable that is cassava and maize based cropping systems as reflected in their nutrients values.

It was also observed that the land use types employed on the study sites were not very compatible with the characteristics of the soil. Thus, it encourages various nutrients status which are always at the declining rate. This in turn affects soil productivity negatively leading to food insecurity at large.

To take adequate care of these deficiencies, and minimize soil nutrients depletion and degradation in the study sites, the following measures are recommended. These include plausible land use approach, multiple cropping, organic mulching, contour ridge and cultivation of cover crops among others.

In addition to the above measures, there is the need to understand the soil adequately through detailed soil survey and land evaluation. When this is carefully done, the soil can then be put to appropriate land use that is cultivate the crops that are most suitable for the land, having known its capability and constraints as well as use the land for the purposes it is best suited for, this will go a long way to improve the productivity of such lands.

### REFERENCES

Adeoye GO, Agboola AA (1985). Critical levels for soil pH, available P, K, Zn and Mn maize content of P, Cu and Mn in sedimentary

soils of south – western Nigeria. Fertilizer Res., 6: 65-71.
Agboola AA, Corey RB (1973). The relationship between soil pH, O. M., Av. P, Exch. K and nine elements in the maize tissue. Soil Sci. 115: 367-375.
Ahn PN (1970). West African Soils. London Oxford University Press, p. 232.
Aruleba JO (2004). Influence of cropping systems and land types on land suitability and degradation in south western Nigeria. A Ph. D. thesis in the Dept. of Agronomy, University of Ibadan, p. 242.
Babalola O, Lal R (1977a). Subsoil gravel horizon and maize root growth I. Gravel concentration and bulk density effects. Plant Soil, 46: 337-346.
Babalola O, Lal R (1977b). Subsoil gravel horizon and maize root growth II. Effect of gravel size, inter gravel texture and natural gravel horizon. Plant Soil, 46: 347-357.
Banin A, Amiel A (1969). A correlative study of the chemical and physical properties of a group of natural soils of Israel. Geoderma, 3: 185-198.
Bouyoucos GH (1951). A recalibration of the hydrometer for making mechanical analysis of soils. Agronomy J._ 43: 434-438.
Bray RH, Kurtz LT (1945). Determination of total organic and available forms of phosphorus in soils. Soil Sci., 59: 22-229.
Dudal R (1982). "Land degradation in a World perspective" J. Soil Water Conserv. 37(5): 245-249.
Fagbami A, Udoh EJ (1982). The Characteristics of 2 Soil Toposequencies in the Basement Complex in the Federal Capital Territory of Nigeria. Ife J. Agric. 4(112): 9-24.
FAO (1978). Report on the agro-ecological zones projects Methodology and results for Africa World Resources Report 48: FAO Rome, p. 127.
FAO (1984a). Guidelines for land evaluation for Rainfed Agriculture. FAO Soils Bulletin No. 52 FAO, Rome, p. 237.
FAO (1984b). Guidelines for land use planning. FAO development series. FAO Rome, p. 96.
FAO (1984c). Keeping the land Alive: Soil Erosion – its causes and cures. (E.F.S). FAO Soils Bull No. 50.
FAO (1984d). Land food and people. Economic and Social Development Series No. 30 FAO, Rome.Jackson, M.L. (1962). Soil chemical analysis. Prentice Hall, New York.
FAO/UNESCO (2006). A Framework for International Classification, Correlation and Communication. World Soil Resources Reports. No. 103. FAO, Rome.
Lal R (1975). Role of Mulching Techniques in Tropical Soils and Water Management. I.I.T.A. Technical Bulletin No. 1.
Lal R (1988). "Erodibility and Erosivity" In: Soil Erosion Research Methods Soil Water Conserv. Soc., pp.141-160.
Lal R (1990). "Soil Erosion in the tropics; Principles and management" McGraw-Hill Inc., pp. 1-250.
Lal R (1992). Soil erosion and land degradation: The global risk. Adv. Soil Sci. 11: 345.
Lal R (1996). Deforestation and land – use effects on soil degradation and rehabilitation in Western Nigeria, 2. Soil chemical properties. J. Land Degradation Dev. (UK), 7(2): 87-98.
Lal R (1997). Deforestation, tillage and cropping systems. Effect on seepage and run off water quality from a Nigeria Alfisol. Soil Tillage Res. 41: 261-284.
Lal R, Stewart BA (1990). Soil degradation, a global threat: Advances in Soil Sci. Springer – Verlag, 11: 345.
Lal R, Kang BT, Moorman FR, Juo ASR, Moomaw JC (1975). Soil management in tropical America. Bornemisza E. and Alvarodo, A. (ed.) North Carolina State Univ. Raleigh, pp. 372-408.
Lal R, Couper DC (1990). A ten years watershed management study on agronomic productivity of different cropping systems in sub – humid regions of Western Nigeria. Topics – in – Applied – Resources – Management – in – the – Tropics. Germany, 2: 61-81.
Majlis MAK (1967). Some factors affecting the production and measurement of colors in montmorillonite and kaolinite clays and natural soils. Ph. D. dissertation, Texas A and M. Univ., College Station, Texas.
Mbagwu JSC (1985). Estimating dry season water requirements from climatological and soil available capacity data in the sedimentary and basement complex areas of south Nigeria. Catena, 12: 201-209.
Mckeague JA, Day JH, Shields JA (1971). Evaluating relationships among soil properties by computer analysis. Canadian J. Soil Sci. 51: 105-111.
Mclean EO (1965). Aluminum: In methods of soil analysis (ed. C. A. Black) Agronomy No. 9 Part 2, Am. Soc. Agronomy, Madison, Wisconsin, pp. 978-998.
Nye PH (1954). Some soil forming processes in the humid tropics I. A. field study of a catena in West African Forest. J. Soil Sci. 5: 7-21.
Onasanya SO (1992). The relationship between topographic location. Soil properties and cultural management and the productivity. Maize – Soybean intercrop. A Ph. D thesis in the Dept. of Agronomy, University of Ibadan, p. 319.
Philipson WR, Drosdoff M (1972). Relationships among physical and chemical properties of Representative soils of the Tropics from Puerto Rico. Soil Science Society of American. Proceedings, 36: 815-819.
Senjobi BA, Peluola CO, Senjobi CT, Lawal IO, Ande OT, Salami BT (2010a). Performance of Cochorus Olitorius as influenced by soil type and organic manure amendments in Yewa North Local Government Area, Ogun State. Afr. J. Biotechnol. 9(33): 5309-5312.
Senjobi BA, Ogunkunle OA (2010b). Effect of Land Use on Soil Degradation and Soil Prodcutivity Decline on Alfisols and Ultisols in Ogun State in Southern Western Nigeria. J. Agric. Conspectus Sci. Faculty of Agriculture, University of Zagreb, Crotia, 75(1): 9-19.
Senjobi BA (2007). Comparative Assessment of the Effect of Land Use and Land Type on Soil Degradation and Productivity in Ogun State, Nigeria. Published Ph.D. Thesis submitted to the Department of Agronomy, University of Ibadan, Ibadan, p. 161.
Senjobi BA, Ayoola AO, Ayanlaja SA (2001). Evaluation of Ogun State University soils for tree and arable crops production. Ogun J. Agric. Sci. 1(2001): 214-224.
Senjobi BA (2001). Parametric and conventional approaches for soil potential evaluation in three ecological zones of southern Nigeria. Moor J. Agri. Res. 2(2): 91-102.
Sharpe CFS (1938). Landslides and related phenomena. Columbia Univ. Press New York.
Smyth AJ, Montgomery RF (1962). Soil and land use in central western Nigeria. Govt. Printer, Ibadan, Western Nigeria, pp. 264.
Sparks DL (1996). Methods of Soil Analysis. Part 3. Chemical Methods. SSSA and ASA. Madison, 1: 551-574.
Stoop WA (1987). Adaptation of Sorghum/Maize and Sorghum Pearl millet inter crop system to the toposequence land type in the north 5 – 2 or West African Savannah. Field Crops Res. 16: 255-272.
United Nations/FAO (1997). Land Quality Indicators and their use in sustainable agriculture and rural development. FAO Land and Water Bulletin 5, co-published with World Bank, UNEP and UNDP, Rome, Italy.
Walkey A, Black IA (1934). Determination of organic matter in soil. Soil. Sci. 37: 549-556.

# Biofertilization of banana (*Musa* spp. L.) with free-living $N_2$ fixing bacteria and their effect on mycorrhization and the nematode *Radopholus similis*

**Adriano-Anaya ML[1], Gutiérrez-Miceli FA[2], Dendooven L[3] and Salvador-Figueroa M[1]***

[1]Centro de Biociencias. Universidad Autónoma de Chiapas. Carretera a Puerto Madero Km 2.0 Tapachula, 30700, Chiapas, México.
[2]Departamento de Ingeniería Química y Bioquímica. Instituto Tecnológico de Tuxtla-Gutiérrez, Tuxtla-Gutiérrez, Chiapas, México.
[3]Laboratory of Soil Ecology, GIB, Dept. Biotechnology and Bioengineering, Cinvestav, México.

**Banana plants (*Musa* spp. L.) cv 'Great dwarf' were amended with free-living $N_2$ fixing bacteria (FLNFB) and arbuscular mycorrhizal fungi (AMF) and the presence of the burrowing nematode *Radopholus similis* was monitored in the field. Five treatments were applied by inoculating banana roots with four strains of FLNFB, that is C1, C2, C3 and C4 isolated from the rhizoplane of the same banana cultivar, or by keeping them uninoculated. The largest number of nematodes was found in the untreated roots and the lowest in the roots inoculated with C4. The largest percent of mycorrhizal colonization was found when banana roots were inoculated with C1 and the lowest in roots that were not inoculated. The number of *R. similis* decreased with increased colonization with AMF. The largest percentage of necrosis of the banana roots was found when banana roots were not inoculated and the lowest in roots that were inoculated with C4. The necrosis of the roots increased with increased number of *R. similis*, but decreased with increased colonization with AMF. It was found that inoculating the banana cv 'Great dwarf' with FLNFB increased the root colonization with AML, and reduced the number of nematodes and root damage.**

**Key words:** Arbuscular mycorrhizal fungi, burrowing or banana-root nematode, free-living $N_2$ fixing bacteria, plant development.

## INTRODUCTION

Plant development is closely related to the interaction with microbiota in the phylosphere and rhizosphere (Green et al., 2006). For instance, 80 and 92% of surveyed plant species and families, respectively, were mycorrhizal with arbuscular mycorrhizal fungi (AMF) being dominant (Wang and Qiu, 2006). Wang and Qui suggested that the origin of arbuscular mycorrhiza coincides with the origin of land plants. In the rhizosphere, AMF and pathogens need to be in equilibrium, so that a normal plant development can occur (Reyes and Valery, 2007). The increase in pathogens populations leads to diseases (Whipps, 2001) thereby hampering plant growth (Siddiqui and Akhtar, 2008). The interaction in the rhizosphere between plant and microorganisms has been studied intensively (Mathesius, 2009) and the importance of certain root exudates in the relationship between plants and microorganisms, beneficial and pathogenic, has been demonstrated (Rojas-Andrade et al., 2003). *Radopholus similis,* the burrowing or banana-root nematode, is one of the most important root parasites of banana plants (*Musa* spp. L.). Although, effective chemical nematicides exist,

*Corresponding autor. E-mail: msalvad@hotmail.com.

but their high cost and toxicity limit their use. Nematode management for resource-poor farmers relies on the use of hot water-treated suckers (Speijer et al., 1995), tissue culture-derived planting material or mulching (Talwana et al., 2003). None of these methods, however, offers complete control of nematodes (Athman et al., 2006). The roots of the banana, as those of most plants, are colonized by AMF and the intensity of colonization depends on fertilizer use and AMF species (Declerck et al., 2002). The hyphae of AMF reduce the entry points of the nematodes and induce physiological changes that increase plant growth (Andrade et al., 2009). Studies have shown that AMF might reduce the damage done to plants by nematodes (Jaizme-Vega et al., 1997), but the synergistic effect of soil fungi and nematodes might break down the protection of AMF against pathogens (Greipsson and El-Mayas, 2002). It is well known that conventional agricultural practices might have a negative effect on soil fertility due to the use of fertilizers, tillage and pesticides (Gregory et al., 2005). Excessive fertilizer use increases soil salinity, tillage disrupts soil structure accelerating soil organic matter decomposition, while pesticide might negatively affect beneficial organisms (Jorgenson and Kuykendall, 2008). New agricultural practices have been developed to reverse those negative effects (Butler et al., 2007). The introduction of beneficial microorganisms often through inoculation upon planning of crops or biofertilization can improve yields without negatively affecting soil fertility (Avis et al., 2008). Application of organisms that are antagonistic against pests or pathogens is another way to improve crop development (Alabouvette and Steinberg, 2006). Therefore, an alternative way to control *R. similis,* instead of using a nematicide, is to inoculate roots with beneficial symbiotic organisms so that through antagonisms or simple competition the pathogen is prevented to infect the plant (Van der Veken et al., 2008). The objective of this work was to investigate the effect of biofertilizing roots of the banana plant with four free-living $N_2$ fixing bacteria isolated from roots of the similar banana cv 'Great dwarf' and to determine how this affected the colonization of the roots with AMF and the presence of *R. similis*.

## MATERIALS AND METHODS

### Free-living $N_2$ fixing bacteria

Four free-living $N_2$ fixing bacterial strains C1, C2, C3 and C4 (*Azospirillum* spp.) were isolated from the rhizosphere of banana plants cultures and selected for their high production of indole-3-acetic acid (IAA).

Bacteria produce IAA under stress conditions promoting plant growth and shoot water content and thus increase plant drought tolerance (Marulanda et al., 2009). The isolated bacterial strains were kept in nFB culture medium and cultivated in nutrient broth (pH 7.0) at 200 rpm and at 30°C before inoculation.

### Plantlets culture and inoculation

Nine hundred banana plantlets were cultivated in the greenhouse in pots containing 5 kg soil from a banana field. The soil contained 320 g $kg^{-1}$ sand, 410 g $kg^{-1}$ loam and 270 g $kg^{-1}$ clay. The plantlets were fertilized with urea (10 g $pot^{-1}$) 60 days after sowing. The plantlets were divided in five groups. Hundred and eighty plants were inoculated with 5 ml of each of the four bacteria ($1 \times 10^8$ cells $ml^{-1}$) eight days after application of urea. The remaining 180 plantlets were not inoculated and served as control.

### Field experiment

The plantlets were transferred to the '*Santa Lucrecia*' farm (14° 40' 21.56" N; 92° 10' 27.19" W), a banana plantation in Suchiate, (Chiapas, México), 18 days after bacterial inoculation. Five 1000 $m^2$ plots (180 plants $plot^{-1}$) were outlined (one for each of the four bacterial strains and one for control). Plots were arranged in a complete randomized block design. Each plant was fertilized montly with 20 g N, 72 g K, 11 g Ca, 4 g Mg and 2 g S. Eight days after the fertilizer was applied, the roots of each plant were inoculated with 30 ml of the previously mentioned bacterial inoculums, i.e. four different strains at $1 \times 10^8$ cells $ml^{-1}$.

### Sampling and assays

After 100, 128, 156, 184, 212 and 240 days, 10 banana plants were selected at random and the rhizosphere collected (30 $cm^3$ soil + roots). The diazotrophic bacteria in the rhizoplane were determined by cultivating them on nFB medium. Nematodes in the roots were extracted with a Baermann funnel as described by Hooper et al. (2005) and counted under a microscope. Nematodes were identified as described by Hunt et al. (2005). The mycorrhizal colonization was determined as described by Giovannetti and Mosse (1980) on previous stained roots as described by Phillips and Hayman (1970). The damage done by the nematodes to the roots was visually determined in three categories: 1) undamaged roots, 2) roots with < 50% damage and 3) necrotic roots with > 50% damage.

Different phenological stages of the banana plants were defined as follows (Soto, 1991). The infantile stage was defined as the time between the bud outset and the first formal leaf formation and lasted from 120 to 160 days. The juvenile stage was defined from the first formal leaf to the floral differentiation stage and took between 50 and 60 days, while the reproductive phase covered the time between floral differentiation stage until the fruits were collected.

### Statistical analysis

The number of root nematodes, AMF colonization and root necrosis was subjected to one-way analysis of variance (Tukey, $p < 0.05$) to test for significant differences between the treatments, i.e. different strains, with InfoStat Professional version 2008.

## RESULTS

The number of free-living $N_2$ fixing bacteria on the roots

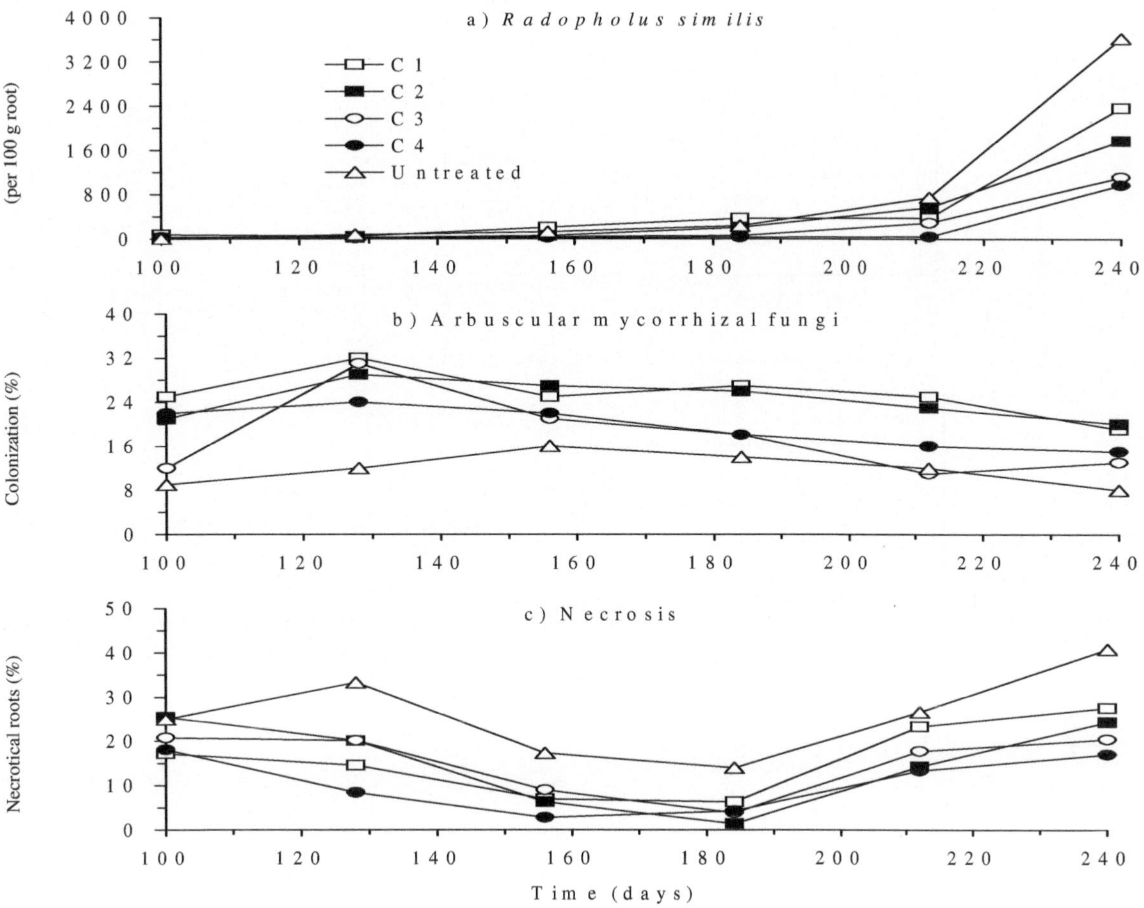


**Figure 1.** The number of nematodes *Radopholus similis* (per 100 g roots), b) the percentage of arbuscular mycorrhizal fungi colonization (%) and c) necrosis (%) of the roots of the banana plant (*Musa* spp. L.) cv 'Great dwarf' inoculated with four different free-living $N_2$ fixing bacteria (C1 (□), C2 (■), C3 (○), C4 (●)) and left untreated (△) grown in the field.

of inoculated banana plants was >$10^8$ $g^{-1}$ roots, while in the non-inoculated plants it was between $10^5$ and $10^6$ $g^{-1}$ roots. The number of nematodes found on the roots remained low until day 128 and started to increase thereafter (Figure 1a). At day 240, the number of nematodes found in the roots was significantly different between the treatments (P<0.05). The largest amount was found in the untreated roots (3618 nematodes 100 $g^{-1}$) and the lowest in the roots inoculated with C4 (969 nematodes 100 $g^{-1}$). The colonization of the banana roots with AMF increased with a maximum found between 128 and 156 days, and then decreased (Figure 1b). The colonization of the banana roots with AMF was significantly different between the treatments (P<0.05). The largest percent colonization was found when banana roots were inoculated with C1 (32%) and the lowest in roots that were not inoculated (16%). The number of *R. similis* decreased significantly and exponentially with increased colonization of the roots with AMF (P<0.05) (Figure 2a).

The necrosis of the banana roots decreased until day 184 in all treatments, except in the roots that were not inoculated which showed an increase between day 100 and 128 (Figure 1c). After day 184, the number of roots that showed necrosis increased in all treatments. The necrosis of the banana roots was significantly different between the treatments (P<0.05). The largest percentage of necrosis of the banana roots was found when they were not inoculated (26%) and the lowest in roots that were inoculated with C4 (11%). The necrosis of the roots increased significantly with an increased number of *R. similis*, but decreased significantly with increased colonization of the roots with AMF (P<0.05) (Figure 2b and c).

## DISCUSSION

### Root colonization with arbuscular mycorrhizal fungi (AMF)

The colonization of banana roots changed with time. In

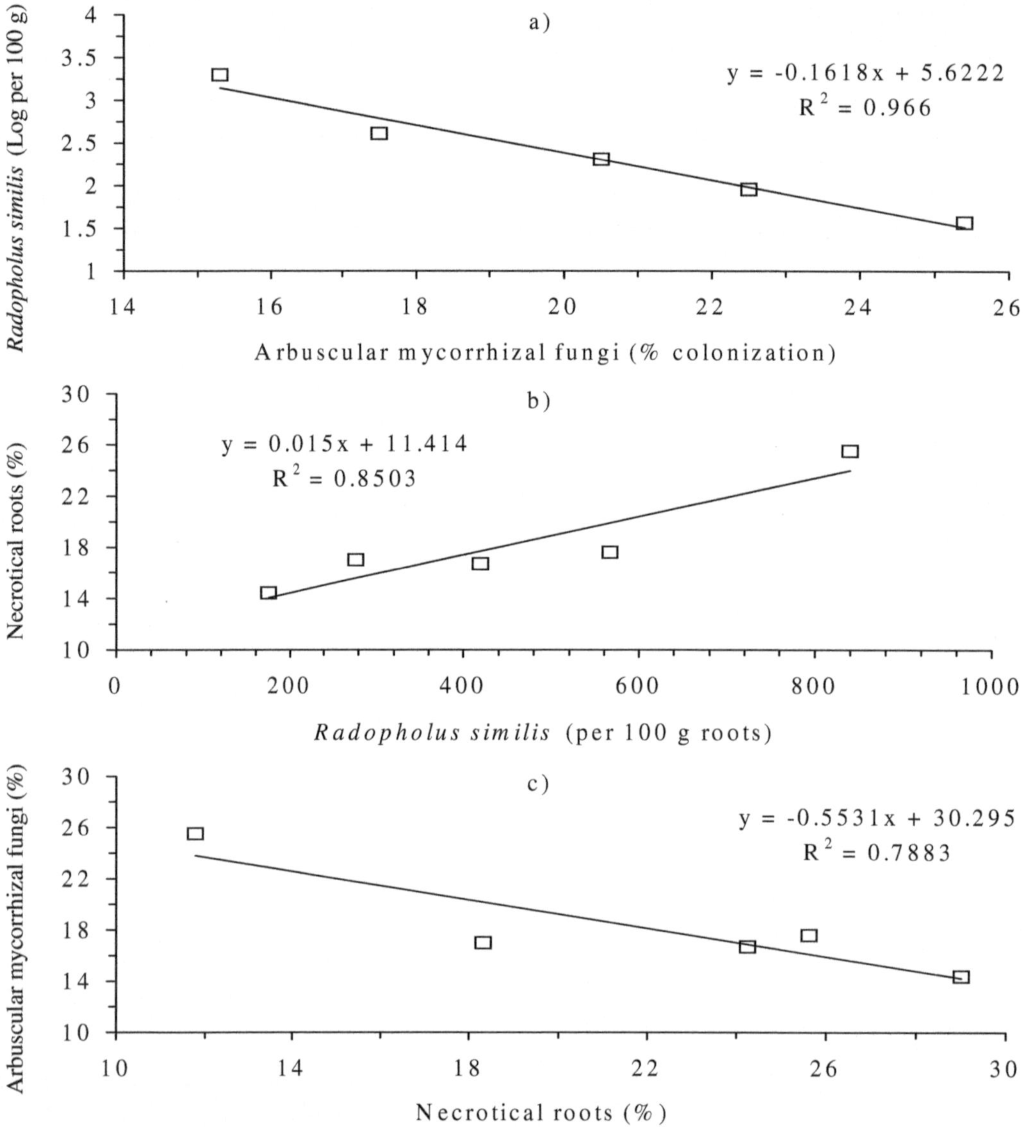


**Figure 2.** The correlation a) between the number of nematodes *Radopholus similis* (per 100 g roots) and the percentage of arbuscular mycorrhizal fungi colonization (%), b) the number of nematodes *Radopholus similis* (per 100 g roots) and necrosis (%) of the roots c) necrosis (%) of the roots and the percentage of arbuscular mycorrhizal fungi colonization (%) of the roots of the banana plant (*Musa* spp. L.) cv 'Great dwarf' inoculated with four different free-living $N_2$ fixing bacteria.

cereals, AMF root colonization is related to the phenological stage of the plant (Gavito and Varela, 1993; Majic et al., 2008). Colonization increased during the vegetative phase and decreased during the reproductive phase. Photosynthetic products are transported to the roots where they are used by the microorganisms in the rhizosphere. The microorganisms proliferate and their metabolites and/or released nutrients are taken up by the plants. When the plant matures and the reproductive phase starts, photosynthetic products are used for flowering and seed development. The plant metabolites released in the rhizosphere decreases, microbial growth slows and the AMF loosen their association with cortical cells.

The level of colonization with AMF found in this experiment for banana plants not inoculated were less than those reported. Adriano (2001) found a 58 to 88% colonization with *Glomus mosseae* for the banana cv 'Great dwarf' during the reproductive phase. Jaizme-Vega and Pinochet (1997) found a 47% colonization of roots of the same banana cultivar with *G. mosseae*, 34% with *G. aggregatum* and 33% with *G. intraradices*.

## Population of *Radopholus similis*

Jaizme-Vega et al. (1997) reported that the banana cv 'Great dwarf' colonized by the AMF *G. mosseae* were less susceptible to infections with the root-knot nematode *Meloidogyne incognita* favouring plant growth. Contrarily, Jaizme-Vega and Rodriguez-Romero (2004) reported that the same banana plants inoculated with *G. mosseae*, *G. aggregatum* and *G. intraradices* had the same number of *Pratylenchus goodeyi* and *M. incognita* than the untreated plants. However, Elsen et al. (2008) found that AMF have the ability to induce systemic resistance against plant parasitic nematodes *R. similis* and *Pratylenchus coffeae* in the root system. AMF reduced both nematode species by more than 50%, even when the AMF and the plant parasitic nematodes were spatially separated. Similar results were obtained with maize (*Zea mays* L.), olive (*Olea europaea* L.) (Castillo et al., 2006) and kiwi (*Actinidia deliciosa*) (Verdejo et al., 1990). The mycorrhiza had neither effect on the number of pathogenic nematodes nor was the number of AML affected by the pathogens.

## Root damage

Jaisme-Vega et al. (1997) reported a 36 to 64% reduction in root galling caused by the root-knot nematode *M. incognita* when micropropagated banana plants cv 'Great dwarf' were inoculated with *G. mosseae* reaching 58 to 88% of AM root colonization. Jaizme-Vega and Rodriguez-Romero (2004) found that banana plants cv 'Great dwarf' inoculated with *G. mosseae*, *G. aggregatum* and *G. intraradices,* and infected with the nematode *P. goodeyi*, had 4, 15 and 13% lesions, respectively, while the plants that were not inoculated had 28%. The colonization with *G. mosseae* was 47%, with *G. aggregatum* 34% and with *G. intraradices* 33%. No differences, however, in nematode populations were found. Elsen et al. (2008) found that the number of lesions in the roots of banana cv 'Williams' amended with ±1850 mycorrhizal spores and infected with 5000 eggs and juvenile *M. javanica* one month later, were 29% lower with *G. mosseae*, 25% with *G.macrosporum* and 16% with *G. caledonium* compared to plants not inoculated with mycorrhizae. They suggested an inhibitory effect of the micorrhiza on the infection with nematodes through changes in the nutrition of the plant, biochemical changes in the plant tissue, e.g. an increase in aminoacids, peroxidases, quitinases, phytoalexins, increases in lignin, stress reduction, changes in the microbial populations in the roots and a more dense root system. Vaast et al. (1998) found a decrease in lesions by *P. coffeae* in the roots of coffee (*Coffea arabica* L.) when inoculated with the mycorrhizae *Acaulospora mellea* or *G. clarum.* Castillo et al. (2006) found similar results when studying the effect of mycorrhiza on the pathogenicity of *M. incognita* and *P. vulnus* on roots of olive tree. Castillo et al. (2006) suggested that the reduction in pathogenicity was not related to a reduction in infection rate, but was related to a higher tolerance towards the nematode.

Our findings conclude that inoculating roots of the banana plant cv 'Great dwarf' with free-living $N_2$ fixing bacteria isolated from roots of the similar plant resulted in a higher colonization of the roots with AMF, a lower number of pathogenic nematodes and healthier roots.

## ACKNOWLEDGEMENTS

We thank FOMIX CONACyT-Estado de Chiapas for funding the research (Grant CHIS-2002-C01-4332).

### REFERENCES

Adriano L (2001). Microbiología de la rizosfera del banano clon "Gran Enano". Tesis Doctoral. Universidad Autónoma de Chiapas, p.107.

Alabouvette C, Steinberg C (2006). The soil as a reservoir for antagonist to plant disease. In: Eilenberg J, Hokkanen HMT (eds) Progress in biological control. An ecological and societal approach to biological control. Springer Netherlands, 2: 123-144.

Andrade SAL, Mazzafera P, Schiavinato MA, Silveira APD (2009). Arbuscular mycorrhizal association in coffee. J. Agric. Sci., 147: 105-115.

Athman SY, Dubois T, Coyne D, Gold C, Labuschagne N, Viljoen A (2006). Effect of endophytic *Fusarium similis* to tissue culture banana plants. J. Nematol., 38: 455-460.

Avis TJ, Gravel VG, Antoun H, Tweddell RJ (2008). Multifaceted beneficial effects of rhizosphere microorganisms on plant health and productivity. Soil Biol. Biochem., 40: 1733-1740.

Butler SJ, Vickery JA, Norris K (2007). Farmland biodiversity and the footprint of agriculture. Science, 315: 381-384.

Castillo P, Nico AI, Azcon-Aguilar C, Rincon CD, Calvet C, Jimenez-Diaz RM (2006). Protection of olive planting stocks against parasitism of root-knot nematodes by arbuscular mycorrhizal fungi. Plant Pathol., 55: 705-713.

Declerck S, Risede JM, Rufyikiri G, Delvaux B (2002). Effects of arbuscular mycorrhizal fungi on severity of root rot of bananas caused by *Cylindrocladium spathiphylli.* Plant Pathol., 51: 109-115.

Elsen A, Gervacio D, Swennen R, De Waele D (2008). AMF-induced biocontrol against plant parasitic nematodes in *Musa* sp.: a systemic effect. Mycorrhiza, 18: 251-256.

Gavito M, Varela L (1993). Seasonal dynamics of mycorrhizal associations in maize fields under low imput agriculture. Agric. Ecosys. Environ., 45: 275-282.

Giovannetti M, Mosse B (1980). An evaluation of techniques for measuring vesicular-arbuscular mycorrhizal infection in roots. New Phytol., 84: 489-500.

Green SJ, Inbar E, Michel FC, Hadar Y, Minz D (2006). Succession of bacterial communities during early plant development: Transition from seed to root and effect of compost amendment. Appl. Environ. Microbiol., 72: 3975-3983.

Gregory OJ, Ingram JSI, Brklacich M (2005). Climate changes and food security. Phil. Trans. R. Soc. B., 360(1463): 2139-2148.

Greipsson S, El-Mayas H (2002). Synergistic effect of soil pathogenic fungi and nematodes reducing bioprotection of Arbuscular mycorrhizal fungi on the grass *Leymus arenarius*. Biocontrol., 47: 715-727.

Hooper DJ, Hallman J, Subbotin A (2005). Methods for extraction, processing and detection of plant and soil nematodes. In: Luc M, Sikora RA, Bridge J (eds) Plant parasitic nematodes in subtropical and tropical agriculture. 2nd Edition. CAB International, Wallingford, UK, pp. 53-86.

Hunt DJ, Luc M, Manzanilla-López RH (2005). Identification, morphology and biology of plant parasitic nematodes. In: Luc M, Sikora RA, Bridge J (eds) Plant parasitic nematodes in subtropical and tropical agriculture. 2nd Edition. CAB International, Wallingford, UK, pp. 11-52.

Jaizme-Vega MC, Pinochet J (1997). Growth response of banana to three mycorrhizal fungi in *Pratylenchus goodeyi* infested soil. Nematropica, 27: 69-76.

Jaizme-Vega MC, Rodriguez-Romero AS (2004). Uso de micorrizas en banano: Logros y perspectivas. XVI Reunión Internacional ACORBAT. Oaxaca, México. Publicación Especial, pp. 143-160.

Jaizme-Vega MC, Tenoury P, Pinochet J, Jaumot M (1997). Interactions between the root-knot nematode *Meloidogyne incognita* and *Glomus mosseae* in banana. Plant Soil, 196: 27-35.

Jorgenson AK, Kuykendall KA (2008). Globalization, foreign investment dependence and agriculture production: pesticide and fertilizer use in less-developed countries, 1990-2000. Soc. Forces, 87: 529-560.

Majic I, Ivezic M, Raspudic E, Gantner V (2008). Effect of soybean cultivar on endoparasitic nematodes and arbuscular mycorrhizal fungi relationship. Cereal Res. Comm., 36:1823-1826.

Marulanda M, Barea JM, Azcón R (2009). Stimulation of plant growth and drought tolerance by native microorganism (AM fungi and bacteria) from dry environment. Mechanism related to bacterial effectiveness. J. Plant Growth Regul., 28: 115-124.

Mathesius U (2009). Comparative proteomic studies of root-microbe interactions. J. Proteomics, 72: 353-366.

Phillips JM, Hayman DS (1970). Improved procedures for clearing roots and staining parasitic and vesicular-arbuscular mycorrhizal fungi for rapid assessment of infection. Trans. Br. Mycol. Soc., 55: 158-161.

Reyes I, Valkery A (2007). Efecto de la fertilidad del suelo sobre la microbiota y la promoción del crecimiento del maíz (*Zea mays* L.) con *Azotobacter* spp. Bioagro, 19: 117-126.

Rojas-Andrade R, Cerda-García-Rojas CM, Friás-Hernández JT, Dendooven L, Olalde-Portugal V, Ramos-Valdivia A (2003). Changes in the concentration of trigonelline in a semi-arid leguminous plant (*Prosopis laevigata*) induced by an arbuscular mycorrhizal fungus during the presymbiotic phase. Mycorrhiza, 13: 49-52.

Siddiqui ZA, Akhtar MS (2008). Synergistic effects of antagonistic fungi and a plant growth promoting rhizobacterium, an arbuscular mycorrhizal fungus, or composted cow manure on populations of *Meloidogyne incognita* and growth of tomato. Biocontrol Sci. Techn., 18: 279-290.

Speijer PR, Gold CS, Kajumba C, Karamura EB (1995). Nematode infestation of 'clean' banana planting materials in farmer's fields in Uganda. Nematologica, 41: 344.

Talwana HAL, Speijer PR, Gold CS, Swennen RL, De Waele D (2003). A comparison of the effects of the nematodes *Radopholus similis* and *Pratylenchus goodeyi* on growth, root health and yield of an East African highland cooking banana (Musa AAA-group). Int. J. Pest Manage., 49: 199-204.

Vaast P, Caswell-Chen EP, Zasoski RJ (1998). Influences of a root-lesion nematode, *Pratylenchus coffeae*, and two arbuscular mycorrhizal fungi, *Acaulospora mellea* and *Glomus clarum* in coffee (*Coffea arabica* L.). Biol. Fert. Soils, 26: 130-135.

Van der Veken L, Win PP, Elsen A, Swennen R, De Waele D (2008). Susceptibility of banana intercrops for rhizobacteria, arbuscular mycorrhizal fungi and the burrowing nematode *Radopholus similis*. Appl. Soil Ecol., 40: 283-290.

Verdejo S, Calvet C, Pinochet J (1990). Efecto de la micorrización en kiwi infestado por los nematodos *Meloidogyne hapla* y *M. javanica*. Biol. San. Veg. Plagas, 16: 619-624.

Wang B, Qiu YL (2006). Phylogenetic distribution and evolution of mycorrhizas in land plants. Mycorrhiza, 16: 299-363.

Whipps JM (2001). Microbial interactions and biocontrol in the rhizosphere. J. Exp. Bot., 52: 487-511.

# Effect of pre-emergence herbicides on weed infestation and productivity of sesame (*Sesamum indicum* L.) in the Sudan Savanna Zone of Nigeria

**E. O. Imoloame[1], S. D. Joshua[2]* and N. A. Gworgwor[2]**

[1]Regional Agricultural Training School, Ngala, Borno State, Nigeria.
[2]Department of Crop Production, Faculty of Agriculture, University of Maiduguri, Borno State, Nigeria.

**Field trials were conducted in 2006 and 2007 cropping seasons at the University of Maiduguri Teaching and Research Farm located in Maiduguri (11°50'N; 13°10'E), Nigeria, to evaluate the effect of different pre-emergence herbicides on weed infestation and productivity of sesame (*Sesamum indicum L.*) The experiment consisted of 22 treatments, which included 4 different pre-emergence herbicides applied at 5 rates each namely butachlor, metolachlor, diuron and pendimethalin at 0.5, 1.0, 1.5, 2.0 and 2.5 kg a.i./ha, weeding at 3 and 6 weeks after sowing (WAS) and a weedy check. The treatments were laid out in randomized complete block design (RCBD) replicated 3 times. Results showed that diuron at 1.5 to 2.5 kg a.i./ha produced significantly (P<0.05) the least weed cover and dry matter at 3 and 6 WAS, which was comparable with two-hoe-weedings at harvest. Both butachlor at 2.0 kg a.i./ha and metolachlor at 1.5 kg a.i./ha produced significantly (P<0.05) low weed cover score and weed dry matter, which were comparable with the least weed cover score in both trials and their combined means at harvest. Among the herbicides tested, butachlor at 2.0 kg a.i./ha and diuron at 1.0 kg a.i./ha produced significantly (P<0.05) the highest grain yield in 2006, while metolachlor at 1.5 kg a.i./ha produced significantly (P<0.05) the highest grain yield in 2007 and the combined means. For effective weed control and higher yield, metolachlor at 1.5 kg a.i./ha and butachlor at 2.0 kg a.i./ha are recommended as an alternative to two hoe-weedings at 3 and 6 WAS for the production of sesame in the Sudan Savanna zone of Nigeria.**

**Key words:** Pre-emergence herbicides, weed infestation, productivity, sesame, savanna, Nigeria.

## INTRODUCTION

Sesame (*Sesame indicum* L.) is thought to be the oldest oil producing plant, and its seeds have the highest oil content close to 50% (Anonymous, 2006). Asia accounts for three quarters of the global production of sesame seeds. The world's total production of the crop estimated by FAO was 3 million metric tonnes in 2005 and since then, Africa's share has been rising steadily. More than 20 African countries produce this crop with Nigeria, Sudan and Uganda as the leading producers and exporters of the crop (FAO, 2006). In Nigeria, sesame is mainly grown in areas of southern Guinea savanna ecological zone, notably in Benue and Kogi States and the Federal Capital Territory, Abuja (Philips, 1977). Recent reports by Iwo and Idowu (2002), indicated that production trend of sesame in Nigeria is on the increase by 7.8%, where production has increased from 15,000 metric tonnes in 1980 to 56,000 metric tonnes in 1994. Further increase in production was seen in 1996 of 64,000 and 67,000 metric tonnes in 1997. They have projected that production will increase to 130,000 metric tonnes by the year 2010. Sesame is an annual up to 1.5 m tall (Roland, 1993). It has a strong tap root and dense, much branched lateral root system which spreads in the surface. The lower leaves are opposite, broad and palmately lobed, while the upper leaves are alternately arranged, narrow and lanceolate (Onwueme and Sinha, 1991). The stems of most varieties are branched. The flowers are white or pink. The fruit is a rectangular,

*Corresponding author. E-mail: asdjoshua@yahoo.com.

deeply-grooved capsules up to 3 cm long with a short beak (Onwueme and Sinha, 1991). The importance of sesame lies in its high content of oil, protein, calcium, iron and methionine, which are very important for pregnant and lactating women (Gupta et al., 1998).

Yermanos et al. (1972) reported that, the seed of sesame contains about 50 to 51% oil, 17 to 19% protein and 16 to 18% carbohydrate and that the oil is edible, odourless and semi-drying, containing oleic, stearic and palmitic acids. Sesame oil serves as antioxidant in the manufacture of margarine and salad creams, and as a fixative in the industries for making of perfumes and cosmetics. It is also used as carriers for fat-soluble substances in pharmaceuticals and insecticide manufacturing (Godin and Spensley, 1971). The traditional methods of weed control, namely, hoe-weeding, are the commonest method of weed control by farmers in the Sudan Savanna Zone of Nigeria. This method is not only labour intensive, expensive and strenuous, but can also cause mechanical damage to growing branches and roots of plants. In addition to high cost, labour availability is uncertain, thus making timeliness of weeding difficult to attain, leading to greater yield loss (Adigun et al., 2003).

Chemical weed control has revolutionized farmers approach to weed control in the world. Chemical weed control is more adapted to large scale crop production than other weed control methods and it is labour saving (Anon, 1994). Also, herbicide use has been reported to be more profitable than hoe-weeding in the production of various crops in Nigeria (Shrock and Monaco, 1980; Okereke, 1983; Usoroh, 1983; Sinha and Lagoke, 1984; Ogungbile and Lagoke, 1986; Adigun et al., 1993). Most of the work done on chemical weed control in sesame has been limited to the Southern Guinea Savanna Zone of Nigeria and other parts of the world with agro-ecology different from that of the semi-arid zone of Nigeria, characterized by low, erratic, variable and unreliable rainfall, which affect the performance of herbicides. For example, Ndarubu et al. (2003), reported highest sesame yield and effective weed control with application of herbicide mixture of metolachlor + metobromuron at 1.0 + 1.0 kg a.i./ha in the Guinea Savanna of Nigeria. As a result of the paucity of data on the use of herbicides in sesame production, it has made it necessary to evaluate some herbicides in this zone to find out not only how safe they are, but their effectiveness in the control of weeds to achieve higher yield in sesame (Imoloame, 2009).

## MATERIALS AND METHODS

Field trials were conducted during the 2006 and 2007 cropping seasons on the university Teaching and Research Farm, Faculty of Agriculture, University of Maiduguri, Maiduguri (11°51'N; 13°15'E; elevation 336 m) located in the Sudan Savanna Zone of Nigeria. The experimental site which measured 12 x 88 m was harrowed, leveled properly using a hand-hoe, and marked out. The plot size was 4 x 4 m leaving a distance of 1.0 and 0.5 m between replications and plots, respectively. NPK 15:15:15 was applied to each plot to provide 50 kg P and 75 kg N. The full rate of P was applied before sowing and incorporated into the soil. The application of N however, was done in 2 equal split dose. The first dose was applied before planting as starter dose and the second dose was applied at 6 WAS which was drilled into the soil 15 cm away from the stand in a continuous band. The experiment consisted of 22 treatments, which included 4 different pre-emergence herbicides at five rates each, namely butachlor, diuron, metolachlor and pendimethalin at 0.5, 1.0, 1.5, 2.0 and 2.5 kg a.i./ha, weeding at 3 and 6 WAS and a weedy check. The treatments were laid out in randomized complete block design (RCBD) with 3 replications. Data collected were subjected to analysis of variance (ANOVA) and the difference between means were determined using Duncan's Multiple Range Test (DMRT). Sesame var. Gwoza Local was sown on 6 August, 2006 and 26 July, 2007 by drilling the seeds at the rate of 6 kg/ha into the soil, at a spacing of 90 cm between the rows. The following day, the pre-emergence herbicides were applied using CP-15 knapsack sprayer at a pressure of 2 kg/cm$^2$ calibrated to deliver 250 l/ha spray volume using a green nozzle. Observations taken included the following:

**Crop vigor score:** This was assessed visually at 3, 6 and 9 WAS, using a scale of 1 to 9, where 1= completely dead plants and 9 = most vigorous plants. The features of the plants used for scoring were, leaf and stem size, greenness and size of the plant (Ishaya et al., 2009).

**Plant height (cm):** This was taken at 4, 8 and 12 WAS. This was done by selecting 5 plants randomly from a plot and then measured their heights from the soil level to the tip of the apical bud with a graduated meter ruler. The mean of the plant heights was recorded as height of plant/plot (Imoloame, 2009).

**Phytotoxicity rating:** This was taken at 2, 4 and 8 WAS by visual observation using a scale of 1 to 9, where 1 = no crop injury and 9 = complete crop kill (Imoloame, 2009).

**Weed dry matter:** This was estimated at 3, 6 WAS and at harvest. This was done by harvesting weed biomass from three (3) randomly located 1 m$^2$ quadrats in each experimental plot. The weeds were oven-dried and weighed. The average weight was recorded (Imoloame, 2009).

**Weed cover score:** This was taken at 3, 6 WAS and at harvest using a scale of 1 to 9, where 1=complete absence of weeds in the plot and 9 =complete coverage of the plot by weeds (Imoloame, 2009).

**Number of pods (capsules)/plant:** This was determined by counting all the pods on 5 randomly selected plants at 10 and 12 WAS and the average recorded (Imoloame, 2009).

**Grain yield/ha:** This was determined after harvest. After threshing the plant, the grain yield/plot was weighed and recorded. This was converted to grain yield per hectare. (Imoloame, 2009). Harvesting was done in November with only the 3 inner rows of each plot harvested (net plot) and the first and last rows were discarded.

## RESULTS AND DISCUSSION

Pre-emergence herbicides had significant phytotoxic effect on sesame crop at 2, 4 and 8 WAS (Table 1). In both years and their combined means at 2, 4 and 8 WAS, all the rates of butachlor, all the rates of metolachlor except at 2.0 and 2.5 at 2 WAS in 2006 and combined means, diuron and pendimethalin at 0.5 kg a.i./ha had significantly ($P<0.01$) least phytotoxicity effect on sesame. The slight phytotoxic effect of metolachlor at the aforementioned rates in 2006 at 2 WAS could be due to the lower rainfall recorded in this year than 2007, which

**Table 1.** Effect of pre-emergence herbicides on sesame phytotoxicity at 2, 4 and 8 WAS in Maiduguri, 2006 and 2007.

| Treatment | Rates (kg a.i./ha) | Phytotoxicity rating[1] | | | | | | | | |
|---|---|---|---|---|---|---|---|---|---|---|
| | | 2 WAS[2] | | | 4 WAS | | | 8 WAS | | |
| | | 2006 | 2007 | Mean[3] | 2006 | 2007 | Mean | 2006 | 2007 | Mean |
| Butachlor | 0.5 | 2.3ghi[4] | 1.8fg | 2.1efg | 1.7e[3] | 1.8c | 1.8de | 1.0e | 1.0e | 1.0f |
| " | 1.0 | 2.7ghi | 2.3d-g | 2.5efg | 2.1de | 1.7c | 1.9de | 1.0e | 1.0e | 1.0f |
| " | 1.5 | 2.7ghi | 2.0efg | 2.3efg | 1.9e | 2.0c | 2.0de | 1.0e | 1.2e | 1.1f |
| " | 2.0 | 2.0hi | 3.0d-g | 2.5efg | 1.6e | 2.2c | 1.9de | 1.0e | 1.3e | 1.2f |
| " | 2.5 | 3.0f-i | 2.0efg | 2.5efg | 2.4de | 1.8c | 2.1de | 1.0e | 1.3e | 1.2f |
| Metolachlor | 0.5 | 2.5ghi | 1.8fg | 2.2efg | 1.8e | 1.7c | 1.8de | 1.0e | 1.0e | 1.0f |
| " | 1.0 | 3.0f-i | 1.5g | 2.3efg | 2.2de | 1.3c | 1.8de | 1.0e | 1.0e | 1.0f |
| " | 1.5 | 3.7e-i | 1.8fg | 2.8edg | 2.5de | 1.7c | 2.1de | 1.0e | 1.0c | 1.0f |
| " | 2.0 | 5.0c-g | 1.8fg | 3.4de | 2.0de | 1.7c | 1.8de | 1.0e | 1.3e | 1.2f |
| " | 2.5 | 5.0c-g | 2.0efg | 3.5de | 4.8cd | 1.8c | 3.3cd | 1.7de | 1.0e | 1.3f |
| Diuron | 0.5 | 2.7ghi | 2.3d-g | 2.5efg | 2.3de | 2.3c | 2.3de | 1.3e | 1.3e | 1.3f |
| " | 1.0 | 4.0d-h | 2.2efg | 3.1def | 6.3bc | 3.2c | 4.8c | 2.3cde | 1.7e | 2.0f |
| " | 1.5 | 6.0a-e | 7.0ab | 6.5bc | 7.8ab | 6.3b | 7.1b | 6.0ab | 5.3cd | 5.7de |
| " | 2.0 | 7.3abc | 6.0bc | 6.7ab | 8.3ab | 5.7b | 7.0b | 7.3ab | 4.3d | 5.8cde |
| " | 2.5 | 8.7a | 8.3a | 8.5a | 9.7a | 8.8a | 9.3a | 9.0a | 7.3bc | 8.2ab |
| Pendimethalin | 0.5 | 3.2f-i | 2.3d-g | 2.8efg | 2.8de | 2.0c | 2.4de | 1.3e | 1.0a | 1.2f |
| " | 1.0 | 5.7b-f | 3.8def | 4.8cd | 7.0abc | 6.5b | 6.8b | 4.7bcd | 5.0d | 4.8e |
| " | 1.5 | 6.7a-d | 3.0d-g | 4.8bcd | 8.0ab | 8.5a | 8.3ab | 6.6ab | 8.0ab | 7.3bcd |
| " | 2.0 | 7.7abc | 4.3cd | 6.0bc | 9.0ab | 9.5a | 9.3a | 9.0a | 9.8a | 9.4a |
| " | 2.5 | 8.3ab | 4.0cde | 6.2bc | 9.3a | 9.8a | 9.6a | 5.3bc | 9.8a | 7.6abc |
| Weeding at 3 and 6 WAS | | 1.0i | 1.3g | 1.2fg | 1.0e | 1.5c | 1.3e | 1.0e | 1.0e | 1.0f |
| Weedy check | | 1.0i | 1.7g | 1.3fg | 1.0e | 1.7c | 1.3e | 1.0e | 1.0e | 1.0f |
| SE (± | | 1.35 | 0.79 | 0.93 | 1.4 | 0.69 | 0.81 | 1.58 | 1.04 | 0.94 |

1 = On a scale of 1 to 10 where 1 = no crop injury and 10 = complete crop kill; 2 = Weeks after sowing; 3 = Combined mean for two years data; 4 = Means followed by the same letter(s) within the same column are not significantly different at 5% level of probability according to Duncan's Multiple Range Test (DMRT).

led to the uptake of more of the desorbed herbicide at higher concentration in a sandy loam soil. Diuron at 1.5 to 2.5 kg a.i./ha and pendimethalin at 1.5 to 2.5 kg a.i./ha were significantly ($P<0.01$) phytotoxic to sesame in both years and combined means at 2,4 and 8 WAS (Table 1). This is similar to the findings of Grichar et al. (2001) who reported the phytotoxicity of pendimethalin at 1.2 kg a.i./ha which reduced sesame plant stands by 8 to 98% compared to untreated check.

Table 2 shows that the pre-emergence herbicides evaluated had significant effect on crop vigour of sesame in both years and the combined means. Generally, crop vigour decreased with increase in herbicide rates for all the herbicides. This could be due to increase in phytotoxicity effect with increase in herbicide rate. In both years and their combined means at 6 WAS, all the rates of butachlor, all the rates metolachlor except at 2.0 and 2.5 kg a.i./ha in 2006 produced significantly vigorous crops which were comparable to hoe-weeding. However, diuron and pendimethalin at 1.0 to 2.5 kg a.i./ha produced significantly the least vigorous crops. At 9 WAS, however diuron and pendimethalin at 1.0 and 1.5 kg a.i./ha in 2006, produced vigorous crops comparable to hoe-weeding. The depressive effect of pendimethalin and diuron at higher rates had on the crop was due to the phytotoxic effect of these herbicides. However, at 9 WAS in 2006, diuron and pendimethalin at 1.0 and 1.5 kg a.i./ha produced vigorous crops because the crops recovered from the depressive effect of these herbicides. While metolachlor at 2.0 kg a.i./ha produced the most vigorous crop in 2007 at 6 WAS, metolachlor at 1.0 and 1.5 kg a.i./ha produced the most vigorous crop in 2007 at 9 WAS.

Plant height of sesame was significantly affected by

**Table 2.** Effect of pre-emergence herbicides on crop vigour at 6 and 9 WAS and plant height (cm) at 8 WAS and at harvest in Maiduguri, 2006 and 2007.

| Treatment | Rates (kg a:i./ha) | Crop vigour | | | | | | Plant height | | | | | |
|---|---|---|---|---|---|---|---|---|---|---|---|---|---|
| | | 6 WAS[2] | | | 9WAS | | | 8 WAS | | | Harvest | | |
| | | 2006 | 2007 | Mean[3] | 2006 | 2007 | Means | 2006 | 2007 | Mean | 2006 | 2007 | Mean |
| Butachlor | 0.5 | 7.7ab[4] | 7.7abc | 7.7ab | 8.5a | 7.3abc | 7.9ab | 79.6a-d | 91.1a-e | 85.4abc | 122.0abc | 134.5c-f | 128.2abc |
| " | 1.0 | 6.7a-d | 8.2ab | 7.4ab | 8.0ab | 8.0a | 8.0ab | 83.0ab | 107.7a | 95.4a | 125.0ab | 139.3b-f | 132.2abc |
| " | 1.5 | 7.7ab | 7.5abc | 7.6ab | 8.2ab | 7.2abc | 7.2abc | 80.4a-d | 87.1b-f | 83.8abc | 135.6a | 144.5a-d | 140.0a-d |
| " | 2.0 | 7.7ab | 6.8abc | 7.3ab | 7.7abc | 7.0a-d | 7.3a-d | 83.8a-d | 98.1a-d | 87.9abc | 127.2ab | 144.0a-e | 135.6abc |
| " | 2.5 | 7.0abc | 7.3abc | 7.2ab | 7.5abc | 7.0a-d | 7.3a-e | 77.3a-d | 82.8def | 80.1abc | 120.1abc | 141.9a-f | 131.0abc |
| Metolachlor | 0.5 | 7.7ab | 8.0ab | 7.8a | 7.8ab | 7.5abc | 7.7abc | 76.9a-d | 102.0ab | 89.5ab | 111.4bc | 133.1c-f | 122.3cd |
| " | 1.0 | 6.7a-d | 8.2ab | 7.4ab | 7.5ab | 8.0a | 7.7abc | 73.5a-d | 104.8ab | 89.1ab | 124.0abc | 141.2a-f | 132.6abc |
| " | 1.5 | 6.3a-e | 8.0ab | 7.2ab | 7.8ab | 8.0a | 7.9ab | 81.7abc | 105.7ab | 93.7ab | 128.7ab | 146.1a-d | 137.4abc |
| " | 2.0 | 5.3bf | 8.3a | 6.8abc | 7.8ab | 7.7ab | 7.8abc | 62.8a-g | 99.1a-d | 81.0abc | 129.5ab | 146.0a-d | 137.8abc |
| " | 2.5 | 4.0efg | 7.0abc | 5.5cd | 7.0abc | 7.0a-d | 7.0a-e | 62.2a-f | 92.1a-e | 79.2abc | 119.3abc | 126.5f | 122.9cd |
| Diuron | 0.5 | 6.7a-d | 6.8abc | 6,8abc | 8.1ab | 6.8a-d | 7.5abc | 70.2a-e | 83.5c-f | 76.8bcd | 129.6ab | 139.4b-f | 134.5abc |
| " | 1.0 | 4.3d-g | 4.3d | 4.3de | 7.3abc | 6.3c-f | 6.8a-e | 52.1c-g | 70.1fg | 61.1de | 118.7abc | 144.5a-d | 131.6abc |
| " | 1.5 | 2.8fgh | 3.0de | 2.9ef | 7.0abc | 5.5efg | 6.3b-e | 41.1e-h | 57.1gh | 49.1ef | 116.3bc | 151.7ab | 134.0abc |
| " | 2.0 | 3.0fgh | 3.3de | 3.2ef | 5.2c | 5.8d-g | 5.5ef | 36.9fgh | 41.9hi | 39.4fg | 118.7abc | 156.8a | 137.8abc |
| " | 2.5 | 1.8gh | 2.0e | 1.9fg | 5.8bc | 5.3fg | 5.6def | 34.9gh | 34.7i | 34.8fg | 105.0c | 141.5a-f | 123.2bcd |
| Pendimethalin | 0.5 | 6.0a-e | 6.3c | 6.2bc | 8.0ab | 6.7b-e | 7.3a-d | 64.8a-g | 77.1ef | 71.0cd | 119.07abc | 139.7b-f | 129.4abc |
| " | 1.0 | 4.7c-f | 2.3e | 3.5ef | 7.5abc | 4.7gh | 6.1cde | 54.2b-g | 43.8hi | 49.0ef | 116.4bc | 127.1ef | 121.8cd |
| " | 1.5 | 3.0fgh | 1.8ef | 2.4f | 7.5abc | 3.7h | 5.6def | 50.4d-g | 30.6i | 40.5fg | 115.4bc | 104.9g | 110.2de |
| " | 2.0 | 1.0h | 0.3f | 0.7g | 7.2abc | 0.7i | 3.9f | 36.6fgh | 10.9j | 23.8gh | 113.3bc | 93.0g | 102.2e |
| " | 2.5 | 0.7h | 0.3f | 0.5g | 2.3d | 0.7i | 1.5g | 16.2h | 9.4j | 12.8h | 36.1d | 26.4g | 31.2f |
| Weeding at 3 and 6 WAS | | 8.3a | 7.3abc | 7.8a | 9.0a | 8.0a | 8.5a | 85.6a | 99.5a-d | 92.6ab | 137.1a | 146.8abc | 142.0a |
| Weedy check | | 7.0abc | 6.7bc | 6.8abc | 6.8abc | 7.3abc | 7.1a-e | 76.5a-d | 75.9efg | 76.2bcd | 113.6bc | 129.6def | 121.6cd |
| SE (± | | 1.23 | 0.37 | 0.80 | 1.32 | 0.58 | 0.90 | 10.54 | 12.04 | 9.53 | 14.50 | 12.58 | 12.22 |

1 = On a scale of 1-10 where 1 = completely retarded growth and 10 =Most vigorous plant growth. 2 = Weeks after sowing; 3 = Combined means for two years data. 4 = Means followed by the same letter(s) within the same column are not significantly different at 5% level of probability according to Duncan's Multiple Range Test (DMRT).

pre-emergence herbicides at 8 WAS and at harvest. Generally, plant height decreased with increase in the herbicide rate irrespective of herbicide (Table 2). This could be due to the phytotoxic effect of the herbicides at higher rates, which might have caused a depressive effect on crop growth. All the treatments produced significantly tall plants which were comparable to hoe-weeding in 2006 at 8 WAS, except diuron and pendimethalin at 1.0 to 2.5 kg a.i./ha which produced significantly the shortest plants. In 2007 and combined means all the rates of pendimethalin and diuron and the weedy check produced significantly the shortest plants (Table 2). The phytotoxic effect of diuron and pendimethalin and the inability of pendimethalin to control weeds effectively suppressed crop growth resulting in significantly short plants. This is similar to the findings of Ndarubu (1997), that the application of a mixture of simazine+ametryn at 0.75+0.75 kg a.i./ha and 1.0+1.0 kg a.i./ha resulted in stunted crop growth probably as a result of the phytoxic effect of this herbicide at the stated rates on sesame. At harvest, similar trend as at 8 WAS was observed; however, diuron at 1.0 and 2.0 kg a.i./ha in both trials and combined means and diuron at 1.0 to 2.5 kg a.i./ha in 2007 produced plant height comparable to the hoe-weeding. This showed that sesame plants were able to overcome the phytotoxicity effect of diuron at the above rates with time, possibly as the herbicide became degraded or leached out (Ashton and Klingman, 1975). Also, diuron at 1.5 and 2.0 kg a.i./ha produced the tallest plants in 2007 at harvest, as a result of the low plant population and the significantly low weed infestation, under this treatments which minimized weed competition for growth resources and allowed the few plants to take up enough moisture and nutrients for better growth.

Among the herbicides, while butachlor at 1.0 and metolachlor at 1.5 kg a.i./ha produced the tallest plants in both years and the combined means at 8 WAS, butachlor at 1.5 kg a.i./ha and metolachlor at 2.0 kg a.i./ha produced the tallest plants in 2006 and combined means at harvest. The pre-emergence herbicides evaluated significantly ($P<0.01$) affected weed cover scores in both trials at 3, 6 WAS and at harvest (Table 3). All the herbicide treatments resulted in significantly lower weed cover score compared to the weedy check in both years and their combined means at 3, 6 WAS and at harvest, except pendimethalin at 0.5 to 2.0 kg a.i./ha in 2007 at 6 WAS and at 1.0 to 2.5 kg a.i./ha in both years and combined means at harvest which gave poor weed control comparable to unweeded check (Table 3). The progressive loss of efficacy of pendimethalin herbicide over time resulting in poor control of weeds could be due to the volatile nature of dinitroanaline herbicides of which pendimenthalin is a member (Akobundu, 1987; Rao, 2000) and the faster degradation of the herbicide through the action of microorganisms in the soil (Zimdahl and Gwynm, 1977). Compared to other treatments at 3 WAS, application of diuron at 1.5 to 2.5 kg a.i./ha consistently resulted in significantly the least weed cover score which is comparable with diuron at lower rate and pendimethalin at 1.5 to 2.5 kg, a.i/ha in both years and combined means, metolachlor at 1.5 kg a.i./ha in 2006, butachlor at 1.5 to 2.5 kg a.i./ha and metolachlor at 2.0 and 2.5 kg a.i./ha in 2007. The ability of diuron to provide season-long effective weed control, could be due to the persistence of this herbicide in the soil and the fact that it is more effective on broad- leaved weeds than grasses (Akobundu, 1987) which were predominant at the site. Generally, at 3, 6 WAS, the weed cover scores were higher in 2007 compared with 2006. This could be due to higher amount of rainfall received in 2007 which might have solubilised the herbicides resulting to leaching. Similarly, weed cover score were higher in 2007 and combined means at 6 WAS compared with that in 2007 and combined means at harvest, possibly as a result of the ability of the sesame crop to suppress and smother weeds due to early canopy closure which shaded off the weeds, depriving them of growth factors of sunlight, moisture and mineral nutrients (Imoloame, 2004). Table 4 shows that pre-emergence herbicides significantly ($P<0.01$) affected weed dry matter in both years and combined means at 3, 6 WAS and at harvest.

All the herbicide treatments reduced significantly the weed dry matter in both years and their combined means compared to the weedy check at 3, 6 WAS and at harvest, except metolachlor at 0.5 and 1.0 kg a.i./ha and pendimethalin at 2.5 kg a.i./ha at 6 WAS in 2006 and 2007, butachlor at 1.5 kg a.i./ha, diuron at 2.0 kg a.i./ha and pendimethalin at 0.5, 1.0, 1.5 and 2.5 kg a.i./ha in 2007 and combined means, which produced high and comparable weed dry matter to the weedy check. In the combined means also at 6 WAS; all the herbicide treatments produced lower weed dry matter except metolachlor at 0.5 and 1.0 kg a.i./ha, pendemethalin at 0.5, 1.0, 1.5 and 2.5 kg a.i./ha which were comparable to the weedy check. At harvest, all the herbicide rates did not differ significantly from the weedy check in 2006. All the rest of the herbicide treatments in 2007 and the combined means significantly produced lower weed dry matter compared to the weedy check but comparable to hoe-weeded control. The effectiveness of metolachlor and diuron at all rates except at 2.0 kg a.i./ha in 2007 at harvest is an indication that, these herbicides can be used in weed control alternative to two hoe-weedings at 3 and 6 WAS. This agrees with the findings of Ibrahim et al. (2008) that pre-emergence application of diuron at 0.960 kg a.i./ha gave excellent broad-leaved weed control in sesame in Egypt. It also agrees with Grichar et al. (2001) who reported that metolachlor among all the herbicides evaluated, provided the best control of weeds and least sesame injury.

The pre-emergence herbicides significantly affected the number of capsules/plant of sesame (Table 5). In 2006, metolachlor at 2.0 and 2.5 kg a.i./ha produced

**Table 3.** Effect of pre-emergence herbicides on weed cover score at 3, 6 WAS and at harvest in Maiduguri, 2006 and 2007.

| Treatment | Rates (kg a.i./ha) | Weed cover score[1] | | | | | | | | |
|---|---|---|---|---|---|---|---|---|---|---|
| | | 3WAS[2] | | | 6WAS | | | Harvest | | |
| | | 2006 | 2007 | Mean[3] | 2006 | 2007 | Mean | 2006 | 2007 | Mean |
| Butachlor | 0.5 | 3.3cd[4] | 6.0cd | 4.7bc | 4.7bc | 7.0a-d | 5.8b-f | 4.8d-h | 3.7c | 4.3c-f |
| " | 1.0 | 3.3cd | 4.7d-g | 4.0bc | 4.5bc | 5.7b-e | 5.1b-f | 5.5d-g | 2.8cde | 4.2c-f |
| " | 1.5 | 3.0cde | 3.7e-i | 3.3c-f | 3.8cde | 5.8b-e | 4.8c-f | 4.7d-i | 2.5c-f | 3.6c-h |
| " | 2.0 | 3.0cde | 3.7e-i | 3.3c-f | 3.7cde | 6.3a-e | 5.0b-f | 3.3f-j | 2.3def | 2.8d-h |
| " | 2.5 | 3.7cd | 3.7e.i | 3.7cde | 4.7bc | 5.7b-e | 5.2b-f | 6.0def | 2.2ef | 4.1c-g |
| Metolachlor | 0.5 | 4.3bc | 5.0def | 4.7bc | 5.0bc | 7.5abc | 6.3b-e | 5.7d-g | 2.5c-f | 4.1c-g |
| " | 1.0 | 3.0cde | 5.3cde | 4.2bcd | 3.7cde | 5.5b-f | 4.6d-f | 6.0def | 2.7c-f | 4.3cde |
| " | 1.5 | 2.7c-f | 4.0d-h | 3.3c-f | 3.8bcd | 5.0c-f | 4.4efg | 5.5d-g | 1.8ef | 3.7c-h |
| " | 2.0 | 3.3cd | 3.0f-j | 3.2c-g | 4.2bcd | 6.0b-e | 5.1b-f | 6.7cde | 2.5c-f | 4.6cde |
| " | 2.5 | 3.7cd | 3.2f-j | 3.4cde | 4.0bcd | 4.0efg | 4.0fg | 7.0bcd | 2.7c-f | 4.8cd |
| Diuron | 0.5 | 2.6c-f | 2.7g-j | 2.3d-h | 4.3bc | 4.2d-g | 4.3efg | 3.0g-j | 2.3def | 2.7d-h |
| " | 1.0 | 1.3ef | 1.8ij | 1.6gh | 2.0def | 2.7fg | 2.3gh | 3.8e-j | 2.0ef | 2.9d-h |
| " | 1.5 | 1.3ef | 1.7ij | 1.5h | 1.5ef | 2.0g | 1.8h | 2.1hj | 2.2ef | 2.1fgh |
| " | 2.0 | 1.0f | 1.7ij | 1.3h | 1.2f | 1.8g | 1.5h | 1.8ij | 2.0ef | 1.9gh |
| " | 2.5 | 1.0f | 1.5j | 1.3h | 1.1f | 1.8g | 1.5h | 3.2f-j | 1.8ef | 2.5e-h |
| Pendimenthalin | 0.5 | 3.3cd | 7.3bc | 5.3b | 4.3bc | 9.0a | 6.7bcd | 7.5a-d | 3.5cd | 5.5bc |
| " | 1.0 | 3.0cde | 3.7e-i | 3.3c-f | 5.3bc | 8.3ab | 6.8bc | 9.3abc | 8.8a | 9.1a |
| " | 1.5 | 2.0def | 2.8g-j | 2.4e-h | 6.0b | 7.5abc | 6.8bc | 9.5abc | 8.0a | 8.8a |
| " | 2.0 | 2.0def | 3.5e-j | 2.8d-h | 6.0b | 8.0ab | 7.0ab | 9.8ab | 7.8a | 8.8a |
| " | 2.5 | 1.3ef | 2.2hij | 1.8fgh | 5.0bc | 4.7c-g | 4.8c-f | 9.8ab | 8.0a | 8.9a |
| Weeding at 3 and 6WAS | | 6.0ab | 9.7a | 7.8a | 2.0def | 6.7a-e | 4.4efg | 1.4j | 1.5f | 1.5h |
| Weedy check | | 6.3a | 9.0ab | 7.7a | 9.7a | 8.3ab | 9.0a | 10.0a | 5.3b | 7.7ab |
| SE (±) | | 0.98 | 1.07 | 0.83 | 1.11 | 1.41 | 1.06 | 1.44 | 0.65 | 1.10 |

1 = On a scale of 1-10 where 1 = no weed cover and 10 = complete weed cover; 2 = Weeks after sowing; 3= Combined mean for two years data; 4 = Means followed by the same letter(s) within the same column are not significantly different at 5% level of probability according to Duncan's Multiple Range Test (DMRT).

significantly the highest number of capsules/plants which were comparable with butachlor at 2.0 kg a.i./ha, diuron at 1.5 and 2.5 kg a.i./ha, pendimethalin at 0.5 and 1.5 kg a.i./ha but significantly higher than the rest of the treatments. This could be due to the phytotoxic effect which the stated herbicide rates had on sesame crop which reduced the plant population and reduced the inter and intra plant competition leading to the utilization of more growth factors for better growth. However, in 2007 and combined means, pendimethalin at 2.0 and 2.5 kg a.i./ha produced significantly, the least number of capsules/plant probably due to the phytotoxicity of this herbicide at the aforementioned rates and its inability to effectively control weeds which competed for nutrients, moisture and other growth resources with sesame resulting in poor performance.

Grain yield was significantly affected by pre-emergence herbicides in both years and combined means (Table 5). In 2006, butachlor at 2.0 kg a.i./ha and diuron at 1.0 and 1.5 kg a.i./ha produced significantly the highest sesame yield which was comparable to hoe-weeded control and other treatments except butachlor at 0.5, 1.0 and 2.5 kg a.i./ha, metolachlor at 0.5, 2.0 and 2.5 kg a.i./ha, diuron at 2.0 and 2.5 kg a.i./ha, pendimethalin at all the rates and weedy check which produced significantly the lowest grain yield. However, in 2007 and combined means, metolachlor at 1.5 kg a.i./ha produced significantly, the highest yield but comparable with hoe-weeded control and other herbicide treatments except butachlor at 0.5 kg a.i./ha, metolachlor at 0.5, 2.0 and 2.5 kg a.i./ha, diuron at 0.5, 1.5, 2.0 and 2.5 kg a.i./ha, pendimethalin at 1.0-2.5 kg a.i./ha and weedy check which produced significantly, the lowest grain yield (Table 5).

Butachlor and diuron at 2.0 and 1.0 kg a.i./ha, respectively produced significantly, the highest yield; although, not significantly different from some other herbicide treatments and hoe-weeded control in 2006, probably due to the low phytotoxicity of butachlor at

**Table 4.** Effect of pre-emergence herbicides on weed dry matter in sesame at 3, 6 WAS and at harvest in Maiduguri, 2006 and 2007.

| Treatment | Rates (kg a.i./ha) | Weed dry matter (kg/ha) | | | | | | | | |
|---|---|---|---|---|---|---|---|---|---|---|
| | | 3 WAS[1] | | | 6 WAS | | | Harvest | | |
| | | 2006 | 2007 | Mean[2] | 2006 | 2007 | Mean | 2006 | 2007 | Mean |
| Butachlor | 0.5 | 7.1c[3] | 6.2cd | 6.6bc | 34.1b-g | 26.2bcd | 30.1b-e | 21.9a | 80.0bcd | 51.0cde |
| " | 1.0 | 20.2bc | 5.0cd | 12..6bc | 15.3e-h | 31.8bcd | 23.5c-f | 28.1a | 42.2d | 35.7cde |
| " | 1.5 | 15.9c | 5.2cd | 10.5bc | 17.1d-h | 35.0a-d | 26.1c-f | 35.4a | 67.3cd | 51.4b-e |
| " | 2.0 | 21.2bc | 9.7bc | 15.4bc | 28.6b-h | 25.8bcd | 27.2c-f | 35..0a | 50.4d | 42.7cde |
| " | 2.5 | 14.5c | 1.8cd | 8.1bc | 22.9c-h | 22.3cd | 22.6c-f | 27.7a | 103.5bcd | 65.3b-e |
| Metolachlor | 0.5 | 42.6b | 5.0cd | 23.8b | 48.9a-d | 33.7bcd | 41.3abc | 42.7a | 96.1bcd | 69.4b-e |
| " | 1.0 | 11.6c | 4.9cd | 8.1bc | 58.2ab | 37.7abc | 48.0ab | 65.6a | 56.3d | 61.0b-e |
| " | 1.5 | 3.9c | 6.2cd | 5.1c | 31.8b-h | 26.0bcd | 28.9b-f | 26.1a | 40.2d | 33.1cde |
| " | 2.0 | 9.5c | 2.5cd | 6.0c | 14.2e-h | 28.7bcd | 21.4c-f | 65.6a | 56.3d | 61.0b-e |
| " | 2.5 | 11.5c | 4.4cd | 8.0bc | 21.4c-h | 18.0cd | 19.7def | 36.0a | 28.2d | 32.5cde |
| Diuron | 0.5 | 11.1c | 2.4cd | 6.7bc | 16.3e-h | 21.6cd | 18.9def | 41.7a | 32.7d | 37.3cde |
| " | 1.0 | 5.1c | 3.6cd | 4.4c | 9.6fgh | 26.4bcd | 18.0def | 44.8a | 29.7d | 37.3cde |
| " | 1.5 | 13.0c | 3.9cd | 8.4bc | 6.3fgh | 16.5cd | 11.4ef | 15.6a | 22.7d | 19.2de |
| " | 2.0 | 4.7c | 2.4cd | 3.6c | 1.6h | 34.8a-d | 18.2def | 17.7a | 156.5ab | 87.1abc |
| " | 2.5 | 1.0c | 0.6d | 0.8c | 2.6gh | 15.2d | 8.9f | 33.3a | 41.2d | 37.3cde |
| Pendimethalin | 0.5 | 17.1c | 6.8cd | 11.9bc | 24.0c-h | 49.9ab | 35.4a-d | 41.5a | 98.2bcd | 69.8a-e |
| " | 1.0 | 18.2c | 4.9cd | 11.6bc | 44.5a-e | 36.5a-d | 40.5abc | 32.3a | 225.2a | 128.7a |
| " | 1.5 | 9.1c | 3.3cd | 6.2bc | 36.9b-f | 44.8ab | 40.9abc | 67.9a | 75.6bcd | 71.7a-d |
| " | 2.0 | 7.5c | 6.4cd | 7.0bc | 30.2b-h | 29.1bcd | 29.7bc | 59.4a | 72.2bcd | 65.8b-e |
| " | 2.5 | 7.1c | 2.4cd | 4.7c | 71.3a | 35.3a-d | 53.3a | 74.0a | 39.4d | 56.7b-e |
| Weeding at 3 and 6WAS | | 16.6c | 15.5ab | 16.1bc | 24.3c-h | 46.0ab | 35.2a-d | 6.7a | 17.6d | 12.1e |
| Weedy check | | 90.6a | 22.4a | 56.5a | 52.7abc | 56.3a | 54.5a | 72.9a | 147.7abc | 110.3ab |
| SE (±) | | 11.17 | 4.13 | 8.89 | 15.8 | 10.85 | 10.10 | 21.96 | 43.4 | 29.92 |

1 = Weeks after sowing; 2 = Combined means for two years data; 3 = Means followed by the same letter(s) within the same column are not significantly different at 5% level of probability according to Duncan's Multiple Range Test (DMRT).

**Table 5.** Effect of pre-emergence herbicides on number of capsules at 12WAS, and grain yield of sesame in Maiduguri in 2006 and 2007.

| Treatment | Rates (kg a.i./ha) | Number of capsules/plant 12 WAS[1] | | | Grain yield (kg/ha) | | |
|---|---|---|---|---|---|---|---|
| | | 2006 | 2007 | Mean[2] | 2006 | 2007 | Mean |
| Batachlor | 0.5 | 35.8cde[3] | 64.1a | 49.9abc | 105.2f-i | 210.9def | 158.0ghi |
| " | 1.0 | 30.7de | 43.9a-d | 37.3b-e | 128.0efg | 392.7abc | 260.4c-g |
| " | 1.5 | 31.8de | 32.8bcd | 32.3cde | 295.9ab | 443.4ab | 369.6abc |
| " | 2.0 | 43.9a-d | 48.9a-d | 46.4a-e | 330.4a | 464.2ab | 397.3ab |
| " | 2.5 | 39.5cd | 29.1d | 34.3cde | 182.7def | 410.3ab | 292.0b-f |
| Metolachlor | 0.5 | 40.3cd | 57.0ab | 48.7a-d | 202.7b-f | 311.8b-e | 257.3c-g |
| " | 1.0 | 40.2cd | 36.5bcd | 38.4a-e | 263.1a-d | 456.4ab | 359.8abc |
| " | 1.5 | 34.1cde | 54.4abc | 44.3a-e | 286.6ab | 529.7a | 408.2ab |
| " | 2.0 | 65.5a | 46.3a-d | 55.9a | 195.2c-f | 225.2c-f | 210.2e-h |
| " | 2.5 | 65.2ab | 45.3a-d | 55.3ab | 189.4c-f | 189.6ef | 189.5fgh |
| Duiron | 0.5 | 38.3cd | 45.1a-d | 41.7a-e | 239.6a-d | 227.0c-f | 233.3d-g |
| " | 1.0 | 39.8cd | 57.1ab | 48.5a-d | 330.4a | 374.1abc | 352.3a-d |
| " | 1.5 | 42.3a-d | 47.1a-d | 44.7a- e | 317.5a | 302.1b-e | 309.8b-f |
| " | 2.0 | 28.9de | 29.7cd | 29.3ef | 68.7ghi | 207.6def | 138.1ghi |
| " | 2.5 | 41.5a-d | 48.0a-d | 44.8a- e | 115.0fgh | 201.1def | 158.1ghi |
| Pendimethalin | 0.5 | 43.5a-d | 25.8d | 34.7cde | 214.4b-e | 430.0ab | 322.2a-e |
| " | 1.0 | 31.1de | 27.5d | 29.3ef | 110.0fgh | 64.1fg | 87.1hij |
| " | 1.5 | 57.5abc | 28.6d | 43.0a-e | 66.0ghi | 7.9g | 36.9ij |
| " | 2.0 | 25.0de | 1.0e | 13.0fg | 19.9hi | 1.0g | 10.5j |
| " | 2.5 | 11.7e | 1.0e | 6.4g | 12.1i | 1.0g | 6.6j |
| Weeding at 3 and 6WAS | | 40.7bcd | 43.1a-d | 41.9a-e | 331.4a | 543.7a | 437.58a |
| Weedy check | | 23.9de | 38.8bcd | 31.4de | 197.3c-f | 198.5ef | 197.9e-h |
| SE (±) | | 12.18 | 12.28 | 9.21 | 1.36 | 86.14 | 62.92 |

1 = Weeks after sowing; 2 = Combined means for two years data; 3 = Means followed by the same letter(s) within the same column are not significantly different at 5% level of probability according to Duncan's Multiple Range Test (DMRT).

2.0 kg a.i./ha and its ability to control weeds effectively throughout the season, which prevented weed competition for growth resources leading to better growth and high yield. Similarly, diuron at 1.0 kg a.i./ha significantly produced high yield, probably as a result of its low phytotoxicity on the crop at 8 WAS, coupled with its ability to significantly reduce both weed cover and weed dry matter throughout the season, preventing weed competition and allowing the crop to utilize more growth resources resulting in higher yield. However, in 2007 and the combined means, metolachlor at 1.5 kg a.i./ha, butachlor at 2.0 and 1.5 kg a.i./ha produced significantly, the highest yield which was comparable to hoe-weeded control and other herbicide treatments, while metolachlor at 2.0 and 2.5 kg a.i./ha, diuron at 1.5 to 2.5 kg a.i./ha, pendimethalin at 1.0 to 2.5 kg a.i./ha and weedy check produced significantly the lowest yield. The significantly higher yields produced by the aforementioned herbicides at the stated respective rates could be due to their little or no phytotoxic effect on sesame, coupled with its high selectivity to weeds. This resulted in better uptake of plant nutrients and moisture which resulted in better growth and higher yield. This is similar to the findings of Grichar et al. (2001) who reported that metolachlor at 3.36 kg a.i./ha increased yield up to 80% over the untreated check. Diuron at 1.5 to 2.5 kg a.i./ha and pendemethalin at 1.0 to 2.5 kg a.i./ha produced significantly low yields due to their phytotoxic effect on sesame.

From this research, it can be recommended that metolachlor at 1.5 kg a.i./ha and butachlor at 2.0 kg a.i./ha, were the best among the herbicides and can be used as an alternative to two-hoe weedings for effective weed control and higher grain yield in sesame in the

Sudan Savanna Zone of Nigeria.

## REFERENCES


Adigun JA, Lagoke STO, Kumar V, Erinle ID (1993). Weed management studies in transplanted tomato in Nigerian savanna. Samaru J. Agric. Res., 10: 29-39.

Adigun JA, Lagoke STO (2003). Weed control in transplanted rainfed and irrigated tomatoes in the Nigeria Savanna. Niger. J. Weed Sci., 16: 24-30.

Ashton FM, Klingman OA (1975). Weed Science: Principles and Practices. Wiley Inter-science Publications New York, USA, pp. 431.

Akobundu IO (1987). Weed science in the tropics. John Wiley and Sons, New York, p. 522.

Anonymous (1994). Weed control recommendations for Nigeria. Series 3, Sponsored by Department of Agriculture, Federal Ministry of Agriculture, Nigeria, p. 1.

Anonymous (2006). Sesame: Past and Future.SPORE. (CTA), Netherland, 125: 5-6.

Godin VJ, Spensley PC (1971). Oils and oil seeds. Tropical Products Institute. Crop and Product Digest, London, England, p. I.

Grichar WJ, Sestak DC, Brewer KD, Besler BA, Stichler CR, Smith DT (2001). Sesame (*Sesamum indicum L.*) tolerance and weed control with soil applied herbicides. Crop Prot., 20: 389-394.

Gupta SK, Yadawa TP, Parkash K, Thakral NK, Kumar P (1998). Influence of date of sowing on oil and fatty acid composition in sesame (*Sesamum indicum L.*) Ann. Biol., 14: 67-68.

Ibrahim AF, EL-Wekjl HR, Yahia ZR, Shaban SA (2008). Effect of Some Weed Control Treatments on Sesame (*Sesamum indicum L.*) and Associated Weeds J. Agro. Crop Sci. 160(5): 319-324.

Imoloame EO (2004). Effects of seed rate and sowing method on weed infestation, growth and yield of sesame(*Sesamum indicum L)* in the semi arid zone of Nigeria M.Sc. Dissertation University of Maiduguri, p. 48.

Imoloame EO (2009). Effects of pre- and post- emergence herbicides on weed infestation and productivity of sesame (*Sesamum indicum* L.) in a Sudan Savanna zone of Nigeria. Ph.D Thesis, Department of Crop Production University of Maiduguri, Maiduguri, p. 145.

Ishaya DB, Agaji AM (2009). Evaluation of cinosulfuron and prosulfuron in sesame (*Sesame indicum* L.) in the Northern Guinea Savanna of Nigeria. Niger. J. Weed Sci., 22: 23-29.

Iwo GA, Idowu AA (2002). Impact of research and development on sesame towards poverty alleviation in Nigeria . Noma 16:15-21.

Ndarubu AA (1997). Effect of chemical weed control treatments and inter-row spacing in the growth and yield of sesame (*Sesamum indicum L.*) M.Sc. Dissertation. Ahmadu Bello University, Zaria, p. 18.

Ndarubu AA, Adejonwo KO, Adigun JA, Chiezey UF (2003). Effects of weed control treatments and interrow spacing on the growth and yield of sesame (Sesamum indicum L.). Niger. J. Weed Sci., 16: 15-23.

Ogungbile AO, Lagoke STO (1986). On-farm evaluation of the economics of chemical weed control in oxen-mechanized maize production in Nigeria savanna. Trop. Pest Manage., 32: 373.

Okereke UO (1983). Weed control in transplanted dry and rainy sesason tomato (Lycopersicon esculentum Mill) crops with gramoxone (paraquat) and sencor. Paper presented at the 5th Annual Conference of Horticultural Society of Nigeria, University of Nsukka, pp. 6-9.

Onwueme IC, Sinha TD (1991). Field crop production in tropical Africa. Principles and Practice. CTA, Wageningen, Netherlands. p. 348.

Philips TA (1977). Sesame. In: An agricultural notebook. New Ed. Longman Group, Ltd. London, pp. 61-63.

Rao VS (2000). Principles of Weed Science. Second Edition Plymouth, U. K., p. 81.

Roland JRJ (1993). Dryland Farming in Africa CTA,Wageningen, The Netherland, p. 337.

Shrock WA, Monaco TJ (1980). Recent development in weed management programme for tomato. Proceedings of 33rd Annual Meeting of Southern Weed Science Society, 33: 90-92.

Sinha TD, Lagoke STO (1984). Weed control in irrigated tomato (Lycopersicon esculentum. Mill) in northern Nigeria. Trop. Pest Manage., 30: 18-25.

Usoroh NJ (1983). Field screening of herbicides for weed control in tomato (Lycopersion esculentum. Mill). Paper presented at the 13th Annual Conference of the Weed Science Society of Nigeria, National Cereal Research Institute, Ibadan, 5 Dec. 1983.

Yermanos D, Hemstreet MJ, Saleed W, Huszar CK (1972). Oil content and composition of seeds in the world collection of sesame introduction. J. Agric. Chem. Soc., 49(1): 20-23.

Zimdahl D, Gwynn SO (1977). Soil degradation of three dinitroanilines. Weed Sci., 25: 247-51.

# Economic assessment of some pre-emergence herbicides in the Sudan Savanna Zone of Nigeria

**E. O. Imoloame[1], S. D. Joshua[2] * and N. A. Gworgwor[2]**

[1]Regional Agricultural Training School, Ngala, Borno State, Nigeria.
[2]Department of Crop Production, Faculty of Agriculture, University of Maiduguri, Borno State, Nigeria.

**Field trials were conducted in 2006 and 2007 cropping seasons at the University of Maiduguri Teaching and Research Farm located in Maiduguri (11°50'N; 13°10'E) to carry out an economic analysis of the use of some pre-emergence herbicides for the production of sesame. The experiment consisted of 22 treatments which included 4 different pre-emergence herbicides applied at 5 different rates viz: butachlor, metolachlor, diuron and pendimethalin at 0.5, 1.0, 1.5, 2.0 and 2.5 kg a.i. /ha, weeding at 3 and 6 weeks after sowing (WAS) and weedy check. The treatment was laid out in randomized complete block design (RCBD) replicated 3 times. Result showed that although two hoe weedings resulted in the highest yield, percentage yield increase over check and income in both years, it was less beneficial and profitable than butachlor at 1.5 and 2.0 kg a.i /ha in 2006 and metolachlor at 1.5 kg a.i./ha in both years. Therefore, metolachlor at 1.5 kg a.i. /ha was more economical, profitable and beneficial than two hoe weedings in the production of sesame in the Sudan Savanna Zone of Nigeria.**

**Key words:** Cropping seasons, pre-emergence herbicides, millet-cowpea intercrop.

## INTRODUCTION

Sesame (*Sesamum indicum L.*) had been grown for the superior edible oil and nutritious food since before 2360BC (Klingman and Martin, 1954). In Nigeria, sesame is mainly grown in areas of southern Guinea Savanna Ecological Zone, notably in Benue and Kogi States and Federal Capital Territory, Abuja (Philips, 1977). Ingawa et al. (1986) reported that most of the north eastern parts of Nigeria are also involved in the production of sesame which include a stretch from Gumel in Kano State extending to Potiskum in Yobe State, Gwoza in Borno State, and Madagali, Lassa, Yola curving northward to Mubi areas of Adamawa State respectively and then into Bauchi State.

The importance of sesame lies in the variety of useful materials that can be extracted from it. The seed contains about 50 - 51% oil, 17 - 19% protein and 16 - 18% carbohydrate and the oil is edible, odourless and semi-drying containing oleic, stearic and palmitic acids (Yermanos et al., 1972). In sesame growing areas of Nigeria, the seeds and leaves are used for the preparation of soups. Sesame seed is eaten raw or fried in the form of cakes, or fried, pounded and mixed with sorghum and millet flour to a pastry or fried, pounded and mixed with water to give a sustaining drink (Van Rheenem, 1973). The constraint imposed by weeds on high yield in sesame had been well documented. Jain et al. (1985) and Sharma (1985) have stressed that early growth of sesame is slow, so it is important to suppress weed growth at an early stage. Balyan (1993) and Sinha et al. (1992) reported weed- induced yield reduction up to 135% and need for a critical weed free period of 50 days after planting.

The traditional method of weed control, namely, hoe-weeding is the commonest method of weed control by farmers in the Sudan savanna zone of Nigeria. This method is not only labour intensive, expensive and strenuous, but can also cause mechanical damage to growing branches and roots of plants. In addition to high cost, labour availability is uncertain thus making timeliness of weeding difficult to attain, leading to greater yield loss (Adigun et al., 2003).

Chemical weed control has revolutionized farmers approach to weed control in the world. It is one of the recent developments in crop production. It is more adapted to large scale crop production than other weed

*Corresponding author. E-mail: asdjoshua@yahoo.com.

control methods and it is labour saving (Anon, 1994). Also, herbicide use has been reported to be more profitable than hoe-weeding in the production of various crops in Nigeria (Shrock and Monaco, 1980; Okereke, 1983; Usoroh, 1983; Sinha and Lagoke, 1984; Ogungbile and Lagoke, 1986; and Adigun et al., 1993). Judicious use of herbicides has been reported to reduce labour requirements for and cost of weed control, increased crop yields by reducing weed competition and consequently increased profitability (Ogungbile et al., 1982; Sinha et al., 1982). Nazeer et al. (2004) recommended Buctril-M herbicide over hoe-weeding for the management of broadleaf weeds in wheat as a result of the ability of the herbicide to produce higher grain yield compared to hoe-weeding and the attractive benefit-cost ratio of the use of this herbicide. Also, Kehinde (2002) had noted that although controlling weeds with pre- and post-emergence herbicides in upland rice was more expensive than two hoe-weeding, it gave the best weed control and had the highest net returns.

Fadayomi (1991) noted that there was adoption of chemical weed control option to large extent in Nigeria. However, it should be noted that many problems are associated with chemical weed control method in developing countries to make it as an alternative to hand weeding. It was observed by Akobundu (1997) and Fadayomi (1991) that a cross-section of Nigerian farmers do not have formal education, secondly, the chemical and spray equipment are largely imported, hence making price control by government difficult. In the case of herbicide use for example, other than perceived high cost and availability of chemicals, ownership and operation of knapsack sprayers are limiting factors to adoption in areas surveyed in the states of Nigeria (Ikuenobe et al., 2005). Adedzwa and Ortese (2004) reported that while farmers in Benue State of Nigeria acknowledge the efficiency of herbicides in the control of *Imperata cylindrical* the adoption of herbicides was limited by availability and cost.

While the beneficial effects of herbicide applications are well known, the risks to man and the other living organisms that he considers beneficial to himself or his environment should not outweigh the benefits associated with the use of these chemicals. Therefore, distribution should be made between the consequences of an abuse and the benefits from a recommended use of chemical (Akobundu, 1987). Some of the risks posed by herbicide use include, spray drift to sensitive crops, contamination of drinking water especially herbicides that are not bio-degradable and easily leached from the surface of light soil, the reduction and killing of the population of soil-inhabiting animals, the development of resistance in weeds due to repeated treatment of weeds with a particular class or family of herbicides, etc. (Akobundu, 1987).

In order to make farming more attractive and economical to the present and future generations of farmers and to increase food production, there is need to carry out an economic assessment of some pre-emergence herbicides to find out the one that will not only give higher yield but that which is most economical and profitable in the production of sesame in the Sudan Savanna Zone of Nigeria, so as to improve the socio-economic well-being of the farmers.

## MATERIALS AND METHODS

Field trials were conducted during the rainy seasons (July - October) of 2006 and 2007 on the Teaching and Research Farm of the Faculty of Agriculture University of Maiduguri, Maiduguri (11°50'N, 15°10'E) located in the Sudan Savanna Zone of Nigeria. The experimental site, which measured 12 × 88 m, was harrowed, leveled properly using a hand hoe, and marked out. The plot size was 4 × 4 m leaving a distance of 1 and 0.5 m between replications and plots respectively. NPK 15:15:15 was applied to each plot to provide 50 kg P and 75 kg N. The full rate of P was applied before sowing and incorporated into the soil. The application of N however, was done in 2 equal split doses. The first dose was applied before planting as starter dose and the second dose was applied at 6WAS by drilling the fertilizer into the soil 15 cm away from the stand in a continuous band.

The experiment consisted of 22 treatments, which included 4 different pre-emergence herbicides applied at 5 different rates viz: butachlor, diuron, metolachlor and pendimethalin at 0.5, 1.0, 1.5, 2.0 and 2.5 kg a.i. /ha, weeding at 3 and 6WAS and a weedy check. The treatments were laid out in randomized complete block design (RCBD) with 3 replications.

Information on the cost of all the cultural practices from land preparation to harvesting and processing was collected from Borno State Agricultural Development Programme (BOSADP), Maiduguri, an agency responsible for extension services in Borno State, Nigeria. The current price of a big Mudu measure was found out from the open market to calculate the selling price or income. The economic assessment was done for different rates compared with hoe-weeded control and weedy check to find out the most cost-effective or profitable rate for the production of sesame.

The following methods (Joshua and Gworgwor, 2001) were used to assess the economics of production of sesame using different pre-emergence herbicides:

### Cost of production

This include the cost of all the cultural practices, seeds, herbicides and transportation per net plot (0.00216 ha) and later converted to per hectare.

### Yield increase over check

This was obtained by subtracting the check yield from yield of each treatment. This result was then divided by the yield of each treatment and multiplied by 100 to give percentage yield increase according to the following equation.

$$\text{Yield increase} = \frac{Yt - Yc}{Yt} \times 100$$

Where, Yt = Yield of grain in kg/ha of treatment.
Yc = Yield of check in kg/ha.

**Selling price (Income)**

Yields obtained from the experiments were sold at the prevailing market price of the commodity in Maiduguri.

**Profit /loss**

This was determined by subtracting the total cost of production from the selling price (Income) as represented by this equation:

Profit/loss = Sp – TCp,

Where Sp = Selling price (Income)
TCp = Total cost of production.

**Cost-benefit ratio**

This was determined by dividing the total cost of production by the selling price (income) as represented by this equation.

$$\text{Cost-benefit ratio} = \frac{TCp}{Sp}$$

Where TCp = Total cost of production
Sp = Selling price.

## RESULTS AND DISCUSSION

The economic assessment of the use of different herbicide in the production of sesame in Maiduguri is presented in Table 1. The table shows the grain yield of sesame in kilograms per hectare produced by different herbicide rate, two hoe-weeding and weedy checks. It also shows percentage grain yield increase over weedy check, total cost of production per hectare, income per hectare, profit and loss per hectare and the cost-benefit ratio of the use of different pre-emergence herbicides vis-à-vis hoe- weeding and weedy check in the production of sesame in 2006 and 2007 cropping seasons. With regards to sesame production, the highest yield (331.4 kg/ha) was produced under hoe-weeding, followed by butachlor at 2.0 kg a.i./ha (330.4 kg/ha), diuron at 1.0 kg a.i./ha (330.4 kg/ha), diuron at 1.5 kg a.i./ha (317.5 kg /ha), butachlor at 1.5 kg a.i./ha (295.9 kg/ha) and metolachlor at 1.5 kg/a.i./ha (286.6 kg/ha), while the lowest yield (12.1 kg/ha) was produced under pendimethalin at 2.5 kg a.i/ha in 2006. Similarly, in 2007, hoe-weeding recorded the highest yield (543.7 kg/ha), followed by metolachlor at 1.5 kg a.i./ha (529.7 kg/ha), butachlor at 2.0 Kg a.i./ha (464.2 Kg /ha), metochlor at 1.0 kg a.i./ha (456.4 kg/ha) and butachlor at 1.5 kg a.i./ha (443.4 kg/ha) compared to the rest of the treatments, while the lowest yield (1.0 kg/ha) was observed under pendimethalin at 2.0 and 2.5 kg a.i./ha. The above results could be due to the little or no phototoxic effect of hoe-weeding and these herbicides on sesame and their effectiveness in the control of weeds which minimized weed competition with the crop and led to better utilization of growth resources (mineral nutrients, moisture and light) and better crop growth compared to the other treatments. The poor yield by pendimethalin herbicides at higher rates could be due to the phytotoxic effects of this herbicide and its ineffectiveness to control weeds (Imoloame, 2009).

In 2006, hoe-weeding produced the highest percentage yield increase (40.5%) over weedy check treatment, followed by butachlor at 2.0 kg a.i./ha and diuron at 1.0 kg a.i./ha (40.3%), diuron at 1.5 kg a.i./ha (37.9%), butachlor at 1.5 kg a.i./ha (33.3%) and metolachlor at 1.5 kg a.i./ha (31.2%) compared to other treatments, while the lowest grain yield increase over check (-1,532.6%) was recorded under pendimethalin at 2.5 kg a.i./ha. Similarly, in 2007, percentage yield increase over weedy check was highest (63.5%) under hoe-weeding, followed by metolachlor at 1.5 kg a.i./ha (62.5%), butachlor at 2.0 kg a.i./ha (57.2%), metolachlor at 1.0 kg a.i./ha (56.5%) and butachlor at 1.5 kg a.i./ha (55.2%), compared with the other treatments, while pendimethalin at 2.0 and 2.5 kg a.i./ha recorded the lowest percentage yield increase over check (-19,753.2%). The above results show the advantage of effective weed control by hoe-weeding and the above herbicides over weedy check and the unsuitability of pendimethalin at high application rates in the production of sesame. This agrees with the findings of Grichar et al. (2001a) that pendimethalin at higher rates reduced sesame plant stands by 8 - 98% when compared with untreated check.

Although hoe-weeding resulted in the highest yield and percentage yield increase over weedy check, it recorded the highest cost of production (₦55, 574.07), followed by pendimethalin at 2.5 kg a.i./ha (₦53, 44.44) and pendimethalin at 2.0 Kg a.i./ha (50, 962.93), while the lowest cost of production (₦35, 574.07), was observed under weedy check, diuron at 0.5 kg a.i./ha (₦37, 648.15), diuron at 1.0 kg a.i./ha (₦38, 513.89), and diuron at 1.5 kg a.i./ha (39, 384.26) compared with the other treatments in 2006. Similarly in 2007, the highest cost of production (₦55, 574.07) was recorded under hoe-weeding, pendimethalin at 2.5 kg a.i. /ha (₦53, 740.74) and pendimethalin at 2.0 kg a.i. /ha (50, 962.93), while the lowest (₦35, 574.07), (₦37, 994.44) and (₦38, 810.19) were recorded under weedy check, diuron at 0.5 and 1.0 kg a.i. /ha respectively compared with the other treatments. These results show the advantage of herbicide application over hoe-weeding in the reduction of cost of production of sesame. This agrees with the finding of Nazeer et al. (2004) that a very lucrative cost-effective gain was recorded due to the application of Buctril-M on wheat and that the gain in yield from the hand weeded plots was nullified with the additional cost of weeding.

Despite its high cost of production, hoe-weeding in 2006, resulted in the highest income (₦77, 981.48), followed by butachlor at 2.0 kg a.i. /ha and diuron at 1.0 kg a.i./ha (₦77, 745.37), diuron at 1.5 kg a.i./ha (₦74, 703.70), butachlor at 1.5 kg a.i./ha (₦69,606.48) and

**Table 1.** Economic assessment of the use of different pre-emergence herbicides on sesame production in Maiduguri in 2006 and 2007.

| Herbicide | Rates kg/ha | Yield (kg/ha) | | %yield increase over check | | Total cost of production per hectare (₦) | | Selling price /ha(Income) per hectare (₦) | | Profit or loss/ha (₦) | | Cost benefit ratio | |
|---|---|---|---|---|---|---|---|---|---|---|---|---|---|
| | | 2006 | 2007 | 2006 | 2007 | 2006 | 2007 | 2006 | 2007 | 2006 | 2007 | 2006 | 2007 |
| Butachlor | 0.5 | 105.2 | 210.9 | -87.5 | 5.7 | 38,976.85 | 39,273.15 | 24,750.00 | 49,550.93 | -14,226.85 | 10,277.78 | 1:1.575 | 1:0.793 |
| " | 1.0 | 128.0 | 392.7 | -54.0 | 49.4 | 41,175.93 | 41,472.22 | 30,129.63 | 92,393.52 | -11,046.30 | 50,921.30 | 1:1.367 | 1:0.449 |
| " | 1.5 | 295.9 | 443.4 | 33.3 | 55.2 | 43,375.00 | 43,671.30 | 69,606.48 | 104,324.07 | 26,231.48 | 60,652.77 | 1:0.623 | 1:0.419 |
| " | 2.0 | 330.4 | 464.2 | 40.3 | 57.2 | 45,574.07 | 45,870.37 | 77,745.37 | 109,199.07 | 32,171.30 | 63,328.70 | 1:0.586 | 1:0.420 |
| " | 2.5 | 182.7 | 401.3 | -8.0 | 50.5 | 47,333.33 | 47,629.63 | 42,986.11 | 94,412.04 | -4,347.20 | 46,782.41 | 1:1.101 | 1:0.504 |
| Metolachlor | 0.5 | 202.7 | 311.8 | 2.5 | 36.3 | 38,444.44 | 38,740.74 | 47,592.59 | 73,370.37 | 9,148.15 | 34,629.63 | 1.0.808 | 1:0.528 |
| " | 1.0 | 263.1 | 456.4 | 25.0 | 56.5 | 39,555.55 | 39,851.85 | 61,916.67 | 107,384.26 | 22,361.12 | 67,532.41 | 1:0.639 | 1:0.371 |
| " | 1.5 | 286.6 | 529.7 | 31.2 | 62.5 | 41,222.22 | 41,518.52 | 67,495.37 | 124,634.26 | 25,162.04 | 83,115.74 | 1:0.611 | 1:0.333 |
| " | 2.0 | 195.2 | 225.2 | -1.0 | 11.8 | 42,333.33 | 42,629.63 | 45,935.20 | 52,986.11 | 3,601.87 | 10,356.48 | 1:0.922 | 1:0.805 |
| " | 2.5 | 189.4 | 189.6 | -4.2 | 41.2 | 44,000.00 | 44,296.30 | 44,564.81 | 79,500.00 | 564.81 | 35,206.70 | 1:0.987 | 1:0.557 |
| Diuron | 0.5 | 239.6 | 227.0 | 17.7 | 12.5 | 37,648.15 | 37,944.44 | 56,370.37 | 53,398.15 | 18,722.22 | 15,453.71 | 1:0.668 | 1:0.711 |
| " | 1.0 | 330.4 | 374.1 | 40.3 | 46.9 | 38,513.89 | 38,810.19 | 77,754.37 | 88,027.78 | 39,240.74 | 49,217.59 | 1:0.495 | 1:0.440 |
| " | 1.5 | 317.5 | 302.1 | 37.9 | 34.3 | 39,384.26 | 39,680.56 | 74,703.70 | 71,064.81 | 35,319.44 | 31,384.25 | 1:0.527 | 1:0.558 |
| " | 2.0 | 68.7 | 207.6 | -186.9 | 4.3 | 40,250.00 | 40,546.30 | 16,175.93 | 48,833.33 | -24,074.07 | 8,267.03 | 1:2.488 | 1:0.830 |
| " | 2.5 | 115.0 | 201.1 | -71.8 | -11.4 | 41,129.63 | 41,425.93 | 27,023.15 | 47,314.81 | -14,106.48 | 5,888.88 | 1:1.522 | 1:0.876 |
| Pendimethalin | 0.5 | 214.4 | 430.0 | 8.0 | 53.8 | 40,250.00 | 40,546.30 | 50,449.07 | 101,175.93 | 10,199.07 | 60,629.63 | 1:0.798 | 1:0.400 |
| " | 1.0 | 110.0 | 64.1 | -79.3 | -209.8 | 43,722.22 | 44,018.51 | 25,893.52 | 15,074.01 | -17,828.70 | -28,944.50 | 1:1.689 | 1:2.920 |
| " | 1.5 | 66.0 | 7.9 | -198.8 | -2,421.0 | 47,194.44 | 47,490.74 | 15,532.41 | 1,851.85 | -31,662.03 | -45,638.89 | 1:3.038 | 1:25.645 |
| " | 2.0 | 19.9 | 1.0 | -893.2 | -19,753.2 | 50,666.67 | 50,962.93 | 4,671.30 | 236.11 | -45,995.37 | -50,726.82 | 1:10.846 | 1:215.843 |
| " | 2.5 | 12.1 | 1.0 | -1532.6 | -19,753.2 | 53,444.44 | 53,740.74 | 2,870.37 | 4671.30 | -50,574.07 | -49,069.44 | 1:18.619 | 1:227.608 |
| Weeding at 3 and 6 WAS | | 331.4 | 543.7 | 40.5 | 63.5 | 55,574.07 | 55,574.07 | 77,981.48 | 127,939.81 | 22,407.41 | 72,365.74 | 1:0.713 | 1:0.432 |
| Weedy check | | 197.3 | 198.5 | - | - | 35,574.07 | 35,574.07 | 46,416.67 | 46,708.67 | 10,842.60 | 11,134.26 | 1:0.766 | 1:0.769 |

*Cost-benefit ratio is based on N235.29 per kg, the prevailing cost of sesame in Maiduguri.

metolachlor at 1.5 kg a.i./ha (₦67,495.37), compared to the rest of the treatments. Similarly in 2007, the income from sesame was highest (₦127, 939.81) under hoe-weeding, followed by metolachlor at 1.5 kg a.i./ha (₦124, 634.26), and lowest (₦235.29) under pendimethalin at 2.0 and 2.5 kg a.i./ha.

In 2006, the highest profit (₦39,240.74) was observed under diuron at 1.0 kg a.i./ha, followed by diuron at 1.5 kg a.i./ha (₦35,319.44), butachlor at 2.0 kg a.i./ha (₦32,171.30), butachlor at 1.5 kg a.i./ha (₦26,231.48) and metolachlor at 1.5 kg a.i./ha (₦25,162.04) compared with hoe-weeding and lowest profit (-₦50, 574.07) was recorded under pendimethalin at 2.5 kg a.i./ha. However, in 2007, the highest profit (₦83, 115.74) was observed under metolachlor at 1.5 kg a.i./ha, followed by hoe-weeding (₦72, 365.74) compared to the rest of the treatments and lowest (-₦50,726.82) and (-₦49,069.44) under pendimethalin at 2.0 and 2.5 kg a.i./ha respectively. The results show that the use of herbicide is more profitable in the production of sesame than hoe-weeding. This is similar to the findings of Warade et al. (1995), Saikia et al. (1997), Kehinde (2002) and Khan et al. (2005) who reported optimum and highest net returns obtained with the application of herbicides compared with the other methods of weed control treatment. It also confirms the report of Shrock and Monaco, 1980; Okereke, 1983; Usoroh, 1983; Sinha and Lagoke, 1984; Ogungbile and Lagoke, 1986; Adigun et al., 1993 that herbicide use is more profitable than hoe-weeding in the production of various crops in Nigeria. However, the above result is contrary to the findings of Chinaka and Enyinnia (1980) who reported highest profit (N1, 528.20) with manual weeding and lowest (N398.10) with the use of herbicides at Umudike in the eastern part of Nigeria.

The cost benefit ratio in 2006 was highest (1:18.619) and (1:10.846) under pendimethalin at 2.5 and 2.0 kg a.i./ha respectively and lowest (1:0.586) under diuron at 1.0 kg a.i./ha, diuron at 1.5 kg a.i./ha (1:0.522), butachlor at 2.0 kg a.i./ha (1:0.586), metolachlor at 1.5 kg a.i./ha (1:0.611), butachlor at 1.5 kg a.i./ha (1:0.623), metolachlor at 1.0 kg a.i./ha (1:0.639) and diuron at 0.5 kg a.i./ha (1:0.688), compared to hoe-weeding and other treatments. Similarly, in 2007, the highest cost: benefit ratio (1:227.608) and (1:215.843) was observed under pendimethalin at 2.5 and 2.0 kg a.i./ha respectively and the lowest (1:0.333) under metolachlor at 1.5 kg a.i./ha, pendimethalin at 0.5 kg a.i./ha (1:0.400), butachlor at 1.5 kg a.i./ha (1:0.419) and at 2.0 kg a.i./ha (1:0.420), respectively compared to hoe-weeding. The lower cost: benefit ratio recorded by the above herbicide rates compared to hoe-weeding in the two years of study signify that the use of herbicides is more beneficial and economical than hoe-weeding in the production of sesame in the Savanna Zone of Nigeria. This is similar to the findings of Nazeer et al. (2005) and Khan et al. (2005) who reported optimum cost: benefit ratio with the application of herbicides compared to weedy check and other weed control methods.

From this study, it can be concluded that hoe-weeding produced the highest sesame grain yield closely followed by metolachlor at 1.5 kg a.i. /ha and butachlor at 2.0 kg a.i./ha in the two years of study. However, hoe-weeding was more expensive and resulted in the highest cost of production of sesame than all the herbicide treatments and weedy check. Metolachlor at 1.5 and 1.0 kg a.i./ha, butachlor at 1.5 and 2.0 kg a.i./ha, diuron at 0.5 - 1.5 kg a.i./ha had lower cost: benefit ratio compared to hoe-weeding in the two years of study. Therefore for lower cost, higher yield, and better economic returns, metolachlor at 1.5 kg a.i./ha and butachlor at 2.0 kg a.i./ha are recommended to the farmers in the Sudan Savanna Zone of Nigeria as an alternative to two hoe-weedings.

## REFERENCES


Adedzua DK, Ortese JT (2004). Participatory approach to sustainable weed management in traditional farming systerm. Nig. J. Weed Sci. 17: 101-118.

Adigun JA, Lagoke STO, Kumar V, Erinle ID (1993). Weed management studies in transplanted tomato in the Nigerian savanna. Samaru J. Agric. Res. 10: 29 – 39.

Akobundu IO (1987). Weed science in the tropics. John Wiley and Sons, New York p. 522.

Akobundu IO (1997). Weed science development in Nigeria yesterday, today and tomorrow. Nig. J. Weed Sci. 10: 61-70.

Anonymous (1994). Weed control recommendations for Nigeria. Series 3, sponsored by Department of Agriculture, Federal Ministry of Agriculture, Nigeria p. 1.

Balyan RS (1993). Integrated Weed management in oil seed crops in India. Proc. Int. Soc. Weed Sci. pp. 317-323.

Chinaka CC, Enyinnia (1980). Profit derived from the use of herbicides." Proc.10$^{th}$ and 11$^{th}$ Ann. Conf. Weed Sci. Soc. Nig. pp. 112-118.

Fadayomi RO (1991). Weed management in Nigerian agriculture in the 90's. The chemical weed control Option. Nig. J. Weed Sci. 14: 63-66.

Grichar WJ, Sestack DC, Brewer KD, Besler BA, Stichler CR, Smith DT (2001a). Sesame (*Sesamum indicum L.*) tolerance and weed control with soil applied herbicides Crop Protection 20: 389-394.

Ikuenobe CE, Fadayomi RO, Adeosun JO, Gworgwor NA, Melifonwu AA, Ayeni AO (2005). State of adoption of improved weed control technologies by farmers in three Agro- Ecological Zones of Nigeria. Nig. J. Weed Sci. 18: 1-19.

Imoloame EO (2009). Effect of pre-and post-emergengency herbicides on weed infestation and productivity of sesame (*Sesamum indicum* L.) in the Sudan Savana Zone of Nigeria. Unpublished Ph.D Thesis, University of Maiduguri, Borno State. Nigeria p. 135.

Ingawa SA, Misari SM,Kaigama BK (1986). Sesame production and marketing in Nigeria. A report of an exploratory investigation. Institute for Agricultural Research Cropping Meeting. In: Legumes and Oilseeds Research, Ahmed Bello University, Zaria, Nigeria.

Jain GL, Singhi SM, Salin MP, Sharma GL (1985). Sesame production in Rajasthan. Constraints and opportunities. In: Srivastava HC, Bhaskaran S, Vatsya B, Menon KKG (Eds). Oil seed production constraint and opportunities. New Delhi, Bombay, Calcutta pp. 181-198.

Joshua SD, Gworgwor NA (2001). Revised. Economic assessment of chemical weed control in cereal-legume inter crop in the Savanna Zone of Nigeria. A Review. Annals of Borno. 17/18 (2000-2001): 247-256.

Khan BA, Bakhtiar G, Muhammad S, Zahid H (2005). Efficacy of different herbicides for controlling weeds in onions in higher altitudes. Weed Science Society of Pakistan 11(1 – 2): 61 – 68.

Kehinde JK (2002). Influence of seed rate and weed management practices on weed control and Performance of upland rice. Nigerian J. Weed Sci. 15: 1-6.

Klingman ML, Martin JA (1954). Present status of sesame breeding in the United States. Agron. J. 46: 24-27.

Nazeer HS, Gul H, Sad UR, Nazir A, Fazle S (2004). Weed management in wheat on Farmers' fields of DASP command Area. Weed Science Society of Pakistan 10(1-2): 25-32.

Ogungbile AO, Ndahi W, Lagoke STO (1982). Economic evaluation of herbicide use in maize: production In: Proceeding of the 11th Annual Conference on Weed Science Society of Nigeria. Eds. Olunuga,I.O, Akobondu and A. Ayemi. Ahmadu Bello University, Zaria, Nigeria p. 45.

Ogungbile AO, Lagoke STO (1986). On-farm evaluation of the economics of chemical weed control in oxen-mechanized maize production in Nigeria Savanna. Tropical Pest Management 32: 273.

Okereke UO (1983). Weed control in transplanted dry and rainy season tomato (*Lycorpersicon esculentum* Mill) crops with gramoxone (Paraquat) and sensor. Paper presented at the 5th Annual Conference of Horticultural Society of Nigeria University of Nsukka, Nov. 6-9, 1985.

Philips TA (1977). Sesame. In: An agricultural notebook. New Ed. Longman Group Ltd. London pp. 61-63.

Saikia S, Saikia A, Shadeque A, Gogoi S (1997). Integrated weed management in onion (Allium cepa L.). Dept. Hort. Assam Agric. Univ. Jorhat, Assam, India. Annals of Biology Ludhiana 13: 139 – 143.

Sharma SM (1985). Constraints and opportunities for increasing the productivity and production of sesame in India. In: Srivastava HC, Bhaskaran S, Vatsya B, Menon KKG (Eds). Oil seed production constraints and opportunities. New Delhi, Bombay, Calcutta. pp:165-175.

Sinha TD, Lagoke STO, Olukosi JO (1982). AR evaluation of the lost of various weed control treatments in irrigated Onior and tomatoes. In: proceedings of the 5th Annual conference of Horticultural society of Nigeria, Ahmadu Bello University, Zaira, November pp. 107-177.

Sinha TD, Lagoke STO (1984). Weed control in irrigated tomato (Lycopersicon esculentum.Mill) in Northern Nigeria. Trop. Pest Manage. 30: 18-25.

Sinha D, Dangor JC, Gangwar B (1992). Infestation by weeds and their management in oil seed crops A review. Agric. Rev. 13: 163-175.

Shrock WA, Monaco TJ (1980). Recent development in weed management programme for tomatoes. Proceeding of 33rd Annual Meeting of Southern Weed Science Society 33: 90-92.

Usoroh NJ (1983). Field screening of herbicides for weed control in tomato (lycopesicon esculentum Mill). Paper presented at the 13th Annual Conference of the Weed Science Society of Nigeria. National Cereal Research Institute Ibadan 5 Dec. 1983.

Van Rheenen HA (1973). Major problems of growing sesame in Nigeria. Communications Agriculture. University of Wagenigen, Holland p. 30.

Warade SD, Desale SB, Shinde KG (1995). Effect of weedicides on weeds. Mahatma Phule Krishi Vidyapeeth, Ind. J. Maharashtra Agric. Univ. 20(3): 462 – 463.

Yermanos DM, Hemstreet, saleeb W, Huxzar C (1972). Oil contents composition of seed in the world collection of sesame introduction J. Agric. Chem. Soc. 49(1): 20-22.

# Performance evaluation of a mixed-mode solar dryer for evaporating moisture in beans

**Ahmed Abed Gatea**

Agricultural Mechanization College of Agriculture, University of Baghdad, Iraq. E-mail: ahmedabd192000@yahoo.com.

**This research includes the design and manufacture of a solar dryer and its performance test on bean. Solar dryer consists of three parts; 1) the air heating section (solar collector), 2) two fins and 3) the drying chamber section with four trays. The drying test was conducted after setting the drying air speed at 1.5 m/s at the drying chamber. The drying process continued until the total beans weight was (100 kg). The highest and the lowest solar collector efficiencies of 61.82 and 45.40% were obtained at 11.00 and 9.00 h, respectively. Higher moisture evaporating from the beans was achieved in the first tray, as moisture content was reduced from 60 to 8% during the 6 h compared with the other three trays.**

**Key words:** Solar dryer, broad–bean, performance, temperature distribution.

## INTRODUCTION

Drying crops by solar energy is of great economic importance the world over, especially in Iraq where most of the crops and grain harvests are lost to fungal and microbial attacks. These wastages could be easily prevented by proper drying, which enhances storage of crops and grains over long periods of time. Iraq lies within the equator and is blessed with abundant solar energy all the year round. This solar energy can easily be harnessed by a proper design of solar dryers for crop drying. This method of drying requires the transfer of both heat and water vapour (Forson et al., 2007). Most of our crops and grain are harvested during the peak period of rainy season and so preservation proves difficult and most of these grains and crops get spoilt. This results in the crops not lasting the year, resulting in subsequent hunger and malnutrition. Solar drying may be classified into direct, indirect and mixed-modes. In direct solar dryers, the air heater contains the grains and solar energy passes through a transparent cover and is absorbed by the grains. Essentially, the heat required for drying is provided by radiation to the upper layers and subsequent conduction into the grain bed. In indirect dryers, solar energy is collected in a separate solar collector (air heater) and the heated air then passes through the grain bed, while in the mixed mode type of dryer, the heated air from a separate solar collector is passed through a grain bed and at the same time, the drying cabinet absorbs solar energy directly through the transparent walls or roof (Gatea, 2010). These crops can be preserved and stored, so that they can be of economic importance both to the farmers and the entire populace. Rural farmers do this by open-air drying. This practice in the rural areas has some obvious disadvantages (Henry et al., 1999). This method is unhygienic since the crops are easily contaminated by animal droppings and consequent infestation by fungi and bacteria. Human health is thus endangered as a result of food poisoning. This method also prolongs drying and may result in the deterioration of the quality of the crops. Moreover, more labour is involved as the crops are being moved frequently in and out during the day and night and from rain. They are also watched in order to prevent physical attacks birds and other animals. It is a well known fact that in rural areas, conventional sources of energy like petrol and electricity are either totally absent or are not readily available to develop active dryers, which have higher rate of performance. A low temperature passive solar dryer has therefore been developed, which will be appropriate for crops and grains during the low temperature and high relative humidity periods of the year. The obvious advantage of low temperature drying is that, it enables crops to be dried without cracking and hence minimizes the exposure of the crops to fungal and bacterial infestation and wastage, this is also suitable for bulk drying for long-term storage (Forson et al., 2007). The objective of this research was to fabricate the solar dryer using indigenous materials to assess their efficacy in drying beans. The fabricated driers utilized no chimney and both direct and indirect modes in the same unit.

## DESIGN CALCULATIONS

### Declination (δ)

This is the angle between the sun's direction and the equatorial plane and is given by Forson et al. (2007) as,

$$\delta = 23.45 \sin [0.9863 (284 + n)] \quad (1)$$

Where (n) is the day in the year which varies from n = 1 to n = 365.

### Optimum collector slope (β)

The optimum collector slope, β, is determined from

$$\beta = \delta + \varphi \quad (2)$$

Where (δ) is the angle of declination for Baghdad, Iraq and (φ) is the latitude of the location.

### Collector efficiency (η)

This is computed from,

$$\eta = \frac{\rho V c_p \Delta T}{A I_c} \quad (3)$$

Where (ρ) is the density of air (kg/m$^3$), ($I_c$) is the insolation on the collector, (ΔT) is the temperature elevation, ($c_p$) is the specific heat capacity of air at constant pressure (J/kg K), (V) is the volumetric flow rate (m$^3$/s), and (A) is the effective area of the collector facing the sun (m$^3$).

### Dryer efficiency ($\eta_d$)

This is given by Henry et al. (1999) as,

$$\eta_d = \frac{ML}{I_c A t} \quad (4)$$

where (L) is the latent heat of vaporization of water, (M) is the mass of the crop, and (t) is the time of drying.

### Heat energy Q needed for crop drying at moderate temperature

This is given by

$$Q = M_w L = \rho c_p V (T_a - T_b) \quad (5)$$

where L = latent heat of vaporization of water, $M_w$ = mass of crop before drying, ρ = density of water, $T_a$ = ambient temperature, $T_b$ = dryer temperature.

### Moisture content (M.C.)

The moisture content is given as:

$$MC(\%) = \left( \frac{M_i - M_f}{M_i} \right) \times 100\% \quad (6)$$

where $M_i$ = mass of sample before drying and $M_f$ = mass of sample after drying.

### Moisture loss (ML)

The moisture loss is given as:

$$ML = (M_i - M_f) \text{ (g)} \quad (7)$$

Here, $M_i$ is the mass of the sample before drying and $M_f$ is the mass of the sample after.

### Length of the day

The length of the day is given by Henry et al. (1999) as:

$$N = (2/15) \cos^{-1}(-\tan\varphi \tan\delta) \quad (8)$$

## THE EXPERIMENTAL SET-UP

The solar dryer with box-type absorber collector was constructed using the materials that are easily obtainable from the local market. Figure 1 shows a section of the solar crop dryer. The dryer has four main features namely: the box-type absorber solar air collector, the drying chamber, the drying rack and two fines electric. The

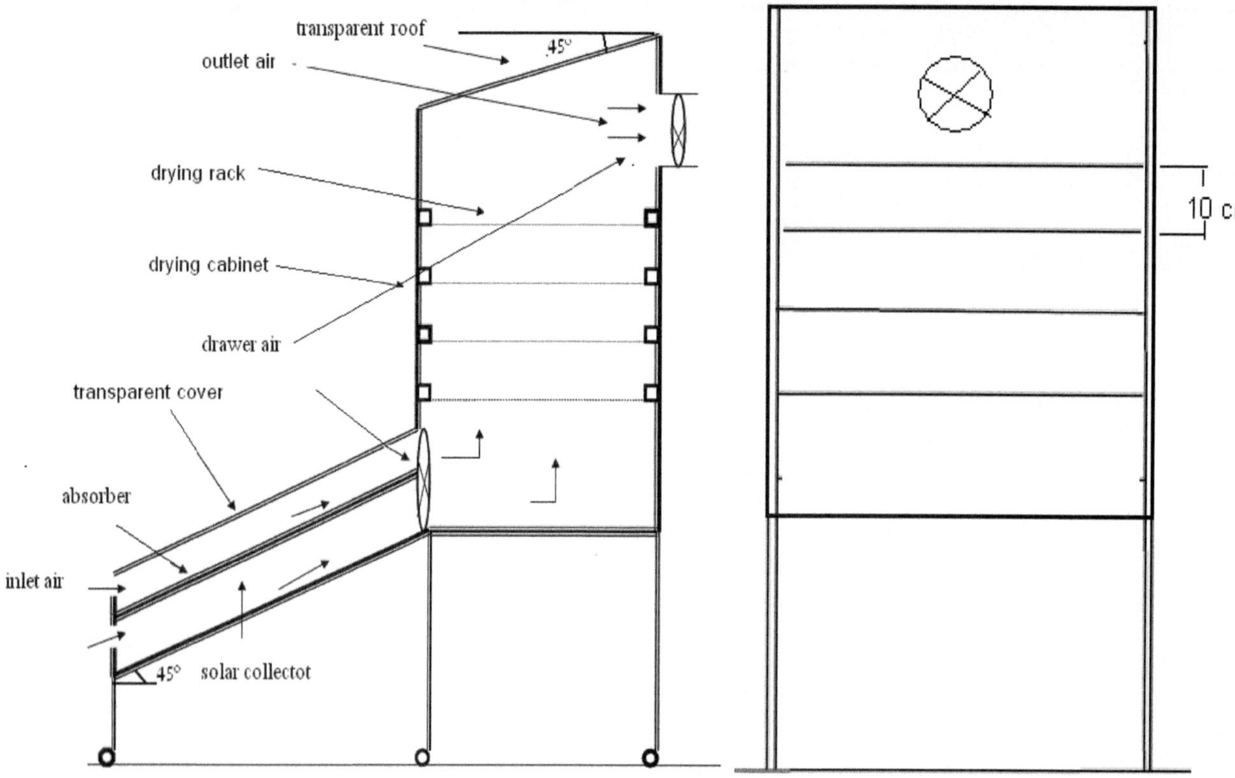


**Figure 1.** Shows section of the mixed-mode solar dryer.

parameters involved in the design of the solar crop-drying systems are shown in Figure 2.

**Collector (solar air heater)**

The heat absorber (inner box) of the solar air heater was constructed using 1 mm thick galvanized plate, painted black, the surface facing sunlight was painted with black paint containing (5%) black chromium powder to increase its absorbing capability (Duffie and Beckman, 1974). The solar collector was insulated with rock wool of about 5 cm thickness and thermal conductivity of 0.04 $Wm^{-1}K^{-1}$ on all sides. The solar collector assembly consists of air flow channel enclosed by transparent cover (glazing). The glazing is a single layer of 4 mm thick transparent glass sheet. It has a surface area of 0.82 by 1.20 cm and of transmittance above 0.86 (Figure 3).

**The drying cabinet and drying racks**

The designing of the drying chamber depends on many factors such as the product to be dried, the required temperature and velocity of the air to dry food material, the quantity of the dried product and the relative humidity of the air passing over the food material. The drying chamber houses four drying racks, between a tray and another tray is 10 cm as shown in Figure 1. Four trays of dimension (0.75 x 0.30 x 0.08 m) were fabricated and stacked uniformly/evenly at distances (0.02 m) apart, for placing of material to be dried. The tray was made from an aluminum wire mesh (0.003 X 0.003 m in size) attached to it. Metal handles (0.076 m) were attached on each tray for ease of handling and sliding the trays inside the chamber through the produce to be dried. The drying chamber was also lined with foam insulation material 5 cm thick to prevent loss of heat (Figure 4).

**Drying mechanism**

In the process of drying, heat is necessary to evaporate moisture from the material and a flow of air helps in carrying away the evaporated moisture. There are two basic mechanisms involved in the drying process: 1) the migration of moisture from the interior of an individual material to the surface, and 2) the evaporation of moisture from the surface to the surrounding air (Youcef-Ali et al., 2001). The drying of a product is a complex heat and mass transfer process which depends on external variables such as temperature, humidity and velocity of the air stream and internal variables which depend on parameters like surface characteristics (rough or smooth surface), chemical composition (sugars, starches, etc.), physical structure (porosity, density, etc.), and size and shape of product.

**The experimental measurements**

Two thermocouples have been positioned to measure the air temperature at the inlet and outlet portion of the air heater with accuracy of ±1. Another four thermocouples have been placed at trays 1, 2, 3 and 4 in order to measure the temperature of trays. Ambient temperature was also recorded during the course of experiments with the help of mercury thermometer. The experiment was conducted at Baghdad / AL.Zafaranya (latitude 33.27° northwards) and the orientation of the solar collector has been fixed towards the south direction, inclined at an angle of 45°. The solar radiation on a horizontal and inclined surface has been recorded

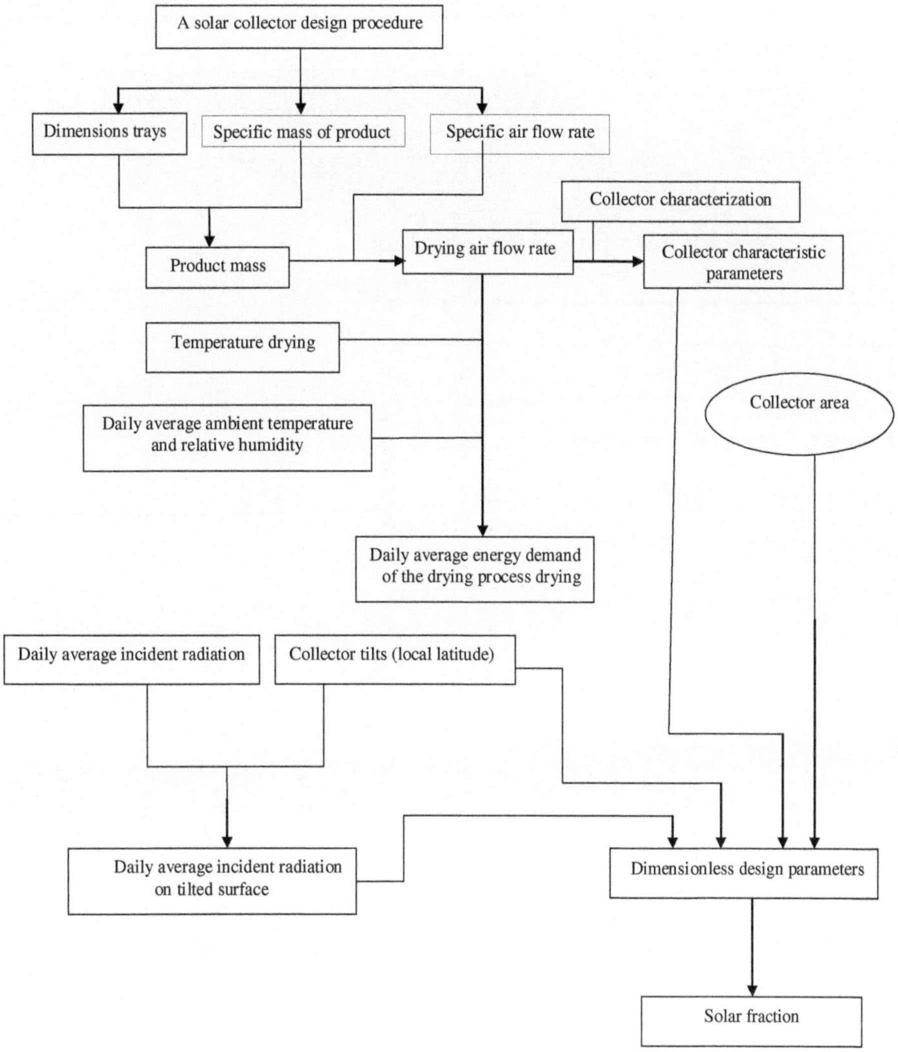


**Figure 2.** Schematic diagram of a mixed-mode solar crop drying system.

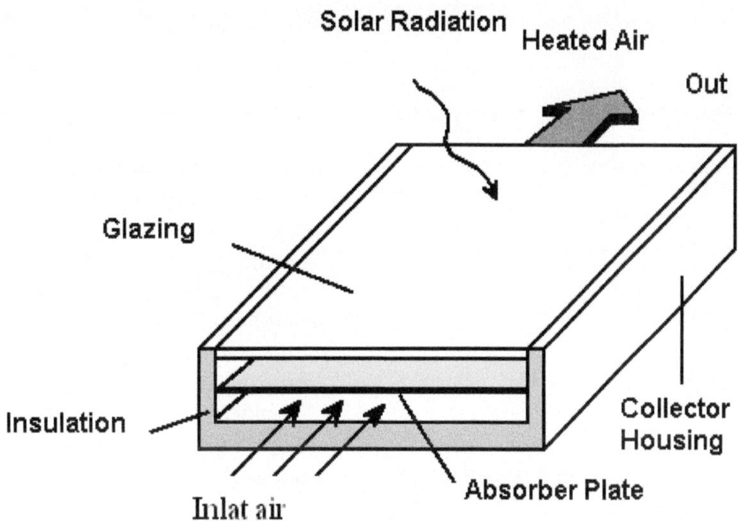


**Figure 3.** Typical solar collector for air heating.

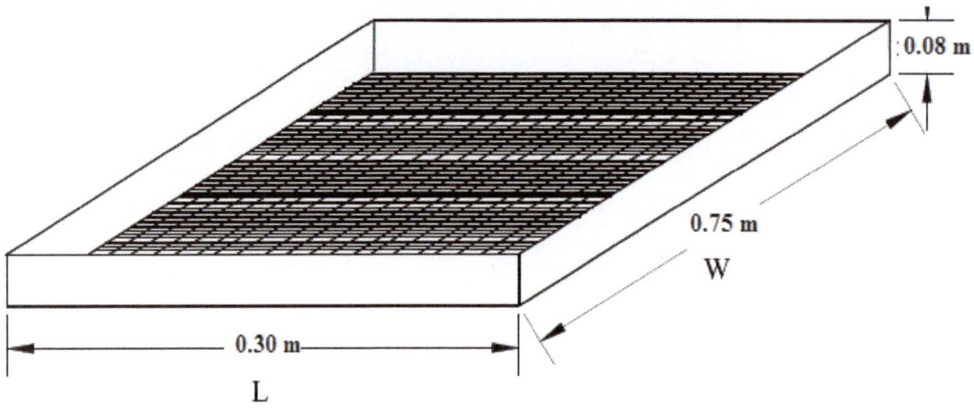


**Figure 4.** Dimension of trays used in the drying chamber.

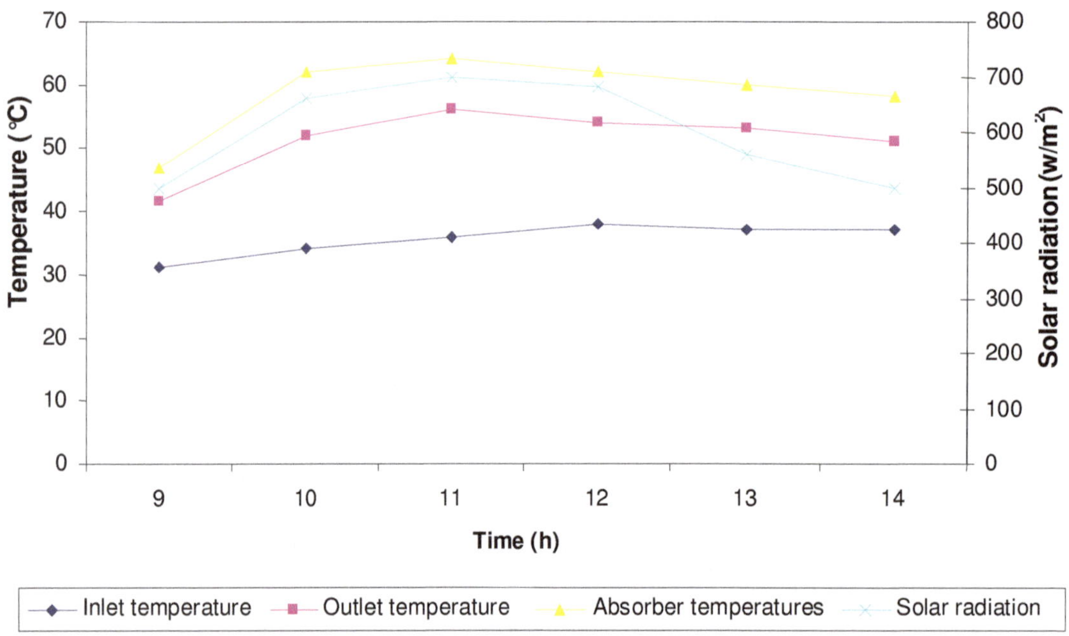


**Figure 5.** Variation of and temperatures of outlet air of collector, absorber, ambient, and radiation during test on June 26, 2009.

during the course of experiments with the help of solar meter. Air velocity or flow rate of air at the inlet position of the drying chamber was measured by anemometer.

## RESULTS AND DISCUSSION

### Solar collector performance

From results of the experiments, the diurnal variation of temperatures of the solar collector outlet air, absorber, ambient, and solar radiation were plotted. One typical day of June is shown in Figure 5. It is observed that the rise in air temperature due to the generated air flow rate in the collector is sufficient for the purpose of most agricultural products drying. The daily solar radiation was relatively high with an average of 600 W/m$^2$. For the inlet air temperature of 38°C, the maximum air temperature at the dryer inlet was recorded as 56°C at a solar irradiance level of 700 W/m$^2$. The mass flow rate of the drying air in the thermo-sphere mode of the collector depends on the prevailing wind conditions, ambient air temperature, incident solar radiation and the collector design. It was observed that, when the wind velocity was more or less uniform throughout the day, the air velocity in the collector shows definite dependence on the stack temperature difference between ambient and collector outlet air.

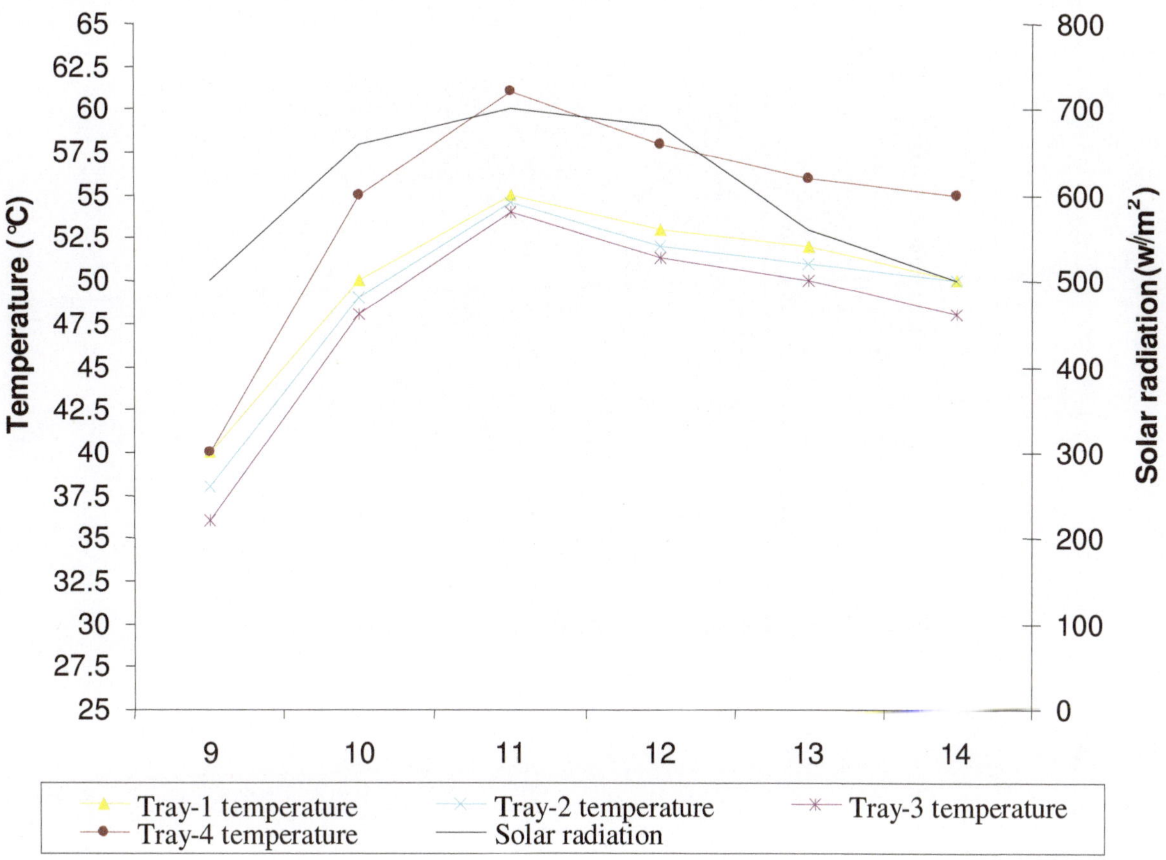


**Figure 6.** Variation of drying air temperatures on each tray, ambient temperature, and radiation during the unloaded experiment with variation of heat sources (June 27, 2009).

## Performance of solar dryer

The performance of the solar dryer depends on several factors, the solar radiation, inlet air temperature of solar dryer and the dryer design factors (Figure 6) note that access to the highest temperature required for drying when tray 4 is from 61°C and is at the highest intensity of the radiation 700 W/m$^2$ at 11. There were no significant differences in temperature at the trays (1, 2 and 3) with the differences ranging from (2 to 5°C).

## Moisture content variation

As shown in Figure 7, the moisture content was reduced from 60% on dry basis to 8% on dry basis through 6 h. Been delivering moisture content of the beans on (trays 1, 2, 3 and 4) to (8, 11, 14 and 18%) respectively.

## Solar collector efficiency

The efficiency of the solar collector depends on the air flow rate and the difference in temperature ($t_o$- $t_i$) and radiation intensity as in the Equation 3 of Figure 8, was obtained at the highest efficiency of the solar collector the amount of 61.82% at midday, and less efficient were obtained from 45.4% at 9 am.

## Conclusions

The study revealed the following conclusions:

1. The maximum drying temperature through the drying time was 61°C.
2. The amount of the moisture content was decreased to 11% in 6 h which depends on drying temperature and the time of exposure to sun radiation. This drying rate was an adequate rate of moisture extraction. The main factor in controlling the drying rate was found to be the drying air temperature and air velocity.
3. The drying time depends on the thermal and physical properties of the biogases pulp and also on the climatic conditions (solar radiation, relative humidity, ambient temperature and wind speed).
4. The highest efficiency solar collector was 61.82% at the 11 o'clock hour and the lowest efficiency was 45.40%

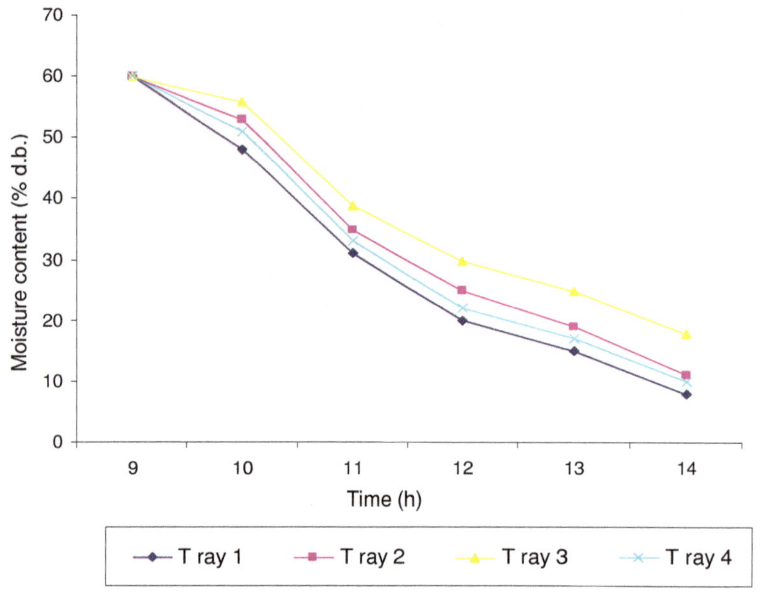


**Figure 7.** Decreasing of moisture content of broad –bean for each tray of the dryer.

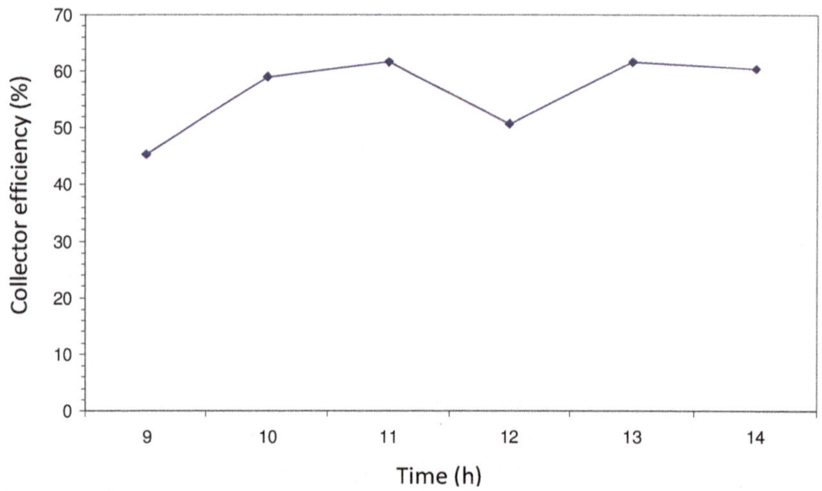


**Figure 8.** Variation of solar collector efficiency.

at the 9 o'clock hour.

## REFERENCES

Duffie JA, Beckman WA (1974). Solar Energy: Thermal Processes. John Wiley Inc., New York, NY.

Henry TS, Price WE (1999). A diffusion model for prune dehydration. J. Food Eng., 42: 167-172.

Forson FK, Nazha MAA, Akuffo FO, Rajakaruna H (2007). Design of mixed-mode natural conbection solar crop dryers: Application of principles and rules of thumb. Renew. Energy, 32: 2306-2319.

Gatea AA (2010). Design, construction and performance evaluation of solar maize dryer. J.Agri. Biotechnol. Sust.Dev., 2(3): 039-046.

Youcef-Ali S, Messaoudi H, Desmons JY, Abene A, Le Ray M (2001). Determination of the average coefficient of internal moisture transfer during the drying of a thin bed of potato slices. J. Food Eng., 48(2): 95-101.

# Investigating Fe and Zn foliar application on yield and its components of soybean (*Glycine max* (L) Merr.) at different growth stages

**A. R. Heidarian[1], H. Kord[1], Khodadad Mostafavi[2], Amir Parviz Lak[2], Faezeh Amini Mashhadi[3]**

[1]M.Sc. Graduate, Department of Agronomy, Islamic Azad University, Mashhad Branch, Iran.
[2]Department of Agronomy and Plant Breeding, Karaj Branch, Islamic Azad University, Karaj, Iran.
[3]Department of Statistics, Khayyam Institute of High Education, Mashhad, Iran.

**In order to investigate Zn, Fe and Zn+Fe foliar application effects on soybean (*Glycine max* L.) yield and its components (number of pods per plant, number of seeds per pod and 1000 grain weight), a field experiment was conducted at Minodasht, Iran during 2009 and 2010 growth season. The experiment was arranged in split plot of a randomized complete block design and replicated three times. Parameters measured were grain yield, harvest index, first pod height, number of pods per plant, number of seeds per pod and 1000 grain weight. Results showed significant effect of Zn+Fe treatment on grain yield, number of pods per plant ($p < 0.05$) and 1000 grain weight ($p < 0.01$). The time of foliar application on number of pods per plant ($p < 0.05$) and 1000 grain weight ($p < 0.01$) was also significant. In general, the highest yield was produced by Zn+Fe combination treatment. There was a significant and positive correlation between grain yield and its components. Leaf area index in 407 growth degree days at the beginning of pod set (6/6), crop growth rate in 272 growth degree day (GDD) at the beginning of flowering (39/52) and net assimilation rate in 272 GDD at 10-leaf stage (3 g $plant^{-1}$ $GDD^{-1}$) reached the highest point, respectively.**

**Key words:** *Glycine max* L., nutrients, seed production, growth indexes.

## INTRODUCTION

Soybean (*Glycine max* (L) Merr.) belongs to Papilionaceae, leguminous family and it is a herbaceous annual legume, normally erect, bushy and leafy that was cultivated in China for the first time which was ranked as an oilseed crop and it provides approximately 50% edible oil of the world (Akparobi, 2009). It has been recognized as a former crop plant since the origin of agriculture (Jandong et al. 2011). Soybean as a meal consists of important minerals, such as Ca, P and Fe. Due to the large amount of macro and micro nutrients, it has been considered as a nutritious food for human needs, livestock, industrial and medicinal purposes (Akparobi, 2009; Berglund, 2002). In Iran, soybean is cultivated after wheat and canola in rotation and soybean seed consists of 18 to 25% oil and 30 to 50% protein. Vahedi et al., (2010) and Salwa et al. (2011) stated that soybean is a crop that compensates shortage of oil and protein of other crops. Furthermore, it is a good source for height energy, protein and essential nutrients to human and animals. Soybean production has been increased to 36% since 2000 in the world. In 1961 to 1965, 28.6 million ton of soybean was produced and it was increased to 217.6 million ton per hectare in 2005 to 2007 (Masuda and Goldsmith, 2009).

Foliar spraying is a new method for crop feeding, which micronutrients in form of liquid are used into leaves (Nasiri et al., 2010). Foliar application of microelements is more beneficial than soil application. Since application rates are lesser as compared to soil application, same application could be obtained easily and crop reacts to nutrient application immediately (Zayed et al., 2011). Undoubtedly, soybean higher yield and quality as well as

*Corresponding author. E-mail: mostafavi@kiau.ac.ir.

its oil will be obtained by microelements foliar spraying (Vahedi, 2011). Foliar spraying of microelements is very helpful when the roots can not provide necessary nutrients (Kinaci and Gulmezoglu, 2007; Babaeian et al., 2011). Moreover, soil pollution would be a major problem by micronutrients soil application. As people are concerned about the environment and plant leaves uptake nutrients better than soil application, foliar spraying was created (Bozorgi et al., 2011). Crop roots are unable to absorb some important nutrients such as zinc, because of soil properties, such as high pH, lime or heavy texture, and in this situation, foliar spraying is better as compared to soil application (Kinaci and Gulmezoglu, 2007). Narimani et al. (2010) reported that microelements foliar application improve the effectiveness of macronutrients. It has been found that microelements foliar application is in the same level and even more influential as compared to soil application. It was suggested that micronutrients could be applied successfully to compensate shortage of those elements (Arif et al., 2006). These authors found that based on soil properties, foliar spraying could be effective 6 to 20 times as compared to soil application. Resistance to different stresses will be increased by foliar application of micronutrients (Ghasemian et al., 2010). Since in field situation, soil features and environmental factors which affect nutrients absorption are extremely changeable, foliar application could be an advantage for crop growth (SeifiNadergholi et al., 2011). Also, effectiveness of foliar spraying is higher and the cost of foliar application is lower as compared to soil application (Yassen et al., 2010).

Salwa et al. (2011) stated that microelements are defined substances that are crucial for crop growth; however, they are used in lower amounts as compared to macronutrients, such as N, P and K. They have a major role in cell division and development of meristematic tissues, photosynthesis, respiration and acceleration of plant maturity (Zeidan et al., 2010). One of the most important roles of micronutrients is keeping balanced crop physiology. Furthermore, these elements play vital roles in $CO_2$ flowing out, improvement in vitamin A and immune system activities (Narimani et al., 2010). Zinc plays a special role in synthesizing proteins, RNA and DNA (Kobraee et al., 2011). Zinc and iron take over different roles in crop, such as formation, partitioning and utilization of photosynthesis assimilates (Sawan et al., 2008). The major role of zinc element in crops is not clear Nasri et al. (2011). These authors found that although zinc is an important element which should be provided for crop growth, but it could be poisonous in large amount. As a matter of fact, the importance of zinc foliar application is due to being given to crop immediately (Alloway, 2003). Gul et al. (2011) claimed that profitability of micronutrients will be obtained in combination with macro elements, such as nitrogen and potassium. Ghasemian et al. (2010) declared that zinc element is essential in chlorophyll production and pollen function. Iron (Fe) contributes in very enzymatic activities, such as cytochromes, ferredoxine, superoxide dismutase (SOD), catalase (CAT), peroxidase and nitrate reductase. It was reported by Kobraee et al. (2011) that growth limitation, symbiosis, nodulation, photosynthesis, dry matter production and plant nutrient disorder were caused by the lack of zinc and iron. Furthermore, reactions of electron transportation are needed for zinc and iron.

Cakmack (2002) found that an important reason of zinc and iron deficiencies around the world is the consumption of cereals which have low amounts of Zn and Fe micronutrients and high levels of bicarbonate in irrigation water increases Fe shortage of soils. The main factors which affect the amount of zinc in soil are pH, carbonate content, organic matter, soil texture and interaction between zinc and other microelements, such as iron (Bukvić et al., 2003). Zinc and iron deficiency in soils could be a restricted factor of yield and extremely decrease crop yield quality (Salwa et al., 2011). Bozoglu et al. (2007) found that in 25 countries, zinc deficiency is a soil common problem. Babaeian et al. (2011) announced that zinc and iron deficiencies are common in 30 and 50% of soils, respectively in the world. These authors claimed that microelements deficiency is mainly due to their low amounts and soil features which hinder crop roots to obtain them. Vahedi (2011) stated that lack of zinc is a major problem in the world and shortage of zinc will reduce crop yield. Generally, dry matter production and its division into different parts of plant will be weakened by lack of micronutrients (Sawan et al., 2008). Lack of zinc in crop plants is due to the fact that this element cannot be solved in soils, and reduction of cell growth and development is one of the symptoms of Zn deficiency (Ghasemian et al., 2010). Kobraee et al. (2011) and Ghasemian et al. (2010) claimed that young leaves chlorosis and plant metabolism disorder will be caused by lack of iron, and in condition of iron stress, the absorption of Fe would be enhanced. Since iron is not soluble in soil, foliar application of this nutrient is necessary. Ai-Qing et al. (2011) and Sliman and Motto (1990) reported that there is an antagonistic interaction between zinc and iron in soybean which influences their absorption, partitioning and utilization. These authors announced that zinc affects absorption and translocation of iron and vice versa. The aim of this study was to investigate zinc and iron foliar application effects on soybean yield and its components (grain yield, number of pods per plant, number of seeds per pod and 1000 grain weight).

## MATERIALS AND METHODS

This study was conducted at Minodasht, Iran during 2009 and 2010 growth seasons. The experiment was arranged in split plot of a randomized complete block design and replicated three times. Main factor were three different stages (10 leaf stage, beginning of flowering and beginning of pod set), sub plots were (control, Zn 116

**Table 1.** Analysis of soil sample taken from 0 to 30 cm depth.

| Sample | Value |
|---|---|
| Clay (%) | 16 |
| Silt (%) | 78 |
| Sand (%) | 6 |
| Fe (ppm) | 7.68 |
| Zn (ppm) | 0.9 |
| Mn (ppm) | 7.0 |
| Cu (ppm) | 2.8 |
| Texture silty loam | - |
| pH | 7.7 |

**Table 2.** Analysis of variance of seed yield, harvest index and first pod height.

| Source of variation | df | Mean of square | | |
|---|---|---|---|---|
| | | Seed yield | Harvest index | First pod height |
| Time (T) | 2 | 188235.111$^{ns}$ | 0.001$^{ns}$ | 4.405$^{ns}$ |
| Fertilizer (F) | 3 | 2676456.4* | 0.003$^{ns}$ | 3.019$^{ns}$ |
| T × F | 6 | 581802.519$^{ns}$ | 0.003$^{ns}$ | 3.739$^{ns}$ |
| Error | 18 | 580269.370 | 0.003 | 2.798 |

**, *, ns: Significant at 1 and 5% probability levels and non significant, respectively.

ppm treatment, Fe 116 ppm treatment and Zn+Fe combination treatment). Minodasht (latitude 160 29 ' N, longitude 260 55 ' E) is situated at 200 m above sea level. Annual rainfall average is 600 mm and the climate is mild-semi humid. Annual temperature average is 11/9°C. Soybean cultivar used in this experiment was Sahar (Persian) which was sown on the 14th July, 2009. Analysis of soil sample taken from 0-30 cm depth explained in Table 1. Seeds were pretreated with *Rhizobium japonicum* and weed control was done manually. The trial was harvested after physiologic maturity. Each plot consisted of six rows with 6 m length and the row spacing was 50 cm, ten selected plants were used to take the data from each plot of each replication. Data were recorded for grain yield, first pod height, harvest index, number of pods per plant, number of seeds per pod and 1000 grain weight. Leaf area index (LAI), crop growth rate (CGR) and net assimilation rate (NAR) was calculated based on growth degree day. Data collected were analyzed statistically using SAS and correlation coefficient by statistical processor system support (SPSS). The means differences among the treatments were compared by least significance difference test (LSD) at 0.05 level of possibility.

LAI = LA/SA

CGR = w2-w1/GDD

$$GDD = \sum \left( \frac{T\max + T\min}{2} - 10 \right)$$

NAR = CGR/LAI

where LA is the leaf area, SA is the ground area that occupied a plant, W is the dry matter and GDD is the growth degree day.

## RESULTS AND DISCUSSION

### Seed yield

Analysis of variance showed that the effect of micronutrients foliar application on grain yield was significant ($p < 0.05$), but the effect of foliar spraying at different growth stages was not significant (Table 2). Mean comparison showed that Zn+Fe combination treatment produced maximum yield which was 15.75 ton/ha. Zn treatment and Fe treatment yields were 25 and 11.41% higher than control treatment, respectively (Table 3). Our results were in line with the findings of Banks (2004) in soybean and Arif et al. (2006) in wheat. In contrast, Mallarino et al. (2001) and Gerwing et al. (2003) showed that zinc foliar spraying did not increase soybean yield. SeifiNadergholi et al. (2011) indicated that ZnSo4 foliar application at flowering and seed set noticeably raised soybean yield components. Berglund (2002) noted that Zn foliar application particularly at vegetative growth stage increased soybean seed yield. Bozoglu et al. (2007) stated that foliar application of micronutrients could be implemented for higher yield and quality. These authors reported that shortage of some micronutrients decreased yield in chickpea. Zinc foliar application enhanced soybean yield by influencing the number of seeds per plant and seed weight (Kobraee et al., 2011). Zeidan et al. (2006) reported that yield components in

**Table 3.** Mean Comparison of seed yield and yield components.

| Treatment | Seed yield (ton/ha) | Pod/plant number | Seed/pod number | 1000 grain weight (g) |
|---|---|---|---|---|
| Control | 6.66$^{c}$ | 24.23$^{b}$ | 1.68$^{b}$ | 144.4$^{c}$ |
| Zn | 7.42$^{b}$ | 35.97$^{a}$ | 2.16$^{a}$ | 167.3$^{b}$ |
| Fe | 8.33$^{ab}$ | 31.31$^{a}$ | 2.23$^{a}$ | 178.2$^{ab}$ |
| Zn + Fe | 15.75$^{a}$ | 36.36$^{a}$ | 2.17$^{a}$ | 200$^{a}$ |

Means, in each column, following similar letter(s) are not significantly different at the 5% level of probability- using Duncan's multiple range test.

**Table 4.** Mean comparison of yield components at different growth stages.

| Treatment | Pod/plant | 1000 grain weight |
|---|---|---|
| 10-leaf | 35.42$^{a}$ | 168.3$^{a}$ |
| Flowering | 33$^{ab}$ | 163.8$^{c}$ |
| Pod set | 31$^{b}$ | 166$^{b}$ |

Means in each column, following similar letter(s) are not significantly different at the 5% level of probability- using Duncan's multiple range test.

lentil are enhanced by foliar application of micronutrients. Due to the enzymatic activity enhancement, micro-elements effectively increased photosynthesis and translocation of assimilates to the seed. Zayed et al. (2011) announced that due to the synergistic effect, zinc + iron treatment as compared to Zn treatment and Fe treatment was more helpful in rice. Arif et al. (2006) reported that higher yield could be achievable by foliar spraying which ensures crop dry matter production. Kobraee et al. (2011) claimed that zinc and iron application at the same time could be lead to higher dry matter and seed yield as compared to using them separately. Kakiuchi and Kobata (2008) revealed that seed production in soybean is lower as compared to the other legumes, and for production, one gram of seed, two gram photosynthesis assimilates will be required. Shiraiwa et al. (2004) declared that pod number and seed number are two major factors which affect soybean seed yield. Variation in dry matter production in seed filling period is a prime factor for different seed yield in soybean genotypes. In soybean, as a result of increasing the ratio of source-sink during flowering stage, seed yield was increased by the seed number enhancement (Cumudini et al., 2011).

## Number of pod per plant

Analysis of variance showed that the effect of micronutrients foliar application and foliar spraying at different growth stages were significant ($p < 0.05$) (Table 5). Mean comparison showed that the highest number of pod per plant was produced at 10 leaf stage (35.42) and the least number was at the beginning of pod set (31) (Table 4). In addition, there was no statistical difference among all the fertilizer treatments and Zn+Fe combination treatment had the maximum number of pod per plant (36.36) and minimum number was produced by control (24.23) (Table 3). Our results were in agreement with Bozorgi et al. (2011) in faba bean, Kobraee et al. (2011) in soybean and SeifiNadergholi et al. (2011) in common bean who showed that the number of pods per plant was enhanced by zinc foliar application. Elballa et al. (2004) revealed that microelements application enhanced the number of pods per plant. Zeidan et al. (2006) found that foliar spray of micronutrients considerably enhanced the number of pods per plant, 1000 seed weight and seed yield. Kakiuchi and Kobata (2006) claimed that one of the most important factors which are determiner in soybean seed yield is pod number per plant. These authors indicated that the rate of pod set in soybean raised with an enhancement in source vigor at the time at which sink/source ratio was altered by flower thinning and defoliation. Furthermore, the ratio of dry matter enhancement in soybean seeds to its shoot is a crucial factor which affects the rate of podding in soybean. SeifiNadergholi et al. (2011) reported that highest pods per plant was produced by foliar spraying at flowering and podding stage and increase of number of pods per plant due to foliar application could be attributed to significant effect of microelements on reproductive organs, such as stamens and pollens. These authors revealed that since soybean is a self-pollinated crop, stamen activity enhances the number of flowers that can

**Table 5.** Analysis of variance_of yield components.

| Source | df | Mean of Square | | |
|---|---|---|---|---|
| | | Pod/plant | Seed/pod | 1000 grain weight |
| Time (T) | 2 | 117.693* | $0.001^{ns}$ | 60.778** |
| Fertilizer (F) | 3 | 294.940* | $0.579^{ns}$ | 2050.185** |
| T × F | 6 | $35.253^{ns}$ | 0.059* | $34.407^{ns}$ |
| Error | 18 | 71.422 | 0.021 | 60.333 |

**, *, ns: Significant at 1 and 5 % probability levels and non significant, respectively.

**Table 6.** The interaction of fertilizer and time of fertilizer application on seed/pod.

| Treatments | 10-leaf stage | Beginning of flowering | Beginning of pod set |
|---|---|---|---|
| Control | $1/7^{de}$ | $1/5^{e}$ | $1/86^{cd}$ |
| Zn | $2/2^{ab}$ | $2/13^{abc}$ | $2/16^{ab}$ |
| Fe | $2/2^{ab}$ | $2/36^{a}$ | $2/13^{abc}$ |
| Zn + Fe | $2/2^{ab}$ | $2/26^{ab}$ | $2/06^{bc}$ |

Means, in each column, following similar letter(s) are not significantly different at the 5% level of probability- using Duncan's multiple range test.

fertile well and as a result, larger number of pods per plant will be produced.

## Number of seeds per pod

Analysis of variance showed that the effect of foliar application and foliar spraying at different growth stages were not significant (Table 5). There was no statistical difference among all the fertilizer treatments (Table 3). Mean comparison revealed that Fe treatment had the maximum number of seed per pod (2.23) and the minimum was produced by control (1.68) (Table 3). The interaction of fertilizer and time of fertilizer application showed that Fe treatment at the beginning of flowering produced maximum number of seed per pod (2.36) and control treatment produced the least (1.5) at the beginning of flowering (Table 6). These results coincided with SeifiNadergholi et al. (2011) in common bean and Banks (2004) in soybean who reported that zinc foliar application increased the number of seeds per pod. This author showed that zinc foliar application in eight-leaf stage with increasing leaf area, dry weight and length of flowering period enhanced the number of seeds per pod. Nasri et al. (2011) reported that it was increased considerably by zinc foliar spraying in *Phaseolous vulgaris*. Ghasemian et al. (2010) declared that maximum number of seeds per plant was produced by zinc application in soybean.

## 1000 grain weight

Analysis of variance revealed that the effect of micronutrient foliar spraying and time of foliar application on 1000 grain weight were significant ($p < 0.01$) (Table 5). Mean comparison showed that maximum 1000 grain weight was found at 10-leaf stage (168.3 g) and the minimum was obtained by the beginning of flowering stage (163.8 g) (Table 4). Zn+Fe combination treatment produced maximum 1000 grain weight (200 g) and minimum 1000 grain weight was obtained by control treatment (144/4 g) (Table 3). The results are in agreement with Arif et al. (2006) who found considerable enhancement in 1000 seed weight by foliar spraying in wheat crop. Nasri et al. (2011) in *P. vulgaris* and Ghasemian et al., (2010) in soybean reported that 100 seed weight was increased considerably by zinc foliar spraying. When photosynthesis assimilates is translocated from vegetative organs to the other parts, seed weight will be enhanced considerably Kakiuchi and Kobata (2008). It was shown that in common bean, microelements affects leaf and as a result larger amounts of assimilate was produced. Since seeds are major source of assimilate, they moves to the seeds and in conclusion larger seeds were produced (SeifiNadergholi et al., 2011).

## Correlation coefficient of yield and its components

Correlation coefficients (Table 7) indicated that soybean yield had positive and significant correlation with 1000 grain weight ($r = 0.711$), number of pod per plant ($r = 0.577$) and number of seed per pod ($r = 0.483$). Correlation of soybean yield with first pod height was negative and significant ($r = -0.433$). Number of pods per

**Table 7.** Correlation coefficients of various agronomic characteristics of soybean (*Glycine max* (L.) Merr).

| | Number of pod per plant | First pod height | Number of seed per pod | 1000 grain weight |
|---|---|---|---|---|
| Seed yield | 0/577** | - 0/433** | 0/483** | 0/711** |
| Number of pod per plant | - | - 0/384* | 0/498** | 0/588** |
| Number of seed per pod | - | - | - | 0/615** |

**, *, ns: Significant at 1 and 5 % probability levels and non significant, respectively.

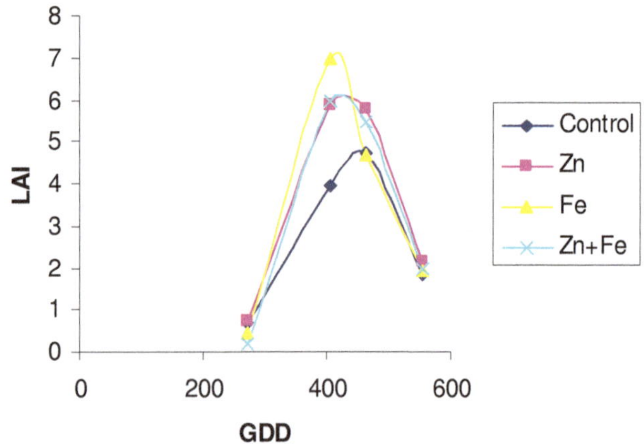


**Figure 1.** Leaf area index at different growth stages.

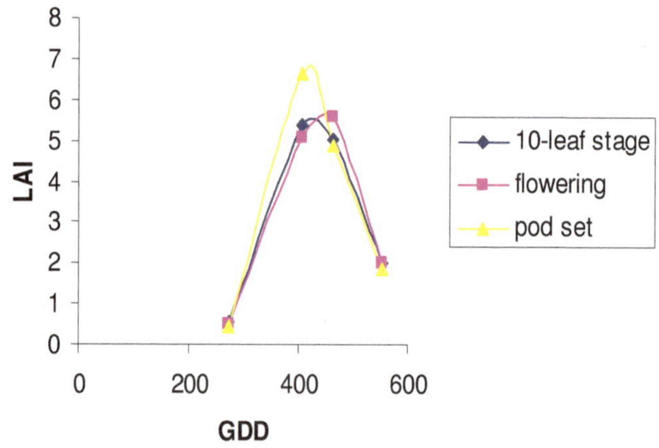


**Figure 2.** Effect of foliar application on LAI.

plant showed positive and significant correlation with number of seed per pod (r = 0.498) and 1000 grain weight (r = 0.588). Furthermore, number of seed per pod had positive significant correlation (r = 0.615) with 1000 grain yield. However, a negative and significant correlation (r = - 0.384) was found between number of pods per plant and first pod height. Our findings were in confirmation with the previous studies carried out by Showkat and Tyagi (2010) and Arshad et al. (2006) who indicated that there was positive and significant correlation between seed yield with number of pods per plant and 100 seed weight. Cumudini et al. (2001) found that there is a significant and positive correlation between seed yield and number of seeds per plant, number of seeds per pod and 1000 seed weight.

## Leaf area index (LAI)

Results of this study indicated that Leaf area index during the experiment decreased at different growth stages so that in 407 GDD and at the beginning of pod set reached to the highest point (6.6) (Figure 1), but after that because of accelerating in leaves, ascending was declined. Leaves ascending decreased 23.73 in LAI. Furthermore, the highest LAI was found by foliar application with Fe (6.9) in 407 GDD and there was no difference among other treatments (Figure 2).

Micronutrients foliar application increases enzymes activities which lead to higher crop production and leaf area (LA) expansion (Zayed et al., 2011). However, Mallarino et al. (2001) found that foliar fertilization could be harmful for leaf area expansion. Zinc foliar application increases tryptophan amino acid and indol acetic acid hormone which were the two main factors in leaf area expansion in *P. vulgaris* (SeifiNadergholi et al., 2011).

## Crop growth rate (CGR)

Maximum CGR was obtained at the beginning of flowering (39.52) in 272 GDD (Figure 3). In addition, maximum CGR was found by zinc and iron combination treatment application (40.05) in 407 GDD and minimum CGR belong to control treatment (5.62) in 553 GDD (Figure 4). Pedersen and lauer (2004) stated that crop growth rate is the most important factor in growth analysis of soybean, because it shows the amounts of canopy assimilate and influences on the amount of dry matter and balances by changing leaf are index (LAI) and net assimilation rate (NAR). These authors showed that optimum CGR will be achievable by LAI 3 to 3.5. It was indicated by Cumudini et al. (2001) that throughout early reproductive stage, there is a significant and positive correlation between crop growth rate and seed number in maize and also, between dry matter production and seed

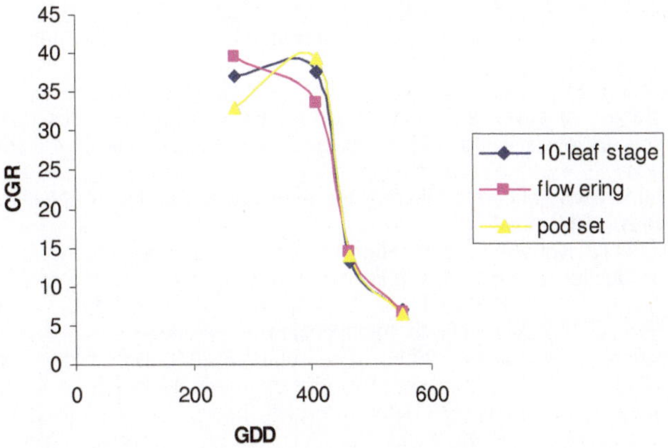


**Figure 3.** Crop growth rate at different growth stages.

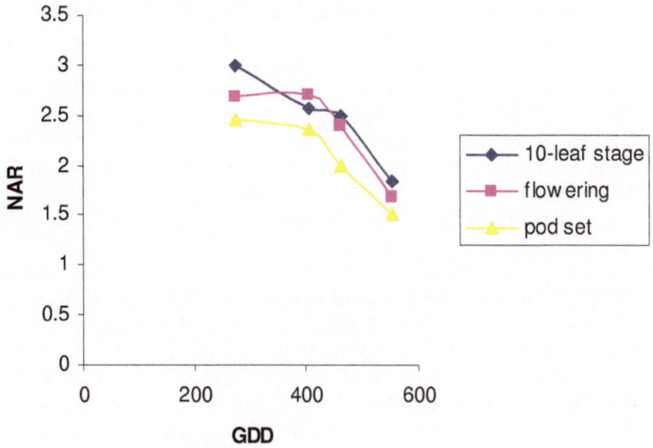


**Figure 5.** Net assimilation rate at different growth stages.

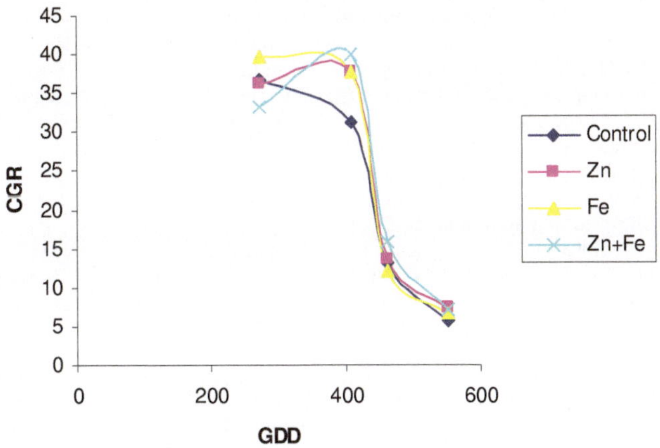


**Figure 4.** Effect of foliar spraying on CGR.

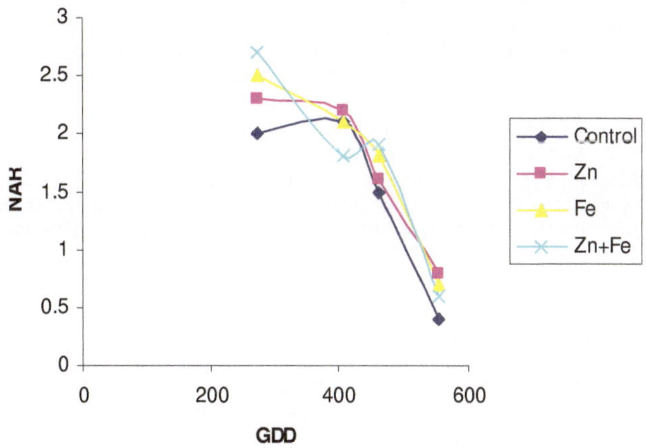


**Figure 6.** Effect of foliar application on NAR.

number in soybean. Ruhul et al. (2009) declared that the highest amount of crop growth rate was observed up to 60 days after sowing and it was decreased at 90 days after sowing. It was reported by Oya et al. (2004) that soybean seed yield was considerably correlated with crop growth rate during the reproductive growth stage. These authors believed that soybean has a special period in which reproductive and vegetative growth stages happen at the same time.

## Net assimilation rate (NAR)

Results of this study revealed that NAR variations during the experiment were descending (Figure 5 and 6). Apparently, NAR is influenced by many factors which are complex and practically are not recognizable. For this reason, results of many researches are different with each other. Maximum NAR was observed at 10-leaf stage (3 g $plant^{-1}$ $GDD^{-1}$) in 272 GDD and minimum NAR belong to pod set stage (1.5 g $plant^{-1}$ $GDD^{-1}$) in 553 GDD (Figure 5). With regard to treatments, Zn+Fe combination treatment (35%) was more than control treatment (Figure 6). So, maximum NAR was found by foliar application of Zn+Fe (2.7 g $plant^{-1}$ $GDD^{-1}$) in 272 GDD and minimum NAR was observed by control treatment (0.4 g plant-1 $GDD^{-1}$) in 553 GDD (Figure 6).

## ACKNOWLEGDMENTS

We thank Dr Mohammad Golbashy for his comments on an earlier draft of this paper. Also we thank the Associate Editor and three anonymous reviewers for their useful comments and suggestions

### REFERENCES

Ai-Qing Z, Qiong-Li B, Xiao-Hong T, Xin-Chun L, Gale W J (2011). Combined effect of iron and zinc on micronutrient levels in wheat

(Triticum aestivum L.). J. Environ. Biol., 32: 235-239. Available online at: http://www.jeb.co.in/journal_issues/201103_mar11/paper_16.pdf

Akparobi SO (2009). Evaluation of six cultivars of soybean under the soil of rainforest agro-ecological zones of Nigeria. Middle-East J. Sci. Res., 4(1): 06-09. Available online at: http://idosi.org/mejsr/mejsr4(1)/2.pdf

Alloway BJ (2003). Zinc in soil and crop nutrition. International Zinc Assocation, p. 114. Available online at: http://sustainablecropnutrition.info/ifacontent/download/10445/151245/version/1/file/2008_IZA_IFA_ZincInSoils.pdf

Arshad M, Ali N, Ghafoor A (2006). Character correlation and path coefficient in soybean *(Glycine max (L.) Merrill)*. Pak. J. Bot., 38(1): 121-130. Available online at: http://www.pakbs.org/pjbot/PDFs/38(1)/PJB38(1)121.pdf

Arif M, Chohan M A, Ali S, Gul R, Khan S (2006) Response of wheat to foliar application of nutrients. J. Agric. Biol. Sci., 1(4). Available online at: http://www.arpnjournals.com/jabs/research_papers/jabs_1106_36.pdf

Babaeian M, Tavassoli A, Ghanbari A, Esmaeilian Y, Fahimifard M (2011) Effects of foliar micronutrient application on osmotic adjustments, grain yield and yield components in sunflower (Alstar cultivar) under water stress at three stages. Afr. J. Agric. Res., 6(5): 1204-1208. Available online at: http://www.academicjournals.org/ajar/pdf/pdf2011/4%20Mar/Babaeian%20et%20al.pdf.

Banks LW (2004). Effect of timing of foliar zinc fertilizer on yield component of soybeans. Australian. J. Exp. Agric. Animal Husbandry, 22(116): 226-231. Doi: http://dx.doi.org/10.1071/EA9820226

Berglund DR (2002). Soybean Production Field Guide for North Dakota and Northwestern Minnesota. Published in cooperative and with support from the North Dakota Soybean Council, 136p. Available online at: http://library.ndsu.edu/tools/dspace/load/?file=/repository/bitstream/handle/10365/5452/a1172.pdf?sequence=1

Bozoglu H, Ozcelik H, Mut Z, Pesken E (2007) Response of chickpea *(Cicer arientinum L.)* to Zink and molybdenum fertilization. Bangladesh J. Bot., 36(2): 145-149. Available online at: http://www.banglajol.info/index.php/BJB/article/download/1503/1468

Bozorgi HA, Azarpour E, Moradi M (2011). The effects of bio, mineral nitrogen fertilization and foliar zinc spraying on yield and yield components of faba bean. World Appl. Sci. J., 13(6): 1409-1414. Available online at: http://www.idosi.org/wasj/wasj13(6)/19.pdf

Bukvić G, Antunović M, Popović S, Rastija M (2003). Effect of P and Zn fertilization on biomass yield and its uptake by maize lines (*Zea mays* L.). Plant Soil Environ., 49(11): 505-510. Available online at: http://www.cazv.cz/2003/PSE11_03/5-bukvic.pdf

Cakmack I (2002). Plant nutrient research: Priorities to meet human needs for food in sustainable ways. Plant Soil, 247: 3-24. Available online at: http://www.harvestzinc.org/pdf/PrioritiestomeetHumanNeeds.pdf

Ghasemian V, Ghalavand A, Soroosh zadeh A, Pirzad A (2010) The effect of iron, zinc and manganese on quality and quantity of soybean seed. J. Phytol., (J Phytol) 2(11): 73-79. Available online at: http://journal-phytology.com/article/download/5066/2565

Cumudini S, Hume DJ, Chu G (2001). Genetic improvement in short season soybean: I. Dry matter accumulation, Partitoning, and Leaf area duration. Crop Sci., 41: 391-398. Available online at: https://www.soils.org/publications/cs/pdfs/41/2/391

Elballa MMA, El-amin AHB, Elamin EA, Elsheikh EAE (2004). Interactive effects of cultivars, foliar application of mirconutrients and rhizobium inoculation on snap bean *(Phaseolus vulgaris L.)* performance. U of K J Agric Sci., 12(3). Available online at: http://staffcv.uofk.edu/agriculture/shayatti/downloads/43.pdf

Gerwing JAB, Gelderman R, Berg R (2003). Foliar nutrient application influence on soybean yield at Aurora and Berestord SD in 2003. Agricultural experiment station plant science department. South Dakota State University. Plant Sci., 315: 65-69. Available online at: http://www.sdstate.edu/ps/research/soil-fertility/reports/upload/03-06-Foliar-Nutrient-Application-Influence-on-Soybean-Yield-at-Aurora-and-Beresford-SD-in-2003-32303E-and-32403W.pdf

Gul H, Said A, Saeed B, Mohammad F, Ahmad I (2011). Effects of foliar application of nitrogen, potassium and zinc on wheat growth. J. Agric. Biol. Sci., 6(4). Available online at: http://www.arpnjournals.com/jabs/research_papers/rp_2011/jabs_0411_265.pdf

Jandong EA, Uguru MI, Oyiga BC (2011). Determination of yield stability of seven soybean *(Glycine max)* genotypes across diverse soil pH levels using GGE biplot analysis. J. Appl. Biosci., 43: 2924-2941. Available online at: http://www.zef.de/module/register/media/ab74_Jandoung%20et%20al%20(2011).pdf

Kakiuchi J, Kobata T (2008). High carbon requirements for seed production in soybeans [*Glycine max* (L.) Merr]. Plant Prod. Sci., 11(2): 198-202. Available online at: http://www.jstage.jst.go.jp/article/pps/11/2/198/_pdf

Kakiuchi J, Kobata T (2006). The relationship between dry matter increase of seed and shoot during the seed-filling period in three kinds of soybeans with different growth habits subjected to shading and thinning. Plant Prod Sci 9(1): 20-26. Available online at: http://ci.nii.ac.jp/els/110004026968.pdf?id=ART0006284350&type=pdf&lang=en&host=cinii&order_no=&ppv_type=0&lang_sw=&no=1316593703&cp=

Kinaci E, Gulmezoglu N (2007). Grain yield and yield components of triticale upon application of different foliar fertilizers. Interciencia, 32(9): 624-628. Available online at: http://redalyc.uaemex.mx/pdf/339/33932909.pdf

Kobraee S, NoorMohamadi N, HeidariSharifabad H, DarvishKajori F, Delkhosh B (2011). Influence of micronutrient fertilizer on soybean nutrient composition. Indian J. Sci. Technol., 4(7). Available online at: http://www.indjst.org/archive/vol.4.issue.7/july11kobraee-9.pdf

Kobraee S, Shamsi k, Rasekhi B (2011). Effect of micronutrients application on yield and yield components of soybean. Annals Biol. Res., 2(2): 476-482. Available online at: http://scholarsresearchlibrary.com/ABR-vol2-iss2/ABR-2011-2-2-476-482.pdf.

Mallarino AP, Haq MU, Wittry D, Bermudez M (2001). Variation in Soybean Response to Early Season Foliar Fertilization among and within Fields. Agron. J., 93: 1220-1226. Available online at: https://www.soils.org/publications/aj/pdfs/93/6/1220

Masuda T, Goldsmith PD (2009). World soybean production: Area harvested, yield and long-term projections. Int. Food Agribus. Manag. Rev., 12(4). Available online at: http://ageconsearch.umn.edu/bitstream/92573/2/20091023_Formatted.pdf

Narimani H, Rahimi MM, Ahmadikhah A, Vaezi B (2010). Study on the effects of foliar spray of micronutrient on yield and yield components of durum wheat. Arch. Appl. Sci. Res., 2(6): 168-176. Available online at: http://scholarsresearchlibrary.com/aasr-vol2-iss6/AASR-2010-2-6-168-176.pdf.

Nasiri Y, Zehtab-Salmasi S, Nasrullahzadeh S, Najafi N, Ghassemi-Golezani K (2010). Effects of foliar application of micronutrients (Fe and Zn) on flower yield and essential oil of chamomile (*Matricaria chamomilla* L.). J. Med. Plants Res., 4(17): 1733-1737. Available online at: http://www.academicjournals.org/jmpr/PDF/pdf2010/4Sept/Nasiri%20et%20al.pdf

Nasri M, Khalatbari M, Aliabadi Farahani H (2011) Zn-foliar Application Influence on Quality and Quantity Features in Phaseolous Vulgaris under Different Levels of N and K Fertilizers. Adv. Environ. Biol., (Adv. Environ. Biol) 5(5): 839-846. Available online at: http://www.aensionline.com/aeb/2011/839-846.pdf

Oya T, Nepomuceno AL, Neumaier N, Farias JRB, Tobita S, Ito O (2004). Drought tolerance characteristics of Brazilian soybean cultivars – Evaluation and characterization of drought tolerance of various Brazilian soybean cultivars in the field. Plant Pro. Sci., 7(2): 129-137. Available online at: http://bioinfo.cnpso.embrapa.br/seca/documentos/Toya.pdf

Pedersen P, Lauer JG (2004). Soybean growth and development in various management systems and planting dates. Crop Sci., 44: 508-515. Available online at: http://corn.agronomy.wisc.edu/Pubs/JL_JournalArticles/44-508.pdf

Ruhul AAKM, Jahan SRA, Karim MF, Hasanuzzaman M (2009). Growth dynamics of soybean (*Glycine max* L.) As affected by varieties and timing of irrigation. Am-Euras J. Agron., 2(2): 95-103. Available online at: http://idosi.org/aeja/2 (2)09/8.pdf

Salwa AIE, Taha MB, Abdalla MAM (2011). Amendment of soil fertility and augmentation of the quantity and quality of soybean crop by using phosphorus and micronutrients. Int. J. Acad. Res., 3(2): part 3. Available online at: http://www.ijar.lit.az/pdf/10/2011(10-127).pdf

Sawan ZM, Mahmoud MH, El-Guibali AH (2008). Influence of potassium fertilization and foliar application of zinc and phosphorus on growth, yield components, yield and fiber properties of Egyptian cotton (Gossypium barbadense L.). J. Plant Ecol., 1(4): 259-270. Available online at: http://jpe.oxfordjournals.org/content/1/4/259.full.pdf

SeifiNadergholi M, Yarnia M, Rahimzade Khoei F (2011). Effect of zinc and manganese and their application method on yield and yield components of common bean *(Phaseolus vulgaris L. CV. Khomein)*. Middle-East J. Sci. Res., 8(5): 859-865. Available online at: http://www.idosi.org/mejsr/mejsr8(5)11/3.pdf

Shiraiwa T, Uneo N, Shimada S, Horie T (2004). Correlation between yielding ability and dry matter productivity during initial seed filling stage in various soybean genotypes. Plant Prod. Sci., 7(2): 138-142. Available online at: http://www.jstage.jst.go.jp/article/pps/7/2/138/_pdf

Showkat M, Tyagi SD (2010). Correlation and Path Coefficient Analysis of Some Quantitative Traits in Soybean (*Glycine max* L. Merrill.). Res. J. Agric Sci., 1(2): 102-106. Available online at: http://www.rjas.info/wp-content/uploads/2010/08/Correlation-and-Path-Coefficient-Analysis-of-Some-Quantitative-Traits-in-Soybean-Glycine-max-L.-Merrill..pdf

Sliman ZT, Motto HL (1990). Differential response of two soybean genotypes to zinc induced iron deficiency chlorosis. J. King Saud Univ. Agric. Sci., 2(1): 71-80. Available online at: http://digital.library.ksu.edu.sa/V2M33R1555.pdf

Vahedi A (2011). The effects of micronutrient application on soybean seed yield and on seed oil and protein content. J. Am. Sci., 7(6). Available online at: http://www.jofamericanscience.org/journals/am-sci/am0706/110_5746am0706_672_677.pdf

Vahedi A (2010) Study of the effects of micronutrient application on the absorption of macro- and micronutrients in the soybean cultivar Telar in the north of Iran. J. Am. Sci., 7(6). Available online at: http://www.jofamericanscience.org/journals/am-sci/am0706/189_5747am0706_1252_1257.pdf

Yassen A, Abou El-Nour EAA, Shedeed S (2010). Response of wheat to foliar Spray with urea and micronutrients. J. Am. Sci., 6(9). Available online at: http://www.jofamericanscience.org/journals/am-sci/am0609/03_2833_am0609_14_22.pdf

Zayed BA, Salem AKM, El Sharkawy HM (2011). Effect of different micronutrient treatments on rice *(Oriza sativa L.)* growth and yield under saline soil conditions. World J. Agric. Sci., 7(2): 179-184 Available online at: http://www.idosi.org/wjas/wjas7(2)/12.pdf

Zeidan MS, Mohamed MF, Hamouda HA (2010). Effect of foliar fertilization of Fe, Mn and Zn on wheat yield and quality in low sandy soils fertility. World J. Agric. Sci., 6(6): 696-699. Available online at: http://www.fspublishers.org/ijab/past-ssues/IJABVOL_3_NO_4/11.pdf

Zeidan MS, Hozayn M, Abd El-Salam MEE (2006). Yield and quality of lentil as affected by micronutrient deficiencies in sandy soils. J. Appl. Sci. Res., 2(12): 1342-1345. Available online at: http://www.aensionline.com/jasr/jasr/2006/1342-1345.pdf

# Proximate analysis of *Sphenostylis stenocarpa* and *Voadzeia subterranean* consumed in South - Eastern Nigeria

**S. Nwodo Chinedu and C. Obinna Nwinyi***

Department of Biological Sciences, School of Natural and Applied Sciences College of Science and Technology, Covenant University, KM 10 Idiroko Road, Canaan Land, PMB 1023, Ota, Ogun State, Nigeria.

**Proximate analyses of two traditional grain legumes consumed in Eastern Nigeria- Bambara groundnut (*Voadzeia subterranean*) and African yam beans (*Sphenostylis stenocarpa*) were carried out. Bambara groundnut was found to contain 2.86 ± 0.02% moisture, 32.40 ± 0.02% protein, 7.35 ± 0.02% fat, 5.78 ± 0.02% ash, 2.68 ± 0.02% crude fiber and 51.78 ± 0.02% total carbohydrates. African yam beans contained following: 1.96 ± 0.02% moisture, 37.21 ± 0.02% proteins, 9.49 ± 0.02% fat, 5.35 ± 0.02 Ash, 3.55 ± 0.02 crude fiber and 44.4 ± 0.02% total carbohydrates. The result indicated that the two underutilized food legumes have high protein content with African yam beans having slightly higher value of the protein nutrient. They also contained varying levels of ash content, crude fiber and carbohydrates. This revealed that bambara groundnut and African yam beans are important food sources that can be exploited particularly in the developing countries where there is shortage in animal protein and undernutrition facing many families in developing countries.**

**Key words:** Bambara groundnut, African yam beans, proximate analysis, undernutrition, protein.

## INTRODUCTION

Currently the need for sustainable development and economic recovery in the developing regions of the world is crucial particularly in the area of access to quality proteinous food and food security. Conversely, the commonly relied source of quality protein which is animal protein is grossly expensive and scarce (Adebowale and Adebowale 2007; Lawal and Adebowale, 2006, Adebowale et al., 2005a, b; Lawal et al., 2005). Thus, legumes are rich sources of plant proteins present as a substitute for the scarce animal proteins. Of the thousand known legumes species, less than 20 are used extensively today as food. Those commonly used include: peanuts, soybeans, peas, lentils, pigeon pea, bambara groundnut and African yam beans. Oils, proteins, gums, resins and medicinal compounds could be derived from this extremely large family. Some leguminous seeds or pulses contain 20 to 40% protein on dry matter basis dry weight, which is nearly twice the value found in cereals and 20 times that of cassava. Only a few grain legumes contain up to 50% proteins. Pulses also contain up to 40 to 60% carbohydrates mainly in the form of starch and some reasonable quantities of niacin, thiamine, riboflavin and ascorbic acid. Pulses are rich in calcium and iron (Maneepun, 2000; Emenalom and Udedibie, 1998).

Bambara groundnuts (*Voadzeia subterranean*) and African yam beans (*Sphenostylis stenocarpa*) are leguminous crops which are cultivated in tropical and subtropical regions. Bambara groundnuts (*V. subterranean*) belong to the family *Fabaceae* which are conspicuously West African crops. Bambara groundnuts are under-utilized or neglected due to storage induced defects and antinutritional factors (toxins and biological active substances) such as trypsin inhibitor, haemoaglutinins, cyanogenic glycosides, anticoagulant, toxic histones (Fadahunsi and Sanni, 2010; Agbede and Aletor, 2003; Vasconcelos et al., 1997). However, most of these toxins are reduced to tolerable levels by simple preparative

*Corresponding author. E-mail: nwinyiobinna@gmail.com.

procedures such as fermentation, germination, roasting, soaking in warm water and thorough cooking which usually requires high fuel input and prolonged cooking (Oluwole and Taiwo, 2009). Groundnuts present good agronomic advantages such as their ability to thrive in drought weather, low nutrient soils and high nutrition potential (Okpuzor et al., 2010; Anchirinah et al., 2001; Azam –Ali et al., 2001). Cultivars are recognized in Africa differing in shape of leaves, sizes, hardness and color of the seed. Bambara groundnuts are known as jugo beans in Swahili. In Zambia, Bambara groundnuts are known as ntoyo, *katoyo* or *mbwiila*. In Nigeria, Hausas refer to it as "Gurjiya" or "Kwaruru". Amongst, the Goemai and Kanuri groups in Nigeria, it is referred to as "Kwam" and "Ngamgala respectively. Bambara groundnut (*V. subterranean*), is a nutritionally balanced seed containing essential amino acids such as isoleucine, leucine, lysine, methionine, phenylalanine, threonine and valine (Ihekoronye and Ngoddy, 1985). According to findings by Adu-Dapaah and Sangwan (2004), the seed of the crop is rich in iron and essential amino acids. Bambara groundnuts are grown for their edible seeds which are used as a nutritional pulse and not as an oil seed due to its low percentage of oil (Minka and Bruneteau, 2000). The fresh seeds are eaten in an unripe state or the pulse, after soaking and boiling as the dried seeds are sometimes hard. The dried seeds are sometimes roasted and ground into flour for preparation of different delicacies (Poulter and Caygill, 2006; Brough et al., 1993).

African yam beans (*S. stenocarpa*) are cultivated both for the seeds and tubers, its valuable and prominent source of plant protein (amino acids –lysine and methionine) in the diets of consumers (Evans and Boulter, 1974). The tuber which is found beneath the ground has resemblance with sweet potatoes or Irish potatoes. Above ground it produces good yields of edible seeds. It could be found in forests, open wooded grasslands, rocky fields, and marshy grounds as weed and cultivated crops (Porter 1992). It grows on a wide range of soils including acid and highly leached sandy soils at altitudes from sea level to 1950 m (Amoaety et al., 2000). In the area of cultivation, it serves as a security crop prior to the new season harvest of main staples such as maize, sorghum, yam and cassava in the tropics. Different studies on nutritional, chemical composition and extraction condition for optimum solubilization of the pulse protein have been studied (Aletor and Aladetimi 1989; Edem et al., 1990; Kine et al., 1991; Akinmutini et al., 2007; Eromosele et al., 2008). African yam beans (*S. stenocarpa*) however have antinutritional factors such as phytin and lectin. The large scale breeding and research on these leguminous crops (Bambara groundnut (*V. subterranean*) and African yam beans (*S. stenocarpa*) has been rather slow. However, evidence and fragmentary research results suggested that it is a crop with much promise. The authors seek to ascertain the proximate composition of these leguminous crops found in the South eastern part of Nigeria with the view of documenting and exploiting them as alternative sources of protein for the emerging population in this zone.

## MATERIALS AND METHODS

### Crop physiology

Bambara groundnut (*V. subterranean*) is an annual herb with short creeping, much branched stems rooting at nodes and with very short internodes thus, giving a bunchy appearance to the plant. Leaves are pinnately trifoliate, petiole long erect, grooved and thickened at base. Seeds are round, smooth, and hard varying in size up to 1.5 cm broad, whitish red, black or variously mottled or blotched hilum white, sometimes surrounded by black eye. The flowers are self pollinated and are cleistogamous. It is grown from seeds either in mixed cultivation or in pure stands with spacing about 12 × 8 inches and matures in 4 months. It is harvested by pulling up the entire plants to which the pods are adhered by tough wiry stalks. The normal yield is about 500 to 1000 seeds per acre.

### African yam beans

*S. stenocarpa* is described as a prostrate or a climbing vine reaching up to 3 m or more in height (Duke et al., 1977). The stems are reddish glabrous or sparsely pulverous. The plant produces spindle shaped starchy tubes smaller in size than those of sweet potato. It takes between 5 to 8 months to mature.

### Collection and preparation

Bambara groundnut (*V. subterranean*) and African yam bean (*S. stenocarpa*) seeds were obtained from a local market (new market) in Enugu, South East of Nigeria. The varieties selected were white black eye typed for the Bambara groundnuts and the greenish type for the yam beans. The seeds were identified and authenticated by a crop scientist and deposited at the department of biochemistry Enugu State University, Enugu. Five hundred and fifty grams (550 g) each of the purchased seeds were washed with distilled water to remove impurities and air dried on aluminum trays to achieve total dryness before pulverization.

Bambara groundnuts (*V. subterranean*) and yam bean (*S. stenocarpa*) seeds were pulverized into powdered flour using a mechanized crown grinding machine Model PHT 94, manufactured in China. This was sieved and used for further work as test samples.

### Chemicals and reagents

All chemicals and reagents used were of analytical grade. Petroleum ether spectrohotometric grade was purchased from Merck. Sodium hydroxide, tetraoxosulphate (VI) acid, boric acid and sodium sulphate were sourced from Fisher scientific, US.

### Determination of proximate analysis

The test sample Bambara groundnut (*V. subterranean*) and African yam beans (*S. stenocarpa*) were subjected to proximate analysis by the methods outlined by AOAC (1984).

### Determination of moisture content

This method is based on moisture evaporation. Here, the aluminum dishes were washed dried in oven and desiccators for cooling. The weight of each dish was taken. 200 g of ground samples of Bambara groundnut (*V. subterranean*) and African yam beans (*S. stenocarpa*) were weighed into a sterile aluminum dish, weight of the dish and weight of undried sample (in duplicate) were taken. This was transferred into an oven set at 80°C for 2 h and at 100°C for 3 h respectively. This was removed and cooled in desiccators. Then the weight was measured using a measuring scale balance. It was transferred back into the oven for another one hour and then reweighed. The process continued until a constant weight was obtained. The difference in weight between the initial weight and the constant weight gained represented the moisture content. The loss in weight multiplied by 100 over the original weight is percentage moisture content. The formula used is presented as:

Moisture content (g/100g) =

$$\frac{\text{Loss in weight (W2-W3) (g)}}{\text{Original weight of sample (W2-W1 (g)}} \times 100$$

Where W1= initial weight of empty crucible, W2 = weight of crucible + food before drying, W3 = final weight of crucible + food after drying.

% Total solid (Dry matter) (%) = 100 - moisture (%)

### Ash content

The ash represents the inorganic component (minerals) of the sample after all moisture has been removed as well as the organic material. The method is a destructive approach based on the decomposition of all organic matter such that the mineral elements may be lost in the process. Twenty grams (20 g) of each of the samples of Bambara groundnut (*V. subterranean*) and African yam beans (*S. stenocarpa*) were weighed into a clean dried and cooled platinum crucible. It was put into a furnace set at 550 to 600°C and allowed to blast for 3 h. It was then brought out and allowed to cool in a desiccator and weighed again. Percentage weight is calculated as weight of ash multiplied by 100 over original weight of the samples used. The formula used is presented as:

Ash content =

$$\frac{\text{Weight of ash (g)}}{\text{Weight of original sample used}} \times 100$$

$$\frac{\text{Loss in weight (W3-W1) (g)}}{\text{W2-W1}} \times 100$$

Where W1 = weight of empty crucible, W2 = weight of crucible + food before drying and or ashing, W3 = weight of crucible + ash.

### Determination of lipid content

The method employed was the Soxhlet extraction technique described by Shir law, (1967). 15 g of the samples were weighed and carefully placed inside a fat free thimble. This was covered with cotton wool to avoid the loss of sample. Loaded thimble was put in the Soxhlet extractor, about 200 ml of petroleum ether were poured into a weighed fat free soxhlet flask and the flask was attached to the extractor. The flask was placed on a heating mantle Such that the petroleum ether in the flask refluxed. Cooling was achieved by a running tap connected to the extractor for at least 6 h after which the solvent was completely siphoned into the flask. Rotary vacuum evaporator was used to evaporate the solvent leaving behind the extracted lipids in the soxhlet. The flask was removed from the evaporator and dried to a constant weight in the oven at 60°C. The flask was then cooled in a desiccator and weighed. Each determination was done in triplicate. The amount of fat extracted was calculated by difference. The formula used is presented as:

Ether extracts (100g) dry matter=

$$\frac{\text{Weight of extracted lipids (g)}}{\text{Weight of dry sample (g)}} \times 100$$

### Protein determination

Total protein was determined by the kjeldahl method as modified by Williams (1964). The analysis of a compound of its protein content by kjeldahl method is based upon the determination of the amount of reduced nitrogen present. About 20 g of the samples Bambara ground nut (*V. subterranean*) and African yam bean (*S. stenocarpa*) were weighed into a filter paper and put into a kjedahl flask, 10 tablets of $Na_2SO_4$ were added with 1 g of $CuSO_4$ respectively. Twenty milliliter (20 ml) of concentrated $H_2SO_4$ were added and then digested in a fume cupboard until the solution becomes colorless. It was cooled overnight and transferred into a 500 ml flat bottom flask with 200 ml of water. This was then cooled with the aid of packs of ice block. About 60 to 70 ml of 40% of NaOH were poured into the conical flask which was used as the receiver with 50 ml of 4% boric acid using 3 days of screened methyl red indicator. The ammonia gas was then distilled into the receiver until the whole gas evaporates. Titration was done in the receiver with 0.1 M $NH_2SO_4$ until the solution became colorless. The formula used is presented as:

The percentage protein =

$$\frac{\text{Vs} - \text{Vb} \times 0.\ 01401 \times \text{N acid (6.25)}}{\text{Original wt of sample used}} \times 100$$

Where Vs = Volume (ml) of acid required to titrate sample; Vb = Volume (ml) of acid required to titrate blank; N acid = normality of acid

### Crude fiber

The bulk of roughages in food is referred to as fiber and is estimated as crude fiber. Twenty grams (20 g) of *V. subterranean* and (*S. stenocarpa*) of the different samples were defatted with diethyl ether for 8 h and boiled under reflux for exactly 30 min with 200 ml of 1.25% $H_2SO_4$. It was then filtered through cheese cloth on a fluted funnel. This was later washed with boiling water to completely remove the acid. The residue was then boiled in a round bottomed flask with 200 ml of 1.25% sodium hydroxide (NaOH) for another 30 min and filtered through previously weighed couch crucible. The crucible was then dried with samples in an oven at 100°C, left to cool in a desiccator and later weighed. This was later incinerated in a muffle furnace at 600°C for 2 to 3 h and later allowed to cool in a desiccator and weighed. The formula used is presented as:

Calculation = Weight of fiber = (C2-C3) y

$$\%\ \text{fiber} = \frac{\text{C2-C3}}{\text{Wt. of original sample}} \times 100$$

**Table 1.** Proximate compositions of Bambara groundnut and African yam bean.

| Nutrient | (*Voadnzeia subterranean*) | (*Sphenostylis stenocarpa*) |
|---|---|---|
| Moisture | 2.86 ± 0.02 | 1.96 ± 0.02 |
| Protein | 32.40 ± 0.02 | 37.21 ± 0.02 |
| Fat | 7.35 ± 0.02 | 9.49 ± 0.02 |
| ash | 5.78 ± 0.02 | 5.35 ± 0.02 |
| Crude fiber | 2.68 ± 0.02 | 3.55 ± 0.02 |
| Carbohydrates | 51.79 ± 0.02 | 44.4 ± 0.02 |

**Table 2.** Percentage total of proximate compositions of Bambara groundnut and yam bean.

| Nutrient | (*Voadnzeia subterranean*) | (*Sphenostylis stenocarpa*) |
|---|---|---|
| Moisture | 2.86 ±0.02 | 1.96±0.02 |
| Protein | 32.40±0.02 | 37.21 ±0.02 |
| Fat | 7.35±0.02 | 9.49 ±0.02 |
| Carbohydrates | 51.79±0.02 | 44.4 ±0.02 |
| Total (%) | 94.40±0.02 | 93.06±0.02 |

**Carbohydrate determination**

Available carbohydrate (%), = 100 - (protein (%) + Moisture (%) + Ash (%) + Fibre (%) + Fat (%). Here, the different samples of Bambara ground nut (*V. subterranean*) and African yam bean (*S. stenocarpa*) were digested with perchloric acid and hydrolyzed starch. The soluble sugar was determined colorimetrically by anthrone method. Duplicate standards were pipetted out using 1 ml dilute glucose standard solutions and 5 ml of freshly prepared anthrone reagents pipetted rapidly to all the tubes and mixed thoroughly. These covered tubes were placed in boiling water bath for exactly 12 min to allow change in color following which sample was allowed to cool at room temperature. The readings of the two cuvettes containing anthrone reagents and deionized distilled water were measured, and difference in readings noted for correction in the calculations. The prepared solutions were transferred to one glass cuvette and their absorbance read at 630 nm against a blank containing 1ml of deionized distilled water and anthrone reagents. Concentrations of the test samples were obtained via interpolation of the absorbance readings. The mass of the glucose was obtained by calculation of the concentration and dilution made. The mass of starch was obtained by calculating the mass of glucose and a factor of 0.9

## RESULTS AND DISCUSSION

The proximate analysis for the legumes is shown in (Tables 1 and 2) and (Figures 1 and 2). In developing countries, there is an increasing mismatch between the rate of population growth and the rate of food production. However, disastrous consequences such as under nutrition often accompanied by diseases such as kwashiorkor or death result, when there is insufficient supply of protein foods. Furthermore, it is known that the brain reaches 80 to 90% of its full weight by age of four; protein malnutrition during this critical period when neurons and neural connections are being formed may cause irreversible brain damage leading to low IQ. This may persist throughout life and may not be remedied by later corrective measures. Women and children in developing countries are more prone to this challenge. The issue of malnutrition alleviation can simply be dealt with by proper understanding and awareness of the food we eat, thus, this research work was aimed at evaluating the nutritive value of the neglected and under-utilized crops in sub Saharan Africa (Bambara groundnut (*V. subterranean*) and African yam beans (*S. stenocarpa*) and to assess their potentials in dealing with protein under- nutrition prevalent in the Eastern Nigeria and other parts of the world. From research studies, legumes are plant foods which are nutritionally rich in high protein content which is twice the percentage found in cereals and thrice of such as found in cassava. However, plant proteins are generally nutritional inferior to those from animal sources due to the protein rich portions of the grains being surrounded by protective husks cellulose and other polysaccharides that are not hydrolyzed by intestinal enzymes. Conversely, high costs of animal protein have made it unavailable in some parts of the world. The combinations of plant proteins can provide a completely adequate and balance mixture of essential amino acids. The content of the essential amino acids is therefore very essential in nutrition, where it provides precursors for protein biosynthesis needed for synthesizing the proteins needed for the maintenance, repair and growth of tissues. From the results in Table 1 the composition of the two legumes showed that they are good sources of protein with the protein content accounting for 32.40 ± 0.02% and 37.21±% for Bambara groundnut (*V. subterranean*) and Africa yam beans respectively. This result compares well with the reports of Okpuzor et al. (2010) who identified a total of ten and twelve different proteins in the malted and dry seeds of

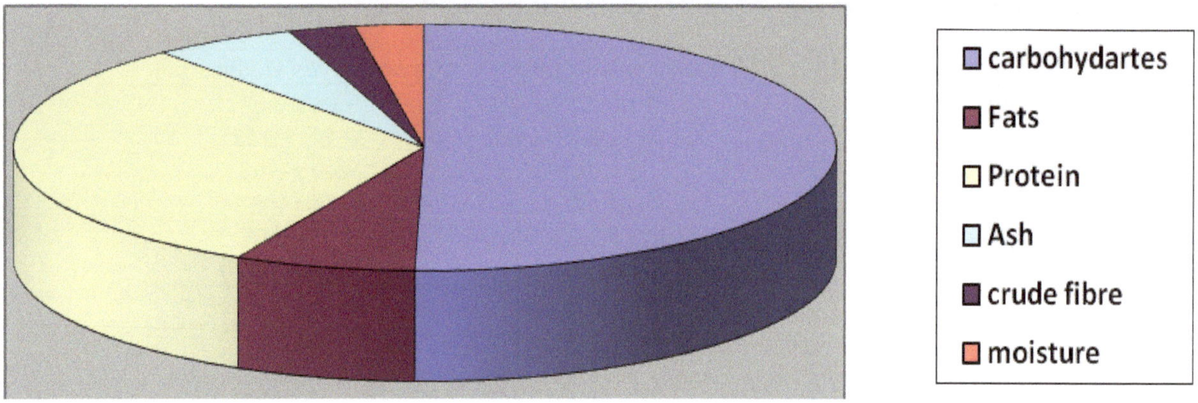


**Figure 1.** Pie chart represents proximate analysis for Bambara groundnut (*Voadnzeia subterranean*).

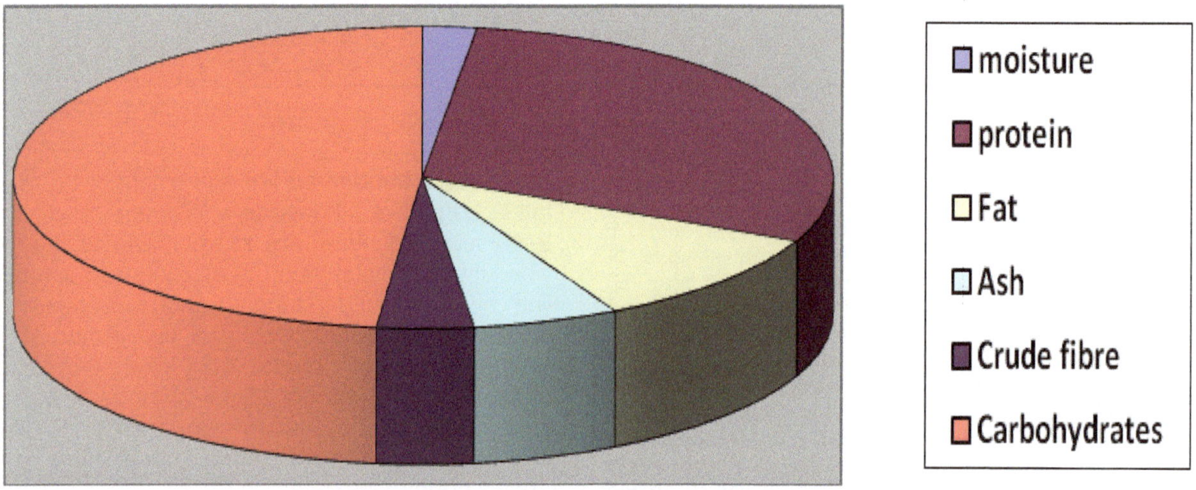


**Figure 2.** Pie chart represents proximate analysis of African yam bean (*Sphenostylis stenocarpa*).

Bambara groundnut (*V. subterranean*). This shows that the seeds have a great potential in alleviating chronic protein deficiency syndromes that are widespread in developing countries such as Nigeria. According to Adu-Dapaah and Sangwan (2004), the seed of Bambara is regarded as a balanced food because when compared to most food legumes, it is rich in iron with high lysine and methionine content. Minka and Bruneteau (2000) reported that Bambara groundnuts (*V. subterranean*) contain about 63% carbohydrates, 18% oil and the fatty acid composition is predominantly linoleic, palmitic and linolenic acids. From our results in (Table 1 and Figures 1 and 2) on the content of carbohydrates and fat in the Bambara (*V. subterranean*) were 51.79 and 7.35% for African yam bean (*S. stenocarpa*) 44.4 and 9.49%. This showed that there are marginal differences between the two legumes in terms of the content of carbohydrates and fats. On comparing the results of the analysis reported by Minka and Bruneteau (2004) on Bambara seed, the difference in result may be due to the processes involved with the handling of the seeds during preparation of different foods from the seeds. In line with this fact, Okpuzor et al. (2010) reported about ten different proteins in the malted seed while the dry seeds were twelve different proteins. The defection in the values may be due to some contributory factors which may include the germination of the seeds; where part of the protein may have been utilized by the growing embryo. It could also be different in mineral content of the soil, climatic factors and the presence of tubercles or nodules of roots that contain nitrogen-fixing bacteria. However, in these observations, the two legume plants studied are relatively very cheap and readily available compared to other legume and protein sources. The nutritional benefits derivable from these legumes can be harnessed by reduction of anti-nutrient contents and formulation of infant foods and diet (Brough et al., 1993; Poulter and Caygill, 2006). According to Adeyeye and Agesin (1997), (*S. stenocarpa*) has a high mineral content which is needful for sustainable health of human life. The mineral composition include:

iron, copper, manganese, phosphorus and calcium. Recent findings reported that seed proteins of bambara (*V. subterranean* have been implicated in anti-hypertensive activities (Yamada et al., 2008); regulation of food intake and body weight in humans, (Anderson and Moore, 2004) and the improvement in plasma and liver lipid profile of rats fed with high cholesterol diet (Yang et al., 2007). The ash content reveals the degree of purity and mineral content of the grains. The low water content of 2.86 and 1.96% for Bambara groundnut (*V. subterranean*) and African yam beans (*S. stenocarpa*) showed that the grain has a high storage potential. Thus, the rate of deterioration of the seeds will be low in view of the little moisture level. Thus, the consumption of the leguminous foods rich in the essential minerals could be an alternative to the solution of protein- energy malnutrition and preference to animal fats which is believed to predispose people to atherosclerosis and coronary heart diseases due to high content of cholesterol. Bambara groundnuts (*V. subterranean*) and African yam beans (*S. stenocarpa*) still exhibit great potential as cash crops which are grown in mixed association with other root tubers. From this study, Bambara groundnuts (*V. subterranean*) and African yam beans (*S. stenocarpa*) offers nutritional and health benefits to the ever –increasing populations of the developing world.

## REFERENCES


Adebowale YA, Adebowale KO (2007). Evaluation of gelation characteristics of mucuna bean flour and protein isolate. Elect. J. Environ. Agric. Food Chem., 6 (8):2243-2262.

Adebowale YA, Adeyemi IA, Oshodi AA (2005a). Variability in the physicochemical, nutritional and antinutritional attributes of six *Mucuna* species. Food Chem., 89: 37-48.

Adebowale YA, Adeyemi IA, Oshodi AA (2005b). Functional and physicochemical properties of flours of six Mucuna species. Afr. J. Biotechnol., 4(12): 1461-1468.

Adeyeye EI, Agesin OO (2007). Dehulling the African Yam Bean (*Sphenostylis stenocarpa* Hochst. ex A. ich) Seeds: Any Nutritional Importance? Bangladesh J. Sci. Ind. Res., 42(2): 163-174.

Adu-Dapaah HK, Sangwan RS (2004). Improving bambara groundnut productivity using gamma irradiation and *in vitro* techniques. Afr. J. Biotechnol., 3:260-265.

Aletor VA, Aladetimi OO (1989). Compositional evaluation of some cowpea varieties and some under- utilized edible legumes in Nigeria (Die Narung), 33 (10) 999-1007.

Anderson GH, Moore SE (2004). Dietary proteins in the regulation of food intake and body weight in humans. J. Nutr., 134:974- 979.

Anchirinah VM, Yiridoe EK, Bennett- Lartey SO (2001). Enhancing sustainable production and genetic resource conservation of Bambara groundnut: A survey of indigenous Agricultural knowledge systems. Outlook on Agric. 30:281-288.

AOAC (1984). Official Methods of analysis, Washington DC USA: Association of Official analytical chemists.

Azam-Ali SN, Sesay A, Karikari KS, Massawe FJ, Aguilar-Manjarrez J, Bannayan M, Hampson KJ (2001). Assessing the potential of an underutilized crop - a case study using bambara groundnut. Exp. Agric., 37:433-472.

Agbede JO, Aletor VA (2003). Comparative evaluation of weaning foods from gliricidia and leucaena leaf protein concentrates and some commercial brands in Nigeria. J. Sci. Food Agric., 84: 21-30.

Brough SH, Azam-Ali SN, Taylor AJ (1993). The potential of bambara groundnut (*Vigna subterranea*) in vegetable milk production and basic protein functionality systems. Food Chem., 47:277-83.

Duke JA, Okigbo BN, Reed CF (1977). African Yam bean (*Sphenostylis sternocarpa*) (Hoscht exA.Rich). Tropical grain legume Bulletin 10:4-5.

Edem DO Amugo CT, Eka OU (1990). Chemical composition of Yam Bean (*Sphenostylis sternocarpa*). Trop. Sci., 30: 59-63.

Eromosele CO, Arogundade LA, Eromosele IC, Ademuyiwa O (2008). Extractability of African yam bean (*Sphenostylis sternocarpa*) protein in acid, salt and alkaline aqueous media. Food Hydrocolloids, 22:1622-1628.

Evans DM, Boulter D (1974). Amino acid composition of seed meals of Yam bean (*Sphenostylis stenocarpa*) and Lima bean (*Phaseotlus lunatus*). J. Sci. Food Agric., 25: 919-922.

Fadahunsi IF, Sanni AI (2010). Chemical and biochemical changes in Bambara nut (*Voandzeia subterranean* (L) (thours) during fermentation to tempeh Elect. J. Environ. Agric. Food Chem., 9 (2): 275-283.

Ihekoronye AI, Ngoddy PO (1985). Integrated Food Science and Technology for the tropics, University of Ibadan Press. p. 283.

Kine BB, Eka OU, Aremu CY, Ifon ET (1991). Chemical evaluation of the nutritive value of African yam bean (*Sphenostylis stenocarpa*). Trop. J. Appl. Sci., 1: 99 - 102.

Lawal OS, Adebowale KO (2006). The acylated protein derivatives of *Canavalia einsiformis* (Jack beans): A study of functional characteristics. Food Sci. Technol., 39: 918-926.

Lawal OS, Adebowale KO, Ogunsanwo BM, Sosanwo OA, Bankole SA (2005). On the functional properties of globulin and albumin protein fractions and flours of African Locust bean (*Parkia biglobossa*). Food Chem., 92: 681-691.

Minka SR, Bruneteau M (2000). Partial chemical composition of Bambara pea (*Vigna subterranean (L. Verde*). Food Chem., 68:273-276.

Oluwole SI, Taiwo RE (2009). Comparison of nutritional composition and anti-nutrient status of fermented, germinated and roasted bambara groundnut seeds (*vigna subterranea*)", British Food J., 111(4): 376 – 386.

Okpuzor J, Ogbunugafor HA, Okafor U, Sofidiya MO (2010). Identification of protein types in bambara nut seeds: perspectives for dietary protein supply in developing countries. EXCLI J., 9:17-28

Poulter NH, Caygill JC (2006). Vegetable milk processing and rehydration characteristics of bambara groundnut [*Voandzeia subterranean* (L.)thouars. J. Sci. Food. Agric., 31:1158-1163.

Vasconcelos IM, Siebra EA, Maia AA, Moreira RA, Neto AF, Campelo GJA, Oliveira JTA (1997). Composition, toxic and antinutritional factors of newly developed cultivars of Brazilian soybean (glycine max)", J. Sci. Food Agric., 75: 419-26.

Yamada Y, Nishizawa K, Yokoo M, Hui Z, Onishi K, Teraishi M, Utsumi S, Ishimoto M, Yoshikawa M (2008). Anti-hypertensive activity of genetically modified soybean seeds accumulating novokinin. Peptides, 29: 331-337.

Yang SC, Liu SM, Yang HY, Lin YH, Chen JR (2007).Soybean protein hydrolysate improves plasma and liver lipid profiles in rats fed high-cholesterol diet. J. Am. Coll. Nutr., 26:416-420.

# 20

# Land use planning regarding sustainable development through agritourism: Şarköy example

**Tuğba Kiper**

Department of Landscape Architecture, Faculty of Agriculture, Namık Kemal University, Tekirdağ/Turkey.
E-mail: tugbadokumaci@hotmail.com.

**In rural areas, where agricultural tourism activities are not dense but having apparent potential, in order to provide sustainable development, it is necessary to maintain a rational and protective planning process. With this point, firstly, landscape potential of a field should be determined, while the activities which can be performed and the implementation areas regarding this potential should be determined. This is because planning decisions made without considering conservation-use balance and providing daily solutions would damage unique sources of the rural areas in an irrecoverable way. At this stage, analysis studies performed to determine land use feasibility are important. Within this scope, this study aims to reveal the capacity of the field, actively regarding sustainable agritourism of Şarköy, which has importance with its variety of agricultural products, the values of natural and cultural resources and its social structure in the Thracian Region. In this research, a trial is made to answer questions such as: "at which level subjects regarding agricultural activities are convenient?" and "what should be done to develop agritourism activities?" Parallel to this, firstly, through forming a database by processing the natural and cultural characteristics of the research field in geographical information system (GIS) environment, evaluation factors regarding use of agritourism (viticulture and olive cultivation) and convenient coefficients were ranked according to their importance. The convenience values obtained were questioned and convenience maps were made and the local people's recommendations were developed within the dimensions of local people and participation. The results obtained in this research will be guiding both the local people and local administrations in the decisions of planning made about agritourism subject and in the implementation phase.**

**Key words:** Agritourism, land use planning, sustainable development, Şarköy/Turkey.

## INTRODUCTION

Agritourism is defined differently by different people and organizations. Some related terms that are used interchangeably with agritourism or that are complementary to agritourism include nature tourism, alternative farming, wildlife enterprises, ecotourism, agritainment, heritage tourism, agrieducation and value added agriculture (Brant and Rhoades, 2007).

Agritourism is defined as travel, which combines agricultural or rural settings with products of agricultural operations all within a tourism experience. Agritourism can be defined as "A range of activities, services and amenities provided by farmers and rural people to attract tourist to their area in order to generate extra income for their businesses". Agritourism is also viewed as, "an alternative enterprise that links value-added or non-traditional agricultural production or marketing with travel to a farm or ranch". Agritourism is defined here as a set of rural activities, including participating in farming activities, exploring local culture, enjoying the landscape and agro-biodiversity, observing organic and conventional agricultural practices and sampling tropical fruits and vegetables. Reynolds (2005) mentioned that agritourism is businesses conducted by farmers whose working agricultural operations for the enjoyment and education of visitors. Agricultural producers are beginning to try their hand in the tourism industry by opening their farm to guests seeking entertainment or educational activities; this type of tourism is known as agritourism (Mcgehee and Kim, 2004). Agro-tourism has the potential to play an important role in a country whose economy is largely dependent on income derived from agricultural exports and recreational tourism. In extensions meaning, agrotourism presents the potential generating farm revenues and increasing profitability. Additionally, visitors of agritourism contact directly with farmers and support the increase of agriculture products indirectly (Catalino and Lizardo, 2004; Reynolds, 2005; Gopal et al., 2008; Maetzold, 2002).

For rural residents not living on farms, it enables them to upkeep large heritage homes and beautifies their surrounding grounds and landscapes. Rural tourism is based on the rural environment in general, whereas agritourism is based on the farm and the farmer. Rural tourism provides many benefits to local areas by bringing visitors to the region, increasing awareness for agricultural products and showcasing produce to the local and regional community. It can also provide incentives for local heritage and conservation groups to preserve unique heritage landscapes and built heritage. Most studies have highlighted the multiple benefits that agritourism brings to the farm, local communities, agriculture heritage and natural resources (Fleischer and Pizam, 1997; Busby and Rendle, 2000; Ventura and Milone, 2000; Sharpley, 2002; Wicks and Merrett, 2003; Hegarty and Przezbórska, 2005; Fleischer and Tchetchik, 2005; Mitchell and De Waal, 2009). Lacking a formal definition, agritourism can be summarized as anything that connects consumers with the heritage, natural resource or culinary experiences unique to the agricultural industry or a particular region of the country's rural areas.

Agritourism activities are limited only by the imagination. Some common examples include (Barbieri and Tew, 2010):

- Special events, including farm festivals and fairs, bonfires, outdoor plays or concerts, dances, rodeos, livestock shows, farm activity demonstrations and classes, farm implement or antique shows and farm tours.
- Direct sale of products, through farm retail markets, roadside stands, pickyourown operations, farmers' markets, community supported agriculture (CSA) and adoptananimal or tree programs.
- Children's activities, school or youth group tours and farm school programs.
- Alternative uses of farmland, including farm fishing, hunting, horseback riding, trap and skeet shooting, canoeing, bird and wildlife viewing, camping, biking, etc.
- Hospitality services, such as cabin or room rentals; restaurant or catered meals; facilities for weddings, corporate retreats and camping; working vacations and picnic areas outdoor recreation.

Agritourism is a segment within the larger sector of rural tourism. Agritourism is a new type of rural tourism, which is generally implemented in disadvantaged agricultural lands. It is developed as a sector with the aim of not only a development instrument for local people who are dependent on agricultural production, but also for sustaining the agricultural lands. In some countries like Italy, Spain, Greece, Austria, Denmark, England, France, Germany, Ireland and Norway it is a growing sector with the policies of the governments aiming to benefit from it as a regional development instrument (Demirtaş Topcu, 2007; Frater, 1983). Agrotourism generates a significant contribution to the rural development process in rural areas and does indeed do so in many cases (Butler and Hall, 2003). Agritourism may be regarded as part of the change in the European model of agricultural development from productivism towards sustainability and multifunctionality (Van der Ploeg, 2008) and it has received great attention in rural/agricultural politics and economics over the last decade (Brandth and Haugen, 2010). In addition, there are opportunities for expanding agritourism-related enterprises in many countries across the state, based on the presence of significant natural amenities or cultural/historical sites and events (Wilson et al., 2006).

Agritourism aids rural economic development by providing alternative use of farmland, increasing revenue of on-farm activities, improving business sustainability and bringing economic revenue to rural areas both on-site and near the operation. Agritourism brings diversify-cation opportunities to farms and ranchers that can help buffer fluctuating markets. It can increase farm revenue and increase community economic activity. It can provide economically feasible ways to care for natural habitats, natural scenic areas, national resources and special places (Keith et al., 2003; Jensen et al., 2006; Geisler, 2008).

Nowadays, in Turkey some recent developments are occurring related to the agritourism development in the political and practical areas, because the rural develop-ment issue is one of the main problems that has to be solved in the process of inclusion to the European Union. Nowadays, there is more emphasis on the legislative and regulative literature about alternative tourism, rural tourism and agritourism as an alternative tool. In the rural development programs and strategy documents of eighty five-year development plan, the existence of rural tourism potential in some rural areas and regions and their potential powers for diversifying the rural economy are emphasized. It is also declared that agritourism will be strongly supported as one of the rural development strategies (Demirtaş Topcu, 2007).

However, when considering that tourism is a sensitive issue and local people would damage the sources of region while using them, it is seen that a sustainable tourism planning which sustains conservation-usage balance unique to the region and uses the sources rationally are required. At this point, asking questions such as "is each part of the geographical place convenient for the development of tourism?", "is it correct to open every place to tourism?" and "which places should be chosen for tourism and how the selection

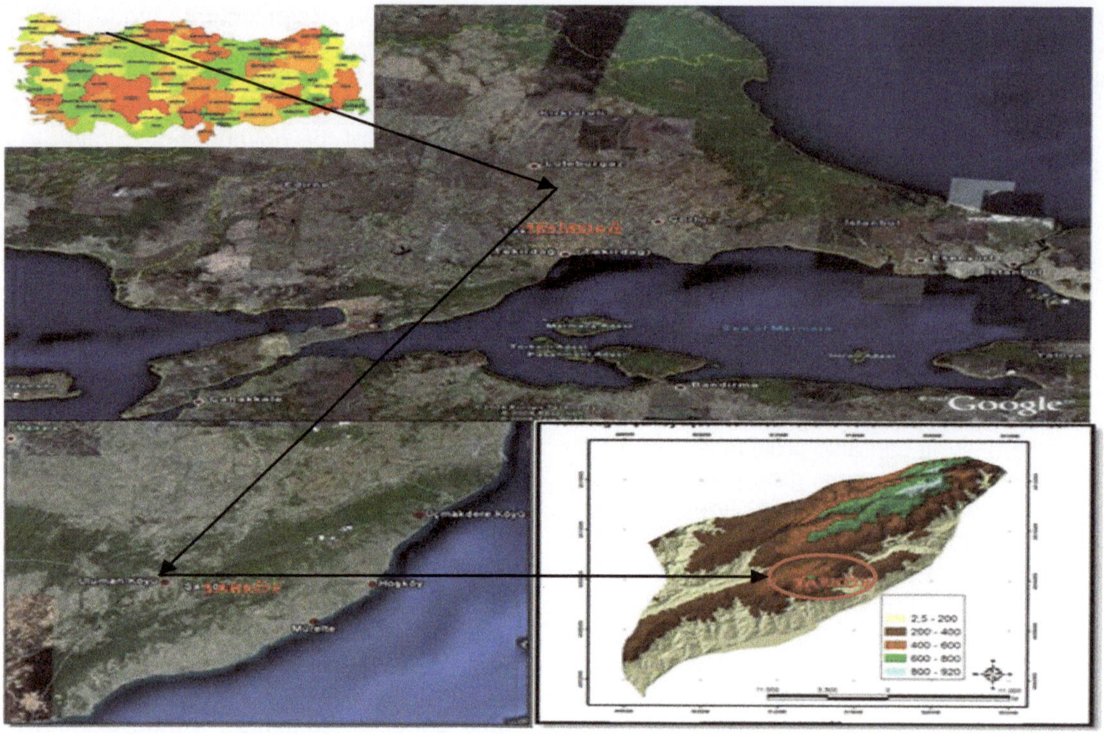


**Figure 1.** Location of the research area.

should be made?" is very critical.

Şarköy district chosen as the sample in the research has a position to create opportunities for visitors to be personally involved in agricultural activities, to be familiar with farm life, to be actively and passively involved in various agricultural activities and to be acquainted with local culture (tasting traditional local food, introducing local handicrafts and participation to various festivals). Although, the research field has agricultural, natural and cultural characteristics, it is not at a sufficient level in terms of agritourism activities. At the same time, even though most of the local people that have an active role in agricultural production, are unpaid family workers.

Accordingly, in the research, the hypothesis of "in order to provide an active rural development in Şarköy agritourism activities should be developed and an appropriate field use planning regarding these activities should be performed" has been developed. From this point of view, the study is important in terms of the development of agritourism of Şarköy district, obtaining an authentic identity and revealing unknown values.

## MATERIALS AND METHODS

Şarköy district of Tekirdag province constitutes the basic material of the study (Figure 1). The research field is a district of Tekirdag province located at the Thracian part of Marmara region on the north-western part of Turkey. The district is surrounded by Malkara in the north, Tekirdag in the north-east, Marmara sea in the south and south-east and Gelibolu in the west. The district has a surface area of 481 km². Şarköy is 84 km away from Tekirdag, 87 km from Çanakkale and 220 km away from Istanbul. It is composed of two towns (Mürefte and Hoşköy) and 26 villages. According to the 2008 census, the population of the city center is 16.121 and the populations of the villages are 13812, and it is 29933 in total (Anonymous, 2009).

The economy of the district depends on agriculture (viticulture, olive), stock-breeding, fishery and tourism. Total surface area of the district is 48.100 ha, 89.40% (43.000 ha) of this area is agricultural land and 10.60% (5.100 ha) of it is non-agricultural land (Anonymous, 2010).

Şarköy district comes first among the districts of Tekirdag province in terms of possession of vineyard and grape cultivation. Grape fields of Şarköy district constitute 79% (52.210 das) of total grape fields of the province and approximately 80% (52.120 tones) of grape production. When types of grapes grown in the district are analyzed, it is noticed that 86.4% of grapes grown are wine grape varieties and 13.6% are table grape varieties (Özdemir and Kiracı, 2006).

According to the environmental condition planning report for Tekirdag (2007), there are forty winemaking enterprises (Anonymous, 2007).

In the district, 3 water product cooperatives, 13 agricultural development cooperatives and 1 irrigation cooperative, 17 cooperatives in total are active (Anonymous, 2010).

Besides, in this study, 1/25.000 scaled topographic maps (11 units), 1/25.000 scaled soil maps (11 units), climate data, population data and literature knowledge were utilized as material. Arc-GIS 9.0 software was used in computerizing and analyzing data and maps.

In the research, with field studies conducted in various times, examination was made about characteristics of natural landscape (geographical position, surface features and natural vegetation) and cultural landscape (history, archeology, socio-economic structure, handicrafts, socio-cultural life style and preserved areas), and natural-cultural landscape values obtained were processed in the GIS and the database was created. A database was formed by preparing some thematic maps concerning the area's geographical position, transport situation, hydrological structure, earth properties, etc., in computer medium in Arcview (9.0) program. Observations performed were supported by literature knowledge and SWOT analysis regarding issues, potential and obstacles was carried out.

Various evaluation factors (for example slope, exposure, height and climate) and sub-units influencing convenience for agritourism land use types notably regarding viticulture and olive cultivation were selected separately using literature and expert opinions.

In terms of viticulture and olive cultivation, evaluation factors and suitability coefficients were ranked as the most suitable (3), suitable (2) and not suitable (1) according to numerical values demonstrating their relative importance (Tables 1 and 2).

In the determination of evaluation factors and measures, literature data (Çelik, 2007; Ağaoğlu et al., 2001; Çelik et al., 1998; Toksarı and Can, 2008; Gündüzoğlu, 2004; Mendilcioğlu, 2002; Ünal et al., 2007; Woo and Yean, 2006), purpose of the study, the characteristics the region possesses, expert opinions, the state of user group and the methods used by Kiper (2006), Mcharg (1969) and Topay (2003) were leading. Convenience maps were constituted regarding use of viticulture and olive cultivation separately by questioning the natural and cultural landscape characteristics via geographical information system (GIS) according to convenience coefficients.

## RESEARCH RESULTS

Viticulture and oil cultivation which play important roles in agritourism activities have always taken an important place within the socio-economic structure of eastern and western civilizations both economically and mystically. For instance, in Europe and in the world, viticulture is very common. Especially France, Spain, Italy, (world's oldest and third biggest winemaker), the USA (California), Australia (New South Wales is the first winemaking center of Europe) and South Africa are the leading countries with their wine yards and winemaking (www.wineguides.co.uk). Turkey is located on the most convenient climate zone for viticulture and at the crossroads of gene centers of grape vines. And because of the position of geography where the grape vine was taken to the culture, it has preserved its position to be one of the cultivated plants which all civilizations prevailing in Anatolia have valued most until today (Anonymous, 2005). Olive (*Olea europeae L.*), has had an important place in both old civilizations of the Mediterrenean basin where is its region of origin and within social and economic life of existing countries (Ünal et al., 2007; Denk, 2004).

Therefore, in the research conducted, in order to contribute to the local people economically, socio-culturally and environmentally, and to introduce the culture of grape vine and olive with all aspects, the best places for cultivation were determined.

At the end of the questionings, it was assessed that proportion of the most convenient areas to be separated for use of grape vine activities aimed at agritourism in the region, and is approximately 17.24 ha. For the use of olive cultivation activities, the proportion of most convenient and convenient areas to be separated were established as "10101.06" and "1814.0616" ha, respectively (Figures 2 and 3).

According to the evaluations of the method, wine-growing areas would be suitable for use for agricultural tourism (2-scoring) could not be identified. Although, research field is suitable in terms of slope, elevation and climate, it is subjected to some limitations in terms of aspect, soil type and soil depth. When it was examined vectoraly, the southern, south-western and south-eastern slopes were determined favorable for wine-growing .

SWOT analysis was performed based on the field observations carried out, the surveys literature review and analysis (Table 3). According to this, it is seen that the field has a strong agritourism potential. However, although the local people lean towards this subject, but this potential has not been utilized. For example, in the 1/100.000 scaled Thracian Sub Region, Ergene Basin Revision of Environmental Order Planning, it was aimed to support sectors which provide or will provide competitive advantages and within this scope Şarköy district was defined as "tourism (agritourism) region" (Ergene Environmental Order Planning Report, 2009).

### Conclusion

Agricultural product demand increases together with the tourism activities in rural areas. The production and the producers are positively affected by such conditions. In addition, employment, additional income, personal income, standard of living, rural migration and public revenues are also positively affected, and the region becomes a center of attraction. In addition, agricultural tourism will provide the visitors to participate in person the harvesting, tillage and other businesses and in nature areas where the local products on display they will feel themselves as a part of this place. According to Ungureanu (2008) agri-tourism represents a chance for the inhabitants of villages, the inhabitants being willing to try a new seasonal (about 100 days per year) activity which would bring them extra incomes.

With this purpose the research, providing a database for natural and cultural landscape values, the field convenience analysis regarding agritourism was conducted. Field convenience analysis was performed to rank which area is the most or the least convenient for agritourism activities (vine cultivation, olive cultivation). In this study, the most suitable and appropriate wine-growing and olive production areas for use of agricultural tourism were identified. In this context, the slopes facing coastal areas and the southeastern and southwestern sections of the research area were determined the most suitable areas for agricultural tourism. This potential will highly increase only if it is combined with the other uses of agricultural tourism.

**Table 1.** Agritourism (viticulture) evaluation factors and criteria of suitability class.

| | Evaluation factors | Evaluation sub-units | | | Point |
|---|---|---|---|---|---|
| Natural Landscape Features | Climate | Temperature (°C) | Annual mean temperature | 11–16 | 3 |
| | | | | <9 , >35 | 1 |
| | | | The average temperature in the growth period | >13 | 3 |
| | | | The average temperature in summer | >20 | 3 |
| | | Precipitation (mm) | Annual total precipitation | 500-600 | 3 |
| | | | | <300<br>>900 | 1 |
| | | Wind speed (m/sn) | The average wind speed | <3 | 3 |
| | Great soil groups | Alluvial soil | | | 3 |
| | | Brown forest soil | | | 2 |
| | | Lime brown forest soil | | | 2 |
| | | Vertisol | | | 2 |
| | Soil depth | Deep | | | 3 |
| | | Medium deep | | | 2 |
| | | Shallow-very shallow | | | 1 |
| | Land use capability Classes | I. - II. Class | | | 3 |
| | | III. - IV. Class | | | 2 |
| | | V. Class and over | | | 2 |
| | Accessibility – Being access to water (m) | Access to streams (m) | 0-100 | | 1 |
| | | | 100-500 | | 3 |
| | | | 500 and over | | 2 |
| | | Access to the shore (m) | 0-100 | | 1 |
| | | | 100 and over | | 3 |
| | Height Groups | 0-750 | | | 3 |
| | | 750-950 | | | 2 |
| | Slope (%) | 0-5% (Flat) | | | 2 |
| | | 5-25% | | | 3 |
| | | 25% and over | | | 2 |
| | Aspect | South-southwest | | | 3 |
| | | West | | | 2 |
| | | East, North | | | 1 |
| Cultural Landscape Featutes | Agri tourism in the local people's perspective | Positive | | | 3 |
| | Traditional settlement pattern | Availability | | | 3 |
| | Traditional social activities (Festivals, Ceremonies) | Availability | | | 3 |
| | Accommodation opportunities | Availability | | | 3 |
| | Communication opportinitics | Positive | | | 3 |

In the evaluation made for agritourism, at the uses of the most convenient areas, the local products can be cultivated without any limitations whereas the convenient areas are subject to some limitations in the field. This may lead to decrease in productivity. Nonetheless, vine and olive can be cultivated in various amounts. Therefore, in the selection of activities to be delivered the field as agritourism activity types, it will be important to take rural and ecological characteristics of "the most convenient and convenient" areas into account and to develop them within the scope of conservation-use understanding.

Some recommendations for the local people and participators have been made to improve agritourism vision of the field in the direction of the research results (Table 4).

According to Bradley (1982) wine tourism can be defined as visitation to vineyards, wineries and wine festivals. In particular, festivals and events based around rural products are growing in popularity (Higham and

**Table 2.** Agri tourism (olive cultivation) evaluation factors and criteria of suitability class.

| | Evaluation factors | Evaluation sub-units | | | Point |
|---|---|---|---|---|---|
| Natural Landscape Features | Climate | Temperature (°C) | Annual mean temperature | 15-20 | 3 |
| | | | | <-7 40< | 1 |
| | | | The average temperature in the growth period | 20 | 3 |
| | | Precipitation (mm) | Annual total precipitation | 700-800 | 3 |
| | | | | 800-200 | 3 |
| | | | | <200 | 1 |
| | | Wind Speed (m/sn) | The average wind speed | <3 | 3 |
| | | | | 10< | 1 |
| | Great Soil Groups | Alluvial Soil | | | 3 |
| | | Brown Forest Soil | | | 3 |
| | | Lime Brown Forest Soil | | | 3 |
| | | Vertisol | | | 3 |
| | Soil Depth | Deep | | | 3 |
| | | Shallow-Very Shallow | | | 1 |
| | Land Use Capability Classes | I. and II. Class | | | 3 |
| | | III. Class and over | | | 2 |
| | Accessibility — Being Access to Water (m) | Access to Streams (m) | 0-100 | | 1 |
| | | | 100-500 | | 3 |
| | | | 500< | | 2 |
| | | Access to the Shore (m) | 0-100 | | 1 |
| | | | 100< | | 3 |
| | Height Groups (m) | 0-500 | | | 3 |
| | | 500-800 | | | 2 |
| | | 800 and over | | | 1 |
| | Slope (%) | 0-25% | | | 3 |
| | | 25%< | | | 2 |
| | Aspect | South-southwest | | | 3 |
| | | West-East-North | | | 2 |
| Cultural Landscape Features | Agri tourism in the Local People's Perspective | Positive | | | 3 |
| | Traditional Settlement Pattern | Availability | | | 3 |
| | Traditional Social Activities (Festivals, Ceremonies, Gastronomy) | Availability | | | 3 |
| | Accommodation Opportunities | Availability | | | 3 |

Ritchie, 2001), as is "agritourism" (Langworthy et al., 2006). Therefore, festivals and days organized related to wine have been increasing the brand value of the region, and have raised the prestige of wine producers. In this sense, fairs and festivals (festivals of vintage which can be set up within the period between August and November when it is grape harvesting period) will bring economic, social and cultural dynamism.

Besides, according to Tefler (2001), the basic motivation factor for wine tourists is to know the culture of wine region, to taste wine and to have the knowledge of it seeing the production phase. For example in the Wine Tourism Strategic Business Plan 2002 to 2005 in Australia, in order to promote wine tourism, visitation to wineries and wine regions were recommended (Anonymous 2002). Within this scope, in Şarköy in addition to wine distilleries, there are wine tasting houses and sales stores. For instance, Kayra, Doluca and Melen winemakers have been organizing tours to wine production plants and vineyards until 15 October. This is an important tool to attract visitors. In this point, while visitors observe the period from grape growing to wine making in natural environment, at the same time it should give the opportunities for them to see remnants of the

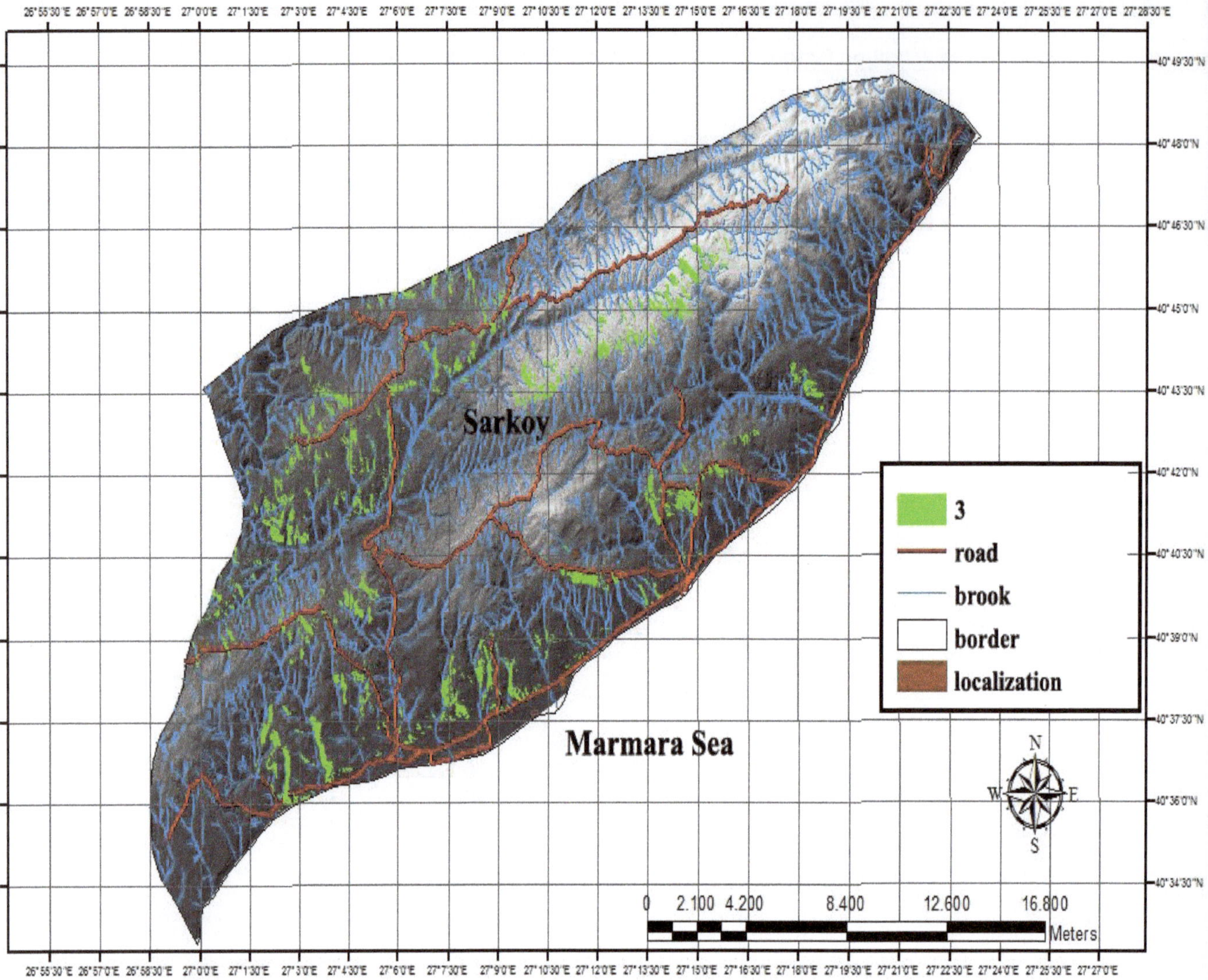


**Figure 2.** Areas suitable for agritourism (viticulture) in research area.

Thracian civilization. For example, today, the area where Gaziköy (Ganos) and Hoşköy (Hora) settlements are located was named GANOHORA as a religious center of Byzantine age. Ganos (Light Mountain) is the highest mountain of the Thrace region with 980 m height; where in the past Christians accepted as "Sacred Mountain" and visited it to be a pilgrim. In the abbey on this region, priests produced their own food, and sold the surplus to other settlements. Wine distributed in amphora (barrel to conserve wine in) on the coast of Ganohora made this region one of the amphora producing centers. Density of wine production is understood via Ganos sealed amphora, too. Within this scope, the historical values of the region and various activities regarding agro tourism support each other and bringing them together in integrating way will be useful to stir up the interest of the visitors.

Furthermore, in order to increase agricultural potential, various steps should be taken. For this, production of all products derived from grape (molasses, sausage, dried fruit rollup, köfter, vinegar, grape juice, brined vine leaf and grape cologne) by the diversification of them, and these should be sold in the local market. For example, currently one day in a week (Friday), local products are being sold under the name of "producer's market". Here, the point required to be emphasized is to develop rural tourism which will contribute to sell these products in place, to attract customers to the villages and to strengthen rural economy in place. The villagers should not come to markets of towns, but city folks should come to villages. Accordingly, brand name products regarding the district should be diversified and subjects of quality, standard, price and logo should be considered.

Some villagers utilize Şarköy district forest by collecting

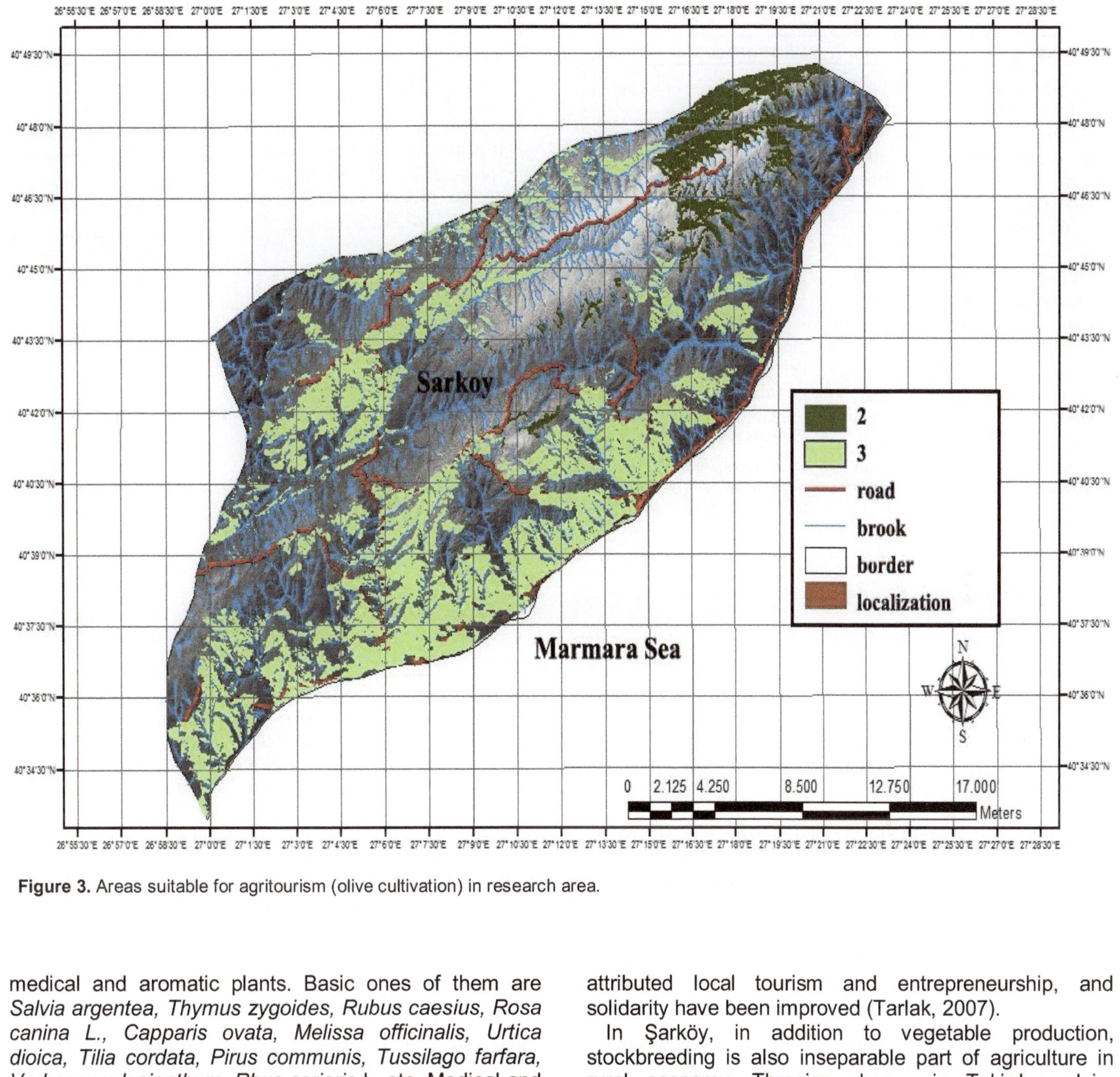


**Figure 3.** Areas suitable for agritourism (olive cultivation) in research area.

medical and aromatic plants. Basic ones of them are *Salvia argentea, Thymus zygoides, Rubus caesius, Rosa canina L., Capparis ovata, Melissa officinalis, Urtica dioica, Tilia cordata, Pirus communis, Tussilago farfara, Verbascum lasianthum, Rhus coriaria* L, etc. Medical and aromatic plants should also be considered within agritourism. Besides, establishing small local health care centers which will sell these plants, inform about them and operate various applications regarding treatments would contribute to the economy as well. For example, according to Rural Europe (2001) against the regressive agriculture in Bormida valley on the piedmonts of Alp in Italy, a cooperative was established to cultivate herbs in the region, a certificate for organic agriculture was obtained for 19 vegetable oil types produced by dried herbs and they were put in sale. Moreover, aromatic therapy courses were taught. As a result, abandoned agricultural lands began to be cultivated again, it has attributed local tourism and entrepreneurship, and solidarity have been improved (Tarlak, 2007).

In Şarköy, in addition to vegetable production, stockbreeding is also inseparable part of agriculture in rural economy. Thracian cheese in Tekirdag, dairy products like goat cheese, skim milk cheese and curd cheese have local taste. For instance, according to Soykan (2000), some of the property owners dealing with stock breeding in France produce cheese, open their factories to visitors and sell their products there. In this sense, villages (Yörgüç and Yayaköy) who produce dairy products in Şarköy can also take part in agro tourism.

There is an educating side of tourism for tourists. In Europe, pedagogic farms are places of learning, observing and implementation for especially students. Within this scope, villages and farms having dairies (Yörgüç), villages producing olive oil and wine (Mursallı, Eriklice, Hoşköy and Güzelköy) can be utilized for

**Table 3.** The result of the SWOT analysis of agritourism.

| Strengths | Weaknesses | Opportunities | Threats |
|---|---|---|---|
| - Climatic values of the field is appropriate,<br>- Positive look at the vegetable production that most of local people's main pursuit,<br>- Rich water resource,<br>- The presence of wine production and olive oil processing plant,<br>- The presence of medicinal and aromatic plants,<br>- To be a branded agricultural products (grapes and wine, olives and olive oil, fish Şarköy specific, cheese etc.).<br>- Boutique to cultivate an organic product compliance (Grapes and wine)<br>Residents' hospitality | - In terms of land-use capability classes for agriculture may be the most suitable I, II and III. Classes of land area to be covered of a certain portion<br>- Lack of organization agricultural products production and marketing,<br>- Lack of young work force,<br>- The lack of publicity in the study will take place in the field of agricultural tourism.<br>- Problems arising from unplanned development of second homes<br>- Underdeveloped physical infrastructure (especially in the Uçmakdere Village) | - To promote of local agricultural products to the participants,<br>- Chemicals like hormones, insecticides etc. little or no use in production. Therefore, the ability to create the potential ecological agriculture,<br>- Rural-tourism, agro-eco-board increase in demand for<br>- The opportunity to participate of local people especially in the training of vineyard and olive cultivation,<br>- Tekirdag province to show diversity in terms of agricultural production in other districts (vineyard, olive),<br>- Growing market economy opens possibilities for the development of local and visiting tourism | - Lack of agricultural cooperativization<br>- Economic difficulties,<br>- Rural– urban migration,<br>- Generally, agricultural lands have multi-partner,<br>- Qualitative and quantitative decadence of<br>-Recreation resources |

**Table 4.** The recommendations for agritourism.

| Activities | For the local people | For the participators |
|---|---|---|
| Buying fresh products and homemade food from the farms and roadside stand booth | Production of local food products in traditional methods, (molasses, couscous (kuskus), fettuccini, paste, dried fruit rollup, jam) should be supported. | - Advertisements should be given regarding foods naturally growing in ecological conditions of the region and production methods and benefits of homemade products via brochures and booklets |
| Joining activities like planting, caring, or collecting of plants such as olive, vine, medical and aromatic plants by participating in garden tours (self-collection activities) and observing the development phases of products grown ecologically (from blossoming to bearing fruit) | - For the continuity of products obtained from natural vegetation a calendar should be prepared for planting, and harvest.<br>- A training program should be organized for farmers on organic production and harvest.<br>- Specialization training about local food production should be provided to the local people.<br>- Calendars should be prepared showing blossoming and bearing fruit seasons of products cultivated in the region ecologically. | - For ones wanting to be in agricultural activities actively, a training program should be arranged about organic production.<br>- Planting and harvest calendars prepared for the continuity of products obtained from natural vegetation should be prepared for participators too.<br>- Especially by making vintage and olive collecting activities entertaining, the participators should be stimulated.<br>- Calendars prepared showing blossoming and bearing fruit seasons of products cultivated in the region ecologically should be prepared for the participators, too. |
| Visitation to wine museums and wineries | The opinions of local people should be taken. | - Visitation to wine museums and wineries should be included in tour programs.<br>- For participators, brochures and booklets including information about wine museums and wineries should be prepared.<br>- For participators, used materials and techniques regarding winemaking from past to present should be presented in cinevision. |

**Table 4.** Cont'd.

| | | |
|---|---|---|
| Visitation to olive oil factories | For farmers, a training program on organic olive oil production should be arranged and the local people should be encouraged in this subject. | - In tour programs, olive oil factories should definitely be included.<br>- For participators, brochures and advertisements about olive oil factories should be prepared.<br>- For participators, used materials and techniques regarding olive oil making from past to present should be presented in cinevision. |
| Learning local agricultural products (olive oil soap and olive cologne) | For farmers, a training program on olive oil soap and olive cologne should be arranged and the local people should be encouraged in this subject. | - Opportunities should be created in order to introduce different local agricultural products to the participators.<br>- Especially introduction of olive oil soap and olive cologne should be provided through enriching them in various ways from the packaging to presentation |
| Having picnic | - The opinions of local people should be taken. | - Activities of natural and cultural products should be maintained without damaging them. |
| Be familiar with farm life | - The opinions of local people should be taken.<br>- Various farm life activities from stock breeding to vegetable cultivation should be taken into account in a program and training system. | - In order to attract the participators' interest to agricultural products cultivated in the farms and stock breeding, the activities should be changed into entertaining games<br>- Introduction of farm life should be sustained within certain programs. |
| Attending to festivals | They should be taken into account within the frame of the collection of products and harvest season and in a program. | Advertisement of vines and seasonal fruits grown naturally in the ecological conditions of the region should be provided within a festival program. |

educational purposes.

In conclusion, Şarköy district which is very rich in terms of natural and agricultural values has value of use to be considered significant in terms of agritourism. However, because of lack of planning and management, only sea and costal tourism are apparent, and this leads to intense use of natural landscape values in certain time. On the other hand, it is possible to carry out tourism land use planning based on realistic data obtained via systematic inventory analysis and on balanced use of them by increasing tourism potential of the region via its implementation actively.

## REFERENCES

Ağaoğlu YS, Çelik H, Çelik M, Fidan Y, Gülşen Y, Günay A, Halloran N, Köksal Aİ, Yanmaz R (2001). General horticulture. Ankara Universty Faculty of Agriculture Publication, 5, Ankara/Turkey.

Anonymous (2002). Wine tourism strategic business plan 2002–2005. Embrace the Challenge, November, available online at www.fao.org/es/esa/eJADE, 1 (1): 87-116.

Anonymous (2005). Analyses of temperature and precipitation fields in Turkey, according to the Best grapes areas Report. General Directorate of the State Meteorological, Ankara/Turkey.

Anonymous (2007). Tekirdag Province Environment Plan Report (2007). T.C. Tekirdag Governorship Provincial Directorate of Environment and Forestry, Tekirdag./Turkey. p. 316.

Anonymous (2009). Turkey Statistical Institute for 2008 Population Data. http://report.tuik.gov.tr/reports. Accessed: (18.01.2009).

Anonymous (2010). Briefing Report of the District Directorate of Agriculture of Tekirdag, 2010.

Barbieri C, Tew C (2010). The economic benefits of agritourism in Missouri farms. Technical Report Developed For The Missouri Department Of Agriculture. Jefferson City, Missouri. Available On-Line At: http://www.Agrimissouri.Com/Pdf/Mda_Specialreport_B September 2010.pdf. Accessed: (15.04.2009).

Bradley R (1982). The small wineries of Australia: A guide to the best winemakers. South Melbourne: McMillan.

Brandth B, Haugen MS (2010). Farm diversification into tourism e implications for social identity? J. Rural Studies Xxx (2010). pp 1-10.

Brant S, Rhoades S (2007). Agritourism series agricultural diversification through agritourism. http://www. Oklahomaagritourism.com./pdf /Agritourism_ Resource_ Manual.pdf.

Busby G, Rendle S (2000). The transition from tourism on farms to farm tourism. Tourism Management. 21 (6): 635–642.

Butler R, Hall CM (2003). Tourism and recreation in rural areas. New York: John Wiley & Sons.

Catalino HA, Lizardo M (2004). Agriculture, environmental services and agro tourism in the Dominican Republic. Ejade electronic J. Agric. Dev. Econ. Agric. Dev. Econ. Division (ESA) FAO.

Çelik H, Ağaoğlu S, Fidan Y, Marasalı B, Söylemezoğlu G (1998). General viticulture. Sun Fidan Professional Books Series, p. 253,

Ankara/Turkey.
Çelik S (2007). Viticulture. Volume I. Extended 2 Print. Avcı Offset İstanbul/Turkey. p. 428.
Demirtaş Topcu E (2007). Agri-tourısm: As a new element of country planning. A Thesis Submitted to the Graduate School of Natural and Applied Sciences Of Middle East Technical University By, 187p.
Denk G (2004). Olive view survey. http://www.tgdf.org.tr/turkce/tr/rapor/ITO/2004/. Accessed: (04.07.2010).
Ergene Environmental Order Planning Report (2009). 1/100.000 scaled Thrace sub-region Ergene basin revision environment planning, Ministry of Environmental and Forest, p.78.
Fleischer A, Pizam A (1997). Rural tourism in Israel. Tourism Manag., 18 (6): 367-372.
Fleischer A, Tchetchik A (2005). Does rural tourism benefit from agriculture? Tourism Manag., 26, 493-501.
Frater JM (1983). Farm tourism in England-planning, funding, promotion and some lessons from Europe. Tourism Management, 4(3), 167-79.
Geisler M (2008). Agritourism profile. Retrieved March 12, 2008, From http://www.Agmrc.Org/Agmrc/Commodity/Agritourism/Agritourism/Agritourismprofile. Accessed: (18.01.2009).
Gopal R, Varma S, Gopinanthan R (2008). Rural tourism development: constraints and possibilities with a special reference to agri tourism a case study on agri tourism destination - Malegaon Village, Taluka Baramati, District Pune, Maharashtra, Conference On Tourism in India – Challenges Ahead, 15-17 May 2008, Irmak.
Gündüzoğlu G (2004). Natural environment analysis in West Anatolia by using GIS method (ın olive example). 3 rd GIS Days in Turkey, Oktober 6-9 2004, Fatih Universty, İstanbul/Turkey.
Hegarty C, Przezbórska L (2005). Rural and agri-tourism as a tool for reorganizing rural areas in old and new member states - a comparison study of Ireland and Poland. Int. J. Tourism Res., 7 (2): 63-77.
Higham JES, Ritchie B (2001). The evolution of festivals and other events in rural southern New Zealand, Event Manag., 7 (1): 39-49.
Jensen K, Lindborg C, English B, Menard J (2006). Visitors to Tennessee agritourism attractions: demographics, preferences, expenditures, & projected economic impacts. Retrieved February 18, 2008, From University of Tennessee. Web Site: http://Web.Utk.Edu/~Aimag/Pubs/Research%20report%20visitors%20surveys3.Pdf. Accessed: (21.05.2009).
Keith D, Rilla E, George H, Lobo R, Tourte L, Ingram R (2003). Obstacles in the agritourism regulatory process: perspectives of operators and officials in Ten California Counties. AIC Issues Brief, Agricultural Issues Center, University of California, p.22, Davis.
Kiper T (2006). An assesment of the landscape potential of Safranbolu-Yörükköyü in the frame of rural tourism. Universty of Ankara Graduate School of Natural and Applied Sciences Department of Landscape Architecture (Ph. D. Thesis ), Ankara /Turkey. p. 367.
Langworthy A, Howard J, Mawson F (2006). Building the relationship between agriculture and tourism-models and benefits of co-operation. Report Prepared for the Centre for Agriculture and Business, Swinburne University of Technology, Melbourne.
Maetzold J (2002). Nature-based tourism & agritourism trends: unlimited opportunities. www.kerrcenter.com/publications/2002_proceedings/agritourism.pdf. Accessed: (10.01.2011).
Mcgehee NG, Kim K (2004). Motivation for agri-tourism entrepreneurship [Electronic Version]. J. Travel Res., p. 43.
McHarg IL (1969). Design with nature. ISBN 0-471-11460-X.
Mendilcioğlu K (2002). Climate subtropical fruits (olive). Ege Universty Faculty of Agriculture Publication, 12/7, Izmir/Turkey.
Mitchell CJA, De Waal SB (2009). Revisiting the model of creative destruction: sty. Jacobs, Ontario, a decade later. J. Rural Stud., 25: 156-167.
Özdemir G, Kiracı MA (2006). Solving the problem of producers in viticulture in the organizational status and opportunities: examples of Tekirdag province Şarköy county, Turkey VII. National Congress of Agricultural Economics, Akdeniz University, Antalya.
Reynolds AK (2005). Consumer demand for agricultural and on-farm nature tourism. Uc Small Farm Center Research Brief, p.7.
Sharpley R (2002). Rural tourism and the challenge of tourism diversification: the case of Cyprus. Tourism Manag., 23(3): 233-244.
Soykan F (2000). Rural tourism and experience earned in Europe. Anatolia: J. Tour. Res., 11 (March- June): 21-33.
Tarlak Ş (2007). Model of rural tourism in Mumcular. Dokuz Eylul University Faculty of Architecture Department of City and Regional Planning, Project Work, Izmir/Turkey p.100,
Telfer DJ (2001).Strategic alliances along the Niagara wine route. Tourism Management. 22: 21-30.
Toksarı M, Can SA (2008). New approach to diversify tourism in the process of globalization: rural tourism. III. Balıkesir the National Tourism Conference, April 17-19 2008, Nobel Publications, p. 47
Topay M (2003). A research on the evaluation of landscape features of Bartın-Uluyayla with respect to recreation-tourism uses. Universty of Ankara Graduate School of Natural and Applied Sciences Department of Landscape Architecture (Ph. D. Thesis), p. 210.
Ungureanu D (2008). Sustaınable development altenatıves by rural tourısm and agritourism in the Romanian mountain rural area. Towards Business Excellence. pp. 207-212.
Ünal HB, Kaya Ü, Gülsoylu E (2007). Perceptible connected olive growing soil and water conservation measures: alternative soil tillage practices. J. Ege Universty, Faculty of Agriculture. 11 (3/4):73- 79.
Van der Ploeg JD (2008). The new peasantries. Earth Scan, London In 2002, p.867.
Ventura F, Milone P (2000). Theory and practice of multi-product farms: farm butcheries in Umbria. Sociologic Rurally. 40 (4): 452-465.
Wicks BE, Merrett CD (2003). Promoting rural development through agri-tourism. Rural Research Report, 13(9): 1-8.
Wilson J, Thilmany D, Suillins M (2006). Agritourism: a potential economic driver in the rural west. Economic Development Report. p. 8.
Woo N, Yean S (2006). Agro-tourism as a rural development strategy in Korea, Lee Seong, J. Rural Dev., 29 (6): 67-83. http://www.wineguides.co.uk Accessed: (09.01.2011).

# Crop-machinery management system for field operations and farm machinery selection

Lotfie A.Yousif[1], Mohamed H. Dahab[2] and Haitham R. El Ramlawi[3]

[1]Agricultural Research Corporation, Agricultural Engineering Research Programme, Gedarif, Sudan.
[2]Department of Agricultural Engineering, Faculty of Agriculture, University of Khartoum, Shmbat, Sudan.
[3]Center of Dry Land Research and Studies, Faculty of Agriculture and Environmental Sciences, University of Gedarif, Gedarif, Sudan.

**The main objective of this study is to develop a computer system for farm management and selection of required farm machinery to perform field operations in time for crops grown in rotations. Excel and Visual basic software were used to develop the program. The input data included 4 crops (sorghum, sesame, sunflower and cotton), 3 field operations (seedbed preparation, seeding, and weeding operations) and 3 farming systems (zero-tillage, conventional, and heavy machines farming systems). In addition, tractor and 6 implements (wide level disk, disk harrow, chisel plow, row crop planter, inter-row cultivator and sprayer) were also used. The system estimates the size and number of machine, power requirement and fuel consumption for the implements and operations. Verification showed that, the system has the ability to estimate the required parameters as soon as input data was entered. System validation indicated no significant differences between predicted results and actual data. The sensitivity analysis showed that, changing of input variables affects the output parameters and consequently selection is possible. The system was applied to estimate the required output variables in the mechanized rainfed agriculture in Gedarif, Sudan. It can be used for proper crop and machinery management as pre-season decision making with great confidence.**

**Key words:** Crop production, machine selection, farm management, computer system.

## INTRODUCTION

Farm mechanization is the use of machines for production process. Crop production involves sequence of actions, operations and other factors that affect production. A good farm management plan has to analyze the whole farming system for maximizing production and minimizing risks. The farm manager has to study the relationships between machines, weather and crop combinations. Management can more or less control machine capacity (machine width, speed, and field efficiency), as well as selection of implement, operation and crop. Computer systems can facilitate the examination of these relationships.

Many computer systems were developed elsewhere to analyze the factors that affect field operations and machinery performance. The purposes of these systems varied from power selection and implement matching (Dahab and Mohamed, 2006; Alam et al., 2001; Bol et al., 2006; Yousif and Dahab, 2010), to systems incorporate farm size, cropping patterns, soil properties and climatic conditions to calculate tractor power, machine width and estimating costs (Isik and Sabanci, 1993; Ismail and Burkhardt, 1994). Other systems deal with special crops (Parmar et al., 1994; Dahab and O'Callagham, 1998).

Management of rainfed agriculture is a complex process. The complexity is due to the nature of various and interrelated factors involve in crop production process. These factors include; selection of crops and agricultural machinery (tractors and implements), the expected working time for field operation, inputs cost and outputs prices. It is believed that, promotion of farm management can be realized by the use of modern techniques like computer and softwares. Computer systems can facilitate the process of planning and decision making. Therefore, a computer system was developed to be used as a tool for crops -machinery management. The goals of the system were to select the number and size of machinery (tractor and implement) required to perform timely seedbed preparation, seeding and weed control operations for 4 crops; sorghum, sesame, sunflower and cotton grown in 1, 2, 3, and 4 crops combination.

## MATERIALS AND METHODS

### Characteristics of the study area

The study was conducted in Gedarif State, which lies in the Eastern part of the Sudan between latitudes 12.67 and 15.75°N and longitudes 33.57 and 37.0°E, where about more than 3 000 000 ha are put under cultivation. The soil is heavy cracking clay soils (Vertisols), which was characterized by shrinking when dried and swelling when moistened. The clay content ranged between 65 and 75%.

### Crops grown and field operations

Sorghum is a dominant grown crop, constitutes about 85% of the cropped area, followed by sesame, while sunflower and cotton are sown in limited areas where rain amount is sufficient for their growth. Land preparation and seeding operations usually start when the accumulative rainfall reaches about 100 to 125 mm, which is sufficient for establishing crops. This usually occurs during the second and third week of July. The wide level disk plow is the main machine used for seedbed preparation and seeding operations. However, deep plowing (chiseling) and shallow plowing (disk harrowing) and row crop planter are sometimes practiced in limited areas. Weed control usually starts 2 to 4 weeks after crops germination. Hand weeding is the common practice, however, due to shortage and high expense of labors, mechanical weed control (Sarwala operation) by WLD as well as herbicides application are recently adopted. Tractors of 75 to 80 ha are the main source of farm power; however, big tractors were recently introduced to operate large and heavy implements for improving timeliness of agricultural operations. According to the used machinery and cultural practices there are three farming system practiced; namely conventional, zero-tillage and heavy machinery farming systems. The mechanized farming system consists of large commercial farms 210 to 420 ha or more.

### Data collection and analysis

The required input data to run the developed system is collected from many sources, such as agricultural engineers, Agricultural Research Corporation (ARC) Reports, Mechanized Farming Corporation (MFC), field observations, ASABE standard, John Deer publications, and agricultural machinery dealers. The collected data include; crops and their operations and type and size of machine, field efficiency, speed, draft requirement, machinery capacity, fuel consumption rates.

Statistical measures employing mean, standard deviation, maximum and minimum, correlation analysis were used as data analysis tools. Also, T-test was used to compare and test the significance between the predicted and actual data. Moreover, the Root mean square error (RMSE) criterion was used to compare the values of the predictions and actual data. The RMSE was calculated as follows:

$$RMSE = \sqrt{\frac{1}{n}\sum_{i=1}^{n=5}\left((predicted)_i - (actual)_i\right)^2}$$

Where: n = number of observation, i = different, predicted = unit predicted values and actual = actual values.

### Computer system development

#### *System structure*

The system was developed in Excel and Visual basic computer softwares. Data entry is a step by step process in specifically designed cells. The user is always given the freedom to use site specific data or use built-in data. Data input can be corrected done directly on screen. The system output can be displayed on the screen or as print out.

#### *System description*

The crop production and farm machinery combination management system was designed to work with a sequence of procedures, crop and implement procedure, farm machinery management procedure and farm costs analysis procedure. All procedures work collectively. The system flow chart was described in Figure 1.

### Crop and implement selection procedure

This is the basic procedure for the whole system, which allows the user to choose crops and required field operations. The designed options of crops were 1, 2, 3, or 4 crop combinations. The designed crops were sorghum, sesame, sunflower and cotton. The system deals with 3 operations; seedbed, seeding, and weed control; via 6 implements namely; chisel plow, disk harrow, wide level disk, row-crop - planter, inter-row-cultivator and sprayer. The wide level disk with the seeder box may be used for seedbed, seeding and weed control (Sarwala) operations. The user has to enter the

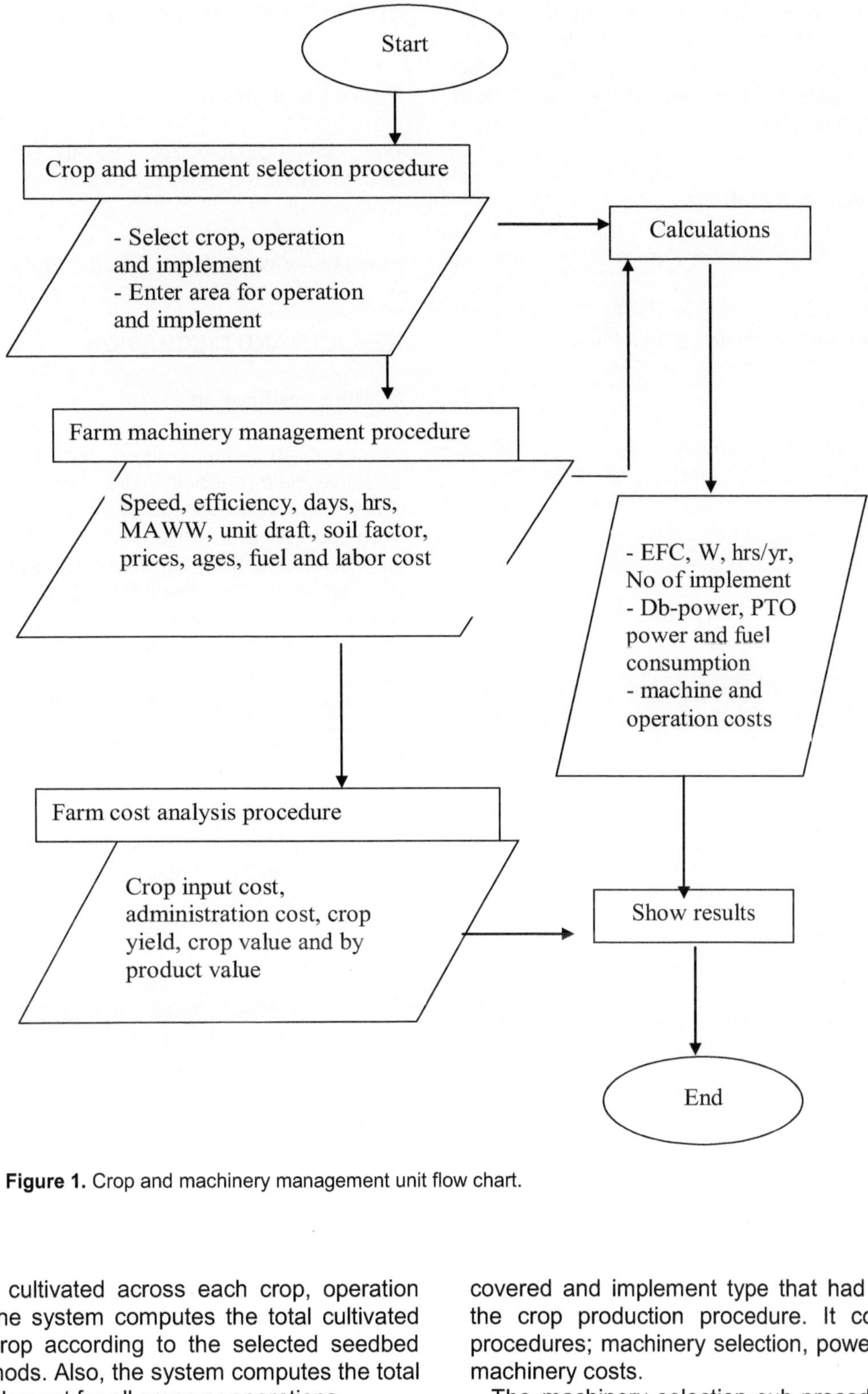


**Figure 1.** Crop and machinery management unit flow chart.

area (ha) to be cultivated across each crop, operation and machine. The system computes the total cultivated area for each crop according to the selected seedbed preparation methods. Also, the system computes the total area of each implement for all crops or operations.

### Machinery management procedure

This procedure deals with machinery (implements and tractor). It uses the intended operations, area to be covered and implement type that had been specified in the crop production procedure. It consists of 3 sub-procedures; machinery selection, power requirement and machinery costs.

The machinery selection sub-procedure computes the size and the number of implements and power units required to complete field operations during specific period of time by computing effective field capacity (ha/h). The required input data include: speed of field operation, field efficiency, available working days, and available working hours per day.

The machinery power requirement sub-procedure determines the required drawbar power (kw), tractor PTO power (kw) and fuel consumption (L/h) for the selected implement. The required input data include: unit draft (kn/m) and soil condition factor.

### Implement selection calculations

#### *Field capacity*

The effective field capacity (EFC) (ha/h) is calculated using total area and total available time as follows:

EFC (ha/h) = A/(D × h) (1)

Where: A = area under operation (ha), D = available working days and hr = available working hours per day.

#### *Required width*

The required working width (W) (m) of a machine is calculated as follows:
W = (EFC × C)/Se (2)

Where: C = conversion factor = 10, S = field speed (km/hr) and e = field efficiency of the machine, decimal.

#### *Number of machine(s) required*

The number of the machine(s) required is calculated as follows:

Number of M = W/MAWW (3)

Where, Number of M = number of machine required and MAWW = maximum available working width (m).

### Machinery power requirement

#### *Drawbar power*

The drawbar power (DBHP), (kw) for an implement is calculated as follows:

DBHP = (S × D × W)/ 3.6 (4)

Where: D = Unit draft (kn/m)

#### *Tractor take – off shaft power*

Tractor PTO required power (TPTO) is calculated using implement drawbar power and soil condition factor as follows:

TPTO = DBHP × soil condition factor × 1.25 (5)

### Fuel consumption

Fuel consumption for diesel engine is calculated according to the method described by (FMO, 1976) as follows:

Fuel consumption (L/hr) = DBHP (kw) × 0.226 (6)

## RESULTS AND DISCUSSION

### System verification

The system was verified for the implement width, effective field capacity, drawbar power, required PTO power and fuel consumption by using published data from Sumner and William (2007) as shown in Table 1. It was observed that, as soon as entering input data the unit displays the results. The predicted results were identical to that obtained by Sumner and William (2007). This means that, the unit is able to calculate the required parameters correctly.

### System validation

The system validity was tested for the five implements to compare between actual and predicted effective field capacity (ha/h), implements width (m) and fuel consumption (L/h) as shown in Table 2. The root mean square of the error criterion was used as a comparison measure. The results showed very low RMSE (0.179), (1.095), and (0.1612) between the predicted and actual data for effective capacity, implements width (m) and fuel consumption (L/h), respectively (Table 2). Moreover, Paired T-test indicates no significant difference (at 5%) between the system predictions and actual data (Table 3). These indicate a high consistency between actual data and the system predictions.

### System Sensitivity analysis

#### *Effect of changing cropped area on number of implement*

Increasing cultivated area from 420 to 3780 ha, the number of machines changed from 1 machine to 3, 6, 9, 9, 5, and 3 machines for wide level disk, disk harrow, chisel plow, row crop planter, inter-row cultivator and sprayer, respectively (Table 4). The results demonstrated that, chisel plow and row crop planter are highly sensitive to the changes in area. This may be due to their effective width, working speed or available working hours per day.

**Table 1.** Verification of the unit with published data.

| Input variable | Input data for Sumner and William (2007) and the model | |
|---|---|---|
| Implement name | Disk harrow | |
| Area, ha | 49 | |
| Days | 5 | |
| Hours/day | 8 | |
| Speed Km/h | 8.1 | |
| Efficiency, decimal | 0.82 | |
| Maximum available width, m | 1.8 | |
| Unit draft, kn/m | 5.37 | |
| **Output parameter** | **Sumner and William (2007)** | **Model prediction** |
| Implement width, m | 1.82 | 1.80 |
| Effective field capacity, ha/h | 1.21 | 1.20 |
| Drawbar power, kw | 21.94 | 21.75 |
| Required PTO power, kw | 46.12 | 46.00 |
| Fuel consumption, L/h | 10.00 | 10.00 |

**Table 2.** Comparison between predicted and actual field capacity (ha/h).

| Machine name | Field capacity (ha/h) | | Fuel consumption (L/h) | | Implement width (m) | |
|---|---|---|---|---|---|---|
| | predicted | actual | predicted | actual | predicted | Actual |
| Wide level disk | 2.7 | 2.9 | 11 | 12 | 3.6 | 3.7 |
| Offset disk harrow | 1.4 | 1.5 | 11 | 13 | 1.6 | 1.7 |
| Chisel plow | 1.3 | 1.4 | 13 | 13 | 1.9 | 2 |
| Row crop planter | 1.9 | 1.8 | 12 | 13 | 3.3 | 3.2 |
| Sprayer | 9.6 | 9.9 | 0.2 | 0.2 | 13.7 | 14 |
| RMSE | 0.179 | | 1.095 | | 0.1612 | |

RMSE = root mean square of the error.

**Table 3.** T-test for the mean difference of the evaluation indicators for system outputs and field data.

| Statistical parameter | Field capacity (ha/h) | Fuel consumption (L/h) | width (m) |
|---|---|---|---|
| Variance of the difference between the means | 0.0044 | 0.1400 | 0.004 |
| Standard deviation of the difference | 0.0663 | 0.3742 | 0.0632 |
| Effective degree of freedom | 4 | 4.0 | 4.0 |
| Probability of t | 0.1447 | 0.0993 | 0.1890 |
| f-calculated | 1.0661 | 1.1578 | 1.0418 |
| T-calculated | -1.8091 | - 2.1381 | -1.5811 |
| T-tabulated | 2.776 | 2.776 | 2.776 |

On the other hand, WLD and sprayer are less sensitive to the changes in cultivated area.

### *Effect of changing annual workdays on implements width*

The system was used to determine optimum machine width when changing annual working days. The results showed that, at specific cultivated area (e.g., 420 ha), changing annual working days changed the required width of each implement. This helps the farm manager to select the suitable working width of a machine according to the available working days and farm size. The results also indicated that, the predicted working width for all studied machines decreased as annual working days

**Table 4.** Effect of changing cropped area on number of machines for different operations.

| Area (ha) | WLD | DH | CP | RCP | IRC | SP |
|---|---|---|---|---|---|---|
| 420 | 1 | 1 | 1 | 1 | 1 | 1 |
| 1260 | 1 | 2 | 3 | 3 | 2 | 1 |
| 2100 | 2 | 3 | 5 | 5 | 3 | 2 |
| 2940 | 2 | 4 | 7 | 7 | 4 | 2 |
| 3780 | 3 | 6 | 9 | 9 | 5 | 3 |

WLD = wide level disk, DH = disk harrow, CP = chisel plow, RCP = row crop planter, IRC = inter-row cultivator and SP = sprayer.

**Table 5.** Effect of changing workdays on machines width.

| Working days | WLD | DH | CP | RCP | IRC | SP |
|---|---|---|---|---|---|---|
| 9 | 3.6 | 3.2 | 3.8 | 8.0 | 5.0 | 7.6 |
| 12 | 2.7 | 2.4 | 2.9 | 6.0 | 3.7 | 5.7 |
| 15 | 2.2 | 1.9 | 2.3 | 4.8 | 3.0 | 4.6 |
| 18 | 1.8 | 1.6 | 1.9 | 4.0 | 2.5 | 3.8 |
| 21 | 1.5 | 1.4 | 1.6 | 3.4 | 2.1 | 3.3 |
| 24 | 1.4 | 1.2 | 1.4 | 3.0 | 1.9 | 2.8 |
| 27 | 1.2 | 1.1 | 1.3 | 2.7 | 1.7 | 2.5 |
| 30 | 1.1 | 1.0 | 1.1 | 2.4 | 1.5 | 2.3 |
| 33 | 1.0 | 0.9 | 1.0 | 2.2 | 1.4 | 2.1 |

*Using total area of 420 ha, WLD = wide level disk, DH = disk harrow, CP = chisel plow, RCP = row crop planter, IRC = inter-row cultivator and SP = sprayer.

**Table 6.** Effect of changing soil factor on power required (kw) for different machines.

| Soil factor | WLD | DH | CP | RCP | IRC |
|---|---|---|---|---|---|
| Firm (1) | 51 | 42 | 58 | 43 | 29 |
| Tilled (2) | 61 | 50 | 69 | 52 | 35 |
| Sandy or soft soils (3) | 71 | 59 | 81 | 60 | 41 |

WLD = wide level disk, DH = disk harrow, CP = chisel plow, RCP = row crop planter and IRC = inter-row cultivator

increased (Table 5). This confirms the fact that, there is a reverse relationship between cultivated area and working time.

### *Effect of changing soil type on implements power requirements*

The system offers the user three choices for changing soil conditions which are firm, tilled and sandy or soft soils, and then the system computes the required drawbar power for the selected implements. Table 6 shows the effect of changing soil condition factor on power requirements for the selected implements. For all tested implements, the power requirements in Kilowatts increased as soil factor changed from firm to tilled and sandy or soft soils.

### *Effect of the cropping system on power requirement and fuel consumption*

The results showed that, conventional farming system gave the smallest values of power required, fuel consumed as well as operation cost per hour and per hectare (Table 7). These findings explain why farmers still hold on to conventional cropping system. In contrast heavy machinery system resulted in the highest values of the mentioned parameters. In this regard, Alam et al. (2001) found that, optimum power level varied with the size of farmland and cropping patterns. However, zero tillage seems to be time effective and can be used as a

**Table 7.** Effect of cropping system on tractor annual working hours, power requirement, fuel consumption and operations cost.

| Cropping system | Tractor annual working hour | Maximum fuel consumption L/ha | Maximum power required (kw) | Operation cost (SDG/h) | Operation cost (SDG/ha) |
|---|---|---|---|---|---|
| CFS | 468 | 11 | 51 | 123.05 | 46.19 |
| ZTFS | 310 | 12 | 52 | 394.08 | 122.81 |
| HMFS | 589 | 13 | 58 | 494.88 | 178.88 |

CFS = conventional farming system, HMFS = heavy machine farming system, ZTFS = zero tillage farming system.

farming system when all requirements for its successes are available. These results can help the user in pre season planning and management.

## Conclusions

A computer system for crop-machinery management was developed. The developed system is user-friendly and could be run on most available computers. The system was validated and statistically analyzed by comparing the predicted output to the actual data and its accuracy was approved. The system can quickly be used to explore the effect of changing one or more of input parameters on output variables, and thus can help in quick decision-making. The system can be used as pre-season planning and management tool.

## REFERENCES

Alam M, Hossain MM, Awal MA (2001). Selection of farm power by using a computer system. AMA (32)1:65-68.

Bol MB, Mohamed HI, Ali MA (2006). Development of a Computer Model for Machinery Selection and Management. University of Khartoum. J. Agric. Sci. 14(2):135-149.

Dahab MH, O'Callagham JR (1998). Simulation Model for Crop – disease Spraying Management. AMA (29)3:27-32.

Dahab MH, Mohamed OE (2006). A Computer Model for Selection of Farm Machinery. U. K. J. Agric. Sci. 14(2):167-181.

FMO (1976). Fundamentals of Machine Operation – Tillage. Deere and Company, Moline, Illiois.

Isik A, Sabanci A (1993). A Computer model select optimum sizes of mechanization planning. AMA 24(3):68-72.

Ismail WI, Burkhardt H (1994). Expert System for Crop Production Machinery system. AMA. 25(3):55-62.

Parmar RS, McClendon RW, Williams EJ (1994). A Computer Simulation Model for Peanut Machinery Management. Applied Engineering in Agriculture, American Society of Agricultural Engineers. 10(4):455- 461.

Sumner PE, Williams EJ (2007). What size farm tractor do I need? Mis. Pub. No. ENG07-003. Available from. http://www.caes.uga.edu/departments/bae/pubs/.../farm%20. last viewed January 2011.

Yousif LA, Dahab MH (2010). Computer Programming for Selection of Boom Width and Tank Capacity of Field Sprayer. U. K. J. Agric. Sci. 18(3):363-1377.

# A comparative analysis of yield performance of maize (*Zea mays* L.) under different tillage methods in Musana communal area, Zimbabwe

**Nyakudya I. W.[1]*, L. Jimu[2], V. Muripira[1] and T. J. Chikuvire[1]**

[1]Department of Agriculture, Bindura University of Science Education, P. Bag 1020, Bindura, Zimbabwe.
[2]Department of Environmental Science, Bindura University of Science Education, P. Bag 1020, Bindura, Zimbabwe.

**In order to reduce the negative effects of conventional tillage there is need to develop or adopt alternative tillage systems. A comparative study on the effects of three tillage methods on the yield performance of a maize cultivar SC633 was conducted in Musiiwa ward of Musana Communal Area, Mashonaland Central Province in Zimbabwe. The experiment was conducted on existing farmers' fields. A randomized complete block design with three treatments (direct seeding, ripping and conventional mouldboard ploughing) replicated four times was used. The data were subjected to analysis of variance using SPSS for Windows, Version 15.0 (2006). Results showed that there was no significant difference ($P>0.05$) in maize grain yield mouldboard ploughing and ripping. Mouldboard ploughing out-yielded ($P<0.05$) direct seeding. Stover yields and harvest indices under the three tillage methods were not significantly different ($P>0.05$). It was concluded that ripping can perform perfectly well just like the conventional mouldboard ploughing. In view of the comparative advantages of conservation tillage in terms of carbon sequestration and the subsequent soil quality improvement, mitigation of greenhouse gas emissions, and reduced soil erosion, adoption of conservation tillage is recommended.**

**Key words:** Tillage, maize, communal area, conservation farming, grain yield.

## INTRODUCTION

Zimbabwe has an agrarian based economy with more than 1.2 million peasant farmers in the communal sector (Matarira et al., 2004). The major field crops grown include maize (*Zea mays*), soybeans (*Glycine max*), cotton (*Gossypium hirsutum*), groundnuts (*Arachis hypogaea*), wheat (*Triticum aestivum*), tobacco (*Nicotiana tabacum*), sorghum (*Sorghum bicolor*) and sunflower (*Helianthus annuus*). Maize is both a staple food and a cash crop. According to Mashingaidze (2006), maize ranks first in terms of producers, area grown and total cereal production in Zimbabwe. Being a strategic crop, maize quantities must be maintained at adequate levels in order to ensure food security and self-sufficiency at both household and national level. However, the past decade has seen a decline in the production of maize in communal areas mainly as a result of inefficient management practices like burning of crop residues and inappropriate tillage practices (Blevens and Fry, 1993). Organic matter is involved in the enhancement of soil quality since it acts on soil structure, nutrient storage and biological activity (Melero et al., 2009). Burning of crop residues therefore implies that its beneficial effects will not accrue to the soil. In addition, burning crop residues leaves the soil bare and pre-disposes it to agents of soil erosion that remove the productive top soil. Throughout the world, soil erosion on cultivated land causes on-site as well as off-site damage, generating profound economic consequences (Raclot et al., 2009). On-site damage includes soil and nutrient loss while the off-site damages include siltation and eutrophication of water bodies and air pollution.

According to Van Straaten (1999), soil degradation has

*Corresponding author. E-mail: pupunege@yahoo.com.

increased dramatically over the last decades, affecting millions of hectares of land, posing a threat to maize productivity particularly in areas with fragile ecosystems. The realization that agricultural production causes alteration of natural ecosystems, producing disturbances in abiotic and biotic components (Melero et al., 2009) should be accompanied by deliberate efforts to minimize negative. In Zimbabwe, smallholder farmers have been practicing conventional tillage through the use of the ox-drawn mouldboard plough. Conventional tillage includes practices that leave less than 15% of the soil surface covered with stubble (Baker et al., 2007). Usually, it involves both primary and secondary tillage, with the former being associated with the inverting of the plough furrow, lifting and moving all the soil in the plough layer, usually to a depth of 0.1 to 0.2 m (Morgan, 1986) and the later, being applied to control weeds. The mould board plough cuts a rectangular slice of soil, turns it over, loosens it and buries weeds and organic matter on or near the surface. However, this method has been blamed for soil degradation and requiring an investment to buy oxen and the plough which most farmers cannot afford. According to Wang et al. (2006), soil carbon loss and tillage-induced $CO_2$ emissions associated with conventional tillage affect not only productive capacity but also global environmental quality. As a result, the farmers have been urged to adopt conservation tillage to save the planet's soil (Grarcia-Torres et al., 2001).

Conservation tillage is a way of farming that allows the soil-water resources to maintain its natural ability to produce crop today and in the future (Papendick et al., 1983). Conservation agriculture and conservation tillage are the collective umbrella terms that usually denote the retention of at least 30% ground cover by residues (Baker, 2007). Conservation tillage has the following multiple functions: carbon sequestration (increase soil carbon storage for carbon trading), mitigation of greenhouse gas emissions and improvement of soil quality for economic crop production (Chatskikh et al., 2008; Sainju et al., 2008). As early as the late eighties, Wilhelm et al. (1987) had noted that in dryland cropping systems, enhanced water storage from conservation tillage results; theoretically, in greater yield potential than conventional tillage systems. Data of grain yield indicates that maize well tolerates simplified tillage systems including no-tillage with retained annual chopped straw on both loamy sand and sandy loam (Pabin et al., 2006). Melero et al. (2009) also showed that long-term soil conservation management improved the quality of both sandy clay loam (Entisol) to heavy clays (vertisols) through enhancing organic carbon fraction and biological status especially at upper layers. This implies that the benefits of conservation tillage are not restricted to specific soil types, thus, widening the scope for experimentation.

The objective of this study was to compare maize grain and stover yield under ripping, direct seeding and conventional ploughing.

## MATERIALS AND METHODS

### Description of study area

The research was conducted in Musiiwa communal area (31° 15' E, 18°15' S; 1250 m.a.s.l) during the 2007/2008 croppi ng season. The area is located at 65 km along Harare-Shamva Road in Bindura district of Mashonaland Central Province of Zimbabwe. Musiiwa is in Natural Region II (b) with an annual rainfall of 700 to 1000 mm and mean annual temperature of 26 to 28°C. The area lies in a dissected plateau with 30 to 50% of the land area comprising rock outcrops of the bornhardt and castle koppie type (Anderson et al., 1993). A catenary association of soils occurs on the pediment slopes. The upperslope section is adjacent to rock outcrops and has well-drained, moderately shallow to deep, coarse-grained sand to sandy loam over yellowish red to red coarse-grained sandy clay loams. Mid to lower slopes have well to moderately well drained, moderately shallow to deep coarse grained sands or loamy sands over strong brown to reddish similar sands often with mottling at about 1m depth (Anderson et al., 1993).

### Experimental design

The experiment was conducted on fields under the Development Aid from People to People (DAPP) Farmers Club. Systematic sampling was used to select four farmer participants and the distance between the furthest farmers was 4 km. A randomised block design comprising three treatments (ripping, direct seeding and conventional ploughing) replicated four times was used. Each farmer's field was regarded as a block and the three treatments were assigned at random to the three plots in each block (Neeley et al., 1991) using random numbers.

### Description of experimental sites

All the four fields were on mid-slope positions with an identical pH ($CaCl_2$) of 4.6 and textural classes ranging from coarse sandy loam to coarse sandy clay loam. During the 2006/2007 cropping season, all the fields were planted to maize under similar treatments to those studied in the 2007/2008 cropping season.

### Field operations

The land was pegged to form blocks measuring 30 × 30 m and plots measuring 30 × 10 m. Footpaths of 1 m were left between plots in each block. Conventional ploughing was done in one plot of each block using a mould board plough at a depth of 0.2 m. Seeding furrows were 0.90 m apart. In the other plot, only the seeding furrows were made using an ox-drawn ripper at a depth of 0.3 m. Hand planting was then done using a hoe. A calibrated planter was used in the final plot to rip lines at a depth of 0.15 m, and plant maize at a spacing of 0.90 × 0.15 m. Planting was done on the 1st of December 2007. Hand seeding at two pips per station was done in mouldboard-ploughed and ripped plots at a spacing of 0.90 × 0.30 m. In the third plot, ripping of lines, sowing, fertilizing and seed covering was done simultaneously at a spacing of 0.90 × 0.15 m and one seed per station was sown. The 0.30 × 0.90 m spacing is commonly used by local farmers when planting manually because there is considerable saving in terms of labour if fewer stations with more plants are used. The 0.90 × 0.15 m spacing is recommended for machine planning because with this method fertilizer application is precise and labour saving is less important. The maize hybrid seed SC 633 was sown in all the plots. The hybrid seed is a long season variety with a yield potential which ranges from 10 to 12 tons per hectare and takes about 136 days to maturity. Basal fertilizer compound D (7% N: 14% $P_2O_5$: 8% $K_2O$) fertilizer was applied at a rate of 176 kg/ha. The fertilizer was banded ±7.5 cm to the side of the seed and covered with soil. The application was according to Grant (1996).

Weed control was done using hoes, slashers and hands. Uprooted weeds were left in between the rows to maintain soil cover. The first weeding operation was done at planting in all the plots. First post emergence weeding was done on the 3rd week in ripped and direct seeded plots only. Second post emergence weeding was carried out on the 7th week when new weeds appeared in all the plots. Lastly, there was pre harvest weeding on the 11th week in all the plots again whereby spot weeding was done after the dough stage of the maize plants. Scouting for pests was done after every fortnight and stalk borer and cutworms were commonly seen. Spot stalk borer control was done through application of dipterex at the rate of 4 kg/ha as soon as the caterpillar feeding damage was seen in the maize funnel (Agritex, 1993). Ammonium nitrate was applied in all the plots at a rate of 222 kg/ha (5 g/station) based on regional recommendations.

#### Rainfall data collection

Rainfall was measured using a rain gauge at Mhumhurwi Primary School. The site was considered to be central with the furthest fields being 2 km away.

#### Harvesting and data collection

Harvesting was done on the 23rd of April 2008. Areas to be harvested were chosen from the centre of each plot to cater for boarder effects during data collection. Four sub-plots were harvested per plot at an interval of 2 m and each of the sub-plots covered 9 $m^2$. Thus, maize plants were harvested from a total area of 36 $m^2$ per plot. Two plants were then drawn randomly from each sub-plot making up a sample of 8 plants per plot. Cobs were removed from the 8 stalks, placed in small plastic bags and the wet weight of the cob was measured. The mass of the 8 stalks was also measured. The samples (8 stalks and 8 cobs) per plot were air-dried separately.Cobs were shelled using hands and the grains were weighed. . In Zimbabwe the maximum moisture content normally recommended for maize storage is 12.5 percent. The weight of maize was adjusted to 12.5% moisture content after determination of the sample grain moisture. All the measurements were done using a balance to the nearest 0.01 kg and were recorded.

#### Determination of harvest indices

Harvest index was calculated using the formula (Shamsi et al., 2003):

$$\text{Harvest index (HI)} = \frac{\text{Economic yield}}{\text{Biological yield}}$$

#### Data analysis

Data were analysed through one way analysis of variance (ANOVA) using SPSS for Windows Version 15.0 (2006). The differences between means were calculated using the least significant difference (LSD) post hoc tests at 5% significance level.

## RESULTS

The rainfall received for the 2007/2008 cropping season was as follows: The rainfall was well distributed during the crop growing season. The mid-season dry spell which normally occurs in January was virtually non-existent. The seasonal total was significantly different ($p<0.05$) from the mean for the nine years of 2000/2001 to 2008/2009. Monthly, totals for all months exceeded the mean monthly totals except for January and February.

### Yield comparisons

Table 2 shows a comparison of maize grain and stover yields produced under ripping, direct seeding and conventional mouldboard ploughing. There were no significant differences ($P>0.05$) on maize grain yield produced through direct seeding and ripping as well as ripping and conventional mouldboard ploughing. However, the grain yield of maize produced through conventional ploughing outperformed ($P<0.05$) those produced through direct seeding. The stover yield of maize plants produced through the three tillage methods did not significantly vary ($P>0.05$).

### Comparison of harvest indices

Maize harvest indices for the three tillage methods were as follows: ripping (0.45 ± 0.10), direct seeding (0.46 ± 0.07) and conventional ploughing (0.44 ± 0.09). The harvest indices were not significantly different at $P = 0.05$.

## DISCUSSION

Maize grain yield produced under conventional mouldboard ploughing out-yielded that under direct seeding because in the first seasons conventional mouldboard ploughing loosens the root zone, allowing plant roots easy access to soil-water and plant nutrients. Direct seeding which involves little disturbance of the soil through opening a rip line with coulters, somehow restricted the lateral and downward movement of the roots (Baker et al., 2007). However, with time it can be assumed that "biological tillage" will improve root growth. Conservation tillage methods like stubble mulch, ripping, strip, and direct seeding result in better returns during dry years compared to conventional tillage (Wang et al., 2006). Thus, the significance of conservation tillage is highly proven in drier seasons than in wetter seasons. In this study, the season was wetter than normal for the area and this could have resulted in inferior performance

**Table 1.** Decadal rainfall totals (mm) for the 2007/2008 rainy season for Mumhurwi Primary School in Musiiwa Communal Area, Zimbabwe.

| Decade of month | October 2007 | November 2007 | December 2007 | January 2008 | February 2008 | March 2008 | April 2008 |
|---|---|---|---|---|---|---|---|
| First | 0.0 | 33.5 | 78.7 | 79.0 | 21.5 | 60 | 20 |
| Second | 0.0 | 58.5 | 242.4 | 35.0 | 58.9 | 100.0 | 0 |
| Third | 0.0 | 84.5 | 175.8 | 160.3 | 19.8 | 0 | 0 |
| Total | 0.0 | 176.5 | 496.9 | 274.3 | 100.2 | 160 | 20 |

**Table 2.** Comparison of maize grain and stover yield produced under direct seeding, ripping and conventional mouldboard ploughing during the 2007/2008 in Musiiwa Communal Area, Zimbabwe.

| Tillage method | Mean yield (SD)(t/ha) | |
|---|---|---|
| | Grain | Stover |
| Ripping | 3.67(0.86)$^{ab}$ | 4.10(2.10)$^{a}$ |
| Direct seeding | 2.43(1.25)$^{a}$ | 3.56(2.06)$^{a}$ |
| Conventional mouldboard ploughing | 4.55(0.06)$^{b}$ | 5.67(1.37)$^{a}$ |

Means in the same column with common superscripts are not significantly different at P = 0.05
SD stands for standard deviation.

of direct seeding to mouldboard ploughing. Rainfall data in Table 1 shows that the January mid-season dry spell is a common feature in most seasons was virtually non-existent with all the first and last decades exceeding 70 mm of rainfall and the seasonal total (1227.9 mm) exceeded the mean (SD) 791 (204 mm) for the seasons 2000/2001 to 2008/2009. Spacing of plants might also have had a bearing on the differences in grain yield obtained between the two tillage methods. According to Strieder et al. (2008), the photosynthetically active solar radiation intercepted ($PAR_{int}$) by the canopy Is one of the main requirements to obtain high yields in the absence of water shortage. Plant arrangement, therefore has influence on crop yields (Strieder et al., 2008).

In the current study, although, the plant population in the two plots was the same (74,074 plants/ ha), the interaction of plants in direct seeded plots with a spacing of 0.90 × 0.15 m (one seed per hill) was different to that in conventionally ploughed plots spaced at 0. 90 × 0.30 m (two seeds per hill). The differences in grain yield between conventional mouldboard ploughing and direct seeding may have been due to the difference of $PAR_{int}$ caused by the different plant arrangements. However, there is not enough evidence that interactions affected the portioning of metabolites between the grain and the stover because the harvest indices are similar in value. Ripping equalled conventional mouldboard ploughing in grain yield because like conventional mouldboard ploughing, ripping also loosens the soil in the root zone allowing easy root growth. In this study, ripping is accompanied by hoeing planting stations, thus, further loosening the root zone.The results support findings by Pabin et al. (2006) who states that reduced and no-tillage in combination with straw retaining can be used without any significant decrease in grain yield of maize grown to medium and coarse textured soils. However, the system is suitable for row crops like maize only (Fenger, 2006). Mouldboard ploughs mostly used by smallholder farmers have been identified as the primary cause of soil erosion although all tillage implements contribute to this problem.

Lobb et al. (1995) estimates soil loss from tillage sequence of mouldboard ploughing and cultivating to be 54 t/ha/year and that a risk of accelerating erosion existed on cultivated land from the moment natural and surface litter is removed. When weighed against these negative impacts of mouldboard ploughing, ripping may be preferred for sustainable agriculture in communal areas. In addition, the adoption of reduced tillage practices may also be able to increase or at least stabilize the carbon content in the previously ploughed soil, thus, reducing respiratory carbon dioxide emissions to the atmosphere. According to McClymont and Winkfield (1984), better yields are unlikely to be experienced under conservation tillage until at least in third and fourth year. Therefore, the early relatively good performance of ripping in this experiment is quite encouraging. Results from this study therefore, imply that farmers can achieve yields similar to those of conventional tillage using ripping. In view of the overwhelming evidence of ecological sustainability of conservation tillage, farmers should be encouraged to adopt it as a viable alternative to conventional tillage in the study area. The lack of significant difference on the maize stover yield produced among the three tillage methods implies that for purposes of fodder production any one of the tillage methods may be used.

## Conclusions

Conventional tillage out yielded direct seeding in grain yield but it was equalled in performance by ripping. There was no variation in stover yield among the three tillage treatments. Further studies are needed in order to establish effects on yield of the different planting arrangements used in this study. We recommend monitoring of the performance of treatments in the medium to long-term in order to establish inter-seasonal performance of the treatments. Multi-season monitoring is important because conservation tillage benefits usually accrue with time as soil quality improves. In view of the comparative advantages of conservation tillage in terms of carbon sequestration and the subsequent soil quality improvement, mitigation of greenhouse gas emissions, and reduced soil erosion, we recommend adoption of conservation tillage.

## ACKNOWLEDGEMENTS

We are thankful to Mr. Mudzvova and Mr. Muzuva, the project leaders of Development Aid from People to People (DAPP) Farmers Club for granting us the permission to undertake our research project in the organization. Direct contributions were also made by the dedicated staff in the organization that assisted in obtaining all the required information for this project.

### REFERENCES

Anderson IP, Brinn PJ, Moyo M, Nyamwanza B (1993). Physical resource inventory of the communal lands of Zimbabwe-An overview. Bulletin 60, Kent, United Kingdom.

Baker C J, Saxton KE, Ritchie WR, Chamen WCT, Reicosky DC, Ribeiro MFC, Justice SE, Hobbs PR (2007). Tillage seeding in conservation agriculture. FAO and CAB International, London, UK.

Blevens RL, Fry WW (1993). Conservation tillage, an ecological approach to soil management. Adv. in Agron. 51: 37-78.

Chatskikh D, Olesen JE, Hansen EM, Elsgaard L, Petersen BM (2008). Effects of reduced tillage on net greenhouse gas fluxes from loamy sandy soil under winter crops in Denmark. Agric. Ecosys. Environ. Elsevier 128:117-126.

Department of Agricultural Technical and Extension Services (AGRITEX) (1993). Farm management handbook. Harare, Zimbabwe, Crops Section, pp 1-35.

Fenger G (2006). Green world actions. The GAIA- Movement Trust Living Earth. Green World Action, India. pp. 44-48

Grant P (1996). Fertilization and liming. In: Gibson IDS (ed). Grain Handbook. Commercial Grain Producers Association of Zimbabwe, Section E, pp E1-E13, Harare, Zimbabwe.

Gracia-Torres L, Benites J, Martinez-Vilela A (2001). Conservation agriculture: A worldwide challenge. In: Proceedings of the First World Congress on Conservation Agriculture, Madrid, Spain. pp. 387.

Hallauer, AR, Russell, WA, Lamkey, KR (1988). Corn Breeding. In: Sprague, GF and Dudley, J.W. (eds). Corn and Corn Improvement. Madison, USA. pp.469-554.

Lobb DA, Kachanoski RJ, Miller MH (1995). Modelling tillage translocation and tillage erosion on shoulder slope landscape positions. Can. J. Soil Sci. 75:211-218.

Mashingaidze K (2006). Maize research and development. In: Rukuni et al. (eds) Zimbabwe's agricultural revolution revisited. University of Zimbabwe Publications. Harare, Zimbabwe, pp. 363-376.

Matarira CH, Makadho JC, Mukahanna-Sangarwe, MM (2004). Vulnerability and adaptation of maize production to climatic change in Zimbabwe. Ministry of Environment and Tourism, Harare, Zimbabwe. pp. 3-64.

McClymont D, Winkfield R (1984). Conservation tillage: A Handbook for Commercial Farmers in Zimbabwe. Laser Print, Harare, Zimbabwe.

Melero S, López-Garrido R, Madejon E, Murillo JM, Vanderlinden K, Ordóñez R, Moreno F (2009). Long-term effects of conservation tillage on organic fractions in two soils in Southwest of Spain. Agric. Ecosys. Environ. Elsevier 133:68-74.

Morgan RPC (1986). Soil erosion and conservation. Longman Group UK Limited, England.

Neeley D, Matata J, Kisyombe F (1991). statistical techniques for analysis of on-farm crop trials: A. Manual. CIMMYT, Harare, Zimbabwe.

Pabin J, Lipiec J, Wlodek S, Biskupski A (2006). Maize response to different straw management and tillage systems under cereal crop rotation. Int. Agrophysics 20:141-146.

Raclot D, Bissonnais Y L, Louchart X, Andrieux P, Moussa R, Voltz M (2009). Soil tillage and scale effects on erosion from fields to catchment in Mediterranean vineyard area. Agric. Ecosys. Environ. Elsevier. 134: 201-210.

Sainju UM, Senwo ZN, Nyakatawa EZ, Tazisong A, Reddy CA (2008). Soil carbon and nitrogen sequestration as affected by long-term tillage, cropping systems, and nitrogen fertilizer sources. Agric. Ecosys. Environ. Elsevier 127:234-240.

Shamsi IH, Shah WA, Bakt J, Masood M (2003). Effect of maize population (As intercrop) on growth of ratoon sugarcane and maize yield. Asian J. Plant Sci. 2(7):532-534.

Strieder ML, da Silva PRS, Rambo L, Sangoi L, da Silva AA, Endrigo PC, Jandrey DB (2008). Crop management sysyemes and maize grain yield under narrow row spacing. Sci. Agric. 65(4):346-353.

Van Straaten P (1999). Cattle manures in Zimbabwe: Methods and minerals for quality improvement. In: Delvin JF, Zettel T (eds) Ecoagrioculture initiatives in Eastern and Southern Africa. Weaver Press. Harare. Zimbabwe, pp. 87-103.

Wang Xiao-Bin, CAI Dian-Xiong, Hoogmoed WB, Oenema O, Perdok UD (2006). Potential effect of conservation tillage on sustainable land use: A Review of global long-term studies. Soil sci. Soc. of China Elsevier. Pedosphere 16(5): 587-595.

Wilhelm WW, Schepers JS, Mielke ML, Doran JW, Ellis JR, Stroup WW (1987). Dryland maize development and yield resulting from tillage and nitrogen fertilisation practices. Soil. Till. Res. 10:167-179.

# Effectiveness of arbuscular mycorrhizal fungi in protection of maize (*Zea mays* L.) against witchweed (*Striga hermonthica* Del Benth) infestation

**Othira, J. O.[1]*, Omolo, J. O.[2], Wachira, F. N.[3] and Onek, L. A.[4]**

[1]Department of Biochemistry and Molecular Biology, Egerton University, P. O. Box 536 – 20115, Egerton, Kenya.
[2]Department of Chemistry, Egerton University, P. O. Box 536 – 20115, Egerton, Kenya.
[3]Tea Research Foundation of Kenya, P. O. Box 820 – 20200, Kericho, Kenya.
[4]Juba University, Private Bag, Juba, South Sudan.

***Striga hermonthica* is one of the most important obligate root hemiparasitic weeds of cereals such as maize and sorghum. This study investigated the impact of arbuscular mycorrhizal fungi (AMF) on the *Striga*-maize host interaction and its potential application in management of *Striga* infestation. Two maize varieties: *Striga*-susceptible Nyamilambo and *Striga*-tolerant KSTP94; three species of AMF; *Glomus etunicatum*, *Scutellospora fulgida*, *Gigaspora margarita* and *Striga hermonthica* seeds were used in a greenhouse experiment. Colonization of maize roots was fastest with *Glomus* spp. (51.98%) than other AMF species; however, *Striga* infestation reduced root colonization but conversely increased relative mycorrhizal dependency (*Glomus* spp. = 6.12%) in both maize cultivars. Inoculation of Nyamilambo maize cultivars with AMF reduced *Striga* plant incidence (9.50) and biomass (5.40 g) to significantly lower (4.90 and 3.82 g) levels. Mycorrhizal Nyamilambo and KSTP94 had higher nitrogen (154.01 and 116.88 mg) and phosphorus (25.33 and 31.07 mg) content compared to non-mycorrhizal cultivars (63.37 mg N and 16.81 mg P) in absence and presence of *Striga* infestation. While mycorrhizal maize had increased height (100.50 to 135.20 cm) and biomass (75.70 to 87.54 cm), their *Striga* damage score (4.07) was significantly lower than with non-mycorrhizal maize (8.10). *Glomus* spp. had most significant positive effect compared to both *Gigaspora* and *Scutellospora* spp. In conclusion, AMF inhibits germination and reduces growth of *Striga hermonthica* while enhancing maize host growth and development.**

**Key words:** Arbuscular mycorrhiza, *Striga*, maize growth.

## INTRODUCTION

World food security depends on ample supply of three major cereals, namely, maize (*Zea mays*), wheat (*Triticum aestivum*) and rice (*Oriza sativa*) (Ferrar, 1995). In Sub-Saharan Africa, maize is of greatest importance in terms of tonnage, consumption and financial value, while sorghum (*Sorghum bicolor*) follows as a very popular food crop. Production of maize and sorghum in Sub-Saharan Africa and Asia are threatened by *Striga* spp. weed infestation. The crop loss was estimated to cost US$ 7 billion per year, directly affecting the welfare and livelihoods of over 10 million people in sub-Saharan Africa (Esilaba and Ransom, 1997). Being a parasitic angiosperm, it attaches to roots of a wide range of tropical cereals and leguminous crops and deprive them of water, soluble mineral salts and metabolites. The host often shows draught symptoms, resulting in total yield loss (Press et al., 1990; Kanampiu and Friesen, 2004). Existing *Striga* control measures include crop rotation, roguing, use of resistant crop varieties, breeding for resistant varieties, biological control, soil fumigation, herbicides, high nitrogen nutrition, synthetic germination

*Corresponding author. E-mail: othirajack@gmail.com.

stimulants and trap cropping (Lagoke et al., 1991; Eplee, 1992; Ejeta et al., 2000; Kabambe et al., 2005). However, these *Striga* control methods have given no conclusive and consistent feasible results for the peasant farmer. They are expensive, labour intensive, may require skilled personnel and, the returns are not immediate on investment, as well as mismatch between technologies and the farmer's socio-economic conditions.

The arbuscular mycorrhizal fungi (Phylum Glomeromycota) are characterized by the production of intracellular absorptive structures (vesicles, arbuscules) and are the most widespread of soil fungi followed by ectomycorrhizal fungi (Koide and Mosse, 2004; Walker and Schüßler, 2004; Schüßler et al., 2001). The external mycelium of arbuscular mycorrhizal fungi (AMF) acts as an extension of host plant roots and serves as a direct link between roots and soil nutrient reserves. Mycorrhizal fungi play a key role in terrestrial ecosystem functioning along with environmental factors such as climate, disturbances, food web interactions, mutualism and ecological history (Wardle and Van der Putten, 2002). In many ecosystems, the major benefits of AMF to symbionts includes enhanced nutrient uptake, increased tolerance to root pathogens, drought resistance, tolerance to aluminium and manganese toxicity and improved soil aggregation and structure (Cardoso and Kuyper, 2006; Xavier and Germida, 1998). Endomycorrhizae formation can also affect water relations, hormonal balance and root colonization by other microorganisms, hence, are important for optimum plant health and development (Gianinazzi and Gianinazzi-Pearson, 1986). In agricultural ecosystems, the composition and vitality of the AMF mycoflora is altered by the introduction of exotic weeds and crops, cultivation and chemicals (Douds et al., 1993). Mutual relations between plant roots and vesicular mycorrhizal fungi have been found to increase plant vigour and consequently crop productivity especially under unfavourable conditions. The mutualistic symbiosis between AMF and crop plants results in increased uptake of phosphorus, potassium, nitrogen, and other nutrients, increased growth at high soil temperatures, better growth in low moisture soils, more efficient water utilization, increased levels of cytokinins, and also increased photosynthetic rates and stomatal conductance (Lendzemo, 2004).

Several species of mycorrhizal fungi have also been shown to increase plant biomass and compensate for damage by *Striga hermonthica*, and their metabolites either stimulate or inhibit weed germination in sorghum, *Sorghum bicolour* (Lendzemo and Kuyper, 2001). This could become a useful biotic interaction for effective and sustainable integrated management of *S. hermonthica* infestation for the resource-poor farming situations. This study investigated the effectiveness of AMF inoculation on growth and development of maize under *S. hermonthica* infestation.

## MATERIALS AND METHODS

### Sources of biological materials

Pure isolates of AMF, Glomus etunicatum Becker and Gerdemann, Scutellospora fulgida Walker and Sanders and, Gigaspora margarita Beck and Hall were obtained from National Museums of Kenya, Nairobi. Field isolates were trapped from farmers' sorghum field soils at Alupe, Kibos, Lambwe, and Oyugis in Western Kenya, as well as Njoro wheat fields in Rift Valley. Mycorrhizal soils obtained from the farmers' fields were analyzed for their physico-chemical characteristics at Egerton University (data not shown). Two maize cultivars used in this study, Striga-tolerant KSTP94 and Striga-susceptible Nyamilambo, provided by KARI's Kakamega Research Station.

### Preparation of mycorrhizal inoculum

To obtain single spore culture, healthy spores of each isolate were placed on root tips of sorghum seedlings freshly germinated in the greenhouse on autoclaved sand. Before spore placement, the seedling was placed in a 2 cm depression made in sterilized growth medium contained in a 7 cm diameter plastic pot. After spore placement, the depression was gently covered with the growth medium taking care not to dislodge the spore from the root tip. Sorghum plants were watered appropriately and supplied with Hoagland's solution minus phosphorus (Murakoshi et al., 1998). General greenhouse sanitation was maintained to ensure purity of the cultures. Potting medium was dried by stopping watering to the pot. After the host plant had wilted, the dried soil was stored and used as mycorrhizal inoculum for greenhouse pot experiments.

### Soil preparation and infestation

In the greenhouse, pots (20 cm diameter, perforated, bottom lined with filter paper) was sterilized and filled with steam sterilized (121°C, 2 h) soil-sand mixture. The physicochemical characteristics of the mixture was determined to be: clay-loamy with 0.6% organic matter, 0.3% total carbon, 0.05% total N, 7 C/N ratio, 12 ppm P-Bray 1 and pH (1:2 $H_2O$) of 5.4. To disinfect fungal propagules, they were incubated in 2% chloramines T, 200 ppm of streptomycin sulphate, and a trace of Tween-20 for 20 min followed by at least five changes of sterile water. A layer of AMF inoculum, which consisted of a soil mixture (10 g/pouch) containing heavily colonized roots of sorghum, spores and mycelium, was put below maize seed and grown under conducive environment for 3 months. Soil inoculation was simultaneously achieved at transplantation using 10 g of inocula for each AMF isolate (Wacheke et al., 2001). Taking desired infestation as 3000 germinable seeds per pot, weight of *S. hermonthica* seeds required per pot = 3000/% germinability × 5 × $10^{-6}$. This weight was mixed with 10 ml of water to infest one pot. The total amount of *S. hermonthica* seeds and amount of water needed to prepare the inoculum for all pots was measured and mixed in a container, from which a 60 ml syringe was used to draw inoculum to infest 6 cm hole at the centre of each pot, then allowed 7 days conditioning period before planting (Berner et al., 1997).

### *Striga*-mycorrhizal maize interaction experiments

Sixty four 20 cm diameter pots were arranged in randomized complete block design with four replicates. The treatments were two contrasting maize cultivars; uninoculated; inoculated with selected AMF species (*Glomus*, *Scutellospora* and *Gigaspora* spp.); and

**Table 1.** Effects of arbuscular mycorrhizal fungi (AMF) inoculation on frequency of colonization and relative mycorrhizal dependency (RMD) of two maize cultivars.

| Maize cultivars | Treatment | Colonization (%) | RMD (%) |
|---|---|---|---|
| Nyamilambo | *Glomus* spp. | 69.03 ±2.19$^{a}$ | 13.53 ±1.26$^{c}$ |
| | *Striga* spp. + *Glomus* spp. | 51.98 ±1.98$^{c}$ | 16.12 ±0.78$^{a}$ |
| | *Scutellospora* spp. | 41.66 ±1.10$^{e}$ | 13.89 ±0.24$^{bc}$ |
| | *Striga* spp. + *Scutellospora* spp. | 39.04 ±1.54$^{f}$ | 9.55 ±0.61$^{e}$ |
| | *Gigaspora* spp. | 43.11 ±1.28$^{e}$ | 14.20 ±0.51$^{bc}$ |
| | *Striga* spp. + *Gigaspora* spp. | 38.01 ±1.03$^{f}$ | 12.11 ±0.63$^{d}$ |
| KSTP94 | *Glomus* spp. | 54.63 ±3.14$^{b}$ | 11.65 ±1.09$^{d}$ |
| | *Striga* spp. + *Glomus* spp. | 46.11 ±2.69$^{d}$ | 14.50 ±1.16$^{b}$ |
| | *Scutellospora* spp. | 38.97 ±1.26$^{f}$ | 12.03 ±1.29$^{d}$ |
| | *Striga* spp. + *Scutellospora* spp. | 34.46 ±1.28$^{g}$ | 7.87 ±0.32$^{f}$ |
| | *Gigaspora* spp. | 31.67 ±1.14$^{h}$ | 9.98 ±0.61$^{e}$ |
| | *Striga* spp. + *Gigaspora* spp. | 23.06 ±1.04i | 11.28 ±0.78$^{d}$ |

Values are mean ± SE; letter show vertical comparisons among treatments at P = 0.05 probability level; means with the same letter are not significantly different from each other.

and inoculated with *S. hermonthica*. Seeds of *Z. mays* were surface sterilized in hydrogen peroxide (15% v/v) for 3 min, then washed in sterile distilled water. The seeds were then incubated in water agar at 28°C in the dark for 2 days. One pre-germinated seed of each variety was planted in pot filled with 1 kg sterilized sand-soil (1:1) mixture. The plants were watered with tap water for the first 2 weeks, then, 100 ml per pot of Steiner universal nutrient solution. Plants were grown in a glasshouse under a day/night cycle of 12 h each, 30/25°C and 60% relative humidity. The pots were weighed daily and water loss compensated for by top watering. Final harvest was done after 4 months and the roots washed clean with tap water (Berner et al., 1997).

#### Data collection and analysis

Percentage root mycorrhizal colonization was assessed by gridline intersect method of Giovannetti and Mosse (1980). The mycorrhizal root system was collected, washed, oven-dried, cleared with 10% KOH at 121°C for 10 min, then rinsed, acidified with 5% HCl for 1 min, stained with 0.01% acid Fuschin for 30 min, dissolved in destaining solution (14:1:1, lactic acid:glycerol:water), drained and destained in water overnight. The stained roots were examined microscopically at between 10 and 100X to observe AM fungal structures. Relative mycorrhizal dependency (RMD) of maize was calculated by expressing the difference between shoot dry weight of mycorrhizal plant and the shoot dry weight of non-mycorrhizal plant as a percentage of the shoot dry weight of the mycorrhizal plant (Waceke et al., 2001). Data on mycorrhizal colonization was subjected to GLM procedures for Windows statistical packages. *Striga* count and emergence were recorded in the course of experiment just as height of maize. Biomass of dry shoot and root of maize and *Striga* were measured 80 days after emergence (DAE) by destructive sampling, and recorded after oven-drying at 65°C for 3 days. Maize damage scores (MDS) were done using *Striga* syndrome rating with a scale of 1 to 9, where 1 = healthy plant and, 9 = dead/dying plant. The MDS are based on leaf, stem and ear symptoms (Berner et al., 1997). Shoot phosphorus and nitrogen mineral content of mycorrhizal maize plants were determined using standard analytical methods: colorimetry for nitrogen after Kjeldahl digestion (Soon and Kalra, 1995) and phosphorus by the molybdenum blue procedure (Boltz and Mellon, 1947).

Data on biomass, maize height, mineral content and emerged *Striga* count were subjected to analysis of variance (ANOVA) with the help of SAS statistical package (SAS System for Windows, 1990) and SPSS v6.1 where applicable (Norusis, 1994). The means of parameters that showed significant differences were separated by Fisher's LSD test at p = 0.05 confidence level. Relationship between *Striga* parameters and AMF types as wells as maize plant development were assessed by regression analysis. All data were checked and transformed appropriately to normalize skewed distributions before statistical analysis.

## RESULTS AND DISCUSSION

### Mycorrhizal colonization of maize roots

Typical AMF structures, arbuscules, vesicles and hyphae were regularly found in stained roots. Thus, establishment of AMF was effective within roots of both maize cultivars. In absence of *S. hermonthica* infestation, Nyamilambo and KSTP94 had 69.0 and 54.63% mycorrhizal colonization, respectively (Table 1). In the presence of *S. hermonthica* infestation, percent mycorrhizal colonization reduced to 51.98 and 46.11% for Nyamilambo and KSTP cultivars, respectively. In the absence of AMF inoculation, all maize plants remained free of mycorrhizal colonization. Generally, all treatments involving Nyamilambo cultivar had higher percent mycorrhizal colonization than those with KSTP94 cultivar. In both maize cultivars, there was higher root colonization after inoculation with *Glomus* spp., than inoculation with either *Scutellospora* or *Gigaspora* spp. Mycorrhizal fungi differ in their ability to infect and colonize roots. *Glomus* spp. has ability to infect and colonize plant roots faster than *Scutellospora* and *Gigaspora* spp., making it highly

**Table 2.** Effects of arbuscular mycorrhizal fungi (AMF) inoculation on emergence and total biomass of *Striga* under two maize varieties.

| Maize cultivars | Treatment | *Striga* count | *Striga* biomass (g) |
|---|---|---|---|
| Nyamilambo | *Striga hermonthica* | 9.50 ±1.36$^{a}$ | 5.40 ±0.26$^{a}$ |
| | *Striga* spp. + *Glomus* spp. | 4.90 ±0.29$^{c}$ | 4.20 ±0.93$^{c}$ |
| | *Striga* spp. + *Scutellospora* spp. | 6.13 ±0.55$^{b}$ | 3.82 ±0.51$^{cd}$ |
| | *Striga* spp. + *Gigaspora* spp. | 5.22 ±0.10$^{c}$ | 4.81 ±0.24$^{b}$ |
| KSTP94 | *Striga* spp. | 3.50 ±0.34$^{d}$ | 2.93 ±0.41$^{e}$ |
| | *Striga* spp. + *Glomus* spp. | 1.81 ±0.06$^{e}$ | 3.53 ±0.20$^{d}$ |
| | *Striga* spp. + *Scutellospora* spp. | 2.93 ±0.32$^{d}$ | 3.84 ±0.39$^{cd}$ |
| | *Striga* spp. + *Gigaspora* spp. | 2.12 ±0.09$^{e}$ | 2.71 ±0.23$^{e}$ |

Values are mean ± SE; I letter show vertical comparisons among treatments at P = 0.05 probability level; means with the same letter are not significantly different from each other.

competitive (Kurle and Pfleger, 1994). The higher mycorrhizal colonization in *Striga*-susceptible maize could be due to strigolactones exuded by host plant roots and taken up by AMF since strigolactones stimulate fungal metabolism and branching (Parniske, 2008).

The relative mycorrhizal dependency (RMD) varied from 7.87% in KSTP94-*Striga*-*Scutellospora* spp. treatment to 16.12% in Nyamilambo-*Striga*-*Glomus* spp. treatment, among the various treatments (Table 1). Both maize cultivars showed greatest dependency with *Glomus* spp. and least dependency with *Scutellospora* spp. However, the calculated RMD was lower than 50% in all the experiments. In both cultivars, RMD was higher in presence of *Striga* (16.12% with *Glomus* spp.) than in absence of *Striga* (13.53% with *Glomus* spp.). Relative mycorrhizal dependency (RMD) refers to the degree of plant responsiveness to mycorrhizal colonization by producing maximum growth or yield at a given level of soil fertility. RMD is related to morphological and physiological properties of root systems (Diop et al., 2003). The significant differences in mycorrhizal colonization as well as RMD also support the fact that response of plants to AMF inoculation depends on fungus-plant combination (Abbot and Robson, 1985). Successful colonization and functional interaction between host plant and mycobiont are based upon exchange of signaling molecules at different stages of symbiosis development. The role of strigolactones as key signaling compounds in the interaction between plants and soil-borne symbiotic AMF has been recently suggested (Soto et al., 2010). Strigolactones, a novel class of phytohormones involved in the regulation of shoot branching in plants are secreted by plant roots and stimulates presymbiotic growth of AMF.

With the identification of strigolactones as the branching factor, not only its production, exudation into the rhizosphere and perception by the AMF but also its specific action in arbuscular mycorrhizal symbiosis (AMS) can now be explored in more details (Bucher et al., 2009)

## Influence of AMF inoculation on *Striga* performance

The *Striga*-susceptibe Nyamilambo maize cultivar had 9.50 and 4.90 *Striga* plants per pot in absence and presence of *Glomus* spp., respectively (Table 2). Non-mycorrhizal KSTP94 maize cultivar had 3.50 *Striga* plants per pot compared to 1.81 with *Glomus* spp. The average number of *Striga* plants emerging per pot significantly differed between the maize cultivars. Fewer shoots of *Striga* plants were observed in pots with KSTP94 compared to those with Nyamilambo (Table 2). The emergence of *Striga* shoots was reduced in all mycorrhizal treatments with *Glomus* spp. being the most significant. Inoculation with AMF reduced *Striga* biomass, especially in treatments with Nyamilambo cultivar than those with KSTP94 cultivars. Nyamilambo cultivar colonized with *Glomus* spp. significantly reduced *Striga* biomass from 5.40 to 4.20 g. The *Striga* biomass significantly differed between the maize cultivars and less biomass was observed in treatments with KSTP94 compared to those with Nyamilambo maize cultivars. The non-significant effects of *Gigaspora* spp. in *Striga*-infested maize cultivars could be due to its low percent colonization.

The ability of AMF to reduce *Striga* count can be explained in three ways: (i) The formation of metabolites, especially strigolactones that are responsible for the induction of *Striga* germination is down regulated upon mycorrhizal colonization; (ii) Plant metabolites, such as cyclohexenones which arise through carotenoid degradation, that are up-regulated upon mycorrhizal colonization inhibit *Striga* germination and; (iii) Mycorrhizal colonization induces mycorrhizosphere effects that negatively impact on *Striga* germination (Lendzemo et al., 2007). There are further roles of AMF that affect the performance of *Striga*. The observation that *Striga* incidence and biomass is reduced in mycorrhizal maize, lends support to rhizosphere changes as a mechanism by which *Striga* damage could be

**Table 3.** Effects of arbuscular mycorrhizal fungi (AMF) inoculation on nitrogen and phosphorus shoot content under *Striga hermonthica* infestation.

| Maize cultivar | Treatment | Nitrogen (mg) | Phosphorus (mg) |
|---|---|---|---|
| Nyamilambo | Without any additive | 63.37 ±1.69[e] | 16.81 ±0.61[d] |
| | *Striga* spp. | 58.41 ±2.85[e] | 10.37 ±0.28[e] |
| | *Glomus* spp. | 154.01 ±6.09[a] | 25.33 ±0.94[a] |
| | *Striga* + *Glomus* spp. | 115.22 ±4.88[b] | 22.08 ±0.71[b] |
| | *Scutellospora* spp. | 110.80 ±3.32[bc] | 22.16 ±0.28[b] |
| | *Striga* + *Scutellospora* spp. | 104.28 ±3.45[c] | 19.31 ±0.69[c] |
| | *Gigaspora* spp. | 96.47 ±2.83[d] | 21.45 ±0.23[b] |
| | *Striga* + *Gigaspora* spp. | 93.75 ±3.09[d] | 19.60 ±0.76[c] |
| KSTP94 | Without any additive | 64.19 ±1.95[e] | 16.81 ±0.28[f] |
| | *Striga* spp. | 61.46 ±1.50[e] | 11.42 ±0.41[g] |
| | *Glomus* spp. | 116.88 ±4.11[a] | 31.07 ±0.33[a] |
| | *Striga* + *Glomus* spp. | 110.64 ±3.28[b] | 23.35 ±0.50[b] |
| | *Scutellospora* spp. | 101.05 ±2.75[c] | 21.11 ±0.19[c] |
| | *Striga* + *Scutellospora* spp. | 98.26 ±2.13[c] | 20.32 ±0.69[c] |
| | *Gigaspora* spp. | 84.31 ±1.99[d] | 18.94 ±0.31[de] |
| | *Striga* + *Gigaspora* spp. | 83.72 ±2.87[d] | 18.52 ±0.22[e] |

Values are mean ± SE; letter show vertical comparisons among treatments at $P < 0.05$ probability level; means with the same letter are not significantly different from each other.

reduced. Upon colonization by AMF, the quality and quantity of root exudates substantially change. Marschner et al. (1997) observed that mycorrhizal colonization simplified the composition of the rhizosphere soil solution. In this study, both a direct simplification of the root exudate composition and a simplification due to changes in the rhizosphere community composition with its specific metabolites could have occurred. Such community-wide changes may also affect the germination of *Striga* seeds. The observed decline in *Striga* in mycorrhizal maize could be attributed to increased nitrogen availability (Mumera and Below, 1993). According to Swift (2001), mycorrhizal fungi exercise a greater benefit when soil phosphorus levels are at or below 50 ppm (50 mg/kg). Other studies have also reported that fertilisers that are high in phosphorus or nitrogen reduce AMF colonisation (Azcón et al., 2003; Xu et al., 2000).

Sood (2003) also showed that plant growth-promoting rhizobacteria prefer mycorrhizal over non-mycorrhizal roots, which feed back to germination of *Striga* seeds. It has also been observed that *Fusarium solani* inhibited *S. hermonthica* seed germination in *Striga*–infested farmlands by 100% due to presence of extracelluar metabolites in the fungal filtrate which most likely inhibits ethylene action rather than its synthesis (Bethlenfalvay, 1993).

## Influence of AMF inoculation on mineral status of maize

In absence of *Striga* infestation, colonization with *Glomus* spp. increased nitrogen content in *Striga*-susceptible Nyamilambo from 63.37/100 to 154.01 mg/100 g and from 64.19/100 to 116.88 mg/100 g in *Striga*-tolerant KSTP94 maize cultivars (Table 3). In presence of *Striga* infestation, mycorrhizal maize of both cultivars had higher nitrogen content above controls. Similarly, mycorrhization with *Glomus* spp. increased phosphorus content in both Nyamilambo and KSTP94 maize cultivars from 16.81/100 and 16.81 mg/100 g to 25.33/100 and 31.07 mg/100 g, respectively. In general, all AMF species studied increased nitrogen and phosphorus content in both maize cultivars in absence and presence of *Striga* infestation. Earlier studies by Bethlenfalvay (1993) have similarly shown that species and strains of AMF differ in their effectiveness in increasing nutrient uptake and plant growth. The function of extra- and intra-radical forms of AMF hyphae could also explain differences in phosphorus

**Table 4.** Effects of arbuscular mycorrhizal fungi (AMF) inoculation on growth parameters of Nyamilambo and KSTP94 maize cultivars under *Striga hermonthica* infestation.

| Maize cultivar | Treatment | Maize height (cm) | DSW (g) | DRW (g) | Syndrome rating |
|---|---|---|---|---|---|
| Nyamilambo | Without any additive | 100.50 ±2.09$^{d}$ | 75.70 ±3.76$^{d}$ | 17.84 ±0.61$^{f}$ | 1.73 ±0.71$^{gh}$ |
| | + *Striga* spp. | 81.30 ±1.83$^{f}$ | 53.24 ±4.10i | 15.41 ±0.78i | 8.10 ±0.42$^{a}$ |
| | + *Glomus* spp. | 135.20 ±3.01$^{a}$ | 87.54 ±5.35$^{a}$ | 22.51 ±1.28$^{a}$ | 1.30 ±0.09$^{h}$ |
| | + *Striga* + *Glomus* spp. | 114.50 ±2.62$^{b}$ | 77.78 ±3.14$^{c}$ | 18.66 ±1.26$^{de}$ | 4.07 ±0.11$^{f}$ |
| | + *Scutellospora* spp. | 107.89 ±2.16$^{c}$ | 78.33 ±3.27$^{c}$ | 18.61 ±1.35$^{de}$ | 1.78 ±0.13$^{g}$ |
| | + *Striga* + *Scutellospora* spp. | 96.01 ±1.29$^{e}$ | 72.13 ±3.45$^{e}$ | 16.96 ±1.03$^{g}$ | 5.98 ±0.29$^{c}$ |
| | + *Gigaspora* spp. | 111.23 ±2.19$^{bc}$ | 80.76 ±4.13$^{b}$ | 19.94 ±1.77$^{c}$ | 1.95 ±0.41$^{g}$ |
| | + *Striga* + *Gigaspora* spp. | 102.50 ±1.98$^{d}$ | 74.23 ±4.57$^{d}$ | 16.33 ±1.62$^{h}$ | 6.67 ±0.22$^{b}$ |
| KSTP94 | Without any additive | 74.75 ±1.12$^{g}$ | 66.13 ±3.32$^{g}$ | 15.73 ±1.53i | 1.65 ±0.20$^{gh}$ |
| | + *Striga spp.* | 61.00 ±1.09$^{j}$ | 61.03 ±2.85$^{h}$ | 13.57 ±0.21$^{k}$ | 5.89 ±0.71$^{cd}$ |
| | + *Glomus spp* | 83.00 ±2.19$^{f}$ | 74.85 ±4.03$^{d}$ | 20.42 ±1.99$^{b}$ | 1.23 ±0.12$^{h}$ |
| | + *Striga* + *Glomus* spp. | 71.00 ±1.16$^{h}$ | 71.38 ±3.55$^{e}$ | 18.21 ±0.24$^{ef}$ | 5.07 ±0.87$^{e}$ |
| | + *Scutellospora* spp. | 80.00 ±2.08$^{f}$ | 72.24 ±3.95l | 18.70 ±1.37$^{d}$ | 3.67 ±0.28$^{f}$ |
| | + *Striga* + *Scutellospora* spp. | 76.50 ±1.18$^{g}$ | 66.24 ±3.17$^{g}$ | 14.33 ±1.15$^{j}$ | 5.53 ±0.51$^{d}$ |
| | + *Gigaspora spp.* | 75.66 ±1.14$^{g}$ | 69.88 ±3.79$^{f}$ | 17.03 ±0.75$^{g}$ | 4.11 ±0.64$^{f}$ |
| | + *Striga* + *Gigaspora* spp. | 64.50 ±0.88$^{i}$ | 68.79 ±3.45$^{f}$ | 15.48 ±0.24$^{i}$ | 5.70 ±0.95$^{cd}$ |

Values are mean ± SE; letter show vertical comparisons among treatments at P = 0.05 probability level; means with the same are not significantly different from each other. Key: DSW = dry shoot weight and DRW = dry root weight.

phosphorus acquisition among the AMF isolates. Hence, the status of the extra-radical mycelium development in the soil appears to be a major determinant of the efficiency of AMF to phosphorus uptake (Rakshit and Bhadoria, 2009). Similar results have been found on soybean cultivars indicating that phosphorus uptake by mycorrhizal plants fluctuate with fungal isolates and genetic variability within cultivars (Diop et al., 2003).

The nitrogen and phosphorus transfer to the maize plants may also be as a consequence of the fungal demand for the nutrients, with both host plant and fungus evolving transporters to take advantage of localized increases in nutrients.

## Influence of AMF inoculation on maize growth

The AMF inoculation had significant positive effect on height of both Nyamilambo and KSTP94 maize cultivars in presence and absence of *Striga* infestation (Table 4). *Glomus* spp. increased height of Nyamilambo maize cultivar from 100.50 to 135.20 and 114.50 cm in absence and presence of *Striga,* respectively. Similarly, in absence of *Striga* infestation, KSTP94 maize cultivar increased in height from 74.75 to 83.00 cm with *Glomus* spp. Analysis of variance showed that both maize shoot and root biomass were positively affected by maize cultivar, *Striga*, and mycorrhiza and the interaction among these factors. In absence of *Striga, Glomus* spp. increased shoot biomass of Nyamilambo (75.70 g) and KSTP94 (66.13 g) maize cultivars to 87.54 and 74.85 g, respectively. The dry shoot biomass of both mycorhizal Nyamilambo and KSTP94 maize cultivars were higher with all the AMF species used. The AMF effectively cancelled out damage by *S. hermonthica* in both cultivars but the tolerant cultivar had significant gain in shoot biomass compared to its control. The dry root biomass of Nyamilambo maize cultivar (17.84 g) was significantly reduced (15.41 g) by *Striga* infestation in absence of mycorrhizal fungi (Table 4). But *Glomus*-colonized Nyamilambo cultivar had increased dry root biomass of 22.51 and

18.66 g in absence and presence of *Striga* infestation. The other mycorrhizal fungal isolates, *Scutellospora* and *Gigaspora* spp. did not improve dry root biomass as much. Based on *Striga* syndrome rating scale, AMF colonization significantly reduced *Striga* damage score on Nyamilambo cultivar from 8.10 to 4.07 with *Glomus* spp, 5.98 with *Scutellospora* spp. and 6.67 with *Gigaspora* spp. With KSTP94 cultivar, reduction in *Striga* damage score was not significantly different from control. In absence of *Striga* infestation, *Glomus* spp. reduced *Striga* damage score from 1.73 and 1.65 to 1.30 and 1.23, with Nyamilambo and KSTP94 maize cultivars, respectively.

The observed differences among treatments with *Striga*-susceptible Nyamilambo and *Striga*-tolerant KSTP94 maize cultivars (Table 4) is an indication that the overall significant effect of treatment obtained for growth parameters were largely attributed to the presence or absence of AMF or *S. hermonthica*. The ability of *Glomus* spp. to significantly improve maize shoot dry biomasses (Table 4) confirms previous reports on the ability of AMF to enhance plant growth. AMF might have enhanced growth of untreated and *Striga* infected plants through enhanced nutrient uptake and synthesis of plant growth promoting hormones, in particular, auxins, cytokinins and gibberellins (Allen et al., 1982). Improved nutrient uptake results from increased absorptive surface of the root system by AMF external mycelia. The fungi might have in addition, improved water uptake from the growth medium, which results indirectly from improved plant nutritional status (Harley and Smith, 1983), increased cytokinin levels and/or increased number of vascular bundles (Allen et al., 1982). The differences in efficacy of AMF isolates to alter maize growth parameters might be further explained by differences in their effectiveness to colonize maize and their ability to grow in the soil and enhance nutrient uptake. *Glomus* spp. was the most effective in promoting maize growth, since high root colonization reflects high degree of AMF effectiveness. The ability of AMF to increase nutrient uptake is related to their ability to form extensive and well distributed hyphae in soil and throughout the developing root system. Additionally, the maize-AMF compatibility, mycorrhizal dependency of maize, as well as fungal inoculum density might have played a significant role in influencing improved maize growth parameters (Abbott and Robson, 1985).

Increased nitrogen and phosphorus uptake could also subsequently lead to increased plant height and biomass and reduced *Striga* syndrome rating, leading to increased productivity. It is evident that AMF could play a role within the *S. hermonthica* cereal pathosystem. This is in line with the observations by Lendzemo and Kuyper (2001) who found out that AMF could increase sorghum output and cancel out damage by *S. hermonthica*. Gworgwor and Weber (2003) observed that AMF do not colonize *S. hermonthica* roots and have the potential to reduce or compensate for damage caused by *S. hermonthica* infestation especially in sorghum varieties that are mycorrhiza-responsive. The highly significant difference in heights of the maize cultivars could be due to the inherent differences in the maize genotypes and not due to AMF inoculants since KSTP94 is a relatively shorter variety than Nyamilambo. Increased dry root biomass of mycorrhizal plants is often due to an increase in the proportion of lignified and suberized higher order roots due to secondary thickening than an increase in root biomass (Rajapakse and Miller, 1992).

The non-significant effects of *Scutellospora* and *Gigaspora* spp. on maize could be due to their low percent colonization and relative mycorrhizal dependency of maize.

## Conclusion

Result of this study indicated that the colonization of maize roots was fastest with *Glomus* spp. than other AMF species. *Striga* infestation reduced root colonization but conversely increased relative mycorrhizal dependency in both maize cultivars. Inoculation of Nyamilambo maize cultivars with AMF reduced *Striga* plant incidence and biomass to significantly lower levels. Mycorrhizal Nyamilambo and KSTP94 had higher nitrogen and phosphorus content compared to non-mycorrhizal cultivars in absence and presence of *Striga* infestation. While mycorrhizal maize had increased height and biomass, their *Striga* damage score was significantly lower than with non-mycorrhizal maize. The extent of influence of AMF depends on maize variety as well as on fungal species and strains used. *Glomus* spp. had most significant positive effect compared to both *Gigaspora* and *Scutellospora* spp. In conclusion, AMF reduces *Striga* growth while enhancing maize host growth and development. All these results are indicative of the effectiveness of AMF in protecting *Striga*-susceptible maize against damage by *Striga* infestation. Hence, mycorrhizal association can be used as a promising strategy to develop tools for *Striga* control.

## ACKNOWLEDGEMENTS

The authors highly appreciate Egerton University and German Academic Exchange Service (DAAD) for financial support to carry out this study and Kenya Agricultural Research Institute (KARI), Kibos Field Station where pot experiments were carried out.

### REFERENCES

Abbot LK, Robson AD (1985). The role of vesicular arbuscular mycorrhizal fungi in agriculture and the selection of fungi for inoculation, Aus. J. Agric. Res. 33: 389-408.

Allen MF, Moore TS, Christensen M (1982). Phytohormone changes in

*Bouteloua gracilis* infected by vesicular-arbuscular mycorrhizae. II: Altered levels of gibberellin - like substances and abscissic acid in the host plant. Can. J. Bot., 60: 468-471.

Azcón R, Ambrosano E, Charest C (2003). Nutrient acquisition in mycorrhizal lettuce plants under different phosphorus and nitrogen concentration. Plant Sci., 165: 1137-1145.

Berner DK, Winslow MD, Awad AE, Cardwell KF, Mohan Raj DR, Kim SK (1997). *Striga* Research Methods. Manual, The Pan-African Striga Control Network (PASCON) and the International Institute of Tropical Agriculture, 2nd edition, pp. 13-20.

Bethlenfalvay GJ (1993). Mycorrhizae in the agricultural plant-soil system. Symbiosis, 14(1-3): 413-414.

Boltz DF, Mellon MG (1947). Determination of Phosphorus, Germanium, Silicon, and Arsenic by the Heteropoly Blue Method. Anal. Chem., 19: 873.

Bucher M, Wegmuller S, Drissner D (2009). Chasing the structures of small molecules in arbuscular mycorrhizal signaling. Curr. Opin. Plant Biol., 12: 500-507.

Cardoso IM, Kuyper TW (2006). Mycorrhizas and tropical soil fertility: Nutrient Management in Tropical Agroecosystems. Agric. Eco. Environ., 116(1-2): 72–84.

Diop TA, Krasova-Wade T, Diallo A, Diouf M and Gueye M (2003). *Solanum* cultivar responses to arbuscular mycorrhizal fungi: growth and mineral status. Afr. J. Biotech., 2(11): 429-433.

Douds DD, Janke RR, Peters SE (1993). VAM fungus spore populations and colonization and colonization of roots of maize and soybean under conventional and low-input sustainable agriculture. Agric. Eco. Environ., 43(3-4): 325-335.

Ejeta G, Mohammed A, Rich P, Melake-Bahan A, Housley TL, Hess DE (2000). Selection for specific mechanism of resistance to *Striga* in Sorghum, pp. 29 -37. In: *B. I. G. Haussmann et al, (eds.) Breeding for Striga resistance in cereals. Proceedings of a Workshop, IITA,* Ibadan, Nigeria, 18-20 August 1999, Margarf, Germany.

Esilaba AO, Ransom JK (1997). *Striga* in Eastern and Central African countries: A literature review. Technical Report Series No. 1, African Highlands Initiative, ICRAF, Nairobi.

Eplee RE (1992). Witchweed (*Striga asiatica*): An overview of management strategies in the USA, Crop Prot., 2: 3-7.

Ferrar P (1995). Forward', in McIntosh, RA, Wellings, CR & Park, RF (eds.) *An atlas of Resistance Genes.* CSIRO Publications, E, Mellbourne, Australia.

Gianinazzi S, Gianinazzi-Pearson V (1986). Progress and headaches in endomycorrhiza biotechnology. Symbiosis, 2: 139-149.

Giovannetti M, Mosse B (1980). An evaluation of techniques for measuring vesicular arbuscular mycorrhizal infection in roots. New Phytol., 84: 489-500.

Gworgwor NA, Weber HC (2003). Arbuscular mycorrhizal fungi-Parasite-host interaction for the control of *Striga hermontica* (Del. Benth) in sorghum (*Sorghum bicolor*) (L.) Moench. Mycorrhiza, 13: 277-281.

Harley JL, Smith SE (1983). Mycorrhizal symbiosis. Academic press, London, p. 634.

Kabambe VH, Kanampiu F, Nampuzi SC, Kauwa AE (2005). Evaluation of herbicide imizapyr and fertilizer application in integrated management of *S. asiatica* in maize in Malawi. Afr. Crop Sci. Proc., 7: 489–493.

Kanampiu F, Friesen D (2004). *Striga* weed control with herbicide-coated maize seed. CIMMYT, Nairobi, Kenya.

Koide RT, Mosse B (2004). A history of research on arbuscular mycorrhizal fungi. Mycorrhiza, 14: 145-163.

Kurle JE, Pfleger FL (1994). The effect of cultural practices and pesticides on VAM fungi. In: F.L. Pfleger and R.G. Linderman (eds.) Mycorrhizae and Plant Health. APS Press, Minnesota, pp. 101-131.

Lagoke STO, Parkinson V, Agunbade RM (1991). Parasitic weeds and control methods in Africa, In: Kim, SK (ed.) Combating striga in Africa. Proceedings of the International Workshop in August 1988, Ibadan, Nigeria, pp. 3-14

Lendzemo VW (2004). The tripartite interaction between Sorghum, *Striga hermonthica* and arbuscular mycorrhizal fungi, Trop. Res. Mgt., p. 55.

Lendzemo VW, Kuyper TW (2001). Effects of arbuscular mycorrhizal fungi on damage by *Striga hermonthica* on two contrasting cultivars of sorghum *Sorghum bicolour.* Agric. Eco. Environ., 87: 29-35.

Lendzemo VW, Kuyper TW, Matusova R, Bouwmeester HJ, Ast AV (2007). Colonization by Arbuscular Mycorrhizal Fungi of Sorghum Leads to Reduced Germination and Subsequent Attachment and Emergence of *Striga hermonthica.* Plant Signal. Behav., 2(1): 58-62.

Marschner P, Crowley DE, Higashi RM (1997). Root exudation and physiological status of a root colonizing fluorescent pseudomonad in mycorrhizal / non-mycorrhizal pepper (*Capsicum annuum* (L,). Plant Soil, 189: 11-20.

Mumera LM, Below FE (1993). Role of Nitrogen in resistance to *Striga* parasitism of maize. Crop Sci., 33: 758-763.

Murakoshi T, Tojo M, Walker C, Saito M (1998). AM fungi on adjacent semi natural grasslands with different vegetations in Japan. Mycoscience, 39: 455-462.

Norusis MJ (1994). *SPSS advanced statistics.* Chicago: SPSS Inc.

Parniske M (2008). Arbuscular mycorrhiza: the mother of plant root endosymbioses, Nature Rev. Microb., 6: 763-775.

Press MC, Graves JD, Stewart GR (1990). Physiology of interaction of angiosperm parasites and their hosts. Plant Cell. Environ., 13: 91-104.

Rajapakse S, Miller-Jr JC (1992). Methods of studying vesicular-arbuscular mycorrhizal root colonization and related root physical properties. *In: Norris, JR,* Read, DJ & Varma, AK (eds.), Methods in Microbiol., Academic Press, London, 24: 301-316.

Rakshit A, Bhadoria P (2009). Influence of arbuscular mycorrhizal hyphal length on simulation of P influx with the mechanistic model, Afr. J. Microb. Res, 3 (1): 001-004.

Schußler A, Schwarzott D, Walker C (2001). A new fungal phylum, the Glomeromycota: phylogeny and evolution. Mycol. Res., 105: 13-1421.

Sood SG (2003). Chemotactic response of plant-growth-promoting bacteria towards roots of vesicular arbuscular mycorrhizal tomato plants. FEMS Micro. Ecol., 45: 219-227.

Soon YK, Kalra YP (1995). A comparison of plant tissue digestion methods for nitrogen and phosphorus analyses. Can. J. Soil Sci. 75(2): 243-245.

Soto MJ, Fernández-Aparicio M, Castellanos-Morales V, García-Garrido JM, Ocampo JA, Delgado MJ, Vierheilig H (2010). First indications for the involvement of strigolactones on nodule formation in alfalfa (Medicago sativa). Soil Biol. Biochem., 42: 383-385.

Swift RS (2001). Sequestration of carbon by soil. Soil Sci., 166: 858–871.

Waceke JW, Waudo SW, Sikora R (2001). Response of *Meloidogyne hapla* to mycorrhiza fungi inoculation on pyrethrum. Afr. J. Sci. Tech., 2(2): 63-70.

Walker C, Schüßler A (2004). Nomenclatural clarifications and new taxa in the glomeromycota. Mycol. Res., 108: 981-982.

Wardle DA, Van der Putten WH (2002). Biodiversity, ecosystem functioning and above-ground- below-ground linkages. In: M. Loreau, S. Naeem and P. Inchausti (eds.), Biodiversity and Ecosystem Functioning: Synthesis and perspectives. Academic Press, New York, pp. 155-168.

Xavier LJC, Germida JJ (1998). Response of spring wheat cultivars to *Glomus clarum* NT4 in a P-deficient soil containing arbuscular mycorrhizal fungi. Can. J. Soil. Sci., 78: 481-484.

Xu H, Wang R, Mridha AU (2000). Effects of organic and chemical fertilisers and arbuscular mycorrhizal inoculation on growth and quality of cucumber and lettuce. J. Crop Prod., 3: 313-324.

# The Qur'anic sustainable remedy to the global food crisis: Nigeria as a case study

**Sulayman Adeniran Shittu**

Department of Arabic and Islamic Studies, University of Ibadan, Nigeria. E-mail: sulaymanshittu@yahoo.co.uk.

**Agriculturalists, economists and experts from different fields of studies have raised alarm on food crisis. All of the theories propounded did not proffer a sustainable remedy to the invisible catastrophe. Therefore, views of the experts of various fields in both advanced and developing countries were examined. It was discovered that a feasible alternative to sustainable remedy to the global crisis not yet explored is that proffered by the Qur'an. The discussion commenced with the Qur'anic narration on the first food crisis ever witnessed on earth. Reasons and factors responsible for the present global food crisis were examined and compared with the Qur'anic submission on the factor that causes the food crisis. As regards the history of food crisis around the world and in Nigeria, steps taken by the Nigerian government as advised by different experts and agencies were also analyzed. The concept of the Qur'an on the effect of food crisis on the citizenry ended this part. Suggestions put forward by experts and the Qur'an on the global food crisis such as the World Bank, public opinion in doubling financial support to the sector, provision of modern storage facilities, preparation of seeds and distribution of fertilizers to appropriate quarters and period. The recommendations on a sustainable remedy as inferred from the Qur'an and agricultural experts which can boost productivity and enhance availability within a short period ended the discussion.**

**Key words:** Nigeria, Quranic sustainable remedy, food crisis.

## INTRODUCTION

Agriculturalists, economists, and experts from different fields of study have raised alarm on the food crisis. Governments and people of different countries have been on red alert on this same issue but all theories propounded by the aforementioned experts are yet to proffer a sustainable remedy to this global crisis. The research examined the theories of the experts in both advanced and developing countries, and it was discovered that the Qur'anic remedy was not explored which could serve as a sustainable remedy to this global crisis. The paper begins with the Qura'nic narration on the first food crisis ever witnessed on earth and this is followed by expositions on the reasons for the food crisis as given by experts and the Quranic submission on the factors that cause the food crisis. A brief overview of the food crisis around the world and Nigeria in particular is given attention while steps taken by the government as suggested by experts of different fields and agencies were critically examined. The effect of the food crisis on the citizenry occupies another main discussion. The Quranic narration on the effects of the food crisis also concludes the discussion. From the analysis of different steps suggested by experts, it is found that the solution proffered by the Qur'an gives a permanent sustainable remedy to the global food crisis in the world at large and Nigeria in particular. Rising of food price have made basic staple food like rice and corn unaffordable for many people, pushing the poor to the extreme poverty because they can no longer get enough to eat (Staff, 2008).

## THE FIRST GLOBAL FOOD CRISIS EVER WITNESSED ON EARTH

Allah says:

**قال الملك إني أرى سبع بقرات سمان يأكلهن سبع عجاف وسبع سنبلات خضر وأخر يابست يآ أيها الملأ أفتوني في رؤياي إن كنتم لرؤياى تعبرون**

Meaning:

The king (of Egypt said: "I do see (invasion) seven fat kine, whom seven lean ones devour, and seven green ears of corn, and seven (others) withered. O ye chiefs! Expound to me my vision if it be that you can interprete visions: Q. 12: 43.

وقالوا أضغاث أحلام وما نحن بتأويل الأحلام بعالمين

Meaning:

"They said 'A confused medley of dreams; and we are not skilled in the interpretation of dreams": Qur'an 12: 44.

يوسف أيها الصديق أفتنا سبع بقرات سمان يأكلهن سبع عجاف وسبع سنبلات
خضر وأخر يابست لعلي أرجع إلى الناس لعلهم يعلمون

Meaning:

O Joseph (he said) O man of truth! Expound to us (the dream) of seven fat kine whom seven lean ones devour, and of seven green ears of corn and (seven) others withered; that I may return to the people and that they may understand: Qur'an 12: 46.

قال تزرعون سبع سنين دأبا فما حصدتم فذروه في سنبلة إلا قليلا مما تأكلون

Meaning:

(Joseph) said: For seven years shall ye diligently sow as is your wont. And the harvests that ye reap, ye shall leave them in the ear, except a little, of which ye shall eat: Qur'an 12: 47.

ثم يأتي من بعد ذلك سبع شداد يأكلهن مما قدمتم لهن إلا قليلا مما تحصنون

Meaning:

Then will come after that (period) seven dreadful (years), which will devour what ye shall have laid by. In advance for them, (all) except a little which ye shall have (specially) guarded: Qur'an 12: 48.

ثم يأتي من بعد ذلك عام فيه يغاث الناس وفيه يعصرون

Meaning:

Come after that (period) a year in which the people will have abundant water and in which they will press (wine and oil) (Abdullahi, 1984): Qur'an 12: 49.

The Qur'anic explanation on the history of the food crisis on earth as mentioned in the Qur'an is self explanatory. One will see that Prophet Yusuf did not only interpret the vision of the king alone by showing what would happen, but also suggests the measures to be taken for dealing with the calamity when it occurs. He said, there would be seven years of abundant harvest with diligent cultivation.

They should take a little for their sustenance and store the rest that will be harvested during the years. It is better to preserve the stored harvests from the pests that attack corn heaps when they have passed through the threshing floor. Then, this would be followed by seven years of dreadful famine which would devour all the stores which they would have laid by in the good years. They must be careful even during the famine to have special arrangement in saving a little for seed, when they should be helpless even when the Nile brought down abundance of water from the rains at its sources. This will be a symbol of a very abundant year, following the seven years of drought. The Nile will bring abundant fertilizing waters and silt from its upper reaches; and there would probably be some rain also in other parts of Egypt. The vine and the dive trees which must have suffered in the drought would now be revived, and yield-their juice and oil seeds. Plants such as linseed, Sesamum, and the castor oil plant, must have grown as there was irrigated land to spare from the abundant grain crops. The people's spirit will now revive to enjoy the inner products of the earth, when their absolute necessities had been more than met in their grain crops.

## THE HISTORY OF GLOBAL FOOD CRISIS

### A brief on food riots around the world

A lot of articles and public opinions were published in the United States of America which highlighted the global food crisis such as the following.

The skyrocketing commodity prices that have made the farm belt one of the most prosperous regions of the United States have had a rather different impact on large areas of the developing world. Prices of food stuffs had escalated. 41% increase in price since October 2007, pushing many people over the line from poverty into privation or even hunger (The Skyrocketing in the Washington Post Company, 2008).

Higher food prices appearing in the United States in an April 2007 report by the Government Accountability office showed that transportation and other overhead costs now consume 65% of U.S. food aid dollars. This was attributed to higher fuel prices and U.S. Laws that require most grain shipment to go out on relatively expensive U.S. flagged cargo vessels. Even some commented "Why does so much food have to travel the high seas in the first place? Because U.S. law requires that the government buys all food donations from U.S. products (The Skyrocketing in the Washington Post Company, 2008: 17).

The Food and Agriculture Organization (FAO), a branch of the United Nations has identified 36 crisis countries, 21 of which are in Africa. The World Food Programme is another United Nations agency and it was estimated that it will need $500 million on top of what donor nations have already pledged to fill what the world food programme calls a global "food gap". Therefore, economists and some relief organizations such as CARE, advised U.S. government to buy the grains from US farmers and donate it to famine zones especially, in Africa. President Bush proposed it to the law makers in the House and Senate rejected it even refusing to include it in their respective versions of the farm bill (The Skyrocketing in the Washington Post Company, 2008: 18). Situation in Haiti was outlined by Spiegel (2008) captioned: "Global Food Crisis: The Fury of the poor", he summarized the situation as following.

For many Haitians, the mud biscuits are their only food. They taste of fat, suck the moisture out of the mouth and leave behind an after taste of dirt".6 It often causes them diarrhea, but they help to numb the pangs of hunger. Some say: "I am hoping one day I will have enough food to eat, so I can stop eating these". Eventually, the shortages triggered revolts in Haiti in April, 2008. A crowd of hungry citizens marched through Port au-Prince, throwing stones and bottles and chanting, "We are hungry: in front of the Presidential palace, tyres were burnt, and people died".

Iraq and Sudan, once the bread baskets of the Arab World, are nowadays dependent on the World Food Programme. The civil war in Sudan made more than 2 million people in refugee camps completely dependent on food aid (Spiegel Online International, 2008: 2). In Dubai, most of the supermarkets have pledged not to raise the prices of 20 staple foods for at least one year. The goal clearly, is to prevent dissatisfaction within the cities among the Indian and Pakistan construction workers, because without them, the enormous hotels, museums and artificial islands with which Dubai is making such a stir in the world would not exist (Spiegel Online International, 2008: 2).

In Egypt, thirty-two million of Egypt's population of 80 million lived on ($1.58) a day and 16 million on even less. The price of cooking oil has risen by 40% within a year. Inflation jumped to above 12% in February, 2008. Mr. Saad Ibrahim, owner of a small snack shop in Cairo in a neighbourhood behind al-Azhar Mosque sells dishes like noodles and chickpeas in tomato sauce, and his shop is in a good location. Nevertheless, most of the faithful worshippers are now quickly passing his shop after Friday prayers and this made him lament that he had fewer customers every day (Spiegel Online International, 2008). In Jordan, which has a modern system of agriculture, the cost of staple foods has increased by 60% within a year. Mr. Hassan Burendi – a vegetable seller said "I can hardly sell my vegetables anymore". The situation made King Abdullah of Jordan say "I fear the return of the 1996 Food riots, when angry citizens clashed with police in the city of Karak" (Spiegel Online International, 2008: 3).

In Nigeria, the inception of the food crisis was traced by economists and experts of different fields of study to 2003 and 2004 when the fuel prices jacked up. The prices of wheat and rice were stable in the late 2003 and early 2004 despite the hike in fuel price. However, in April 2004, rice prices moved higher by 56% and in April, 2005 by 61%. The outbreak of bird flu also led to the high price of chickens. The flu made some of the poultry farmers suffer enormous financial loss and this led to the high price of all the livestock (Okunmadewa, 2009). Professor Okunmadewa (2009) drew attention to Food Riots across the world and said that, in America, 75,000 people demonstrated in food riots likewise in Cameroon, Senegal, and Somalia but Nigerians persevered, as no food riot had never been recorded in the history of the Nigerians.

## Reasons for the global food crisis: Economists and other experts' perspective

There are several reasons for the global food crisis. The World Bank President Robert Zoelick (Spiegel Online International, 2008) and several Agricultural Economists agree with the following reason for global food crisis:

1. The world population is growing constantly, while the amount of arable land is declining. Climate change is causing a loss of agricultural land, irreversible in some cases, as result of droughts, floods, storms and erosion.
2. Because of changing eating habits, more and more arable land and virgin forests are being turned into pasture for livestock. The yield per acre in calories of land given over to pasture is substantially lower than that of arable land.
3. The World Bank wants developing countries to introduce market reforms, including the abolition of protective tariffs, a move that often causes massive damage to local agriculture.
4. The resulting high oil prices leads to energy crops being cultivated instead of grain crops for food or animal feed.
5. Millions of people displaced by civil wars need food, and yet they themselves are no longer capable of producing food.

Okunmadewa (2008) added further the following reasons for Nigeria's food crisis to be:

i. Rising of oil price
ii. Depreciation of dollars
iii. The Indian ban on rice exportation
iv. Unrest and political instability
v. Lip-services which had been paid on storage of agricultural products by Nigeria Government.

## The Qur'anic perspectives on the reasons for food crisis

وضرب الله مثلا قرية كانت آمنة مطمئنة يأتيها رزقها رغدا من كل مكان فكفرت

بأنعم الله فأذاقها الله لباس الجوع والخوف بما كانوا يصنعون

Meaning:

God set forth a parable, a city enjoying security, quiet and abundantly supplied with sustenance from every place, yet was ungrateful for the favors of God, so God made it taste hunger and terror (in extremes) (closing in on it) like a garment (from every side), because of the (evil) which (its people) wrought (Abdullahi, 1984): Qur'an 16: 112.

The fundamental reason mentioned by the Qur'an for the food problem is ingratitude to God's favour. Gratitude to God would be shown in various ways according to Yusuf Ali as follows:

i. Forgetting or refusing to acknowledge the true source of the favour
ii. Misusing or misapplying the bounty, as by committing excess in things lawful or refusing to share them with God's other creatures when the need arises.
iii. Falsely ascribing to God any prohibitions we may set up for ourselves for special reasons or because of our special idiosyncracies (Abdullahi, 1984).

In the Qur'an, we can see the reasons for the food crisis in Nigeria and the world at large. Price increase of fuel caused the high price of food in Nigeria which eventually led to serious inflation. The crude oil on which the Federal government relies, and which increase the Nigerian foreign reserves are not used for the benefit of the people. In spite of the net gain accrued from crude oil, the citizens buy the product at a high rate, while the wind fall hanged on the three tiers of Nigerian government. The proposed removal of government subsidy on fuel price termed deregulation worsens the negative impact of the high price on other sectors even before the implementation of the policy. The Nigerian government imports food from developed countries such as maize and meat from USA and rice from India/Pakistan. These developed countries had ceased the exportation, because they are turning these products into other product for their domestic utility. For example, America is turning corns into Bio fuel instead of wasting their money on crude-oil on which Nigeria's budget relies heavily.

The effects of the food crisis as the Qur'an stated are: i) Hunger and ii) Terror.

In Nigeria, a nation of 140 million inhabitants and with $60 Billion in its foreign reserve, her citizens and government can only boast of the following:

i. High Inflation of all commodities including fuel and other related commodities.
ii. Increasing poverty, which widens the gap between the high and the low, thus, wipes out the middle class due to the low-income of its civil servants.
iii. Worsening malnutrition of the citizens. The economists said, "Nigeria has really reached the apex of 99% inflation" (Okunmadewa, 2009).
iv. Windfall out of oil price has reached $60 billion and as a result, the government is planning to withdraw $10 billion to be shared among the three tiers of government in the country while the citizens are wallowing in abject poverty.

**Terror**

Terror grips the public office holders in Nigeria and wealthy men also suffer from arm bandits and even the poor innocent citizens fell victims during their operations. Bad governance of previous Nigerian governments have led to misappropriation of Allah's bounty while a negligible percent of its citizen stash away billions of Naira in their private accounts by not carrying out projects already paid for and by importing subsidized fertilizers which is distributed to wrong hands. The government claims that there is an increase in food production through various viable mismanaged agencies but food productivity growth is declining. The result is that the wealthy men are not free from the attack of the hoodlums, youths infested by hunger engaged in arm robbery, and kidnapping of innocent poor ones in order to belong to the high class in Nigeria. Nigerians are desperately looking for livelihood and therefore, trained and untrained thugs who erupt daily to exploit the politicians and public office holders whom they believe did not have welfare for the masses in their hearts.

As the poor ones are wallowing in abject poverty, the high classes are being enshrouded with terror, fear, unrest, and witch hunting either by the citizens or by the Economic and Financial Crimes Commission (EFCC). Without proper management of good resources in the nation, we may continue to witness the food crisis.

**Steps taken by various countries to curb global food crisis**

Since most of the developed and underdeveloped countries rely on World Bank (2008), the 10 points plan for overcoming the global food crisis (http://www.worldbankorg/.htmextrd/foodcrisis, 2008) are as follows:

i. Since high fuel costs have resulted to higher agricultural costs, falling of food stocks and shifting of the use of land from food production to bio-fuels production, the world Food Programmes should fully fund the emergency needs of people.
ii. Safety nets should be initiated and this involves distributing food in schools, offering food for works to quickly help those in severe distress.
iii. Seeds and fertilizers should be made available

promptly to farmers for the next planting seasons.
iv. Double the spending on agricultural research and development over the next five years through the Consultative Group on International Agricultural Research (CGIAR).
v. Agricultural business: There should be more investment in agricultural business so that the private sector's ability to work across the value chain would be exploited.
vi. Risk management: development of innovative risk management tools and crop insurance to protect poor farmers and help build food security.
vii. Bio fuels: Ease policies to subsidies, mandates and tariffs on bio fuels derived from corn and oil seeds should be initiated. Policy makers should consider (safety valves) to ease these policies when prices are high. The choice is not limited to bio fuel or food.
vii. Export ban: Removal of export bans that had led to even higher world prices. Twenty-eight countries have imposed such controls. Removing this ban could have a dramatic effect.
viii. World Trade Organization: This organization should consider the removal of agricultural subsidies and Tariffs and create a more efficient and fair global food trade.
ix. Collective action: Countries should work together to counter the global food crisis.

Among the steps taken by the World Bank to curb the Global Food Crisis through new deal on Global Food Policy are:

i. Creating a $1.2 billion rapid financing facility to speed up assistance to the neediest countries.
ii. Boosting overall agricultural lending to $6 billion over the next year.
iii. Launching of risk management tools and crop insurance to protect poor countries and small-holders.
iv. Nearly doubling agricultural lending to Africa from $450 million to $800 million and to Latin America from $250 million to $400 million.
v. Supporting over $1 billion in new projects in agriculture and rural development in South Asia.
vi. Doubling lending for social protection, nutrition and food security, and social risk mitigation to $800 million over the next year.
vii. Providing $100 million to Bangladesh to help address the food crisis.
viii. Providing $100 million to hard hit Burkina Faso, Burundi, Cote d'Ivoire, Ghana, Madagascar, Mali and Niger to meet additional expenses of food imports and to buy seeds for the new season.
ix. Providing grants to Djibouti ($5 million), Haiti ($10 million) and Liberia ($10 million) to feed poor children and other vulnerable groups.
x. Providing grants to Togo, Tajikistan, and Yemen over the coming season.
xi. Working on irrigation and water management in Ethiopia, fertilizer use in Malawi, market access for small holders in Senegal, and crop diversification in Mali and Uganda.
xii. Preparing a food crisis related emergency operation to help Honduras tackle effects of rising food prices.
xiii. Providing financial assistance to Kyrgyz and Tajikistan for nutritional supplements to pregnant women, lactating mothers, infants and small children.
xiv. Conducting rapid needs assessments for countries impacted by crisis, including Burkina-Faso, Burundi, Eritrea, Guinea, Guinea-Bissau, Kenya, Liberia, Malawi, Mali, Mauritania, Niger, Sierra-Leone and Togo.
xv. Urging major grain producing countries to lift or refrain from bans on food exports.
xvi. Working with other donors and agencies on the Africa Union's New Partnership for Africa's Development in Sub-Saharan Africa (http://www.econ.worldbank.org/external, 2008).

The plans had not really addressed some causes of the food crisis such as crop short fall from natural disasters. The Qur'an bluntly put man's effort as commendable but no matter how serious, with labour, and application of modern equipments, the power of productivity belongs to Allah where He says:

أفرأيتم ما تزرعون ءأنتم تزرعونه أم نحن الزارعون ولو نشاء لجعلناه حطاما
فظلتم تفكهون إنا لمغرمون

Meaning:

See ye the seed that you sow in the ground? Is it you that cause it to grow, or are we the cause? Were it our will, we could crumble it to dry powder, and you would be left in wonderment. (Saying): we are indeed left with debts (for nothing): indeed are we shut out (of the fruits of our labour) (Abdullahi, 1984): Qur'an 56: 63 to 67.

The aforementioned quoted verses of the Qur'an expatiate one of the reasons given by the World Bank President and several agricultural economists for this global food crisis that "Climate change is causing a loss of agricultural land, irreversible in some cases, as a result of droughts floods, storms and erosion" (Robber, 2008: 2). Going through all the solutions proffered by the experts of different fields of studies, one can see that none of them gives a consideration to the divine happenings and its solution. A geographer could predict that rain will fall, geologists or agriculturalists could identify a pasture land but who cause the farm products to grow and come forth with profitable productivity except the Creator who is in possession of life and death.

The aforementioned quoted Qur'anic verses made it clear that no matter how we adopt ultra-modern agricultural equipment, the extent of farm productivity lies ultimately with Allah. One of the incidents I witnessed in

Offa, Kwara State, Nigeria in 1990 confirmed this assertion. A colleague and an Agricultural science teacher in a secondary school cultivated about 2000 heaps, sowed guinea corn and applied fertilizer for an appropriate period as an expert. But he abandoned the Mosque and forgot to follow his labour with prayer, thinking the hard labour without Allah could yield a fruitful productivity. I admonished him to work and at the same time to give Allah his due such as observation of five obligatory prayers, and special supplication prayer on the farm so that Allah can crown his efforts but he ignored my advice. Eventually, the guinea corn started growing vegetative instead of bearing seeds. It grew up higher than normal. This is how the efforts of our brothers became fruitless. Nigerians are increasing in Agricultural production but less in productivity.

Going through the verse and reasons given by the experts, when Allah intends to punish a nation or community, He sends down to them drought, or excess of rain and floods, erosion or storm which will destabilize their properties including their farms. Allah confirms that no matter the agricultural methods applied, the productivity of any enterprise is with His permission. Even on the livestock, where we contact the bird flu. What are the solutions we proffer for future occurrences for these catastrophes? There is no solution yet except the one proffers by the Divine Lord in the Qur'an.

## NIGERIAN AND NIGERIA GOVERNMENT ON THE FOOD CRISIS

Many countries took emergency steps on the food crisis such as restructuring food exports, setting limits on food prices, rigorous, extensive analysis and consultations on way out, food safety net programmes, reduction in import duties, as well as, lifting of ban on rice importation among others. The Agricultural Machineries and Equipment Fabrications Association of Nigeria (AMEFAN) (The Nation Newspaper, 2008a: 26) raised alarm and advised the Federal and State Governments to evolve long-term measures for solving the food crisis. The association is ready to partner with government to produce basic farm machineries and urges the government to discourage rice importation and other food items. The association feels that the remedy to food crisis could be found in the 'promotion' of local fabrication of basic agricultural machineries and processing equipment coupled with discouragement of food importation.

Another group, the Oodua People's Congress (OPC) also suggests other steps to be taken for a short-time and long-term measures. In a conference organized in Lagos at the Century Hotel, the group urges the Federal government to draft soldiers, policemen and Youth Corps members into agriculture to boost food output. The group also suggests that adequate funding of the agricultural sector by those in authority would help to avert a looming economic crisis. The group advised the government to show seriousness on re-distribution of wealth from the rich to the poor. Incessant power failure makes the industries-produce at sub-optimal level and small scale enterprises to go under fizzle. A majority of Nigeria's civil servants spend 70% of their income on food.

Therefore, the time has come to release the country's grain stock and food reserve into the market with urgent measures to be taken to force down the prices of food stuff in the market place. The association submits that for a short-term measure, the Federal and State government must work out subsidies on food, food imports and food production ingredients such as fertilizer (The Nation Newspaper, 2008b).

### Nigerian Government's response to the food crisis

The government ordered for the immediate release of part of the country's grain stock and food reserve into the market. The government stood firm on awarding licenses to individuals on cement importation and tried to force down the price to ₦800.00 per bag when the importation materialized. This forced the skyrocketed cement price to come down even before the ₦800.00 per bag policy was implemented. Federal government distributed the fertilizers to states and local governments at subsidized rate, though, failed to ascertain if it went to appropriate quarters and hands at the affordable subsidized rate or not. Federal government has started sounding a warning on raising the price of fuel in the country from next year. National Assemblies of Nigeria also raised alarm on the food crisis and promised to have special debates on it. Both their short and long term measures are yet to be disclosed. So far, the Nigerian government focuses on consumers of food but neglect its producers. Kwara State Government prioritizes food production, but the impact is yet to be felt at the market. Although, most of the Agricultural agencies owned by the government are aware of this global food crisis, none of them has given the public an insight into the short and long term measures to avert this crisis.

### Qur'anic remedy on the food crisis

Qur'an 12:47 indicates the solution to the crisis where he says:

قال تزرعون سبع سنين دأبا فما حصدتم فذروه في سنبلة إلا قليلا مما تأكلون

"For seven years shall you diligently sow as is your want and harvest a little, of which you shall eat". The government is advised to pump more funds to agricultural production while monitoring closely the resultant mass production obtained with modern equipment. The suggestion of the Oodua People Congress on driving soldiers, police and corps members into farms seems to be useful. It was done and found workable. It forced down the price of food stuff during the military regime of Bukhari/Idiagbon when all the public civil servants irrespective of cadre and discipline had at least a farm. During this time, secondary and primary school farms were also improved and yielded a fruitful result. Price of food stuff drastically decreased without any government's enforcement.

The Qur'an urges Nigeria government and Nigerians to repeat the same for both short and long term measures and if this is done for three years, Nigeria will be exporting a lot of food stuff. The Qur'an also says that the harvests that you reap shall last for years. We should eat a little out of our production.

There is one aspect of farming that needs improvement. All the peasant farmers and large scale farmers harvest at the same period. Because of lack of modern storage facility, much of the farm produce could not be preserved for many years. In more advanced countries, government buys the farm products from farmers and stores them for distribution even as food aids to needy countries. If majority of Nigerians embark on farming, applied modern equipment sublet to them at cheaper rate, provided prepared seeds and fertilizers at appropriate period, large farm produce will be the result at the same time. If the government buys their farm products at a reasonable rate and keeps in modern silos for preservation, it will definitely encourage the farmers and within few years, the problem of food crisis will be overcome. Few years later, Nigeria will be subsequently among the largest exporter of food stuff in Africa if not in the world because of the natural gift of Allah, that is, pasture land.

The Qur'an clears it that no matter how the earth or land is fertile, no doubt, it will be inflicted with climate changing, which may lead to drought or storm, where He says:

ثم يأتي من بعد ذلك سبع شداد يأكلهن مما قدمتم لهن إلا قليلا مما تحصنون

Then will come after that (period) seven dreadful (years), which will devour what you shall have laid by. In advance form them, (all) except a little which shall have (special) guarded: Qur'an 12: 48.

The oil boom with an unexpected high price of crude oil made the Nigerian government neglect agriculture because the few wealthy ones feel that the high price of the food stuff gulps only 0.2% of their income and forget that majority are suffering.

Qur'an 16 verse 112 spelt out the way out for Nigerian government and its citizens by saying:

وضرب الله مثلا قرية كانت آمنة مطمئنة يأتيها رزقها رغدا من كل مكان فكفرت
بأنعم الله فأذاقها الله لباس الجوع والخوف بما كانوا يصنعون

God sets forth a parable, a city enjoying security and quite abundantly supplied with sustenance from every place, yet it was ungrateful for the favour of God, so God made it taste of hunger and terror (in extremes) (closing in on it) like a garment (from every side), because of the (evil) which (its people) wrought. The verse describes Nigeria as a city, which Allah abundantly blessed with all kinds of mineral resources, enjoying security but injudiciously distribute the wealth of Allah in their possession. We see this in recent situations in the country:

i. The president of the present administration declared his entire asset when he was a Governor while others who promulgated the law declined. When he was elected as the President of Nigeria in 2007, he also declared his asset. His colleagues in political offices are yet to make declaration of their assets. Is it because, they possess unlawful wealth or what?

ii. Federal government imports fertilizer and distributes it at a subsidized rate but most of these political office holders give the distribution paper to their relatives who in turn sell the articles to peasant farmers at quadruple the price. The question is how many among these farmers could afford the price?

## Distribution from excess crude oil

The Federal government proposed to withdraw $10 billion from the foreign reserve to be shared among the three tiers of government but is yet to receive the approval. It is the same government that refuses to use out of this external reserve to subsidize fuel to its citizens. Everything in Nigeria relies on fuel and if the price of fuel comes down, the traders, farmers, civil servants and common citizens would enjoy the dividends of democracy. But instead, our government is giving another red alert to increase the fuel price by January, 2009 despite the advice and reasons agreed upon by experts of both advanced and developing countries that the global food crisis centres on the high fuel price of each country. If the proposal of the Federal Government got approved at least each state will share not less than ₦50 billion and eventually as usual we may not see any beneficial project carried out with this money.

The same government who caters for political office holders, and reviews their allowances and their gratuity for a period of four years finds it difficult to cater for teachers at all levels, civil servants and the masses that get their livelihood from the aforementioned cadres of workers through patronizing them at the market. The government finds it difficult to give the workers minimum wage of ₦50, 000.00 per month. Is the government telling us that a country does not have enough money to cater for the workers while the chairman of a political party in a local government earns ₦120, 000.00 for just being local government political meeting coordinator?

Nigerians are graduating every year from our tertiary institutions. The government creates a lot of schemes which can make these graduates to have self reliant job. How many innocent brilliant graduates benefit from these schemes? Instead, the dividends are distributed at ward level of each local government. Award of contracts is based on familiarity not competency. Where is the justice in Nigeria, unqualified people get into political offices; a citizen who had been alleged of corruption in the last few years got re-appointed to political offices in the country. Although, the government claim that previous office holders no longer enjoy immunity but only few who run foul of their political mentors had been prosecuted.

Nigeria is on the verge of having a food crisis which can be quickly averted if a just redistribution of the nation's wealth favours the masses. 10% of the population is stinking rich while 90% are wallowing in abject poverty. A decade ago, Nigerians could be classified into high class, middle class and masses. But recently, the gap between middle and high classes has increased so much that the middle class does no longer exist. A grade level 12 officers in the civil service was among the middle class but the high cost of food stuff, transportation and accommodation has gulped his income and pushed him into the class of the masses. The masses are hungry, the wealthy men are fearful of the masses' as a reaction to their unlawful accumulations. For this, the wealthy men consequently lodge their money into foreign accounts from which these developed countries develop their country. Is it not an immoral act for these developed countries that are aware of the debts owed by Nigeria as a country and still they keep billions of Naira in political officers' account in their country without returning the money to government treasury for development?

Due to incessant power outage which dragged some of the talented artisans and craftsmen to be commercial motorcycle riders, while few graduates from our tertiary institutions turned to commercial motor drivers. Foreign investors signified interest in Nigeria but most of the existing industries that improve the livelihood of our youth are breaking down due to the aforementioned reasons. Nigerian governments have been planning since a decade ago to overcome the problem in the energy sector but to no avail. The situation has not improved because people who do not know anything about the sector, such as a state chairman of the ruling party, gets a contract to supply items such as transformers while qualified graduates in electrical/electronic engineering are ignored.

Nigeria is endowed with brilliant, diligent youths who could handle the agricultural sector properly if the government involves them in improving farm production. Most of Agricultural Research resources centres owned by both Federal and State governments carry out research every day, and they produce seeds to be planted but how many sincere peasant farmers have access to this government effort?

To avert the food crisis, justice needs to be done by the Federal government not only pumping money to agricultural sectors but also assisting the peasant farmers with modern equipment and giving a thorough supervision to the government fund pumped into the sector as the national dailies' editorial comment advised the Federal government in the caption: "The idea behind the ₦200 Billion agricultual fund will be laudable if corruption and red tape do not stand in the way". The Federal government through the Debt Management Office (DMO) and Central Bank of Nigeria (CBN) has floated a ₦200 Billion credit for commercial agriculture. The package was described as part of government's response to the current global economic/food crisis. The United Bank for Africa and the First Bank of Nigeria as the fund's managers. The new scheme targets corporate/large scale integrated commercial farms with assets of ₦350 Million excluding land and capacity to grow such asset to ₦500 Millions in three years while Nigerian Agricultural Corporative and Rural Development Bank (NACRDB) targets small and medium-scale farmers.

Factors responsible for the failure of the agricultural sector to produce enough food for the people are multifarious such as the following; the gross neglect of agriculture over the years, poor funding, poor extension services and lack of loans to the practical peasant farmers and successive government have invariably demonstrated ambivalence towards research institute in general. For instance, less than one percent of Gross National Product (GNP) is spent on research annually. Results of numerous agricultural researchers are gathering dust in the various cabinets of Nigeria ministries because there is a clear skilled extension worker to impart to the farmers. Various river basins and rural development authorities have not made any appreciable impact on local food production almost 33 years after their inception and despite the huge investments in irrigation schemes, most of the equipment at the basin are not working such as one of the biggest Ogun/Osun River Basin in Itoikin along Ikorodu/Ijebu-Ode road, Nigeria. Its dam can irrigate several hectares of farmland but the place is now a parody of its former eminence. These basins had failed to supply water to over 90 farmers in each of their areas.

The Qur'an gives a remedy to another major cause of food crisis that is, drought, erosion, storm, climate change and so on. Allah says:

فقلت استغفروا ربكم إنه كان غفارا يرسل عليكم مدرارا ويمددكم بأموال وبنين
ويجعل لكم جنات ويجعل لكم أنهارا ما لكم لا ترجون لله وقارا وقد خلقكم أطوارا

Meaning:

Saying 'Ask forgiveness from your Lord, for He is off-forgiving; He will send rain to you in abundance give you increase in wealth and sons; and bestow on you rivers (of flowing water) what is the matter with you, that you place not your hope for kindness and long-suffering in God. Seeing that, it is He that has created you in diverse stages: Qur'an 71: 10 to 14.

The enumerated system of governance in Nigeria glaringly shows the great extent of oppression and unjust distribution of Allah's abundant resources blessed to the country. There are some parts of the country specifically in Southwestern Nigeria who need not to cultivate heap or apply fertilizer before they reap mass productivity, and some even plant thrice in a year but yet these areas are not having food surplus. Because Federal government scheme on agriculture did not reach the appropriate quarters.

The World Bank seems to regard Nigeria as a nation that is blessed with everything. Because going through the Programme of the International Monetary Fund and World Bank, both have no plan for Nigeria. The immorality of the citizens also contribute to the food crisis, for instance, a farmer, be it large or small scale who produce vegetable at large quantity prefer to export the farm produce at getting exorbitant profit than selling it to its citizens at cheaper rate. Unless we return to Allah, we cannot seek forgiveness for sins committed which our conscience and Allah know better.

Some agricultural experts suggest four salient points that Saudi government adopted as a method on agricultural sector which helped the country to overcome the food crisis and was able to transfer from a food importer to an agriculturally self-sufficient nation being able to feed its population, accommodate the teeming millions of pilgrims who flop to the kingdom during 'hajj' and 'umrah' without causing a food crisis and these salient points are:

i. Saudi government created hundreds of purpose-built farms. Several boreholes are sunk and diesel pumps and irrigation implementation installed to ensure that crops are adequately watered even during a drought.
ii. The areas to be cultivated are prepared by clearing and leveling the land so that water from rain or irrigation will not wash away the top soil but sink directly into the ground.
iii. Access roads are built to connect the farms to major roads to help in evacuating farm products to major markets. Comfortable houses are built for farmers and their families, and powerful generating sets are provided for the farmsteads.
iv. The farmsteads are allocated to professional farmers (and not politicians) on a soft loan basis.

With the aforementioned views of expert, the following additional recommendations will definitely cater for short term and long term measures for averting the food crisis in Nigeria and at any place that boldly adopts the recommendations:

i. Fuel price for the Nigerian citizens must be subsidized to the minimum level such as N40.00 per litre while smuggling fuel into the neighbouring countries should attract the same penalty as cocaine pushers.
ii. Lift ban on importation of food stuff. Let our men at Nigerian borders stop harassing small scale traders for importing goods into the country. It will force the high prices of these commodities down. For instance, a bag of rice got to ₦10, 500.00 by April, 2008 but now the price seems to be stagnant at ₦7,000.00/₦6,700.00 as from July, 2008 (Interview with Alhaja Salamat Lawal of shop 129A, 2008). With massive production of our professional farmers in modern way both local and imported rice will be at par in cost.
ii. The government should encourage all parastatal, government workers, and cooperative societies to embark on farming either on a large or small scale that will take care of immediate needs and to put the agricultural sector among the priority of the three tiers of Nigeria government by having adequate funding. The young brilliant graduates of agricultural economics will be employed in the six geo-political zones of the country to identify certain farm products which consume less energy and give high productivity.

The genuine experts posted to these geographical zones will be employed to monitor the following:

- Safety nets: Such as distributing the released grains from the country reserve to the masses.
- To prepare seeds and fertilizers: For the coming planting season. Not just delivery alone but to monitor how they were planted and how they apply the fertilizers at appropriate period.
- Risk management: The government should use these experts to develop innovative risk management of tools and crop insurance to protect poor farmers and help to build food security. However, the government needs to double its financial allocation to research.
- Storage facilities: This can preserve our farm products for at least five years. For instance, yam can only be preserved for a year. They can still carry out more research on these seasonal food stuff and crops. We refresh ourselves with fresh fruits at anytime in Makkah. How do they preserve it? Why can we not have fresh mango, banana and orange throughout the year in Nigeria? We can take an example from the Saudi government who use the Qur'an as a pivot of governance.

If AMEFAN promised to produce cassava flour as a substitute to wheat flour, what of yam, rice, beans, millets, guinea corn and so on. Therefore, our agricultural research institute should investigate what those items can be turned into as substitute. Nigeria government can resuscitate community boards at these six geo-political zones with new method and not the old method used in olden days for cocoa board in the Southern part of the country. This will give an open policy to the farmers and each zone will produce massively, our government should govern its subject in transparency and accountability when it comes to buying and paying for the farm products by the board. FADAMA, a community based society on Agricultural production, could also be properly funded and monitored.

Our experts or Agriculturalist should be sponsored for training in and outside the country to expose them to modern equipment and methods on how such application could fit into their own locality. Since Nigeria pasture land can be used to plant varieties of seeds, the peasants and the professional farmers should be tasked and assisted on how to improve production through irrigation in order to harvest farm produce throughout the year. Nigeria is a nation blessed with a lot of resources. As crude oil had been discovered in many geo-political zones which make Nigeria government to be rest assured of buoyant foreign exchange reserves, we can still embark on agricultural sector which will double our status as a giant of Africa when we started granting Food Aids to developing countries and even developed countries in the nearest future. Welfare of farmers at the rural areas needs to be improved. They should have permanent social amenities, such as good roads to transport their products to urban areas at cheaper transportation rate, good schools equipped with adequate teachers, and with access to what goes on in the world, access to modern farm equipment, opportunity to sell their product at minimum profit range. If all these are done, surely, the farmers who have chosen farming as their profession will be ready to serve humanity.

In conclusion, it is unfortunate that malnutrition has affected a large proportion of our large population. About 38% of our population is affected and this has a serious effect on the life expectancy of our people which is 44 years for females and 52 years for males. Let us adopt the remedy and solutions supplied by the Qur'an and nations and abide by the Qur'anic views (Saudi Arabia) eventually; our nation will surely become a major exporter of food stuff in the nearest future. Therefore, let the government make the seeds, agricultural inputs and equipment available, accessible and affordable to all interested Nigerians.

**REFERENCES**

Abdullahi YA (1984). The Holy Qur'an Translation and Commentary Lahore: Islamic Propagation Centre Int., pp. 566-568.

http://www.econ.worldbank.org/external (2008). Food Crisis downloaded on August 9, 2008.

http://www.worldbankorg/.htmextrd/foodcrisis (2008). Downloaded on August 9, 2008. Over coming the Crisis 10 Point Plan.

Okunmadewa F (2009). Food Price Crisis and Sustainable Development in Nigeria. In L Popoola, O Olorunnisola and O Ademowo (Ed.): Contemporary issues in sustainable development: lessons for, and challenges to Nigeria. Postgraduate School, University of Ibadan, Ibadan. 2: 31-79

Interview with Alhaja Salamat Lawal of shop 129A (2008); Grains Dealer at Bodija Market, Ibadan, Oyo State, Nigeria on 25/8/2008.

Robber Z (2008): in and Several Agricultural Economists as in Der Spiegel: Consequence of Food Crisis, op.cit.,p. 2

Spiegel Online International (2008). Food Riots around the World, Part 2, "People are dying before our Eyes", p. 2-3.

Spiegel S (2008). Online International Global Food Crisis, The Fury of the Poor Der Spiegel, Article of April 14, 2008, part 1, p. 1.

Staff S (2008). Global Food Crisis: The Fury of the Poor, Spiegel Online's Daily Newsletter of 14th April, 2008, Part 1, p. 1.

The Nation Newspaper (2008a). Agriculture Mechanized Farming an Antidote to Food Crisis, Friday June 20, p. 26.

The Nation Newspaper (2008b), OPC "How to Avert Food Crisis, Friday June 20, p. 26.

The Skyrocketing in the Washington Post Company, Friday, March 14, (2008), pp. A16-18.

World Bank (2008): Addressing the food crisis: the need for rapid and coordinated action. Report preentedto Group of 8 meeting of finance ministers, Osaka, June 13-14.

# Role of arbuscular mycorrhizal fungus (*Glomus intraradices*) – (fungus aided) in zinc nutrition of maize

K. S. Subramanian[1]*, V. Tenshia[1], K. Jayalakshmi[1] and V. Ramachandran[2]

[1]Department of Soil Science and Agricultural Chemistry, Tamil Nadu Agricultural University, Coimbatore 641 003, India.
[2]NA & BTD, Bhabha Atomic Research Centre, Trombay, Mumbai 220 085, India.

**A greenhouse experiment was conducted to assess the nutritional improvement of maize plants (variety COHM5) inoculated with and without arbuscular mycorrhizal (AM) fungus (*Glomus intraradices* Schenck and Smith) under varying levels of Zn concentration (0, 1.25, 2.5, 3.75 and 5.0 mg kg$^{-1}$). Growth (leaf area index and root morphological attributes) and nutrient status (Zn and P concentrations) were measured at 45 and 75 days after sowing (DAS). Maize grains were analyzed for Zn and tryptophan concentrations. The increasing levels of Zn fertilization have no impact on the total mycorrhizal colonization, but the hyphal growth decreased slightly. Mycorrhizal inoculation consistently increased the root morphological attributes regardless of Zn fertilization. Mycorrhizal inoculation and Zn fertilization exhibited prominent increase in the leaf area index and chlorophyll content of plants and the response to mycorrhizal inoculation was higher at lower levels of Zn fertilization. The mycorrhizal dependency decreased gradually with increase in Zn levels. Mycorrhizal plants had higher P and Zn concentrations in both the stages in comparison to non-inoculated plants. Maize grains of AM+ plants had higher Zn and tryptophan concentrations. The plant available Zn concentration in mycorrhizal fungal inoculated soil was higher than non-inoculated soils, which directly contributed to increase in Zn status of mycorrhizal plants. In conclusion, mycorrhizal symbiosis enhances Zn supply to the host plants by extensive root development enabling the plant to maintain higher nutritional status and produce grains with rich tryptophan concentrations.**

**Key words:** Arbuscular mycorrhiza, chlorophyll, maize, nutritional quality, zinc deficiency.

## INTRODUCTION

Soil Zn deficiency is a world wide problem in the crop production, affecting the growth of the crops in over 50% of agricultural lands (Alloway 2001). Zinc deficiency is closely associated with 80% grain yield reduction, beside reduction in grain Zn content and other nutrients in cereal crops (Cakmak et al., 1998). The high dependency on foods cooked with reduced Zn content brings out malnutrition and Zn deficiency among people, who are estimated to be over two billion globally (Welch and Graham 2002). In order to ensure nutritional security of human beings, the production of cereals with rich grain nutritional quality should be enhanced. Zinc fertilization, a common remedial practice have use efficiency rarely exceeding 1%, while remaining Zn is fixed in the soil due to alkaline pH and higher free lime status (Mandal and Mandal 1986). However, the total Zn content in majority of the arable lands is reported high, but mostly they are present in the unavailable form. The magnitude of Zn deficiency is high in calcareous soils of arid and semi arid tropical regions, where major portion of added Zn gets fixed in the soil (Marschner 1995). The imbalanced use of fertilizers and non-addition of organic manures also cause widespread soil Zn deficiency (Singh et al., 2005). There has to be some mechanism to transforming tightly bound Zn into plant available forms.

Arbuscular mycorrhizal (AM) fungi are known to improve the availability of nutrients especially phosphorrus, moving them to the root surface by diffusion (Jakobsen et al., 1992). It is also reported that mycorrhizal symbiosis enhances absorption of relatively immobile micronutrients such as Zn and Cu (Faber et al., 1990; Kothari et al., 1991; Li et al., 1991; Liu et al., 2000; Ryan and Angus, 2003). The micronutrient improvement in mycorrhizal plants is always associated with rhizosphere

*Corresponding author. E-mail: kssubra2001@rediffmail.com.

acidification (Dodd et al., 1987), more external mycelium in the soil (Jakobsen et al., 1992) and soil biochemical changes (Subramanian and Charest, 2007). Besides, host plants retain the large green leaf area (Subramanian et al., 1997) and chlorophyll concentration (Subramanian and Charest, 1995; Augé, 2001) under the water deficit conditions. Liu et al. (2000) have indicated that the external mycelium of mycorrhizal fungus contribute significantly to the tune 60% towards host plant Zn nutrition under deficient conditions. These data suggest that mycorrhizal association may play an important role both under drought and nutrient deficient conditions. It is proved that mycorrhizal symbiosis improves Zn nutrition in maize as a secondary consequence of P nutrition. We hypothesised that the maize's responses to mycorrhizal inoculation differ with respect to varying degrees of Zn deficiency. To address this question, the changes in nutrient status, morphological and physiological attributes of maize plants that were inoculated with and without mycorrhizal fungi and subjected to varying levels of Zn fertilization, were measured at 45 and 75 days after sowing (DAS).

## MATERIALS AND METHODS

### Experimental soil

The physical characters of experimental soil were Alfisol, sandy loam in texture with neutral pH (7.4) and free from salinity (0.04 $dSm^{-1}$). The soil had less quantity of nutrients, which include organic carbon (0.4%), available N (1.23g $kg^{-1}$), available ($NaHCO_3$-extractable) P (0.058 g $kg^{-1}$), available K (1.6 g $kg^{-1}$) and available Zn (DTPA extractable) (0.63 mg $kg^{-1}$). The experimental soil was collected from the field that was kept fallow for more than three years to avoid indigenous mycorrhizal population and propagules in the root portions of previous crop. The examination of indigenous mycorrhizal status in the experimental soil revealed very less indigenous viable spore population (<10 spores in 100 $g^{-1}$ soil).

### Greenhouse experiment

The greenhouse experiment was conducted at the Department of Soil Science and Agricultural Chemistry, Tamil Nadu Agricultural University, Coimbatore. The climatic conditions viz. temperature (24 - 28°C), light intensity (800 - 1000 µmols provided by natural light), relative humidity (60 - 65%) and 12 h photoperiod were maintained in the greenhouse. The experiments had 10 treatments with six replications in a randomized block design. In six replications, each three were kept for sampling at 45 and 75 DAS. One plant per pot was maintained throughout the experiment. Maize (var. COHM5) plants were grown in earthen pots of 10 kg capacity (30 cm diameter and 30 cm height) with and without Arbuscular Mycorrhiza (AM) inoculation. The mycorrhizal strain (*Glomus intraradices* Schenck and Smith; TNAU-03-06) was cultured in maize plants and propagules comprised of infected root bits and spores were blended in sterile vermiculite. The vermiculate based mycorrhizal inoculum was applied 5 cm below the seeds prior to sowing at the rate of 10 g per pot as a thin layer. Zn fertilization was done at five concentrations (0, 1.25, 2.5, 3.75 and 5.0 mg $kg^{-1}$) in the form of $ZnSO_4$ in addition to regular NPK dose of 0.068:0.03:0.03 g NPK $kg^{-1}$ soil. The pots were watered daily to maintain the moisture content of the soil to the saturation. The roots and shoots were sampled from each plant at 45 and 75 DAS and analyzed for nutrient concentrations. The root architecture of mycorrhizal (AM+) and non-mycorrhizal (AM-) plants was evaluated measuring root length (vertical), spread (horizontal) and volume. Leaf area index (LAI) to assess the leaf area of the AM+ and AM- plants was calculated using leaf area meter (LICOR Model LI 8000). The leaf chlorophyll content was analysed extracting 500 mg fresh leaves with 50 ml acetone, and measuring the absorbance at 663 nm (Bruinsma 1963).

### Mycorrhizal colonization

The stages of mycorrhizal colonization were described as functional and vigorous colonization at 45 and 75 DAS respectively. Maize plants were pulled out from the pots without damaging the roots. The roots were repeatedly washed with tap water until they are free from dirts and soil particles. The root segments of 1 cm size in 100 numbers were cut per treatment, and estimated for mycorrhizal colonization following Dalpé (1993). Before mounting the root segments on slides, they were bleached with 2.5% KOH, acidified in 1% HCl and stained in 0.05% tryphan blue solution (tryphan blue 0.5g, glycerol 500 ml, 1% HCl 50 ml and distilled water 450 ml) and destained. Root segments were observed under the 10 x lens microscope for the presence of any of the mycorrhizal structures such as arbuscules, vesicles, external hyphae and spores.

### Nutrient analysis

Plants: The shoots and residual roots were dried in hot air oven at 70°C for 48 h. The collected plant samples weighing 0.5 g were digested in triple acid mixture (9:2:1 nitric acid: sulphuric acid: perchloric acid) for nutrient analyses. The colorimetric examination of plant extract to measure the phosphorous (P) level was taken following vanadomolybdo phosphoric acid yellow colour method (Piper, 1966). The plant samples diluted to 50 ml using distilled water were fed to an Atomic Absorption Spectrometer (Varian Spectra AA 220, Australia) to determine Zn concentration. Similarly, Zn level in grains was also analyzed. The total Zn and P levels in the plant were calculated multiplying the nutrient concentration in the samples with plant dry mass. The powdered grain sample was prepared to measure the concentration of tryptophan following the protocol described by Sadasivam and Manickkam (1996). Mycorrhizal dependency, a measure of plants response to mycorrhizal inoculation in accumulating Zn in their roots was calculated using the formula from Hetrick et al. (1992).

$$\text{Mycorrhizal dependency (\%)} = \frac{\text{Zn content (AM+)} - \text{Zn content (AM-)}}{\text{Zn content (AM+)}} \times 100$$

Soil: Soil nutrient analyses were conducted to find the concentrations of Zn and Phosphorous in the soils. The DTPA soil test was done with DTPA extractant (13.1 ml triethanolamine, 1.967 g DTPA and 1.47 g $CaCl_2$ are mixed together, made up to 1 L with pH adjusted to 7.3) to extract Zn from the soil (Lindsay and Norvell, 1978). Available phosphorus in the soils was estimated by extraction with sodium bicarbonate ($NaHCO_3$) following Olsen et al. (1954).

### Statistical analysis

A one-way analysis of variance (ANOVA) was done for all data and comparisons among means were made using LSD (least square difference) test, calculated at $P \leq 0.05$. Linear correlation

**Table 1.** Percentage of mycorrhizal colonization examined in the arbuscular mycorrhiza inoculated (AM+) and non-inoculated (AM-) root segments of maize crop plants at 45 and 75 days after sowing (DAS) under different Zn levels. Values in parentheses indicate the standard error. The levels of significance for ANOVA, * = P ≤ 0.05; ** = P ≤ 0.01; NS = Not significant.

| Zn levels (mg $kg^{-1}$) | 45 DAS | | | | | | 75 DAS | | | | | |
|---|---|---|---|---|---|---|---|---|---|---|---|---|
| | Arbuscules | | Vesicles | | External mycelium | | Arbuscules | | Vesicles | | External mycelium | |
| | AM- | AM+ | AM- | AM+ | AM- | AM+ | AM- | AM+ | AM- | AM+ | AM- | AM+ |
| 0 | 17.8 | 40.2 | 22.4 | 67.0 | 15.2 | 46.8 | 13.0 | 33.5 | 26.0 | 43.7 | 10.5 | 41.2 |
| | (1.35) | (0.86) | (0.92) | (1.54) | (1.52) | (1.24) | (0.812) | (0.583) | (1.88) | (1.06) | (0.860) | (1.07) |
| 1.25 | 18.2 | 38.8 | 17.0 | 62.6 | 15.0 | 44.4 | 8.5 | 30.7 | 17.5 | 48.0 | 11.0 | 42.0 |
| | (1.56) | (1.15) | (1.00) | (0.871) | (1.00) | (2.03) | (0.509) | (0.927) | (1.58) | (0.58) | (0.509) | (1.02) |
| 2.5 | 21.2 | 40.2 | 30.0 | 64.0 | 23.0 | 40.4 | 7.2 | 36.0 | 16.5 | 48.0 | 10.0 | 42.5 |
| | (0.96) | (0.86) | (1.30) | (1.67) | (1.58) | (1.60) | (0.447) | (0.509) | (1.14) | (2.42) | (1.02) | (0.509) |
| 3.75 | 17.8 | 45.2 | 26.2 | 63.8 | 17.0 | 40.4 | 8.2 | 29.5 | 26.7 | 37.0 | 9.5 | 35.5 |
| | (0.66) | (1.28) | (1.49) | (1.24) | (2.00) | (1.35) | (1.07) | (0.860) | (1.42) | (3.27) | (0.860) | (0.707) |
| 5.0 | 18.2 | 45.2 | 26.2 | 63.4 | 16.8 | 38.2 | 5.5 | 32.7 | 18.5 | 52.2 | 11.5 | 33.5 |
| | (1.20) | (1.28) | (1.31) | (1.56) | (2.37) | (0.927) | (0.860) | (0.871) | (0.707) | (3.31) | (1.39) | (1.01) |
| Mean | 18.6 | 41.9 | 24.4 | 64.2 | 17.4 | 42.0 | 8.48 | 32.5 | 21.0 | 45.8 | 10.5 | 34.9 |
| ANOVA: M (Mycorrhizal treatment), Z (Zn levels) | | | | | | | | | | | | |
| M | ** | | ** | | ** | | ** | | ** | | ** | |
| Z | ** | | ** | | ** | | ** | | ** | | ** | |
| M x Z | NS | | NS | | NS | | NS | | NS | | NS | |

coefficients (r) were calculated between Zn and P content in the roots and shoots of the maize plants. Statistical procedures were carried out with the IRRI stat (IRRI, Manila Philippines) software package.

## RESULTS AND DISCUSSION

### Mycorrhizal colonization

A significant increase in the proportion of mycorrhizal colonization in Arbuscular Mycorrhiza inoculated roots (AM+) from non-inoculated (AM-) maize plants at all the five levels of Zn fertilisation (Table 1). The higher proportion of mycorrhizal colonization in the AM+ maize plants was consistent across mycorrhizal structures such as arbuscules, vesicles and external mycelium, and across time of sample collection (45 and 75 days after sowing (DAS)). However, a significant reduction in the mycorrhizal colonization was recorded in the samples collected 75 days after sowing. The increase of soil Zn application had no influence with mycorrhizal colonization in both the AM+ and AM- maize plants, where as negative correlation between Zn concentrations and mycorrhizal colonization was recorded elsewhere (Gildon and Tinker 1983; Boyle and Paul 1988). Ortas et al. (2002) proved that such negative correlation was confined to P alone, and an increase of Zn up to 5 mg $kg^{-1}$ had no effect on mycorrhizal colonization in the roots. The readily available Zn from applied fertilizer (2002) proved that such negative correlation was confined to P alone, and an increase of Zn up to 5 mg $kg^{-1}$ had no effect on mycorrhizal colonization in the roots. It was also reported that Arbuscular Mycorrhizal colonization increased even at the highest level level of contamination of 250 mg Zn $kg^{-1}$ soil suggesting the potential of mycorrhiza in reduction of metal contamination (Audet and Charest 2006).

Nevertheless, the development of mycorrhizal structures such as arbuscles and vesicles had no changes at varying Zn levels, but the growth of external mycelium was decreased at higher levels of Zn application in AM+ plants. The readily available Zn from applied fertilizer could

**Table 2.** Root morphological attributes viz. root length, spread and volume measured in the arbuscular mycorrhiza inoculated (AM+) and non-inoculated (AM-) maize crop plants at 45 and 75 days after sowing (DAS) under different Zn levels. Values in parentheses indicate the standard error. The levels of significance for ANOVA, * = $P \leq 0.05$; ** = $P \leq 0.01$; NS = Not significant.

| Zn levels (mg $kg^{-1}$) | 45 DAS | | | | | | 75 DAS | | | | | |
|---|---|---|---|---|---|---|---|---|---|---|---|---|
| | Root length (cm) | | Root spread (cm) | | Root volume (ml) | | Root length (cm) | | Root spread (cm) | | Root volume (ml) | |
| | AM- | AM+ | AM- | AM+ | AM- | AM+ | AM- | AM+ | AM- | AM+ | AM- | AM+ |
| 0 | 39.2 | 52.1 | 10.0 | 13.5 | 21.3 | 26.4 | 44.3 | 58.4 | 14.2 | 17.6 | 29.3 | 35.3 |
| | (0.577) | (1.73) | (0.230) | (0.288) | (0.750) | (1.09) | (2.30) | (1.73) | (0.536) | (0.440) | (1.45) | (0.881) |
| 1.25 | 44.1 | 53.3 | 10.6 | 14.3 | 23.0 | 28.2 | 49.2 | 60.1 | 14.9 | 17.3 | 31.0 | 38.3 |
| | (2.30) | (1.73) | (0.288) | (0.230) | (0.866) | (2.07) | (1.73) | (2.30) | (0.200) | (0.491) | (1.73) | (1.45) |
| 2.5 | 44.1 | 53.4 | 11.1 | 14.8 | 24.5 | 30.7 | 51.5 | 62.2 | 15.6 | 17.2 | 34.3 | 40.3 |
| | (1.73) | (2.30) | (0.230) | (0.346) | (0.866) | (1.32) | (2.30) | (2.88) | (0.702) | (0.450) | (1.45) | (2.02) |
| 3.75 | 45.3 | 54.2 | 11.8 | 15.2 | 26.7 | 33.0 | 53.2 | 63.4 | 16.2 | 18.7 | 37.3 | 43.0 |
| | (1.73) | (2.30) | (0.288) | (0.404) | (0.981) | (1.15) | (1.73) | (2.30) | (0.416) | (0.448) | (2.02) | (1.73) |
| 5.0 | 45.3 | 54.2 | 12.1 | 16.0 | 28.6 | 34.2 | 53.2 | 65.3 | 16.3 | 19.1 | 40.3 | 45.0 |
| | (1.15) | (1.73) | (0.346) | (0.346) | (1.03) | (0.405) | (2.30) | (3.46) | (0.731) | (0.378) | (2.02) | (2.88) |
| Mean | 43.6 | 53.4 | 11.1 | 14.7 | 24.8 | 30.6 | 50.3 | 61.8 | 15.5 | 18.0 | 34.4 | 40.4 |
| ANOVA: M (Mycorrhizal treatment), Z (Zn levels) | | | | | | | | | | | | |
| M | ** | | ** | | ** | | ** | | ** | | ** | |
| Z | NS | | ** | | ** | | * | | NS | | ** | |
| M x Z | NS | | NS | | NS | | NS | | NS | | NS | |

cause plants not to invest more on hyphal growth for foraging immobile nutrients.

## Root morphology

The measurements of root architecture variables: root length (vertical), spread (horizontal) and volume in the AM+ plants were significantly higher than non-mycorrhizal (AM-) maize plants (Table 2). Graded levels of Zn addition increased the root masses of both AM- and AM+ plants. At higher levels of Zn, the response to added Zn was more pronounced in AM- plants than AM+ plants.

The increase in root architecture measurements from mycorrhizal non-inoculated (AM-) plants to mycorrhizal inoculated (AM+) maize plants was found in both the root samples collected at 45 and 75 DAS. The extensive root growth of AM plants can be attributed to the improved P nutrition of host plants. The enhanced supply of P by mycorrhizal symbiosis has been unequivocally demonstrated (Jakobsen et al., 1992; Asmah, 1995; Hetrick et al., 1996; Smith and Read, 1997; Subramanian et al., 2006). The experimental soil had extremely low available P (0.58 mg $kg^{-1}$) which is a constraint to root growth where as the improved P availability to mycorrhizal plants would have resulted in vigorous root growth. Zinc applica- nutrition of host plants. The enhanced supply of P by mycorrhizal symbiosis has been unequivocally demonstrated (Jakobsen et al., 1992; Asmah, 1995; Hetrick et al., 1996; Smith and Read, 1997; Subramanian et al., 2006). The experimental soil had extremely low available P (0.58 mg $kg^{-1}$) which is a constraint to root growth where as the improved P availability to mycorrhizal plants would have resulted in vigorous root growth. Zinc applica-tion significantly increased ($P \leq 0.01$) all the root morphological attributes except root length at 45 DAS and spread at 75 DAS which is in line with the observation of Nable and Webb (1993) who reported that Zn is essential for root development. The synergistic interaction between P and Zn has resulted in extensive root growth in colonized plants.

Mycorrhizal inoculated plants produced significantly ($P \leq 0.01$) higher root masses irrespective of

**Table 3.** Dry weight of roots and shoots measured in the arbuscular mycorrhiza inoculated (AM+) and non-inoculated (AM-) maize crop plants at 45 and 75 days after sowing (DAS) under different Zn levels. Values in parentheses indicate the standard error. The levels of significance for ANOVA, * = P ≤ 0.05; ** = P ≤ 0.01; NS = Not significant.

| Zn levels (mg kg$^{-1}$) | 45 DAS | | | | 75 DAS | | | |
|---|---|---|---|---|---|---|---|---|
| | Roots (mg kg$^{-1}$) | | Shoots (mg kg$^{-1}$) | | Roots (mg kg$^{-1}$) | | Shoots (mg kg$^{-1}$) | |
| | AM- | AM+ | AM- | AM+ | AM- | AM+ | AM- | AM+ |
| 0 | 1.95 | 2.62 | 3.38 | 3.50 | 2.70 | 3.37 | 4.56 | 4.90 |
| | (0.088) | (0.103) | (0.233) | (0.202) | (0.155) | (0.129) | (0.141) | (0.101) |
| 1.25 | 2.15 | 2.81 | 3.73 | 3.85 | 2.92 | 3.55 | 4.64 | 5.67 |
| | (0.057) | (0.129) | (0.116) | (0.202) | (0.129) | (0.155) | (0.181) | (0.202) |
| 2.5 | 2.25 | 2.97 | 3.96 | 4.08 | 3.10 | 3.78 | 5.27 | 5.91 |
| | (0.083) | (0.072) | (0.233) | (0.116) | (0.103) | (0.181) | (0.080) | (0.131) |
| 3.75 | 2.40 | 3.16 | 4.43 | 4.55 | 3.28 | 3.91 | 5.29 | 6.23 |
| | (0.090) | (0.116) | (0.233) | (0.202) | (0.077) | (0.155) | (0.121) | (0.080) |
| 5.0 | 2.52 | 3.20 | 4.55 | 4.66 | 3.51 | 4.00 | 5.85 | 5.11 |
| | (0.132) | (0.129) | (0.202) | (0.116) | (0.129) | (0.103) | (0.060) | (0.060) |
| Mean | 2.25 | 2.95 | 4.01 | 4.13 | 3.10 | 3.72 | 4.32 | 5.56 |
| ANOVA: M (Mycorrhizal treatment), Z (Zn levels) | | | | | | | | |
| M | ** | | NS | | ** | | ** | |
| Z | ** | | ** | | ** | | * | |
| M x Z | NS | | NS | | NS | | NS | |

Zn levels at 45 and 75 DAS (Table 3). Graded levels of Zn addition increased the root masses of both AM- and AM+ plants. At higher levels of Zn, the response to added Zn was more pronounced in AM- plants than AM+ plants. At early stage of crop growth, AM fungal inoculated plants showed an increase in root mass while the shoot masses were similar. This may be attributed to the utilization of carbon for establishment of functional symbiosis (Fitter, 1988; Jakobsen and Rosendahl, 1990). The data are in conformity with the findings of Kothari et al., 1991) who obtained no response in terms of shoot growth of maize to mycorrhizal inoculation in the early stages. Zinc nutrition favoured the dry matter production in the early stages than in later stages. Zinc by way of activating C metabolism enzymes such as carbonic anhydrase (Ohki, 1976) and amino acids, improved the photosynthetic ability (Marschner, 1986) and protein synthesis of plants and thereby plant growth.

## Leaf area index and chlorophyll content

Leaf area index (LAI), measured in AM+ plants at 45 and 75 DAS were significantly (P ≤ 0.01) higher than AM- plants (Table 4 and Figure 1) at both stages of observation. Mycorrhizal colonization would have stimulated leaf expansion as a result of improved Zn nutritional status (Cakmak et al., 1998) or better water use efficiency (Subramanian and Charest 1995). As expected, zinc application progressively increased the LAI of both inoculated and uninoculated plants. The chlorophyll content of AM+ plants at 45 DAS was also significantly higher than AM- plants (Table 4 and Figure 2). The response to added Zn was more pronounced in AM- plants than AM+ plants. The increase in chlorophyll content of AM plants can be attributed to enhanced Zn nutrition as Zn promotes the development of photosyn thetic pigments (Misra et al., 2005).

## Plant nutritional status

Mycorrhiza-inoculated maize plants had significantly (*P* ≤ 0.01) higher Zn and P concentrations than uninoculated (AM-) plants (Table 5 and 6). The increase in Zn and P concentration in AM inoculated plants can be attributed to the extensive root development and hyphae that reduce the distance for diffusion of nutrients thus enhancing the nutrient absorption. AM+ shoots had higher Zn concentration (30.2 and 34.7 mg kg$^{-1}$) than AM- shoots (29.1 and 33.9 mg kg$^{-1}$), at 45 and 75 DAS, respectively. The enhanced acid phosphatase activities in mycorrhizal plants (Dodd et al. 1987) assist in releasing P which in turn transported by the external mycelium (Jakobsen et al. 1992) and resulted in enhancement of nutritional status of the host plants. The hyphal transport of Zn to the host plant has been reported as 60% (Liu et al. 2000). Increasing levels of Zn significantly (P ≤ 0.01) increased the Zn concentrations of roots and shoots at 75 DAS and roots alone at 45 DAS.

There was a strong positive relationship between Zn and P contents in the roots (r = 0.95; *P* < 0.001) and the

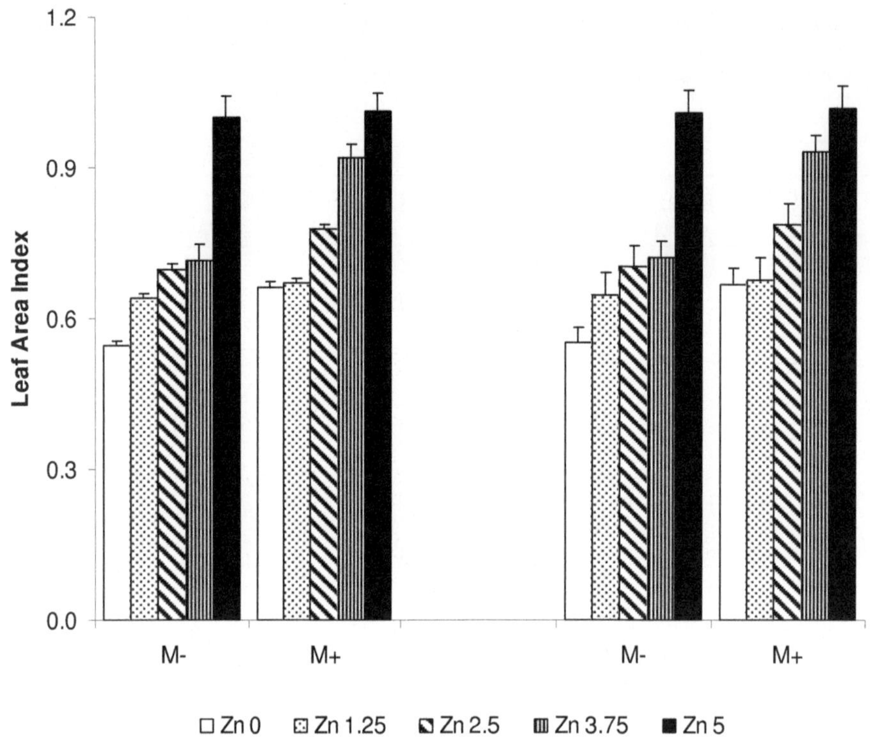


**Figure 1.** Leaf area index (LAI) measured in the arbuscular mycorrhiza inoculated (AM+) and non-inoculated (AM-) maize crop plants at 45 and 75 days after sowing (DAS) under different Zn levels. Error bars represent standard errors of three replications

**Table 4.** The levels of significance for ANOVA for Leaf Area Index and chlorophyll content (mg $g^{-1}$) under different Zn levels (Z) with (AM+) or without (AM-) arbuscular mycorrhizal fungal inoculation. ** P ≤ 0.01; NS, not significant.

| | M | Z | M x Z |
|---|---|---|---|
| **Leaf Area Index** | | | |
| 45 DAS | ** | ** | ** |
| 75 DAS | ** | ** | ** |
| Cholorophyll content | | | |
| 45 DAS | ** | ** | NS |

shoots (r = 0.96; P < 0.001), which confirms the synergistic interaction among P and Zn aided by mycorrhizal colonization (Figure 3). Mycorrhizal dependency based on Zn contents was reported to be 19.7, 16.0, 16.2, 17.6 and 14.7% in Zn fertilization levels from 0, 1.25, 2.5, 3.75 to 5.0 mg $kg^{-1}$ soil respectively.

AM inoculated plants produced grains with significantly (P ≤ 0.01) higher Zn concentration in comparison to non-noculated plants (Table 7). Mycorrhizal plants assimilated greater amounts of Zn and translocated to grains which assisted in enrichment of tryptophan in grains. Grains of AM inoculated plants had significantly higher concentrations of tryptophan than non-inoculated plants (Table 7). The role of Zn in tryptophan biosynthesis has been well established (Cakmak et al., 1989; Brown et al., 1993).

## Soil nutrient status

Overall, available Zn concentration was recorded high in the soil collected from mycorrhiza-inoculated maize plants (AM+) (Table 8). With the progression of plant growth stages, AM- soil had lower DTPA Zn status while AM+ soil had consistently higher available Zn. Mycorrhizal inoculation increased the DTPA-Zn by 74% in comparison to AM- soil even at the highest dose of Zn applied (5 mg $kg^{-1}$) in the soil. The soil from mycorrhizal inoculated maize plants (AM+) had more available phosphorous (P) than from (AM-) maize plants. The increase in P concentration was independent of Zn levels. Mycorrhiza-inoculated plants were reported to release organic acids that reduce rhizospheric pH, and facilitate nutrients availability (Koide and Kabir, 2000).

## Conclusion

To conclude, mycorrhizal inoculation in the maize plants improves the root architectural attributes, plant nutritional

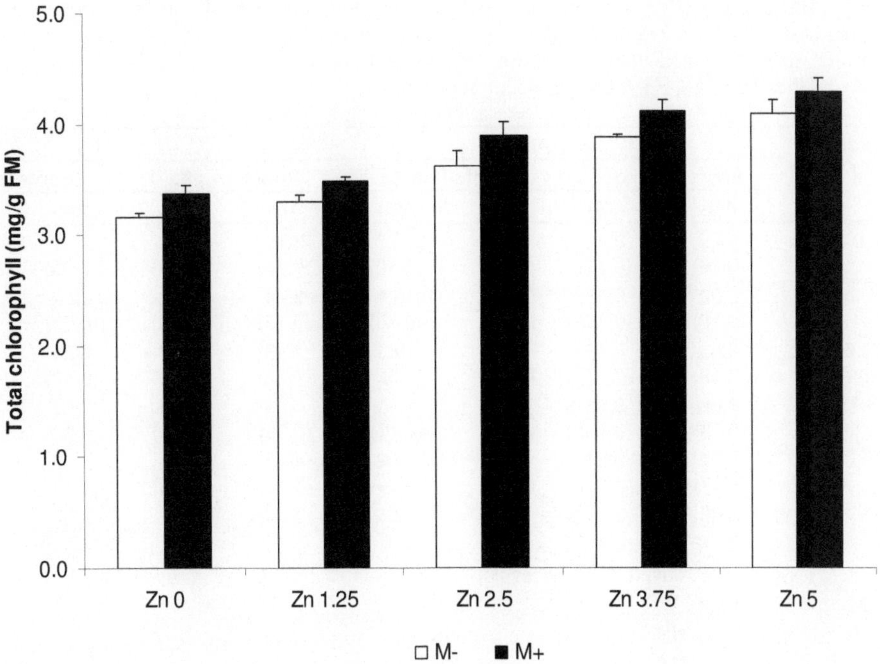


**Figure 2.** Total chlorophyll content in the arbuscular mycorrhiza inoculated (AM+) and non-inoculated (AM-) maize crop plants at 45 and 75 days after sowing (DAS) under different Zn levels. Error bars represent standard errors of three replications.

**Table 5.** Zinc (Zn) concentration examined in the roots and shoots of arbuscular mycorrhiza inoculated (AM+) and non-inoculated (AM-) maize crop plants at 45 and 75 days after sowing (DAS) under different Zn levels. Values in parentheses indicate the standard error. The levels of significance for ANOVA, * = P ≤ 0.05; ** = P ≤ 0.01; NS = Not significant.

| Zn levels ($mg\ kg^{-1}$) | 45 DAS | | | | 75 DAS | | | |
|---|---|---|---|---|---|---|---|---|
| | Roots ($mg\ kg^{-1}$) | | Shoots ($mg\ kg^{-1}$) | | Roots ($mg\ kg^{-1}$) | | Shoots ($mg\ kg^{-1}$) | |
| | AM- | AM+ | AM- | AM+ | AM- | AM+ | AM- | AM+ |
| 0 | 33.3 (0.86) | 35.5 (0.0029) | 27.8 (0.577) | 28.5 (0.520) | 33.5 (0.346) | 35.9 (0.520) | 32.6 (0.404) | 33.7 (0.404) |
| 1.25 | 33.6 (0.009) | 35.9 (0.0029) | 28.4 (0.693) | 29.1 (0.520) | 33.9 (0.520) | 36.4 (0.693) | 33.1 (0.346) | 34.1 (0.577) |
| 2.5 | 34.3 (0.0029) | 35.6 (0.0029) | 28.9 (0.751) | 29.9 (0.693) | 34.5 (0.866) | 37.1 (0.635) | 33.9 (0.404) | 34.9 (0.520) |
| 3.75 | 35.3 (0.0029) | 37.2 (0.0029) | 29.8 (0.520) | 31.6 (0.462) | 35.2 (0.289) | 37.8 (0.462) | 34.8 (0.231) | 35.4 (0.231) |
| 5.0 | 35.6 (0.0029) | 37.2 (0.666) | 30.4 (0.808) | 31.7 (0.404) | 35.6 (0.520) | 38.1 (0.577) | 35.0 (0.520) | 35.6 (0.289) |
| Mean | 34.4 | 36.3 | 29.1 | 30.2 | 34.5 | 37.1 | 33.9 | 34.7 |
| ANOVA: M (Mycorrhizal treatment), Z (Zn levels) | | | | | | | | |
| M | ** | | ** | | ** | | ** | |
| Z | NS | | ** | | ** | | ** | |
| M x Z | NS | | NS | | NS | | NS | |

**Table 6.** Phosphorous (P) concentration examined in the roots and shoots of arbuscular mycorrhiza inoculated (AM+) and non-inoculated (AM-) maize crop plants at 45 and 75 days after sowing (DAS) under different Zn levels. Values in parentheses indicate the standard error. The levels of significance for ANOVA, * = P ≤ 0.05; ** = P ≤ 0.01; NS = Not significant.

| Zn levels ($mg\ kg^{-1}$) | 45 DAS | | | | 75 DAS | | | |
|---|---|---|---|---|---|---|---|---|
| | Roots ($mg\ kg^{-1}$) | | Shoots ($mg\ kg^{-1}$) | | Roots ($mg\ kg^{-1}$) | | Shoots ($mg\ kg^{-1}$) | |
| | AM- | AM+ | AM- | AM+ | AM- | AM+ | AM- | AM+ |
| 0 | 6.45 (0.16) | 8.12 (0.12) | 16.8 (0.23) | 18.7 (0.40) | 38.9 (0.46) | 41.1 (0.35) | 21.1 (0.35) | 23.2 (0.29) |
| 1.25 | 6.52 (0.18) | 8.33 (0.16) | 17.4 (0.35) | 19.1 (0.35) | 39.4 (0.17) | 41.8 (0.40) | 22.2 (0.23) | 24.1 (0.35) |
| 2.5 | 6.63 (0.19) | 8.42 (0.17) | 17.7 (0.23) | 19.3 (0.52) | 40.3 (0.52) | 42.3 (0.23) | 23.1 (0.40) | 24.5 (0.23) |
| 3.75 | 6.68 (0.18) | 8.52 (0.12) | 18.1 (0.52) | 19.0 (0.17) | 40.6 (0.35) | 42.6 (0.29) | 23.5 (0.23) | 24.8 (0.40) |
| 5.0 | 6.71 (0.17) | 8.54 (0.17) | 18.2 (0.35) | 19.2 (0.29) | 40.5 (0.40) | 42.4 (0.23) | 23.7 (0.35) | 24.6 (0.23) |
| Mean | 6.60 | 8.39 | 17.6 | 19.1 | 39.9 | 42.0 | 22.7 | 24.2 |
| ANOVA: M (Mycorrhizal treatment), Z (Zn levels) | | | | | | | | |
| M | ** | | ** | | ** | | ** | |
| Z | NS | | NS | | ** | | ** | |
| M x Z | NS | | NS | | NS | | NS | |

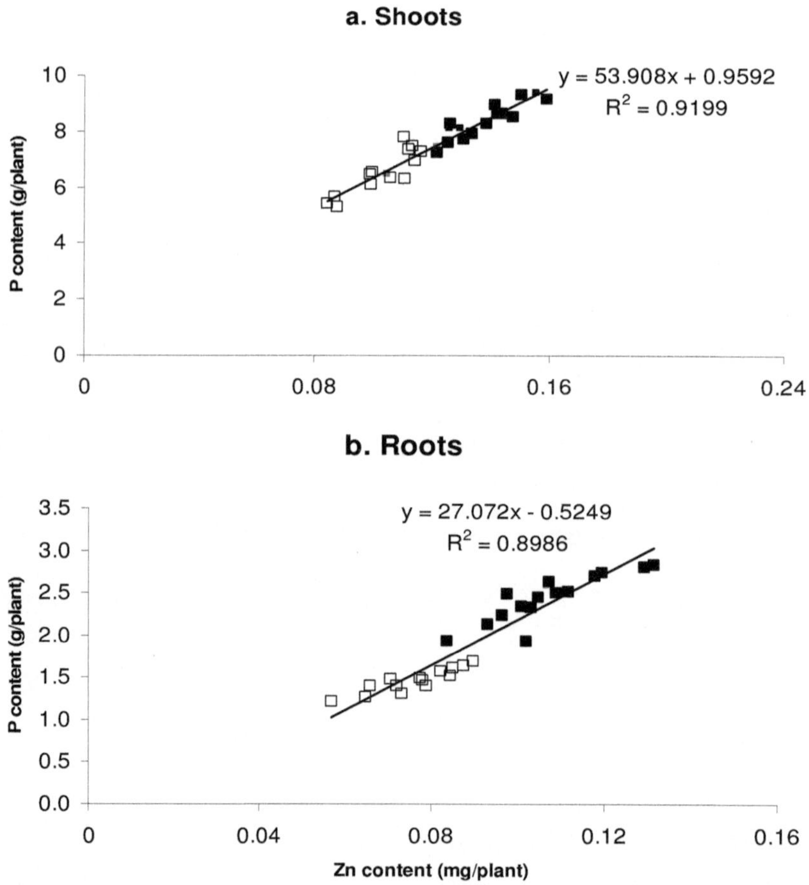


**Figure 3.** Correlations between Zn and P content in the shoots (a) and roots (b) of the arbuscular mycorrhiza inoculated (AM+) and non-inoculated (AM-) maize crop plants.

**Table 7.** Concentrations of Zn and tryptophan measured in the grains of arbuscular mycorrhiza inoculated (AM+) and non-inoculated (AM-) maize crop plants under different Zn levels. Values in parentheses indicate the standard error. The levels of significance for ANOVA, * = P ≤ 0.05; ** = P ≤ 0.01; NS = Not significant.

| Zn levels (mg kg$^{-1}$) | Grain Zn concentration (mg kg$^{-1}$) | | Tryptophan concentration (mg kg$^{-1}$) | |
|---|---|---|---|---|
| | AM- | AM+ | AM- | AM+ |
| 0 | 39.8 (0.404) | 41.5 (0.289) | 76.0 (1.04) | 80.2 (1.27) |
| 1.25 | 40.4 (0.231) | 42.0 (0.408) | 78.0 (0.577) | 80.8 (0.462) |
| 2.5 | 41.2 (0.208) | 42.7 (0.404) | 78.6 (1.27) | 84.2 (1.039) |
| 3.75 | 42.2 (0.340) | 43.8 (0.305) | 80.0 (0.693) | 86.0 (0.924) |
| 5.0 | 42.6 (0.289) | 44.1 (0.346) | 81.0 (0.577) | 88.4 (1.39) |
| Mean | 41.2 | 42.8 | 78.7 | 83.9 |
| ANOVA: M (Mycorrhizal treatment), Z (Zn levels) | | | | |
| M | ** | | ** | |
| Z | ** | | ** | |
| M x Z | NS | | NS | |

**Table 8.** Available zinc (Zn) and phosphorus (P) concentrations examined in the soils of arbuscular mycorrhiza inoculated (AM+) and non-inoculated (AM-) maize crop plants at 45 and 75 days after sowing (DAS) under different Zn levels. Values in parentheses indicate the standard error. The levels of significance for ANOVA, * = P ≤ 0.05; ** = P ≤ 0.01; NS = Not significant.

| Zn levels (mg kg$^{-1}$) | Available Zn (mg kg$^{-1}$) | | | | Available P (kg ha$^{-1}$) | | | |
|---|---|---|---|---|---|---|---|---|
| | 45 DAS | | 75 DAS | | 45 DAS | | 75 DAS | |
| | AM- | AM+ | AM- | AM+ | AM- | AM+ | AM- | AM+ |
| 0 | 1.08 (0.219) | 1.43 (0.029) | 1.07 (0.110) | 1.44 (0.092) | 5.93 (0.029) | 6.19 (0.058) | 5.98 (0.046) | 6.25 (0.040) |
| 1.25 | 1.73 (0.202) | 1.85 (0.069) | 1.72 (0.075) | 1.86 (0.139) | 5.94 (0.035) | 6.22 (0.052) | 6.00 (0.046) | 6.28 (0.075) |
| 2.5 | 2.67 (0.353) | 3.23 (0.116) | 2.62 (0.104) | 3.25 (0.115) | 5.94 (0.052) | 6.24 (0.035) | 6.20 (0.040) | 6.30 (0.052) |
| 3.75 | 3.13 (0.433) | 3.87 (0.060) | 3.02 (0.156) | 4.01 (0.173) | 5.96 (0.017) | 6.26 (0.023) | 6.30 (0.035) | 6.31 (0.035) |
| 5.0 | 3.23 (0.245) | 5.63 (0.364) | 3.06 (0.222) | 5.79 (0.179) | 6.25 (0.023) | 6.25 (0.058) | 6.21 (0.023) | 6.30 (0.035) |
| Mean | 2.37 | 3.20 | 2.29 | 3.27 | 5.94 | 6.23 | 6.14 | 6.29 |
| ANOVA: M (Mycorrhizal treatment), Z (Zn levels) | | | | | | | | |
| M | ** | | ** | | ** | | ** | |
| Z | ** | | ** | | NS | | ** | |
| M x Z | ** | | ** | | NS | | * | |

status and availability of soil nutrients at Zn fertilization levels. While such favourable changes enhance maize plants to survive during Zn deficient conditions, they assure nutritional security providing nutrients enriched grains.

## ACKNOWLEDGMENTS

Authors thank the Department of Atomic Energy (DAE) and Board of Research on Nuclear Sciences (BRNS), Mumbai, India for financially supporting this project (2005/35/30/BRNS/2810). They also thankful Mr. Arthur Selwyn for his scholarly editing of the manuscript.

## REFERENCES

Alloway BJ (2001). Zinc the vital micronutrient

for healthy, high-value crops. International Zinc Association, Brussels, Belgium.

Asmah AE (1995). Effect of phosphorus source and rate of application on VAM fungal infection and growth of maize (*Zea mays* L.). Mycorrhiza 5: 223-228.

Audet P, Charest C (2006). Effects of AM colonization on wild tobacco plants grown in zinc contaminated soil. Mycorrhiza 16: 277-283.

Augé RM (2001). Water relations, drought and vesicular arbuscular mycorrhizal symbiosis. Mycorrhiza 11: 3-42.

Boyle M, Paul EA (1988). Vesicular arbuscular mycorrhizal associations with barley on sewage –amended plots. Soil Biol. Biochem. 26: 1117-1124.

Brown PH, Cakmak I, Zhang LQ (1993). Form and function of zinc in plants. In: Robinson AD (ed) Zinc in soils and plants. Kluwer Publications, Dordrecht. Netherlands pp. 90-106.

Bruinsma J (1963). The quantitative analysis of chlorophyll a and b in plant extracts. Photochem. Photobiol. 72: 241-249.

Cakmak I, Marschner H, Bangerth F (1989). Effect of zinc nutritional status on growth, protein metabolism and levels of indole-3-acetic acid and other phytohormones in bean (Phaseolus vulgaris). J. Exp. Bot. 40: 405-412.

Cakmak I, Tourn B, Erenoglu B, Ozturk L, Marschner H, Kalayci M, Ekiz H, Yilmaz A (1998). Morphological and physiological differences in the response of cereals to zinc deficiency. Euphutica 100: 349-357.

Dalpe Y (1993). Vesicular-arbuscular mycorrhiza. In: Carter MR (eds) Soil sampling and methods of analysis. Lewis Publishers, Boca Raton pp. 287.

Dodd JC, Burton CC, Jeffries P (1987). Phosphatase activity associated with the roots and the rhizosphere of plants infected with vesicular-arbuscular mycorrhizal fungi. New Phytol. 107:163-172.

Faber BA, Zasoski RJ, Burau RG, Uriu K (1990). Zinc uptake by corn affected by vesicular-arbuscular mycorrhizae. Plant Soil 129: 121-131.

Fitter AH (1988) Water relations of red clover *Trifolium pretense* L. as affected by VA mycorrhizal infection and phosphorus supply before and during drought. J. Exp. Bot. 39: 595-603.

Gildon A, Tinker PB (1983). Interactions of vesicular arbuscular mycorrhizal infections and heavy metals in Dodd plants. II. The effects of infection on uptake of copper. New Phytol. 95: 263-268

Hetrick BAD, Wilson GWT, Cox TS (1992). Mycorrhizal dependence of modern wheat varieties, landraces, and ancestors. Can. J. Bot. 70: 2032-2040.

Hetrick BAD, Wilson GWT, Todd TC (1996). Mycorrhizal response in wheat cultivars: relationship to phosphorus. Can. J. Bot.74: 19-25.

Jakobsen I, Abbott LK, Robson AD (1992). External hyphae of vesicular-arbuscular mycorrhizal fungi associated with *Trifolium subterraneum* L. 1. Spread of hyphae and phosphorus inflow into roots. New Phytol. 120: 371-380.

Jakobsen I, Rosendahl L (1990). Carbon flow into soil and external hyphae from roots of mycorrhizal cucumber plants. New Phytol. 115: 77-83.

Koide RT, Kabir Z (2000). Extraradical hyphae of the mycorrhizal fungus *Glomus intraradices* can hydrolyse organic phosphate. New Phytol. 148: 511-517.

Kothari SK, Marschner H, Romheld V (1991). Contribution of the VA mycorrhizal hyphae in acquisition of phosphorus and zinc by maize grown in calcareous soil. Plant Soil 131: 177-185.

Li XL, Marschner H, Romheld V (1991). Acquisition of phosphorus and copper by VA-mycorhizal hyphae and root to shoot transport in white clover. Plant Soil 136: 49-57.

Lindsay WL, Norvell WA (1978). Development of DTPA soil test for zinc, iron, manganese and copper. Soil Sci. Soc. Am. J. 42: 421-428.

Liu A, Hamel C, Hamilton RI, Ma BL, Smith DL (2000). Acquisition of Cu, Zn, Mn and Fe by mycorrhizal maize (*Zea mays* L.) grown in soil at different P and micronutrient levels. Mycorrhiza 9: 331-336.

Mandal LN, Mandal B (1986). Zinc fractions in soils in relation to zinc nutrition of low land rice. Soil Sci. 142: 141-148.

Marschner H (1995). Mineral nutrition of higher plants. Academic Press, Boston, USA p. 889.

Misra A, Srivastava AK, Srivastava NK, Khan A (2005). Zn acquisition and its role in growth, photosynthesis, photosynthetic pigments and biochemical changes in essential monoterpene oils of *Pelargonium graveolens*. Photosynthetica 43: 153-155.

Nable RO, Webb MJ (1993). Further evidence that zinc is required throughout the root zone for optimal plant growth and development. Plant soil 150: 247-253.

Ohki H (1976). Effect of zinc nutrition on photosynthesis and carbonic anhydrase activity in cotton. Physiol. Plant. 38: 300-304.

Olsen SR, Cole CL, Watanabe FS, Dean DA (1954). Estimation of available phosphorus in soils by extraction with sodium bicarbonate. USDA, Circ. p. 939.

Ortas I, Ortakci D, Kaya Z, Cinar A, Onelge N (2002). Mycorrhizal dependency of Sour orange in relation to phosphorus and zinc nutrition. J. Plant Nutr. 25: 1263-1279.

Ryan MH, Angus JF (2003). Arbuscular mycorrhizae in wheat and field pea crops on a low P soil: increased Zn uptake but no increase in P uptake or yield. Plant Soil 250: 225-239.

Piper CS (1966) Soil and Plant Analysis. Hans Publishers, Bombay, India pp. 59-74.

Sadasivam S, Manickam A (1996). Biochemical Methods. New Age International, New Delhi, India pp. 112-113.

Singh B, Senthil Kumar A, Natesan B, Singh K, Usha K (2005). Improving zinc use efficiency of cereals under zinc deficiency. Curr. Sci. 88: 36-44.

Smith SE, Read DJ (1997). Vesicular-Arbuscular Mycorrhizas. In: Mycorrhizal symbiosis. (II ed) Academic Press, New York, USA pp. 9-126.

Subramanian KS, Charest C (1995). Influence of arbuscular mycorrhizae on the metabolism of maize under drought stress. Mycorrhiza 5: 273-278.

Subramanian KS, Charest C, Dwyer LM, Hamilton RI (1997). Effects of mycorrhizas on leaf water potential, sugar and P contents during and after recovery of maize. Can. J. Bot. 75: 1582-1591.

Subramanian KS, Charest C (2007). Arbuscular mycorrhiza and drought management of crops. In: Montano NM, Camargo Ricalde SL, Garcia.

Sanchez RA, Monroy- Ata (eds) Micorrizas arbusculares en ecasistemas aridos y semiaridos (Arbuscular mycorrhizae in arid and semiarid ecosystems), Mundi-Prensa SA of CV, Instituto Nacional de Ecologia - Semarnat, VAM- Izatapalapa, FES-Zaragoza-UNAM. Distrito Federal, Mexico pp. 78-109.

Subramanian KS, Santhanakrishnan P, Balasubramanian P (2006). Responses of field grown tomato plants to arbuscular mycorrhizal fungal colonization under varying intensities of drought stress. Sci Hort. 107: 245-253.

Welch RM, Graham RD (2002). Breeding crops for enhanced micronutrient content. Plant Soil 245: 205-214.

# Influence of canopy structure on yield of cassava cultivars at various toposequences of an inland valley agro ecosystem

**M. T. Lahai[1], I. J. Ekanayake[2,3] and J. P. C. Koroma[1]**

[1]Department of Crop Science, Njala Campus, Njala University, PMB, Freetown, Sierra Leone.
[2]Crop Improvement Division, International Institute of Tropical Agriculture (IITA), Oyo Road, PMB 5320, Ibadan, Nigeria.
[3]IBRD, 1850, I-street, Washington DC 20043, USA.

**A two-year trial to evaluate differences in canopy parameters in relation to productivity of cassava was conducted in inland valley ecology in a 4 × 4 Latin square design. Canopy parameters differed between sites, toposequence positions and years due to large differences in water table depth and weather conditions. Leaf number, leaf formation rate, leaf area index, canopy area and stay green ability correlated positively with yield. The improved cultivars with higher values for these parameters had higher yields than the landraces. However, TMS 4(2)1425 with the highest values had super-optimal, while the landrace produced sub-optimal canopy parameters; both of which reduced the yield. TMS 91/02324 and TMS 91/02327 with intermediate values maintained optimal canopy parameters resulting in high yield. Shallow water table (0.20 to 0.45 m below mound top) and drought reduced the canopy parameters and yield, while well distributed rainfall, high humidity and moderate temperatures (19 to 20°C) improved the canopy parameters and yield. Canopy parameters of low yielding landrace were more susceptible to shallow water table and drought than the high yielding improved cultivars. Selection of cultivars with good maintenance of canopy parameters during mid-season drought and late season excess moisture can contribute to increased tuber yield of cassava in inland valley ecology.**

**Key words:** Cassava cultivars, water table depth, drought, excess moisture stress, canopy parameters, tuber yield.

## INTRODUCTION

Cassava is the world's sixth most important crop (Lebot, 2009) and constitutes a staple food for over 700 million people (Njoku et al., 2010). It is a major source of calories in tropical Africa (Burns et al., 2010) due to its adaptability to marginal soils and erratic rainfall, high productivity per unit of land and labour and possibility of supply throughout the year (Nweke et al., 2002). The adaptation to different edapho-climatic conditions (Adeniji et al., 2011) makes cassava a favourite dry season crop grown in inland valleys in West and Central Africa (Lahai and Ekanayake, 2009).

Inland valleys are first and second order small streams that are found at the upper reaches of drainage basins. They are widely distributed in the agro-ecological zones of sub-Saharan Africa and occupy about 130 million hectares of land in inter-tropical Africa (Jalloh, 2003) and

25 million hectares in West Africa (Andriesse et al., 1994). Hydrologic regimes of inland valleys influence cropping patterns. During the rainy season, water table rises above the soil surface and farmers grow rice at the valley bottoms. In the dry season, the water table recedes and residual moisture content is inadequate to support a second rice crop and cassava is often grown as a dry season crop (Ekanayake et al., 1994; Jalloh, 2003). As a result of water table fluctuations, the soil undergoes three critical moisture conditions during the growing season of cassava. The water table is above the soil surface at planting period in December to January; recedes deep below the soil surface during the peak of the dry season in March to April; and again rises above the soil surface at harvesting in May to June (Mohamoud, 1994; Lahai and Ekanayake, 2009). Thus, cassava growing in inland valley is subjected to excess moisture stress during early and late season growth stages and to moisture deficit in mid-season. However, cassava has highly responsive stomata, deep root system and effective carbon fixation system for enhanced photosynthesis under drought stress (El-Sharkawy, 1993; Lahai and Ekanayake, 2010). Cassava can therefore tolerate long periods of drought (Tafur et al., 1997; Lebot, 2009), but it is highly susceptible to excessive water (Ande, 2011).

Excess soil moisture resulting in anaerobic conditions leads to restriction in rooting zones, nutrient deficiency, tuber rotting and subsequent poor growth and tuber yield of cassava (Izac et al., 1991; Ande, 2011). Selection of cassava cultivars for adaptation to inland valley ecology gives an indication of tolerance to both excess moisture and drought stress. Such information could be useful in breeding for cultivars that adapted to both upland and inland valley soil conditions.

Study of canopy development is a must to estimate the degree of adaptation of a genotype to the environment in which it is grown. Leaf area index (LAI) is the usual measure of canopy development (Sheriff and Muchow, 1984; Lebot, 2009). LAI depends on the rate of leaf formation per apex, apex number per unit area and leaf size and longevity (Lebot, 2009). High positive correlations exist between root yield and LAI, rate of leaf formation and leaf life (Lahai et al., 1999). However, cassava has simultaneous development of leaf area and tuber, and supply of assimilate is partitioned between growth of leaves and tubers (Lebot, 2009). These results in a delicate, balance between top and tuber growth for maximum yield, with maximal tuber growth occurring at LAI values of 2.5 to 3.5 at which optimum light interception and utilisation occur (Ramanujam, 1985; Lebot, 2009).

Research on cassava canopy structure in relation to yield had focused mainly on cassava grown on the upland where tubers are harvested after 12 months of growth. The results may be different in the case of dry season cultivation in inland valley where dry spell and high light intensity in mid-season and excess moisture in late season are constraints to production. The objective of this study was therefore to evaluate differences in canopy parameters in relation to root yield of four cassava cultivars under inland valley conditions.

## MATERIALS AND METHODS

### Site description

Two small inland valleys, one located at the International Institute of Tropical Agriculture (IITA) research farm at Ibadan (7° 30' N and 3° 54' E, 243 m above sea level (a.s.l.)) and the other on a farmer's field at Alabata (4° 46' N and 2° 34' E, 210 m a.s.l.) in the forest-savanna transition zone in South-western Nigeria were the test sites. During the trial period, mean minimum temperature (°C), relative humidity (%) and rainfall (mm) were higher in 2001 (19.8°C, 85%, 955.9 mm) than in 2002 (19.1°C, 75%, 642.6 mm), but mean maximum temperature (°C), evaporation (mm) and solar radiation (MJ/m/day) were higher in 2002 (28.4°C, 962.8 mm, 13.49 MJ/m/day) than in 2001 (27.5°C, 891.1 mm, 13.09 MJ/m/day). The valley bottom soils are classified as Tropaquents (Moormann et al., 1975) having loamy sand at both sites. At Alabata, soil pH ($H_2O$), organic C, N, P and K were 3.91, 0.99%, 0.11%, 7.43 mg/kg and 0.18 cmol/kg in 2001, respectively; the corresponding values in 2002 were 4.60, 0.46%, 0.13%, 6.80 mg/kg and 0.20 cmol/kg. The low organic carbon in 2002 as compared to 2001 reflected reduction in organic carbon as a result of decomposition due to continuous cultivation of the same sites for two consecutive years. At Ibadan soil pH, organic C, N, P and K were 4.87, 1.66%, 0.15%, 3.75 mg/kg and 0.08 cmol/kg in 2001 and 5.40, 0.53%, 0.13%, 2.78 mg/kg and 0.10 cmol/kg in 2002, respectively.

### Experimental design, plant establishment and water table depth measurement

At each site, 3 positions were marked out along the toposequence and designated valley fringe, valley intermediate and valley bottom (Figure 1). Each toposequence position had 16 flat-top mounds arranged in a 4 × 4 Latin square design. Each mound measuring for a plot was 2 × 2 m and 0.6 m high and was spaced 1 m apart. Three 2 m-long polyvinyl chloride (PVC) pipes of 0.05 m in diameter, serving as modified piezometers were buried to a depth of 1.5 m in furrows between mounds in each position to monitor water table depth on weekly basis. Six stem cuttings each 0.25 m long with 8 to 10 effective nodes, from mature healthy plants were planted on each mound at a spacing of 1 m between rows and 0.75 m between plants. Planting was done at Ibadan and Alabata on 12 February and 12 March 2001 in the first season and on 18 and 24 December 2001 in the second season, respectively. Three improved cultivars (TMS 4(2)1425, TMS 91/02324 and TMS 91/02327) and a landrace (Isunikankiyan) were used. Table 1 describes the genetic characters of the cultivars. Hand weeding was done as required but no fertilizer was applied. The non-application of fertilizer was to emulate the local farmers who normally do not apply fertilizer to cassava in inland valleys.

### Leaf area, canopy development and stay-green score

Number, rate of formation (change in number between measurements) and total leaf area (using leaf area meter LI-3000) were used to monitor leaf area development on four tagged plants per plot at bi-weekly intervals. Canopy development was monitored on the same plants using canopy area. Leaf area index (LAI) was calculated as leaf area ($m^2$)/ground area ($m^2$). The combined effects

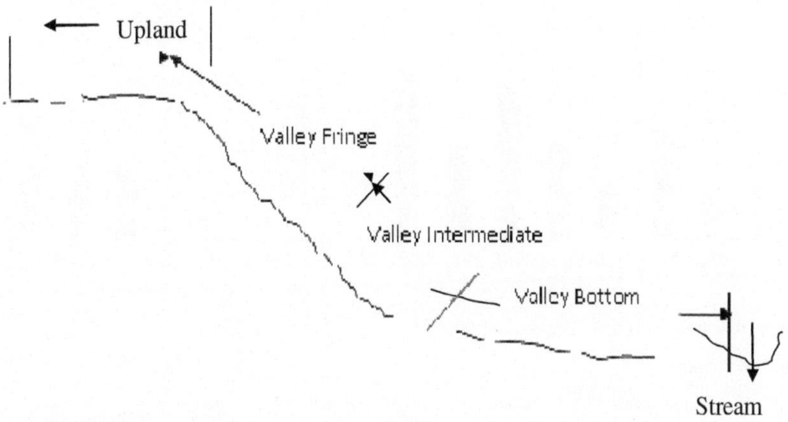


**Figure 1.** Topographic positions of inland valleys used as test sites.

**Table 1.** Tuber and vegetative characteristics of cassava cultivars used in the trial.

| Parameter | Characteristics | | | |
|---|---|---|---|---|
| | **TMS 4(2)1425** | **TMS 91/02324** | **TMS 91/02327** | **Isunikankiyan** |
| **Tuber characteristics** | | | | |
| Fresh yield (t/ha) | 12.98 - 84.42 | 14.84 - 28.16 | 13.00 - 24.58 | 2.00 - 10.00 |
| Dry matter (%) | 28.46 - 39.67 | 21.37 - 37.40 | 27.34 - 36.75 | 30.46 - 40.05 |
| CNP (ppm) | 2.69 - 12.84 | 6.81 - 21.75 | 4.08 - 25.83 | 2.58 - 6.88 |
| Starch (%) | 61.64 - 73.97 | 61.64 - 75.15 | 24.58 - 73.54 | 27.90 - 60.08 |
| Protein (%) | 1.60 - 4.87 | 1.57 - 7.20 | 1.23 - 5.50 | 1.20 - 4.89 |
| Outer root skin colour | Cream | Cream | Cream | Pale brown |
| Inner root skin colour | White/cream | White/cream | White/cream | White |
| Root flesh colour | White | White | White | White |
| Root neck length | Short | Short | Short | Medium |
| **Vegetative characteristics** | | | | |
| Unexpanded leaf colour | Green purple | Green purple | Green purple | Green |
| 1st fully expanded leaf colour | Green purple | Green purple | Green purple | Green |
| Pubescence of young leaf | Moderate | Moderate | Moderate | Low |
| Central leaf lobe shape | Elliptic | Lanceolate | Lanceolate | Lanceolate |
| Petiole colour | Light green | Dark green | Green purple | Red |
| Stem colour | Silver green | Dark brown | Dark brown | Grey |
| Growth habit of stem | Zigzag branching | Straight | Straight | Straight |

of leaf ageing, drought or excess moisture condition, low night temperatures and harmattan winds were visually rated using stay green score (SGS) on a scale of 1 to 9; where 1 = plant with a full canopy, retaining most formed leaves and 9 = complete defoliation with a candlestick appearance of the stem (Ekanayake, 1996; Lenis et al., 2005).

### Harvesting and tuber yield determination

The 2001 and 2002 trials were harvested on 15 August 2001 and 23 July 2002 at Ibadan and on 13 September 2001 and 28 July 2002 at Alabata, respectively. All the plants in each plot were harvested. Storage root biomass weights were recorded and sub-samples of fresh materials (250 g) were dried in a ventilated oven at 70°C for 48 h to determine dry-matter content. Dry tuber yield was expressed as tons $ha^{-1}$.

### Statistical analyses

The data were analysed using mixed model procedure (PROC MIXED) of the statistical analysis system for Microsoft Windows (SAS Institute, 1991). Row and column were taken as random effects and cultivar, toposequence position, site and year as fixed effects. Significant differences between treatments were compared

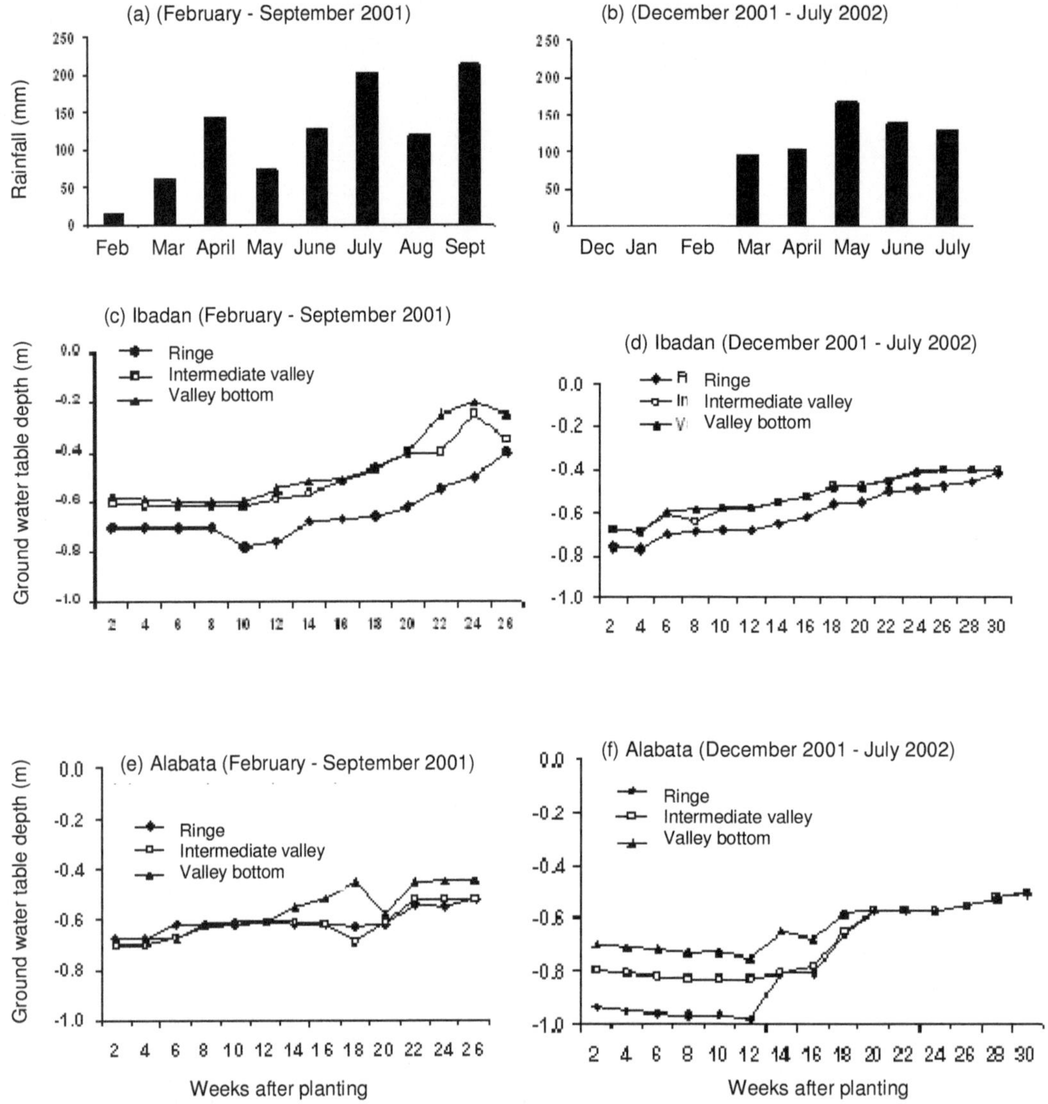


**Figure 2.** Variations in rainfall (at Ibadan) and groundwater table depth during 2001 and 2002 cropping seasons in inland valley at Ibadan and Alabata, Nigeria.

using standard error of the mean. Relationships between canopy parameters and root yield, water table depth and weather variables were explored using conventional correlation approach.

## RESULTS

### Groundwater table depth at both sites

At both sites in both years, water table depth decreased as rainfall increased. The water table was deeper at both sites in 2002 with lower rainfall than in 2001. It was also deeper in the fringe than the other positions at both sites and years. The shallowest water table depth attained at Ibadan ranged from 0.15 to 0.40 m in 2001 and from 0.35 to 0.45 m in 2002, while at Alabata it ranged from 0.45 to 0.50 m in 2001 and from 0.51 to 0.58 m in 2002 below the 0.6 m-high mound surface (Figure 2).

### Dry tuberous root yield

Improved cultivars had significantly higher root yields than the landrace across toposequence positions at both sites in both years. However, TMS 91/02324 and TMS 91/02327 had similar but significantly higher yields than TMS 4(2)1425. Across cultivars in 2001, the highest yield was obtained in valley bottom and the lowest in valley fringe at Alabata, while in 2002 at Alabata and in both years at Ibadan the highest yield was produced in the fringe and the lowest in valley bottom. This led to signi-

**Table 2.** Influence of cassava cultivars and toposequence positions on dry tuber yield (t $ha^{-1}$) and number of leaves $plant^{-1}$ at two inland valley locations in 2001 and 2002 (Nigeria).

| Location (Loc) | Cultivar | 2001 | | | | 2002 | | | |
|---|---|---|---|---|---|---|---|---|---|
| | | Toposequence (Topo) | | | | Toposequence (Topo) | | | |
| | | Fringe | Inter-mediate | Bottom | Cultivar Mean | Fringe | Inter-mediate | Bottom | Cultivar mean |
| | | **(a) Dry tuberous root yield (t $ha^{-1}$)** | | | | | | | |
| Alabata | TMS4(2)1425 | 5.04 | 6.16 | 16.94 | 9.38 | 3.28 | 2.24 | 3.01 | 2.84 |
| | TMS91/02324 | 4.68 | 4.71 | 17.24 | 8.88 | 4.12 | 4.01 | 4.66 | 4.26 |
| | TMS91/02327 | 5.31 | 5.50 | 17.21 | 9.34 | 5.15 | 3.60 | 3.12 | 3.96 |
| | Isunikankiyan | 4.17 | 4.98 | 15.23 | 8.13 | 1.86 | 1.97 | 0.95 | 1.59 |
| | s.e ± | 0.64 | 0.64 | 0.64 | 0.32 | 0.64 | 0.64 | 0.64 | 0.32 |
| | Topo mean (s.e ± 0.32) | 4.80 | 5.34 | 16.66 | - | 3.60 | 2.96 | 2.94 | - |
| | Year mean (s.e ± 0.17) | 8.93 | - | - | - | 3.17 | - | - | - |
| | Loc mean (s.e ± 0.13) | 6.05 | - | - | - | - | - | - | - |
| Ibadan | TMS4(2)1425 | 4.56 | 4.22 | 3.23 | 4.00 | 3.18 | 2.30 | 1.67 | 2.38 |
| | TMS91/02324 | 7.77 | 4.13 | 5.04 | 5.65 | 5.48 | 2.52 | 2.62 | 3.54 |
| | TMS91/02327 | 6.12 | 4.51 | 4.41 | 5.01 | 5.14 | 2.38 | 2.04 | 3.19 |
| | Isunikankiyan | 3.83 | 2.99 | 3.15 | 3.32 | 2.58 | 1.35 | 1.56 | 1.83 |
| | s.e ± | 0.64 | 0.64 | 0.64 | 0.32 | 0.64 | 0.64 | 0.64 | 0.32 |
| | Topo mean (s.e ± 0.32) | 5.50 | 3.96 | 3.96 | - | 4.10 | 2.14 | 1.97 | - |
| | Year mean (s.e ± 0.17) | 4.47 | - | - | - | 2.74 | - | - | - |
| | Loc mean (s.e ± 0.13) | 3.61 | - | - | - | - | - | - | - |
| | | **(b) Number of leaves per plant** | | | | | | | |
| Alabata | TMS4(2)1425 | 131.4 | 100.7 | 116.3 | 116.1 | 76.1 | 63.7 | 53.7 | 64.5 |
| | TMS91/02324 | 82.2 | 84.6 | 86.8 | 84.5 | 72.8 | 46.7 | 56.6 | 58.7 |
| | TMS91/02327 | 100.2 | 81.6 | 68.5 | 83.5 | 70.6 | 57.6 | 42.4 | 56.9 |
| | Isunikankiyan | 107.5 | 81.9 | 67.8 | 85.7 | 49.9 | 38.6 | 36.0 | 41.5 |
| | s.e ± | 8.6 | 8.6 | 8.6 | 3.8 | 8.6 | 8.6 | 8.6 | 3.8 |
| | Topo mean (s.e ± 2.1) | 105.3 | 87.2 | 84.9 | - | 67.4 | 51.7 | 47.2 | - |
| | Year mean (s.e ± 2.6) | 92.5 | - | - | - | 55.4 | - | - | - |
| | Loc mean (s.e ± 2.4) | 74.0 | - | - | - | - | - | - | - |
| Ibadan | TMS4(2)1425 | 135.7 | 117.7 | 96.8 | 116.7 | 107.3 | 55.5 | 53.6 | 72.1 |
| | TMS91/02324 | 117.1 | 64.0 | 61.2 | 80.8 | 82.5 | 51.1 | 51.4 | 61.7 |
| | TMS91/02327 | 101.2 | 65.7 | 57.6 | 74.8 | 94.9 | 46.9 | 56.6 | 66.1 |
| | Isunikankiyan | 72.9 | 52.3 | 56.9 | 60.7 | 67.3 | 45.8 | 46.1 | 53.1 |
| | s.e ± | 8.6 | 8.6 | 8.6 | 3.8 | 8.6 | 8.6 | 8.6 | 3.8 |
| | Topo mean (s.e ± 2.1) | 106.7 | 74.9 | 68.1 | - | 88.0 | 49.8 | 51.9 | - |
| | Year mean (s.e ± 2.6) | 83.2 | - | - | - | 63.2 | - | - | - |
| | Loc mean (s.e ± 2.4) | 73.2 | - | - | - | - | - | - | - |

s.e. = standard error of means.

ficant toposequence position × site, toposequence position × year, site × year and toposequence position × site × year interactions. Root yield was about 55% higher in 2001 than 2002 and 46% higher at Alabata than Ibadan (Table 2a).

## Number of leaves

On average, the improved cultivars had higher number of leaves than the landrace, with TMS 4(2)1425 producing significantly more leaves than the other improved cultivars in all toposequence positions at both sites in both years. Plants in the fringe recorded significantly higher number of leaves than the other positions at both sites in both years. In all cases plants in valley bottom produced the lowest number of leaves. Leaf number was about 33% higher in 2001 than 2002. Site × year interaction was significant because leaf number was higher at Alabata than Ibadan in 2001, but in 2002, it was higher at Ibadan than Alabata (Table 2b).

**Rate of leaf formation**

The landrace, TMS 4(2)1425 and TMS 91/02324 had similar rate of leaf formation, but the landrace and TMS 4(2)1425 formed leaves at faster rate than TMS 91/02327 across toposequence positions, sites and years. Rate of leaf formation was higher at Ibadan (27%) than Alabata and was also higher in 2001 than in 2002, but this was more pronounced at Alabata (83%) than at Ibadan (24%). Rate of leaf formation was affected by toposequence position × site interaction as it was lowest in valley fringe and highest in valley bottom at Alabata, while at Ibadan it was highest in the fringe and lowest in valley bottom. Toposequence position × year interaction was significant due to the fact that rate of leaf formation in 2001 was highest in valley bottom and lowest in valley fringe, but the reverse was the case in 2002 (Table 3a).

**Leaf area index**

TMS 4(2)1425 had the highest leaf area index (LAI), followed by TMS 91/02324 and TMS 91/02327, while Isunikankiyan had the lowest LAI across toposequence positions at both locations in both years. Plants in valley fringe produced higher LAI than those in other positions, with plants in valley intermediate and bottom registering similar LAI at both sites and in both years. LAI was about 60% higher in 2001 than in 2002 and was 41% higher at Alabata than Ibadan. Cultivar x site and cultivar x year interactions were significant because at Alabata, TMS 91/02327 had lower LAI than the landrace but the reverse was recorded at Ibadan. Similarly in 2001, TMS 4(2)1425 had larger LAI than TMS 91/02324 but in 2002 the reverse was the case (Table 3b).

**Canopy area**

Generally, in each toposequence position at both sites in both years, the three improved cultivars had significantly larger canopy area than the landrace cultivar. Plants in valley fringe had larger canopy area than the other positions, with the other two positions producing similar canopy area at both sites in both years. Canopy area was about 40% higher in 2001 than in 2002 and about 34% larger at Alabata than Ibadan. The interaction effects on canopy area were not directional since cultivars changed their rankings only slightly over toposequence positions, sites and years (Table 4a).

**Stay-green score**

Stay-green score (SGS) of the landrace was significantly higher than the other cultivars, with the improved cultivars having similar SGS across toposequence positions at both sites in both years. The lower the SGS the better the stay green ability, indicating that SGS of the improved cultivars was better than that of the landrace. Plants in valley fringe had better SGS than those in valley bottom. Also, SGS in 2001 was better than in 2002 at Alabata, but the reverse was noted at Ibadan. Mean SGS at Alabata was better than that at Ibadan. Toposequence position × site × year interaction was noted. At Alabata, in both years, the fringe had the lowest, while valley bottom had the highest SGS. At Ibadan, in 2001, the fringe had the lowest and valley bottom had the highest, but in 2002, valley bottom had the lowest and the fringe had the highest score (Table 4b).

**Associations of canopy parameters with yield, water table depth and weather variables**

Number and rate of leaf formation, LAI and canopy area correlated positively with yield at both sites. SGS related negatively with yield at Alabata. At Ibadan, water table depth had negative relationship with leaf number, rate of leaf formation, LAI and canopy area, but positive association with SGS, whereas at Alabata, SGS correlated negatively with water table depth. Leaf number, rate of leaf formation, LAI and canopy area related positively with rainfall, minimum temperature and relative humidity, but negatively with evaporation, solar radiation and maximum temperature (Table 5).

## DISCUSSION

**Number of leaves per plant**

Plants obtain their energy and structural materials by photosynthesis and the main photosynthetic organ is the leaf (Jarvis and Morison, 1982; Lenis et al., 2005; Lebot, 2009). This indicates that the higher the number of leaves the higher the photosynthate produced. Leaf number correlated positively with root yield at both sites. Thus, the high tuber yields in 2001 and also of the improved cultivars were partly due to the production of higher number of leaves. The results agreed with Adeniji et al. (2011) who reported that cassava leaves are important determinant of tuber yield. However, TMS 4(2)1425 with the highest leaf number yielded lower than the other improved cultivars. TMS 4(2)1425 branches or forks early, about two months after planting and continues branching until harvest. Due to this branching habit, it produced higher number of leaves than the other cultivars. Shoot apices and storage roots compete for carbohydrate in cassava (Tan and Cock, 1979a). The high number of branches most likely increased competition for available carbohydrate, which may have reduced root yield in TMS 4(2)1425 as compared to the other improved cultivars.

Plants in the fringe with deep water table produced the

**Table 3.** Influence of cassava cultivars and toposequence positions on rate of leaf formation ($day^{-1}$ $plant^{-1}$) and leaf area index (LAI) at two inland valley locations in 2001 and 2002 (Nigeria).

| Location (Loc) | Cultivar | 2001 | | | | 2002 | | | |
|---|---|---|---|---|---|---|---|---|---|
| | | Toposequence (Topo) | | | | Toposequence (Topo) | | | |
| | | Fringe | Inter-mediate | Bottom | Cultivar Mean | Fringe | Inter-mediate | Bottom | Cultivar Mean |
| | | **(a) Rate of leaf formation ($day^{-1}$ $plant^{-1}$)** | | | | | | | |
| Alabata | TMS4(2)1425 | 1.6 | 1.7 | 1.9 | 1.7 | 0.2 | 0.4 | 0.3 | 0.3 |
| | TMS91/02324 | 1.0 | 1.7 | 2.1 | 1.6 | 0.3 | 0.3 | 0.4 | 0.3 |
| | TMS91/02327 | 1.1 | 1.5 | 1.7 | 1.4 | 0.2 | 0.4 | 0.3 | 0.3 |
| | Isunikankiyan | 1.5 | 2.0 | 2.3 | 1.9 | 0.2 | 0.3 | 0.2 | 0.2 |
| | s.e ± | 0.3 | 0.3 | 0.3 | 0.1 | 0.3 | 0.3 | 0.3 | 0.1 |
| | Topo mean (s.e± 0.07) | 1.3 | 1.7 | 2.0 | - | 0.2 | 0.4 | 0.3 | - |
| | Year mean (s.e ± 0.05) | 1.7 | - | - | - | 0.3 | - | - | - |
| | Loc mean (s.e ± 0.04) | 1.0 | - | - | - | - | - | - | - |
| Ibadan | TMS4(2)1425 | 2.2 | 1.7 | 2.1 | 2.0 | 1.5 | 0.9 | 1.1 | 1.2 |
| | TMS91/02324 | 1.7 | 1.0 | 1.7 | 1.5 | 1.6 | 1.2 | 1.4 | 1.4 |
| | TMS91/02327 | 1.4 | 1.7 | 1.2 | 1.4 | 1.8 | 1.3 | 1.3 | 1.5 |
| | Isunikankiyan | 2.1 | 2.0 | 1.6 | 1.9 | 1.4 | 1.3 | 1.4 | 1.4 |
| | s.e ± | 0.3 | 0.3 | 0.3 | 0.1 | 0.3 | 0.3 | 0.3 | 0.1 |
| | Topo mean (s.e ± 0.07) | 1.9 | 1.6 | 1.7 | - | 1.6 | 1.2 | 1.3 | - |
| | Year mean (s.e ± 0.05) | 1.7 | - | - | - | 1.4 | - | - | - |
| | Loc mean (s.e ± 0.04) | 1.6 | - | - | - | - | - | - | |
| | | **(b) Leaf area index (LAI)** | | | | | | | |
| Alabata | TMS4(2)1425 | 6.6 | 5.0 | 5.4 | 5.7 | 3.1 | 2.6 | 1.8 | 2.5 |
| | TMS91/02324 | 4.6 | 4.7 | 4.9 | 4.7 | 3.2 | 2.0 | 2.3 | 2.5 |
| | TMS91/02327 | 4.2 | 3.3 | 3.5 | 3.7 | 2.4 | 1.7 | 1.4 | 1.8 |
| | Isunikankiyan | 5.1 | 4.5 | 3.5 | 4.4 | 2.1 | 1.8 | 1.3 | 1.7 |
| | s.e ± | 0.5 | 0.5 | 0.5 | 0.2 | 0.5 | 0.5 | 0.5 | 0.2 |
| | Topo mean (s.e ± 0.1) | 5.1 | 4.4 | 4.4 | - | 2.7 | 2.0 | 1.7 | - |
| | Year mean (s.e ± 0.1) | 4.6 | - | - | - | 2.1 | - | - | - |
| | Loc mean (s.e ± 0.1) | 3.4 | - | - | - | - | - | - | - |
| Ibadan | TMS4(2)1425 | 6.9 | 3.9 | 2.7 | 4.5 | 3.0 | 0.9 | 1.0 | 1.6 |
| | TMS91/02324 | 6.7 | 1.3 | 1.7 | 3.2 | 3.2 | 0.8 | 1.1 | 1.7 |
| | TMS91/02327 | 5.8 | 1.4 | 1.5 | 2.9 | 3.3 | 0.5 | 1.2 | 1.7 |
| | Isunikankiyan | 3.0 | 1.7 | 1.6 | 2.1 | 1.7 | 1.0 | 0.7 | 1.1 |
| | s.e ± | 0.5 | 0.5 | 0.5 | 0.2 | 0.5 | 0.5 | 0.5 | 0.2 |
| | Topo mean (s.e ± 0.1) | 5.6 | 2.1 | 1.9 | - | 2.8 | 0.8 | 1.0 | - |
| | Year mean (s.e ± 0.1) | 3.2 | - | - | - | 1.5 | - | - | - |
| | Loc mean (s.e ± 0.1) | 2.4 | - | - | - | - | - | - | - |

s.e = standard error of means

highest leaf number, while valley bottom with shallow water table had the lowest number. These results corroborate those of Lahai et al. (1999) and Jalloh and Palada (1989) who have noted better performance of cassava in valley fringe than valley bottom. Also, at shallow water table site at Ibadan, water table depth correlated negatively with leaf number, but the relationship was not significant at Alabata with deep water table. This indicated that shallow water table depth reduced root yield partly by decreasing leaf number. Lahai et al. (1999) and Mohamoud (1994) reported similar results. In 2001, with relatively well distributed

**Table 4.** Influence of cassava cultivars and toposequence positions on canopy area ($m^2$) and stay-green score (SGS) at two inland valley locations in 200 and 2002 (Nigeria).

| Location (Loc) | Cultivar | 2001 | | | | 2002 | | | |
|---|---|---|---|---|---|---|---|---|---|
| | | Toposequence (Topo) | | | | Toposequence (Topo) | | | |
| | | Fringe | Inter-mediate | Bottom | Cultivar Mean | Fringe | Inter-mediate | Bottom | Cultivar Mean |
| | | | | | (a) Canopy area ($m^2$) | | | | |
| Alabata | TMS4(2)1425 | 2.2 | 1.6 | 2.3 | 2.0 | 1.9 | 1.0 | 1.0 | 1.3 |
| | TMS91/02324 | 1.6 | 1.2 | 1.4 | 1.4 | 0.9 | 0.7 | 0.8 | 0.8 |
| | TMS91/02327 | 1.3 | 1.1 | 1.2 | 1.2 | 1.0 | 0.7 | 0.5 | 0.7 |
| | Isunikankiyan | 1.4 | 1.0 | 1.0 | 1.1 | 0.8 | 0.7 | 0.6 | 0.7 |
| | s.e ± | 0.09 | 0.09 | 0.09 | 0.06 | 0.09 | 0.09 | 0.09 | 0.06 |
| | Topo mean (s.e ± 0.03) | 1.6 | 1.2 | 1.5 | - | 1.2 | 0.8 | 0.7 | - |
| | Year mean (s.e ± 0.04) | 1.4 | - | - | - | 0.9 | - | - | - |
| | Loc mean (s.e ± 0.03) | 1.2 | - | - | - | - | - | - | - |
| Ibadan | TMS4(2)1425 | 1.5 | 1.2 | 1.2 | 1.3 | 1.0 | 0.4 | 0.5 | 0.6 |
| | TMS91/02324 | 1.5 | 0.7 | 0.7 | 1.0 | 0.8 | 0.5 | 0.6 | 0.6 |
| | TMS91/02327 | 1.3 | 0.6 | 0.7 | 0.9 | 0.9 | 0.4 | 0.6 | 0.6 |
| | Isunikankiyan | 1.0 | 0.6 | 0.6 | 0.7 | 0.7 | 0.5 | 0.5 | 0.6 |
| | s.e ± | 0.09 | 0.09 | 0.09 | 0.06 | 0.09 | 0.09 | 0.09 | 0.06 |
| | Topo mean (s.e ± 0.03) | 1.3 | 0.8 | 0.8 | - | 0.9 | 0.5 | 0.6 | - |
| | Year mean (s.e ± 0.04) | 1.0 | - | - | - | 0.6 | - | - | - |
| | Loc mean (s.e ± 0.03) | 0.8 | - | - | - | - | - | - | - |
| | | | | | (b) Stay-green score (SGS) | | | | |
| Alabata | TMS4(2)1425 | 2.1 | 2.1 | 2.4 | 2.2 | 3.0 | 3.0 | 3.3 | 3.1 |
| | TMS91/02324 | 2.0 | 2.0 | 2.2 | 2.1 | 2.5 | 2.2 | 3.5 | 2.7 |
| | TMS91/02327 | 2.0 | 2.0 | 2.1 | 2.0 | 2.5 | 2.2 | 4.1 | 2.9 |
| | Isunikankiyan | 2.4 | 2.4 | 2.7 | 2.5 | 3.2 | 3.0 | 4.0 | 3.4 |
| | s.e ± | 0.3 | 0.3 | 0.3 | 0.2 | 0.3 | 0.3 | 0.3 | 0.2 |
| | Topo mean (s.e ± 0.15) | 2.1 | 2.1 | 2.4 | - | 2.8 | 2.6 | 3.7 | - |
| | Year mean (s.e ± 0.08) | 2.2 | - | - | - | 3.0 | - | - | - |
| | Loc mean (s.e ± 0.06) | 2.6 | - | - | - | - | - | - | - |
| Ibadan | TMS4(2)1425 | 3.0 | 3.3 | 3.9 | 3.4 | 3.0 | 2.2 | 2.8 | 2.7 |
| | TMS91/02324 | 3.0 | 3.5 | 3.6 | 3.4 | 3.3 | 2.8 | 2.5 | 2.9 |
| | TMS91/02327 | 2.8 | 3.5 | 3.8 | 3.4 | 3.0 | 2.7 | 2.8 | 2.8 |
| | Isunikankiyan | 3.7 | 4.5 | 4.6 | 4.3 | 3.5 | 3.0 | 3.0 | 3.2 |
| | s.e ± | 0.3 | 0.3 | 0.3 | 0.2 | 0.3 | 0.3 | 0.3 | 0.2 |
| | Topo mean (s.e ± 0.15) | 3.1 | 3.7 | 4.0 | - | 3.2 | 2.7 | 2.8 | - |
| | Year mean (s.e ± 0.08) | 3.6 | - | - | - | 2.9 | - | - | - |
| | Loc mean (s.e ± 0.06) | 3.3 | - | - | - | - | - | - | - |

s.e = standard error of means; SGS rated on a scale of 1 to 9 where: 1 = a normal plant with a full canopy, retaining most formed leaves which are green, turgid and photosynthetically active; 3 = 30% of the leaves have dropped; less than 50% of the remaining older leaves are droopy, partially wilted or dry; young leaves have reduced greenness; 5 = 50% reduction of the leaf number as compared to full canopy; most of the older leaves are droopy, wilted and dry; most of the young leaves have reduced greenness; 7 = 80% reduction of the leaf number as compared to full canopy; more than 75% of the remaining are wilted or brown; young leaves have reduced greenness; 9 = complete defoliation of the stems, with a candlestick appearance and some dieback of the stems evident.

rainfall and moderate temperature, evaporation and solar radiation and high relative humidity, leaf number was higher than in 2002 with low rainfall and relative humidity, but high temperature, evaporation and solar radiation.

**Table 5.** Correlations between selected leaf and stem characters of cassava and tuber yield (n = 32), groundwater table depth (n = 32) and weather variables (n = 64) in inland valley.

| Canopy parameter | Tuberous root yield (t $ha^{-1}$) | | | | Groundwater table depth (m) | | | | Weather variable | | | | | | | | | | | |
|---|---|---|---|---|---|---|---|---|---|---|---|---|---|---|---|---|---|---|---|---|
| | Alabata | | Ibadan | | Alabata | | Ibadan | | Rainfall (mm) | | Evaporation rate (mm) | | Solar radiation (MJ/m/day) | | Minimum temperature (°C) | | Maximum temperature (°C) | | Relative humidity (%) | |
| | r | p | r | P | r | P | r | P | r | p | r | p | R | p | r | p | r | p | r | p |
| Number of leaves | 0.44 | 0.0117 | 0.60 | 0.0002 | 0.27 | 0.1337 | -0.67 | 0.0001 | 0.57 | 0.0001 | -0.57 | 0.0001 | -0.57 | 0.0001 | 0.57 | 0.0001 | -0.57 | 0.0001 | 0.57 | 0.0001 |
| Rate of leaf formation | 0.79 | 0.0001 | 0.35 | 0.0468 | -0.18 | 0.3303 | -0.50 | 0.0036 | 0.69 | 0.0001 | -0.69 | 0.0001 | -0.69 | 0.0001 | 0.69 | 0.0001 | -0.69 | 0.0001 | 0.69 | 0.0001 |
| Leaf area index (LAI) | 0.54 | 0.0015 | 0.72 | 0.0001 | 0.21 | 0.2587 | -0.64 | 0.0001 | 0.60 | 0.0001 | -0.60 | 0.0001 | -0.60 | 0.0001 | 0.60 | 0.0001 | -0.60 | 0.0001 | 0.60 | 0.0001 |
| Canopy area | 0.48 | 0.0055 | 0.69 | 0.0001 | 0.18 | 0.3359 | -0.63 | 0.0001 | 0.51 | 0.0001 | -0.51 | 0.0001 | -0.51 | 0.0001 | 0.50 | 0.0001 | -0.51 | 0.0001 | 0.51 | 0.0001 |
| Stay green score (SGS) | -0.46 | 0.0076 | -0.21 | 0.2415 | -0.47 | 0.0061 | 0.50 | 0.0036 | -0.04 | 0.7406 | 0.04 | 0.7406 | 0.04 | 0.7406 | -0.04 | 0.7406 | 0.04 | 0.7406 | -0.04 | 0.7406 |

This was supported by the positively with root yield at both sites. Similar results were obtained in the upland (El-Sharkawy et al., 1992; Indira, 1996). Photosynthetic rate of cassava leaves decreases markedly with leaf age (Aslam et al., 1977; Lebot, 2009). Also, new leaves formed increase leaf number or replace fallen ones. Increase in the number coupled with high photosynthetic rate of new leaves likely contributed to high root yield. Rate of leaf formation was related to branch number and rate of leaf retention. Irikura et al. (1979) reported that early branching increases leaf formation rate, while Okogbenin et al. (1999) noted that cultivars with high branching characteristics produced more leaves than non-branching ones and that planting methods that favoured leaf production profusely shed more leaves than planting methods with less leaf formation. Results of this study are consistent with these findings in that TMS 4(2)1425, a branching cultivar, TMS 91/02324 with high stem number and the landrace with low leaf number had higher leaf formation rate than TMS 91/02327 with high rate of leaf retention.

Correlation results with weather variables for rate of leaf formation were similar to those for leaf number, indicating that increase in rate of leaf formation increases leaf number. Similarly, water table depth correlated negatively with both number and rate of leaf production at the shallow water table site at Ibadan, but in correlations at Alabata, the deep water table site were not significant, indicating that shallow water table partly reduced root yield by reducing leaf production. Rate of leaf formation at Alabata was lower in valley fringe than valley bottom, whereas at Ibadan it was higher in the fringe than valley bottom. Water table appeared to be deeper in the fringe, likely causing drought stress at Alabata, while the valley bottom at Ibadan had shallower water table and both situations reduced leaf production. With high rainfall in 2001, both number and rate of leaf formation were higher at Alabata than Ibadan, which was partly due to shallower water table depth at Ibadan than Alabata. With low rainfall, water table depth increased markedly and drought stress likely contributed to the drastic reduction in number and rate of leaf production at Alabata in 2002. At Ibadan, with also deeper water table in 2002 than in 2001, but at the same time shallower than at Alabata, both parameters were higher than at Alabata.

## Leaf area index

High yielding cultivars have high LAI (Lahai et al., 1999; Adejinj et al., 2011). The improved cultivars with high LAI produced high root yield, indicating higher photosynthetic capacity of these cultivars than the landrace. Marked positive correlation existed between LAI and tuber yield, stressing the importance of LAI in yield determination. Lahai et al. (1999) and Lebot (2009) reported similar relationship. However, TMS 4(2)1425 with the highest LAI did not give the highest yield. This suggests that other parameters of canopy function, such as leaf carbon fixation and assimilate partitioning could interact significantly in yield formation. TMS 4(2)1425 had lower photosynthetic efficiency (Lahai and Ekanayake, 2010) and partitioned lower dry matter into roots, but higher dry matter into leaves and stems than the other improved cultivars (Lahai and Ekanayake, 2009), partially explaining the lower yield of TMS 4(2)1425 even though it had the highest LAI. LAI development took about four months to reach maximum values at both sites, parallel with gradual decrease in water table depth, but between 0.58 to 0.65 m below the mound tops. Then, LAI decreased rapidly, which corresponded to rapid decrease in water table depth to between 0.20 to 0.55 m below mound tops. Plants in valley fringe had higher LAI than those in other positions and LAI correlated negatively with water table depth at Ibadan. The lowest water table depth at Ibadan was 0.20 and 0.45 m below mound top in 2001 and 2002, respectively, which was within the range (0.2 to 0.4 m) when water table depth is stressful to cassava (Lahai and Ekanayake, 2009). However, the lowest depth at Alabata (0.50 m in 2001 and 0.58 m in 2002 below mound top) was above this range. The high LAI in the fringe as well as at Alabata as compared to other toposequence positions and Ibadan was partly due to deep water table. Similarly, the drastic reduction in LAI at 4 months after planting (MAP) at both sites and years was partially due to decrease in water table depth during this period. Lahai et al. (1999) also noted drastic reduction in LAI after 4 MAP as water table depth decreased in inland valley.

LAI increases with increase in temperature from 20 to 24°C due to early branching and increased leaf formation and size, but above 24°C branch number and leaf life are reduced (Irikura et al., 1979). The minimum and maximum temperatures were 19.8 and 27.5°C in 2001 and 19.1 and 28.4°C in 2002, respectively. The minimum temperatures (19.1 to 19.8°C) were within the minimum range that promotes LAI development, while the maximum temperatures (27.1 to 28.4°C) were above the maximum that favours high LAI, supporting the positive and negative correlations of LAI with minimum and maximum temperatures, respectively. However, the minimum and maximum temperatures in 2001 were closer to the optimum temperature range that increase LAI than in 2002, which partly contributed to the higher LAI in 2001. Leaf size is greatly reduced when water is withheld (Lenis et al., 2005) which reduces LAI. Thus, the low LAI and tuber yield in 2002 was partly due to reduced leaf size, because of drought during the first three months of growth. However, when LAI was reduced for all cultivars in 2002, TMS 91/02324 and TMS 91/02327 maintained higher LAI at both sites, suggesting that they were relatively less affected by drought than the other cultivars, which likely contributed to their high root yields. Lahai et al. (1999) noted that LAI of clone 87/29 that gave the highest yield was also less affected by changes in environmental conditions in inland valley.

In cassava, leaf area and tuber develop simultaneously, and assimilate produced is partitioned between growth of leaves and tubers (Lebot, 2009). This results in a delicate balance between top and tuber growth for maximum yield, with maximal tuber growth occurring at LAI of 2.5 to 3.5 at which optimum light interception and utilization occur. Low yielding cultivars maintain either sub-optimal (less than 2) or supra-optimal (more than 4) LAI during the major growth period, which reduce light utilization efficiency (Ramanujam, 1985). Our results are in line with these findings in that in 2002 when LAI was less than 2, root yield was drastically reduced as compared to 2001 when LAI was above 2. Also, at Alabata in 2001, TMS 91/02327 with LAI of 3.6 yielded higher than the other cultivars with LAI above 4. However, yield increased when LAI was 3.6 to 5.6 at Alabata in 2001. Generally, cassava intercepts 90 and 99% incident radiation at LAI 3 and 6, respectively (CIAT, 1980; Lebot, 2009). Cassava is grown in inland valley in the dry season when solar radiation is at its maximum in the tropics, meaning that at high LAI, high radiation was intercepted, which likely translated into high root yield at Alabata in 2001. This suggests that in inland valley LAI of 4 to 6 may not significantly adversely affect yield due to high solar radiation during this period.

### Canopy area

TMS 4(2)1425 had the largest and the landrace the smallest canopy area. In canopy with full ground cover, over 90% of incident radiation was intercepted by the top 50%, while canopy with partial ground cover intercepted less radiation (Fukai et al., 1984). Canopy area correlated positively with tuber yield, indicating that the high yield of the improved cultivars was partly due to higher light interception with larger canopy area than the landrace.

This agreed with Lebot (2009) who reported that crop canopy is a yield determinant. However, TMS 4(2)1425 with the largest canopy area had lower yield than the other improved cultivars, indicating that large canopy area may reduce yield. The finding was in line with Fukai et al. (1984) who noted that canopy development and root yield are competing components. However, the marked positive correlation between canopy area and yield in the study indicates that the competition was low, likely due to high solar radiation during the growing season in inland valley.

Plants in valley fringe and Alabata had larger canopy area than those in valley bottom and Ibadan. Also canopy area correlated negatively with water table depth at Ibadan, suggesting that the shallow water table depth at Ibadan and generally in valley bottom reduced canopy area. With a more evenly distributed rainfall, moderate temperature and high relative humidity in 2001, canopy area and LAI were higher than in 2002, with drought and high temperature stresses. Therefore, both shallow water table depth and drought stress may reduce canopy area, LAI and consequently yield in inland valley.

### Stay green score

Strategy to maintain high yield during waterlogging or

drought is to maintain photosynthetically active leaf area duration (LAD). LAD is indirectly monitored as stay-green score (SGS) (Ekanayake et al., 1996; Lenis et al., 2005). SGS was lower for the improved cultivars and in the fringe than the landrace and other positions. The lower the SGS, the higher the LAD, indicating that the improved cultivars and plants in the fringe had higher LAD than the landrace and those in other positions. High yield depends on the maintenance of an optimum LAI for much of the growing season (Ramanujam and Indira, 1983; Lebot, 2009). SGS correlated negatively with tuber yield, indicating that tuber yield increases with LAD.

LAD depends on individual leaf life. Leaf life in upland was found to be higher for high yielding improved cultivars than the low yielding landraces (Okogbenin et al., 1999; Lebot, 2009). This agreed with results of our study in that the improved cultivars with high LAD yielded higher than the landrace with low LAD. LAD is influenced by water table depth in inland valley (Lahai et al., 1999). In the valley fringe where water table was deeper, LAD was higher than in valley bottom with shallow water table. This partly explains the reason for the high root yield in the fringe at Ibadan with the shallow water table.

Similarly, SGS at Alabata with deep water table was generally better than at Ibadan with shallow water table. SGS correlated negatively with water table depth at Alabata, but positively at Ibadan. This indicated that as water table depth decreased, LAD decreased at Ibadan, whereas at Alabata LAD increased. At Ibadan, with shallow water table, decreasing water table depth caused waterlogging, which causes chlorosis and subsequent death of leaves (Mohamoud, 1994), thereby reducing LAD. However, at Alabata, decreasing water table likely improved soil moisture status, which increased LAD. Okogbenin et al (1999) noted on the upland that higher soil moisture at high water table site stimulated leaf production. At Alabata, in 2001 with high rainfall and relatively shallow water table, SGS was low indicating high LAD, whereas in 2002, with low rainfall and deep water table, SGS was higher and LAD lower than in 2001. On the contrary, at Ibadan, in 2001 LAD was lower than in 2002, indicating the adverse effect of decrease in water table depth on stay green ability at a shallow water table site. Thus, SGS may be a good indicator of overall performance of cassava in inland valleys.

There were differences between sites, toposequence positions and years for canopy parameters, mainly due to differences in water table depth and weather conditions. Both supra-optimal and sub-optimal canopy parameters reduced tuber yield, but sub-optimal reduced yield more than supra-optimal parameters largely due to high solar radiation during the growing season in inland valleys. High apex and stem and low leaf numbers increased leaf formation rate, while high leaf retention reduced the same. Shallow water table and drought reduced canopy parameters and yield, while well distributed rainfall, high humidity and moderate temperature 19 to 20℃ improved them. Canopy parameters of low yielding landrace were more susceptible to shallow water table and drought than high yielding TMS 91/02324 and TMS 91/02327. Therefore, cultivars with good maintenance of canopy parameters during mid-season drought and late season excess moisture can increase tuber yield when combined with other yield determinants in inland valley ecology.

## ACKNOWLEDGEMENT

This study was funded by IITA. The senior author is grateful for the research fellowship awarded by IITA to study for a PhD degree at Njala University, Sierra Leone.

### REFERENCES

Adeniji OT, Odo PE, Ibrahim B (2011). Genetic relationships and selection indices for cassava root yield in Adamawa State, Nigeria. Afr. J. Agric. Res. 6:2931-2934.

Ande OT (2011). Soil suitability evaluation and management for cassava production in the derived savanna area of Southwestern Nigeria. International J. Soil Sci. 6:142-149.

Andriesse W, Fresco LO, Van Duivenbooden N, Windmeizer PN (1994). Multi-scale characterization of Inland Valley agro-ecosystems in West Africa. Netherlands J. Agric. Sci. 42:159-179.

Aslam M, Lowe SB, Hunt LA (1977). Effect of leaf age on photosynthesis and transpiration of cassava. Can. J. Bot. 55: 2288-2295.

Burns A, Gleadow R, Cliff J, Zacarias A, Cavagnoro T (2010). 'Cassava: The Drought, War and Famine Crop in a changing World'. Sustainability 2:3572-3607.

CIAT (1980). Cassava Program: Annual Report 1979. Centro Internacional de Agricultura Tropical, AA 67-13, Cali, Colombia.

Ekanayake IJ (1996). Procedures for growth analysis of cassava. Procedures Manual, IITA, Ibadan, Nigeria. First edition. P. 28.

Ekanayake IJ, Dixon AGO, Asiedu R, Izac A-M N (1994). Improved cassava for Inland valley agro-ecosystems. In: Akoroda MO (ed.) Root Crop for Food Security in Africa: Proc. of fifth Triennial Symp. of the Intern. Soc. for Tropical Root Crops- Africa Branch held at Kampala, Uganda. Ibadan: IITA/ International Society for Tropical Root Crops, pp. 204-208.

Ekanayake IJ, Dixon AGO, Porto MCM (1996). Performance of various cassava clones in the dry savannah region of Nigeria. In: Kurup CT, Palaniswamy MS, Potty VP, Padmaja G, Kabeerathumara S, Pillai SV (eds.), Tropical Tuber Crops: Problems, prospects and future strategies, Lebanon, USA: Science Publishers, Inc pp. 207-215.

El-Sharkawy MA (1993). Drought-tolerant cassava for Africa, Asia and Latin America. BioSci. 43:441-451.

El-Sharkawy MA, Hernandez ADP, Hershey C (1992). Yield stability of cassava during prolonged mid-season water stress. Exp. Agric. 28: 165-174.

Fukai S, Alcoy AB, Llamelo AB, Patterson RD (1984). Effects of solar radiation on growth of cassava (Manihot esculenta Crantz). I. Canopy development and dry matter growth. Field Crops Res. 9:347-360.

Indira P (1996). Leaf area index and tuber yield in cassava as influenced by the time of application of nitrogen. In: Kurup CT, Palaniswamy MS, Potty VP, Padmaja G, Kabeerathumara S, Pillai SV (eds.), Tropical Tuber Crops: Problems, prospects and future strategies, Lebanon, USA: Science Publishers, Inc., pp. 219-226.

Irikura Y, Cock JH, Kawano K (1979). The physiological basis of genotype-temperature interactions in cassava. Field Crops Res. 2: 227-239.

Izac A-MN, Swift MJ, Andriesse W (1991). A strategy for inland valley agro-ecosystems research in East and Central Africa. RCMP Research Monograph No. 5, Ibadan, Nigeria: IITA, P. 24.

Jalloh A (2003). The potential of inland valley swamps in contributing to food security in Africa. FoodAfrica, 5-9 May 2003, Cameroon.

Jalloh M, Palada MC (1989). On-farm trials in Makeni, Sierra Leone. In: Resource and Crop Management Program (RCMP) 1987 Annual Report. Ibadan, Nigeria: IITA. pp. 177-179.
Jarvis PG, Morison JIL (1982). The control of transpiration and photosynthesis by the stomata. In: Jarvis PG, Mansfield TA (eds.) Society of Experimental Biology Seminar 8. Cambridge: Cambridge University Press, pp. 247-279.
Lahai MT, Ekanayake IJ (2009). Accumulation and distribution of dry matter in relation to root yield of cassava under a fluctuating water table in inland valley ecology. Afr. J. Biotech. 8:4895-4905.
Lahai MT, Ekanayake IJ (2010). Transpiration rate and leaf water potential as indices for cassava yield in inland valley ecology. J. Agric. Sci. Tech. 4:34-44.
Lahai MT, George JB, Ekanayake IJ (1999). Cassava (Manihot esculenta Crantz) growth indices, root yield and its components in upland and inland valley ecologies of Sierra Leone. J. Agro. Crop Sci. 182:239-247.
Lebot V (2009). Tropical root and tyber crops: cassava, sweet potato, yam and aroids. CABI publication. Amazon.com. P. 413.
Lenis JI, Calie F, Jaramillo G, Perez JC, Ceballos H, Cock JH (2005). Leaf retention and cassava productivity. Field Crops Res. 95:126-134.
Mohamoud YM (1994). Effect of mound height and cassava cultivar on cassava performance under a fluctuating water table. Agric. Water. Managt. 26:201-211.
Moormann ER, Lal R, Juo ASR (1975). The soils of IITA. IITA Tech. Bull. No.3, Ibadan, Nigeria.
Njoku DN, Afuape SO, Ebeniro CN (2010). Growth and yield of cassava as influenced by grain cowpea population density in Southeastern Nigeria. Afr. J. Agric. Res. 5:2778-2781.
Nweke FI, Spencer DS.C, Lynam JK (2002).The Cassava Transformation: Africa's best-kept secret. East Lansing: Michigan State University Press. P. 273.
Okogbenin E, Ekanayake IJ, Porto MCM (1999). Effect of planting materials and soil moisture on cassava performance in the semi-arid Sudan Savannah belt of Nigeria. Afr. Crop Sci. J. 1:21-33.
Ramanujam T (1985). Leaf density profile and efficiency in partitioning dry matter among high and low yielding cultivars of cassava (Manihot esculenta Crantz). Field Crops Res. 10:291-303.
Ramanujam T, Indira P (1983). Canopy structure on growth and development of cassava (Manihot esculenta Crantz). Turrialba 33:321-326.
SAS Institute (1991) SAS/STAT User's Guide, Version 6, Fourth Edition. Cary, NC, USA: SAS Institute.
Sheriff DW, Muchow RC (1984). The water relations of crops. In: Goldsworthy PR, Fisher NM (eds.), Physiology of Tropical Field Crops. New York: John Wiley □ Sons Ltd.
Tafur SM de, El-Sharkawy MA, Cadavid L.F, De-Tafur SM (1997). Response of cassava (Manihot esculenta Crantz) to water stress and fertilisation. Photosynthetica 34:233-239.z
Tan SL, Cock JH (1979b). Cassava plant forms and their associated morpho-physiological characters. MARDI Res. Bull. 7:55-69.
Tan SL, Cock JM (1979a). Branching habit as a yield determinant in cassava. Field Crops Res. 2:281-289.

# Effects of nitrogen and phosphorus fertilizers and cotyledon removal on establishment and growth of yeheb nut bush (*Cordeauxia edulis* Hemsl.) outside its natural habitat

**Bertukan Mekonnen[1]*, Kebede Woldetsadik[1], Tamado Tana[1] and Asha Yahya[2]**

[1]Department of Plant Sciences, Haramaya University, P. O. Box: 138, Dire Dawa, Ethiopia.
[2]Department of Crop Production Ecology, Swedish University of Agricultural Sciences, Sweden.

**Yeheb nut bush (*Cordeauxia edulis*) is a multipurpose but endangered shrub native to arid south-Eastern Ethiopia and Central Somalia. Establishment of this species was studied in two sets of experiments during 2005 to 2007. In the first set of experiment, field establishment was conducted at Dire Dawa, Ethiopia, which is located 725 km north west of the natural habitat (Bokh). Nitrogen at 0, 46 and 92 level and phosphorus at 0, 4 and 8 kg ha$^{-1}$ were applied 84 days after transplanting. Application of 92 N/8 P kg ha$^{-1}$ increased seedling height by 134%; 46 N/4 P kg ha$^{-1}$ increased leaf area by 152% and dry weight of seedlings increased by 309% at 46/8 kg ha$^{-1}$ NP compared to the unfertilized plants. Nitrogen and phosphorus fertilizers significantly altered soil as well as plant tissue N and P contents. Leaf and root N was significantly highest at 46 and 92 N kg ha$^{-1}$, respectively while leaf P was maximised at 8P kg ha$^{-1}$. In the second set, the impact of cotyledon removal was assessed in the greenhouse, Haramaya University by excision of cotyledon at 7 days interval for ten consecutive weeks and leaving the cotyledon with plant as a control. Removing cotyledon at 7, 14 and 35 days after germination (DAG) significantly reduced most of the growth parameters while seedlings with intact cotyledon had the highest dry matter accumulated. Thus, 46 to 92 kg N and 8 kg P ha$^{-1}$ would help the establishment and growth of *C. edulis* in the field and seedlings should retain cotyledon during early establishment. However, more rates of N and P as well as methods and frequency of applications deserve further investigation.**

**Key words:** *Cordeauxia edulis*, cotyledon, Ethiopia, nitrogen and phosphorus fertilizers, seedling establishment.

## INTRODUCTION

Yeheb nut bush (*Cordeauxia edulis*) is leguminous, multi-stemmed and evergreen shrub limited to arid south-eastern Ethiopia near Bokh and central Somalia (Seegeler, 1983; Booth and Wickens, 1988; Zimsky, 1990). It typically experiences erratic rainfall as low as 100 to 200 mm year$^{-1}$, average annual temperature of 28°C and grows at altitudes of 300 to 1000 m.a.s.l (Drechsel and Zech, 1988). *C. edulis* adapted to nutrient impoverished soil (Seegeler, 1983; Drechsel and Zech, 1988) and frost free areas. It is the only species within the genus *Cordeauxia*.

The seed of *C. edulis* is nutritious and tasty with a chestnut flavour. They contain 37% starch, 24% low molecular weight sugar, 13% proteins, 11% fats and various mineral salts (Gutale and Ahmed, 1984). The nut is the staple food for the local people which is liked to the common diet of rice and dates (Kazmi, 1979). The nut has also medicinal value and economic importance at

*Corresponding author. E-mail: bertukan.mekonnen@yahoo.com.

local market and even across the continent (Booth and Wickens, 1988). In general, *C. edulis* has multiple uses: as fodder, firewood, house construction, bee food supply, mulch, soil conservation, live fence, dyes etc. (Azene et al., 1993). Moreover, it is a very drought tolerant and often represents the last existing perennial. The palatable leaves are dry season browse for animals (Drechsel and Zech, 1988). Further, *C. edulis* has the potential to be a reliable source of protein and valuable food crop in hot and dry regions (Zimisky, 1990). It could be particularly important for places where rainfall is scares for growing more conventional crops (Vietmeyer, 1985).

However, *C. edulis* is threatened by extinction in its natural habitat because of wars, recurrent drought, over-harvesting and heavy grazing (Miége and Miége, 1978; Zimsky, 1990). Thus, introduction of this multipurpose shrub to arid and semiarid tropics is greatly in need to minimize the threatening of this species.

One of the conditions which influence establishment and growth of plants is soil fertility. Nitrogen and phosphorus are among the major essential nutrient elements for improving plant growth and development. Olaniyi (2008) reported that various levels of individual and combined N and P fertilizers significantly improved growth and seed yield of *Citrullus lanatus*. Mhango et al. (2008) showed the importance of NPK fertilizer in improving seedling growth of *Uapaca kirkiana*. Similarly, *Picea glauca* growth and nutrient uptake increased by application of N fertilizer during its seedling stage (Mcalister and Timmer, 1998). *C. edulis* native area is characterized by red coloured sandy soil which is extremely low in nitrogen (Drechsel and Zech, 1988; Zimisky, 1990). According to Sassaki and Felipe (1998) *Dalbergia miscolobium* Benth which is occurring on a very poor soil responded positively to the addition of mineral nutrients when the plants were grown in pots. In another study, seedling growth of *Acacia auriculiformis* was enhanced significantly with the application of phosphorus fertilizers (Uddin et al., 2007). There might be a possibility of increasing establishment of *C. edulis* through supplying fertilizer.

Rate of fertilizers is considered as one of the factors that affect plants growth. According to Hossain and Hamid (2007), growth and yield of groundnut exhibited improved growth and yield with increased rate of N and P application. Jeppsson (2000) reported that vegetative growth and yield of *Aronia melanocarpa* increased with a combined application of NPK fertilizer in different rates.

The other factor which affects establishment of plants is the seed nutrient reserve (cotyledon). Large seeded species, including *C. edulis* (1.8 to 2.4 g dwt seed$^{-1}$; Andersson et al., 2007) rely heavily on nutrient support from the cotyledon than from the soil (Milberg and Lamont, 1997; Vaughton and Ramsey, 2001) to survive under unfavourable conditions and hazards during the establishment phase (Frost and Rydin, 1997; Kidson and Westoby, 2000). However, loss or damage of cotyledon could be brought by herbivores, insect pests and diseases. Large numbers of seedlings of *C. edulis* are destroyed by goats and camels during browsing (Johnson, 1996) and re-growth of grazed seedlings may be dependent on nutrient reserve. Milberg and Lamont (1997) reported that the effect of cotyledon removal was least in the smallest–seeded *Eucalyptus loxophleba* Benth and greatest in the largest-seeded, *Haeka psilorrhyncha* R. M. Barker. Taylor et al. (2006) animal browse protection is considered indispensable for successful seedling establishment of *Quercus*.

It is a prerequisite to know the stage at which cotyledon loss or damage might bring significant impact on seedling establishment of *C. edulis* to enable take measure at that specific time or phase. Therefore, the aim of this study was to determine the influence of NP fertilizers on *C. edulis* growth and to assess the importance of cotyledon to *C. edulis* establishment.

## MATERIALS AND METHODS

### Experimental locations and seed material

The field experiment (Experiment 1) was conducted at Dire Dawa, Ethiopia (09° 31'N, 41° 51' E), about 500 km east of the capital, Addis Ababa, and 725 km west of the natural habitat at an altitude of 1160 m.a.s.l. The experimental site is characterized as semi-arid tropical climate with average annual rainfall of about 644 mm (NMA, 2006). The rainfall has a bi-modal distribution pattern with small rains from March to May and long and moderate rains from June to September. The experimental soil is alluvial with a sandy clay loam to sandy loam texture and a pH of 8.6. The physic-chemical characteristics of the experimental and the native soils are shown in Table 1.

Plots were selected for the field experiment from a large piece of grazing land covered with grass during the previous nine years and fertilized with animal manure for four years preceding the study. The experimental plots were prepared by digging the soil to one meter depth preceding planning.

Experiment 2 was conducted in a greenhouse at Haramaya University, Ethiopia (9°26' N, 42°3' E).

The fruits of *C. edulis* were collected from natural vegetation in Gambarey village near Bokh town (7°22`N, 46° 35`E), Warder Zone of Eastern Ethiopia, 725 km south east of the study site, Dire Dawa. The fruits were gently dried in shade after harvest; the pericarps were removed and the seeds were kept in cotton-cloth bag at room temperature until used for the experiments.

### Experiment 1: Nitrogen and phosphorus fertilizer supply

The treatments consisted of 3 levels of N (0, 46, 92 kg ha$^{-1}$) and 3 levels of P (0, 4, 8 kg ha$^{-1}$) using urea (46% N) and triple super phosphate (20% P) as a source of N and P, respectively. Nine treatment combinations (0N0P, 0N4P, 0N8P, 46N0P, 46N4P, 46N8P, 92N0P, 92N4P and 92N8P kgha$^{-1}$) were laid out in a randomised complete block design in three replications.

Seedling were raised from ca 1400 seeds which were peeled fourteen days after harvest and sown on 20 June 2005 in a greenhouse. One seed was sown per polyethylene bag in partially-buried position after the bag was filled with sieved river sand and well watered. The bags were open-ended on both sides and had a size of 1.05 L by volume with a height of 0.12 m. After sowing, the

**Table 1.** Physicochemical characteristic of the experimental soil and the native soil.

| Parameter | Experimental soil (Dire Dawa) | | | | Native soil |
|---|---|---|---|---|---|
| | depth (cm) | | | | |
| | 0-25 | 25-75 | 75-123 | 123-200 | 0-20 |
| Texture | Sandy clay loam | Sandy loam | Loamy sandy | Sandy loam | Sandy |
| pH ($H_2O$) | 8.70 | 8.20 | 8.90 | 8.70 | 8.50 |
| EC (μs/cm) | 136.90 | 504.80 | 137.30 | 274.50 | 74.10 |
| Total nitrogen (%) | 0.042 | 0.042 | 0.014 | 0.028 | 0.014 |
| Available P (mg/kg soil) | 2.39 | 3.05 | 0.35 | 1.15 | 0.13 |
| Organic carbon (%) | 0.85 | 0.77 | 0.57 | 0.66 | 0.51 |
| CEC (meq/100 g soil) | 21.80 | 21.00 | 20.80 | 20.80 | 21.60 |
| Exchangeable bases mg (cmol (+) kg/soil | 2.89 | 1.50 | 0.77 | 1.88 | 1.18 |
| Exchangeable bases Na (cmol (+) kg/soil | 0.23 | 0.15 | 0.13 | 0.44 | 0.07 |
| Exchangeable bases Ca (cmol (+) kg/soil | 7.81 | 31.39 | 16.39 | 20.14 | 3.04 |
| Exchangeable bases Mg (cmol (+) kg/soil | 2.89 | 1.5 | 0.77 | 1.88 | 1.18 |
| Exchangeable bases K (cmol (+) kg/soil | 0.35 | 0.17 | 0.10 | 0.18 | 0.32 |
| $CaCO_3$ (%) | 6.70 | 7.00 | 4.70 | 6.70 | 0.90 |

polyethylene bags were covered with plastic sheets to maintain the moisture and they were removed after majority of the seeds had germinated.

Twelve seedlings of one month old were transplanted per plot (2.4 $m^2$) spaced 50 cm between rows and 40 cm between plants. The seedlings were planted in shallow moistened holes, exposing the cotyledon at the surface without covering with soil. Dead or weak seedlings were replaced with vigorous seedlings of similar age a week after transplanting. The nitrogen and phosphorus fertilizers were applied in splits: 11 weeks after transplanting and 44 weeks after the first application. Weeding and hoeing were done as required. Watering was done through furrows along the edges of each experimental plot and between plants for a few weeks then during the remaining experimental period, watering was continued by showering each plot twice a day. Eighty weeks after transplanting, plant height, stem diameter, root diameter, number of leaves, leaf area, fresh weight and dry weight were recorded on four randomly harvested plants per plot.

The amount of NP adsorbed by the soil was assessed by taking soil samples from the experimental field three times: Before planting, ten months and six months after the first and second fertilizer application, respectively. All the samples were taken from 0 to 30, 30 to 60 and 60 to 90 cm depths from each plot. The soil samples were air-dried, sieved with a 2 mm mesh size and analyzed for the total nitrogen and available phosphorus. Total nitrogen was determined according to the procedure of Kjeldahl (1883) and available phosphorus was determined according to Olsen et al. (1954) method.

To determine the amount of NP taken by the plant, four randomly harvested plants per plot were separated into leaf, stem, roots and dried at 70℃ until constant weighs was attained. The dried biomass was analyzed for nitrogen and phosphorus contents. The dry matter was decomposed through wet ashing using the procedure of Kjeldahl (1883). Total N was determined by distillation of an aliquote from the digest with NaOH, collecting the distillate in boric acid and titrating with 0.1 $H_2SO_4$ to the end point of the mixed indicator. Determination of phosphorus was carried out on digest aliquots obtained through wet digestion. The phosphorus in the solution was determined colorimetrically using molybdate and metavanadate for color development. The reading was made at 460 Nm wavelength. Both analyses were done at the soil laboratory of Haramaya University.

### Experiment 2: Date of cotyledon removal

Treatments comprised of ten cotyledon removal dates: 7, 14, 21, 28, 35, 42, 49, 56, 63 and 70 days after germination (DAG) and seedlings with cotyledon left intact were used as a control. The treatments were arranged in a completely randomised design with four replications. Ten seeds per treatment and two seeds per pot (5 L) were sown after filled with sieved moistened river sand on 17 June 2006 in a greenhouse. The pots were covered with plastic sheets until most of the seeds per pot had germinated. Plant height, root length, stem diameter, root diameter, leaf number, leaf area, fresh weight and dry weight of the seedlings were measured and recorded on three plants per replication after five months.

### Statistical analysis

The effect on plant height, root length, stem diameter, root diameter, leaf area, dry weight, soil and leaf, stem, root NP content was tested. Analysis of variance for both experiments was carried out using the General Linear Model of Minitab Statistical Software and when the ANOVA showed significant treatment effects, the means were compared using the least significant difference test at 5% level of significance.

## RESULTS AND DISSCUSION

### Experiment 1: Nitrogen and phosphorus fertilizer supply

Application of nitrogen and phosphorus fertilizers had significant effect on the growth of *C. edulis* (Table 2) but did not affect number of survived seedlings (data not shown). Plant height and stem diameter were maximised at the application rate of 92 kg $ha^{-1}$ nitrogen and 8 kg $ha^{-1}$

**Table 2.** Effect of NP fertilizer on growth of *C. edulis* seedlings.

| Treatment | | Plant height (cm) | Stem diameter (mm) | Root diameter (mm) | Leaf area ($cm^2$) | Dry weight of seedling (g) |
|---|---|---|---|---|---|---|
| N (kg $ha^{-1}$) | P (kg $ha^{-1}$) | | | | | |
| 0 | 0 | 10.5 | 2.7 | 2.4 | 34.1 | 3.3 |
| 0 | 4 | 11.6 | 3.1 | 2.8 | 49.3 | 4.6 |
| 0 | 8 | 11.2 | 2.6 | 2.4 | 36.3 | 3.9 |
| 46 | 0 | 13.4 | 2.9 | 2.8 | 36.1 | 5.9 |
| 46 | 4 | 17.2 | 4.1 | 4.2 | 86.0 | 12.8 |
| 46 | 8 | 13.4 | 4.0 | 3.8 | 53.8 | 13.5 |
| 92 | 0 | 12.7 | 3.6 | 3.3 | 34.7 | 8.7 |
| 92 | 4 | 18.1 | 3.9 | 3.7 | 84.9 | 12.6 |
| 92 | 8 | 24.6 | 4.2 | 3.7 | 67.4 | 12.9 |
| N | | ** | ** | ** | ** | *** |
| P | | * | ** | ** | *** | ** |
| LSD (5%) N*P | | 1.17 | 0.52 | 0.52 | 16.61 | 2.91 |

Ns, Non- significant,*, **,***, significant at P = 0.05, P = 0.01 and P = 0.001, respectively.

phosphorus which were about 134 and 55%, respectively, over the unfertilized plot. Beside flower development of *C. edulis* was recorded at this rate. In agreement with this result, biomass of *Sarcobatus vermiculatus* seedlings was doubled by fertilizer application compared to controls (Breen and Richards, 2008). Hashimi and Hughes (2010) also showed that enhanced growth of *Glochidion obscurum* and *Lagerstroemia speciosa* at high rate of NPK. Root diameter and leaf area were highest at 46 kg $ha^{-1}$ N and 4 kg $ha^{-1}$ P while maximum dry weight of seedlings were recorded at 46 kg $ha^{-1}$ N and 8 kg $ha^{-1}$ P. Nitrogen and phosphorus elements have tremendous roles in plant nutrition in which phosphorus stimulates the formation of lateral and fibrous roots that increases the absorbing surface for nutrients while nitrogen increases the cation exchange capacity of plant roots to use more phosphorus including potash and calcium (Rai, 2002). According to Graciano et al. (2006), soil nitrogen absorption efficiency increased with P fertilization in *Eucalyptus grandis*. All plots that received nitrogen and phosphorus fertilizers resulted in better seedlings growth than those that received either nitrogen or phosphorus.

The influence of nitrogen and phosphorus fertilizers application was evident on the content of soil N and P at different depths (Table 3). Soil phosphorus content was significantly affected at all depths while soil nitrogen was affected only at 60 and 90 cm depths. Phosphorus fertilizer tended to increase soil P at all depths and higher rate of P increased soil N to the maximum. Application rate of 46 kg N $ha^{-1}$ with 4 kg P $ha^{-1}$ and 46 kg N $ha^{-1}$ with 8 kg P $ha^{-1}$ tended to reduce soil P at 30 and 60 cm. By contrast application rate of 92 kg N $ha^{-1}$ with 4 kg P $ha^{-1}$ gave the highest soil P at all depth. Rate of nitrogen might be affected by the available phosphorus by varying soil pH. As nitrogen is the mineral nutrient that is taken up at the highest rate by most plant species (Marschner, 1995), it occurs in the soil as various species that bear different charges and influence soil pH. The highest soil nitrogen was also recorded at 92 kg N $ha^{-1}$ with 8 kg P $ha^{-1}$ at 60 cm which gave the least soil P at 90 cm. Similar trend was also shown at the application rate of 92 kg N $ha^{-1}$/and 4 kg P $ha^{-1}$ at 90 cm whereby soil N was least and soil P was highest. Soil nitrogen content is affected by many factors including soil texture, structure and hydraulic conductivity as this influence the movement of nitrogen in the soil. As a result, the applied nitrogen might be accumulated in the soil or move downward or else lost to the air.

Moreover, NP content of the leaf, stem and root was significantly varied by nitrogen and phosphorus fertilizers (Table 4). Application of nitrogen and phosphorus fertilizer increased the nitrogen and phosphorus content of *Eucalyptus globules* wood (Raymond and Muneri, 2000). Rate of nitrogen affected the level of nitrogen in the different organs of *C. edulis*. The highest leaf N was recorded at 46 kg $ha^{-1}$ N whereas the highest stem and root nitrogen were indicated at 92 kg $ha^{-1}$ N. Hanway and Weber (1971) reported that the amount of soil nitrogen affected the relative distribution of nitrogen in the different plant parts of soybean. Allocation of nitrogen to the leaves was prior to stem and root as leaves are the principal photosynthetic organs of the plant. The least leaf and stem nitrogen were recorded at 46 kg $ha^{-1}$N and 8 kg $ha^{-1}$P but an increasing trend in leaf nitrogen was shown as nitrogen applied at 92 kg $ha^{-1}$ combined with 4 and 8 kg $ha^{-1}$P; however root nitrogen showed a decreasing trend at these rates. Combined application of nitrogen and phosphorus fertilizers or phosphorus alone at low rate, reduced leaf and root phosphorus content significantly. However, leaf phosphorus was maximized as the applied P increased to 8 kg $ha^{-1}$. *Podocarpus urbanii* and *Clethra occidentalis* foliar P concentrations was increased by P fertilization (Tanner et al., 1990).

**Table 3.** Effect of nitrogen and phosphorus fertilizers on nitrogen and phosphorus content of the experimental soil at different depths.

| Treatment | | Soil depth (cm) | | | | | |
|---|---|---|---|---|---|---|---|
| | | 30 | | 60 | | 90 | |
| N (kgha$^{-1}$) | P (kgha$^{-1}$) | Soil N (%) | Soil P (mgkg$^{-1}$) | Soil N (%) | Soil P (mgkg$^{-1}$) | Soil N (%) | Soil P (mgkg$^{-1}$) |
| 0 | 0 | 0.037 | 10.1 | 0.026 | 15.63 | 0.040 | 14.4 |
| 0 | 4 | 0.040 | 15.8 | 0.037 | 15.87 | 0.040 | 13.51 |
| 0 | 8 | 0.042 | 16.35 | 0.049 | 17.10 | 0.048 | 14.9 |
| 46 | 0 | 0.044 | 15.1 | 0.041 | 13.43 | 0.037 | 7.2 |
| 46 | 4 | 0.037 | 8.9 | 0.043 | 8.6 | 0.037 | 10.3 |
| 46 | 8 | 0.049 | 8.4 | 0.047 | 11.33 | 0.044 | 10.1 |
| 92 | 0 | 0.033 | 11.8 | 0.040 | 10.9 | 0.042 | 9.6 |
| 92 | 4 | 0.035 | 17.9 | 0.037 | 17.27 | 0.035 | 17.4 |
| 92 | 8 | 0.051 | 16.2 | 0.049 | 16.9 | 0.047 | 7.4 |
| N | | ns | * | ns | * | ns | ** |
| P | | ns | * | ** | ns | ** | ns |
| N*P | | ns | ** | ns | ns | ns | ns |
| LSD (5%) N*P | | ns | 5.05 | ns | ns | ns | ns |

ns, Non- significant,*, **, significant at P = 0.05 and P = 0.01, respectively.

**Table 4.** Effect of nitrogen and phosphorus fertilizers on nitrogen and phosphorus content in the different organs of *C. Edulis*.

| Treatment | | Nitrogen (%) | | | Phosphorus (%) | | |
|---|---|---|---|---|---|---|---|
| N(kg/ha$^{-1}$) | P(kg/ha$^{-1}$) | Leaf | Stem | Root | Leaf | Stem | Root |
| 0 | 0 | 2.35 | 1.135 | 0.737 | 0.06 | 0.017 | 0.043 |
| 0 | 4 | 1.619 | 1.335 | 1.43 | 0.029 | 0.051 | 0.037 |
| 0 | 8 | 3.40 | 2.165 | 1.645 | 0.136 | 0.048 | 0.050 |
| 46 | 0 | 3.95 | 1.44 | 1.296 | 0.083 | 0.035 | 0.049 |
| 46 | 4 | 1.625 | 1.44 | 2.565 | 0.079 | 0.047 | 0.017 |
| 46 | 8 | 0.955 | 0.96 | 2.165 | 0.074 | 0.047 | 0.039 |
| 92 | 0 | 1.185 | 2.63 | 2.79 | 0.080 | 0.058 | 0.074 |
| 92 | 4 | 2.34 | 2.42 | 2.305 | 0.053 | 0.040 | 0.099 |
| 92 | 8 | 3.195 | 2.485 | 1.915 | 0.047 | 0.055 | 0.056 |
| N | | ns | *** | *** | *** | ns | * |
| P | | ns | ns | ns | *** | ns | ns |
| N*P | | ** | ns | ** | *** | ns | ns |
| LSD (5%) | | 1.88 | ns | 0.75 | 0.01 | ns | ns |

ns, Non- significant,*,**,***, significant at P = 0.05, P = 0.01 and P = 0.001, respectively.

Similarly, foliar phosphorus concentrations of *Zenobia pulverulenta* and *Lyonia lucida* increased in response to enhanced phosphorus availability (Simms, 1987). In another study, addition of NP nutrient solution to *Citrus limonia* increased leaf nitrogen and phosphorus content to a higher level (Ruschel et al., 2004). A decreasing trend in the amount of leaf phosphorus was also found as N rate increased possibly due to increased seedling growth and resulting dilution effect. Likewise, fertilization of *Dendropanax pendulus* and *Hedyosmum arborescens* with N resulted in lower P concentrations (Tanner et al., 1990). On the other hand, roots' phosphorus was highest at 92 kg ha$^{-1}$ N and 4 kg ha$^{-1}$ P while the least was at 46 kg ha$^{-1}$ N and 4 kg ha$^{-1}$ P. In *Cicer arietinum,* phosphate uptake was affected more by nitrogen than phosphorus and severely reduced in low nitrogen condition (Das and Pen, 1981).

**Experiment 2: Date of cotyledon removal**

A weekly interval of removing cotyledon after germination significantly affected most of the growth parameters measured (Table 5) except number of survived seedlings

**Table 5.** Effect of date of cotyledon removal on shoot and root growth of *C. edulis* seedlings.

| Day | Plant height (cm) | Root length (cm) | Stem diameter (mm) | Root diameter (mm) | Leaf area ($cm^2$) | Dry weight of seedlings (g) |
|---|---|---|---|---|---|---|
| 7 | 3.5 | 15.0 | 1.9 | 2.2 | 9.9 | 0.3 |
| 14 | 4.1 | 11.6 | 2.0 | 2.0 | 18.4 | 0.5 |
| 21 | 5.5 | 13.3 | 2.1 | 2.4 | 17.7 | 0.7 |
| 28 | 6.2 | 13.2 | 2.0 | 2.2 | 22.4 | 0.8 |
| 35 | 3.4 | 26.7 | 2.2 | 2.1 | 10.0 | 0.4 |
| 42 | 5.5 | 17.9 | 2.2 | 2.4 | 20.9 | 0.9 |
| 49 | 5.9 | 15.2 | 2.1 | 2.3 | 17.9 | 0.6 |
| 56 | 5.7 | 20.7 | 2.0 | 2.1 | 20.0 | 0.9 |
| 63 | 5.3 | 17.1 | 2.1 | 2.3 | 19.3 | 0.7 |
| 70 | 5.6 | 15.0 | 2.1 | 2.2 | 17.4 | 0.6 |
| Control | 5.5 | 24.8 | 2.3 | 2.4 | 35.4 | 1.0 |
| Significance | *** | ** | ns | * | ** | *** |
| LSD | 1.37 | 9.47 | ns | 0.37 | 11.11 | 0.35 |

ns = non-significant, *,**,***= significant at P = 0.05, P = 0.01 and P = 0.001.

(Data not shown).

Removal of cotyledon 7 DAG reduced plant height, leaf area and dry weight of *C. edulis* significantly compared to the control. Along the same line, García-Cebrián et al. (2003) showed significant reduction of seedling growth of *Quercus robur* L when the cotyledon was removed at emergence and 7, 14 and 21 days after shoot emergence.

Plant height increased with delay in cotyledon removal up to 28 DAG where the highest seedling shoot length was recorded. Cotyledon removal on 35 DAG depressed shoot growth of *C. edulis* whereas this treatment resulted in the longest root growth of the seedlings. From the data, it appears that cotyledon contribute more to growth of the shoot during the first four weeks after germination while more nutrient allocation goes to the roots thereafter. Generally, nutrients supply by cotyledon serve for limited time (Krigel, 1967). For both hypogeal and epigeal species, the timing of maximum relative growth, the exhaustion of cotyledon reserves and the attainment of independence from cotyledon all roughly coincided (Hanley et al., 2004). Similarly, cotyledon abscission of *Caesalpinia peltophoroides* took place at 35 days after sowing (Corte et al., 2008).

The highest leaf area and dry weight of seedlings were recorded in plants with intact cotyledon possibly indicates that plants with cotyledon have a better start and can use the available resource efficiently thus increase biomass whereas plants with detached cotyledon, particularly at 7 and 14 DAG have less developed roots and shoot and this might affect their performance at later stage.

## Conclusion

Result of this study indicated that growth of *C. Edulis* has increased with application of nitrogen and phosphorus fertilizers. Application of 92 kg N and 8 kg P $ha^{-1}$ increased height of seedlings by about 134% compared to those of the unfertilized plants. The highest leaf area and total dry biomass of seedlings were obtained by the treatment combinations of 46 kg N with 4 and 8 kg P $ha^{-1}$ that showed increments of about 152% in leaf area and 309% in total dry biomass compared to the unfertilized control plants. With the treatment combination of 92 kg N and 8 kg P $ha^{-1}$ NP, the leaf area and dry biomass weight of the seedlings showed slight decrease in leaf area (22%) and the dry biomass yield (4%) compared to 46/4 kg $ha^{-1}$ and 46/8 kg $ha^{-1}$ NP treatment combination. Thus, the use of 46 and 92 kg $ha^{-1}$ N with 8 kg $ha^{-1}$ P could help to improve growth of *C. edulis* seedlings in areas with low nitrogen and phosphorus status. However, increased levels and split application of the fertilizers particularly that of nitrogen, deserves further investigation in order to enhance growth and establishment of this valuable but very slow growing species.

Moreover, it was evident that *C. edulis* seedlings with attached cotyledon established better and showed enhanced growth compared to early removal of cotyledon. Hence, *C. edulis* cotyledon should be given special attention during early phase of seedling growth for its successful establishment.

## ACKNOWLEDGEMENT

The authors highly appreciate Haramaya University and Sida/SAREC for financial support to carry out this study and also the owner of Hamdail farm, Mr. Mohammed Shito, for allowing his plot of land to be used to conduct the field experiment.

## REFERENCES


Andersson L, Yahya A, Johansson S and Liew J (2007). Seed desiccation tolerance and storage behaviour in *Cordeauxia edulis*. Seed Sci. Technol., 35: 660-673.

Azene BT, Birnie A,Tengnas B (1993). Useful trees and shrubs for Ethiopia, identification, propagation and management for agricultural and pastoral communities. Regional Soil Conservation Unit /SIDA, RSCU, Nairobi, Kenya. p. 474.

Booth FEM, Wickens GE (1988). Non –Timber Uses of Selected Arid Zone Trees and Shrubs in Africa. Conservation Guide, FAO, Rome.

Breen AN, Richards JH (2008). Seedling growth and nutrient content of two desert shrubs growing in amended soil. Arid Land Res. Manag., 22: 46-61.

Corte VB, de Limae Borges EE, Ventrella MC, de Almeida Leite IT, Braga AJT (2008). Histochemical aspects of reserves mobilization of *Caesalpinia peltophoroides* (Leguminosae) seeds during germination and seedlings early growth. Rev. Vore, p.32.

Das BK, Sen SP (1981). Effect of nitrogen, phosphorus and potassium deficiency on the uptake and mobilization of ions in Bengal gram (*Cicer arietinum*). J. Biosci., 3: 249-257.

Drechsel P, Zech W (1988). Site conditions and nutrients status of *Cordeauxia edulis* (Caesalpinioideae) in its natural habitat in Central Somalia. Econ. Bot., 42: 242-249.

Frost I, Rydin H (1997). Effects of competition, grazing and cotyledon nutrient supply on growth of Quercus robur seedlings. Oikos., 79: 53-58.

García-Cebrián F, Martínez JE and Pelegrín EG (2003). Influence of cotyledon removal on early seedling growth in *Quercus robur* L. Ann. For. Sci., 60: 69 -73.

Graciano C, Goya JF, Frangi JL and Gulamet JJ (2006). Fertilization with phosphorus increases soil nitrogen absorption in young plants of *Eucalyptus grandis,* For. Ecol. Mgmt., 236 (2-3): 202-210.

Gutale SF, Ahmed MA (1984). *Cordeauxia edulis* pigment, cordeauxiaquinone, is deposited on bones and may stimulate hemopoiesis in rats. Riv. Tossicol.Sperim. Clin.,14: 57-62.

Hanley ME, Fenner M, Whibley H, Darvill B (2004). Early plant growth: Identifying the end point of the seedlings phase. New Phytol., 163: 61-66.

Hanway JJ, Weber CR (1971). N, P and K percentages in soybean (*Glycine max* (L.) Merrill) plant parts. Agron. J., 63: 286-290.

Hashimi NR, Hughes FMR (2010). The responses of secondary forest tree seedlings to soil enrichment in Peninsular Malaysia: an experimental approach. Trop. Ecol., 51: 173-182

Hossain MA, Hamid A (2007). Influence of NP fertilizer application on root growth, leaf photosynthesis and yield performance of groundnut. Bangladesh. J. Agric. Res., 32: 369-374.

Jeppsson N (2000). The effects of fertilizer rate on vegetative growth, yield and fruit quality, with special respect to pigments, in black chokeberry (*Aronia melanocarpa*) cv. 'Viking'. Sci. Hort., 83: 127-137.

Johnson MB (1996). Vanishing Legumes. Aridus, 8: 1-3.

Kazmi SMA (1979). Yicib- *Cordeauxia edulis*. An important indigenous plant of Somalia which has many uses. Somalia Range Bull., 7: 13-17.

Kidson R, Westoby M (2000). Seed mass and seedling dimensions in relation to seedling establishment. Oecologia, 125: 11-17.

Kjeldahl J (1883). A new method for the determination of nitrogen in organic matter. Z. Anal. Chem., 22: 366 382.

Krigel I (1967). The early requirement for plant nutrients by subterranean clover seedlings (*Trifolium subterrneum*). Aust. J. Agric. Res., 18: 879-886.

Marschner H (1995). Mineral nutrition of higher plants. 2nd edn. Academic Press, London, UK, p. 889.

Mcalister JA, Timmer VR (1998). Nutrient enrichment of white spruce seedlings during nursery culture and initial plantation establishment. Tree. Physiol., 18: 195-202.

Mhango J, Akinnifesi FK, SA Mng'omba and Sileshi G (2008). Effect of growing medium on early growth and survival of tMüell Arg. seedlings in Malawi. Afr. J. Biotechnol., 7: 2197-2202.

Miége J, Miége MN (1978). *Cordeauxia edulis*- A Caesalpinioideae of arid zones of east Africa. Caryologic, blastogenic and biochemical features. Potential aspects for nutrition. Econ. Bot., 32: 336-345.

Milberg P, Lamont BB (1997). Seed /cotyledon size and nutrient content play a major role in early performance of species on nutrient- poor soils. New Phytol., 137: 665-672.

NMA (National Meteorological Agency) (2006). Agromete. Bull., pp.16: 24.

Olaniyi JO (2008). Growth and seed yield of egusi melon to nitrogen and phosphorus fertilizers application. Am. Eurasian J. Sustain. Agric., 2: 255-260.

Olsen SR, Cole V, Watenabe FS and Dean LA (1954). Estimation of available phosphorus in soils by extraction with sodium bicarbonate. USDA Cir. p. 939.

Rai MM (2002). Principles of soil science. 4th E. D. Macmillan, India. p. 400.

Raymond CA, Muneri A (2000). Effect of fertilizer on wood properties of *Eucalyptus globules*. Can. J. For. Res., 30: 136-144.

Ruschel J, Carmello QAC, Bernardi ACC, Carvalho SA and Mattos D (2004). Leaf nutrient contents of rangpur lime rootstock as affected by N, P, K, Ca and S fertilization. Sci. Agric., 61 (5).

Sassaki RM, Felipe GM (1998). Response of *Dalbergia miscolobium* Benth. Seedlings, a cerrado tree species, to mineral nutrient supply. Rev. Bras. Bot., 21: 1.

Seegeler CJP (1983). Oil plants in Ethiopia, their taxonomy and agricultural significance. Center for Agricultural Publishing and Documentation, Wageningen, Netherlands, pp. 110-111.

Simms EL (1987). The effect of nitrogen and phosphorus addition on the growth, reproduction and nutrient dynamics of two ericaceous shrubs. Oecologia, 71: 541-547.

Tanner EVJ, Kapos V, Freskos S, Healey JR and Teobald AM (1990). Nitrogen and phosphorus fertilization of Jamaican montae forest trees. J. Trop. Ecol., 6: 231-238.

Taylor TS, Loewenstein EF (2006). Effect of animal browse protection and fertilizer application on the establishment of planted Nuttall oak seedlings. New For., 32: 133-143.

Uddin MB, Mukul SA, Khan MASA and Hossain MK (2007). Effects of phosphorus fertilizer on seedlings growth and nodulation capabilities of some popular agro-forestry tree species of Bangladesh. J. For. Res., 18: 283-286.

Vaughton G, Ramsey M (2001). Relationship between seed mass nutrients, and seedling growth in *Banksia Cunninghami* (Proteaceae). Int. J. Plant. Sci., 162: 599-606.

Vietmeyer N (1985). In praise of shrubs. Ceres. - FAO- Rev. Agric. Dev., 18: 28-32.

Zimsky M (1990). Using Nitrogen fixing Trees for Human Food. NFTA NEWS. No 11.

# Innovation dynamics in cassava production systems in Uganda

Deborah Wendiro[1], Watu Wamae[2], Anne Kingiri[3], Margaret Dhabangi[1] and Paul Alex Wacoo[1]

[1]Microbiology and Biotechnology Centre of Excellence, Department of Product Development, Uganda Industrial Research Institute, P. O. Box 7086, Kampala, Uganda.
[2]The Open University Faculty of Mathematics, Computing and Technology, TheOpen University, Milton Keynes, MK7 6AA, UK.
[3]African Centre for Technology Studies P. O. Box 45917 – 00100 Nairobi, Kenya.

**Traditional knowledge has made appreciable contributions to people's sustenance and livelihoods. Its contribution to science and technology is however not recorded, codified, stored or systematized to spur knowledge sharing and science and technology development. It continues to be ordinary, couched and associated with low prestige rural life. An innovation systems framework was used to study the dynamics and mechanism for product, process and organizational innovations in the cassava production systems. The research study revealed that though some traditional knowledge driven innovations may be risky to health and environment; many made a positive contribution to people's sustenance and livelihoods through production of innovative goods and services, improved livelihoods, sustenance, food safety and wholesomeness. The main argument in this study was that innovation strategies rooted in the traditional knowledge systems were socially inclusive and augurs sustainable development. The study underscored the value of creating systemic linkages useful in integrating traditional and modern knowledge systems to develop crop production systems.**

**Key words:** Cassava production systems, rural livelihoods, social inclusion, systems linkages, traditional knowledge.

## INTRODUCTION

There has been an obsession with external sources of innovation embodied in imported technologies (Hall and Nahdy, 1999) and machinery as a panacea to development problems; as is reflected in the strategy documents such as the Plan for Modernization of Agriculture (PMA) (MAAIF, 2000; MAAIF and MoFPED, 2000) and the National Agriculture Advisory Services (NAADS, 2001). Top-down innovation processes have dominated development efforts to fix social, economic problems such as food security (Bolwig et al., 2004), health (Buyinza, 2008) and environment degradation (Buyinza, 2010), and agricultural productivity with varying levels of success (Friis-Hansen, 2005; Friis-Hansen et al, 2004; Bibangambah, 2002). The main argument in this paper is that a lot of innovation activity occurs within the traditional systems that should be identified and nurtured to benefit the wider community and also for policy development. The paper is organized as follows: the introduction setting the agenda; description of the research methodology used to study the traditional innovation systems; description of results and findings and; conclusion.

At onset we argue that ignorance of innovation activities within the traditional knowledge systems constitute missed opportunities for intensive social, cultural and economic transformation. It is further shown that subsistence agricultural production systems draw on traditional knowledge during production activities and have made significant contribution to community sustenance and livelihood, and even thrived. This is based on the premise that the prosperity of any production system depends on its productivity and value created, and that innovation plays a central role in productivity enhancement and value creation. This means that innovation occurs in subsistence agricultural production systems. What is puzzling is why these innovations are depicted as ordinary, couched and associated with low prestige rural life, and are portrayed as inferior. It is suggested that one of the reasons for this is the unobtrusive profile of traditional STI which impacts their utility and affects knowledge sharing and development.

This state of affairs should not continue to be overlooked. In order to understand the dynamics of innovation in the traditional knowledge systems therefore, a holistic approach is essential. The major issue of contention is that traditional innovation systems are a neglected yet critical part of the innovation system that despite disregard has been blooming and hold potential for cultural, social and economic transformation. In order to support the argument, this paper reviews the innovation dynamics in the cassava production systems highlighting novel practices that could be enhanced and replicated elsewhere and used as foundation for policy development which is an organic outcome of the social economic milieu.

The agenda for discussion of this concept has just been set. Next is a presentation of the innovation systems framework as a holistic framework useful for texturing the profile of traditional STI.

## The innovation systems framework

The innovation systems framework identifies the factors enabling knowledge transactions in the innovation processes. An innovation system is defined by the International Bank for Reconstruction, (The World Bank, 2006) as a network of organizations, enterprises, and individuals focused on bringing new products, new processes, and new forms of organization into economic use, together with the institutions and policies that affect their behavior and performance. The innovation systems concept embraces not only the science suppliers but the totality and interaction of actors involved in innovation. It extends beyond the creation of knowledge to encompass the factors affecting demand for and use of knowledge in novel and useful ways. The innovative performance depends not only on how the individual institutions (e.g. organizations, research institutions, universities) perform in isolation, but on how they interact with each other as elements of a communal system of knowledge creation and use, and on their interplay with social institutions e.g., values, norms, and legal frameworks.

Any attempt to bridge the two systems should look at the actors, the processes of knowledge creation and interaction, and the knowledge itself, which is the essence of an innovation systems framework (Bell, 2006). The innovation systems framework therefore provides us with an analytical framework to examine types of relations (Hall and Yoganand, 2004); the network of relationships among firms and the broader institutional setting that supports their innovative activities. The dynamic and cumulative nature of the innovative process is emphasized, since it is based on tracing the flows of knowledge among institutions constituting the system (Wolfe, 2000 in Adam and Wolfe, 2000 Eds.), and analyzing civic uniqueness in the innovation process in a wider economy (Sharif, 2005).

Applying the innovation systems framework, a textured description of the organization and patterns of activity that contribute to innovative behavior is done. Institutions and actors who play decisive roles in the innovation processes would be identified. Yet various factors affect choice of production possibilities (Gans and Stern, 2003; Natario et al., 2011), including the social, economic, cultural setting of the community, physical environment, land use knowledge systems (Mikkelsen and Langohr, 2004); and technological innovation capacity[a] (Sharif, 2005; Asheim, 2007; Borras, 2011; Natario et al., 2011).

Several factors affect conversion of knowledge into innovation as discussed conclusively by Furman et al. (2002) Gans and Stern (2003), and Natario et al. (2011). Mikkelsen and Langohr (2004) moreover argued that the traditional sustainable approach takes great care of the

[a] National innovative capacity is a country's potential—as both a political and economic entity—to produce a stream of commercially relevant innovations. This capacity is not simply the realized level of innovation but also reflects the fundamental conditions, investments, and policy choices that create the environment for innovation in a particular location or nation. National innovative capacity depends in part on the technological sophistication and the size of the scientific and technical labor force in a given economy, and it also reflects the array of investments and policy choices of the government and private sector that affect the incentives for and the productivity of a country's research and development activities (Porter and Stern, 2001).

particular needs of plants and animals and is closely adapted to an instable shifting physical environment. Therefore, the actors, the processes of knowledge creation and interaction, and the knowledge itself (Bell, 2006) have to be examined by tracing the flows of knowledge among institutions constituting the system (Wolfe, 2000 in Adam and Wolfe, 2000 Eds.), and analyzing civic uniqueness in the innovation process in a wider economy (Sharif, 2005) using an intensive approach, which will be discussed later.

The above definition of innovation systems framework shows that in order to understand and prove their worth therefore, information is required to demonstrate the value of traditional STI and constraints, and shade light on how to integrate traditional and modern knowledge for the benefit of marginalized communities; and get answers to the following questions. How the traditional knowledge is applied upon natural and social phenomena for sustenance and livelihoods? What opportunities exist for integration with modern knowledge? What linkages should be established and how this should be done? The information presented in this paper was obtained through case study of cassava production systems in Arua district of Uganda.

## Background and justification

Despite a lot of investment in biotechnology to increase its productivity (Abele et al. 2007; NARO, 2000), cassava production activities have continued to thrive on knowledge transactions within traditional knowledge systems. Using an innovation systems framework, the innovation processes are review-ed in this article. Innovation is the key to development through production of goods and services (Watkins and Ehst Eds.) with increasing scientific and technological content, which transcend local confines into national, regional and even global frontiers (Furman et al., 2002; Gans and Stern, 2003; Natario et al., 2011).

Furman et al. (2002) discussed determinants of national innovative capacity into three categories: the common pool of institutions, resource commitments, and policies that support innovation across the economy; the particular innovation environment in the nation's industrial clusters; and the linkages between them. Natario et al. (2011) defined it as the real and potential capabilities of a system to convert knowledge into innovation. It is based on scientific knowledge base.

This capacity is determined by several factors discussed conclusively by Furman et al. (2002); Gans and Stern (2003), and Natario et al. (2011). In order to understand the dynamics of innovation in the traditional knowledge systems therefore, a holistic approach is essential. The holistic approach is precisely described by the innovation systems framework.

The innovation systems framework aims to identify the factors enabling knowledge transactions in the innovation processes. In this article, information was obtained through case study of cassava production systems in Arua district of Uganda as discussed later. Next is a discussion of the methodology for generating the information.

## METHODOLOGY

The study was done to identify innovations and dynamics of innovation in the traditional knowledge systems using an innovation systems framework. A multistage methodology was used involving literature search, key informant interviews, questionnaire, focus group discussions, and observation. Information was collected in the frontier district of Arua which was selected due to its closeness to regional markets and hence demand opportunities, and poverty and vulnerability issues.

The district leaders and key informants were interviewed using a questionnaire and information generated from them use to develop a sampling frame (Figure 1). The district leaders including the District Chairperson or Chairperson of Local Council Five[b] (LCV), Chief Administrative Officer (CAO), Production Officer, Resident District Commissioner[c] (RDC), and any other government officers or political leaders refereed by them were interviewed using a questionnaire. The sampling frame was developed as summarized (Figure 1).

This methodology generated background information to establish the context, identify innovation spots, innovators and production activities. The district leaders or key informants identified innovation spots by sub-county or village, and innovators by name. In this way a list of names and innovations which constituted the sampling frame was developed from which a sample was selected for interview. The decision on which product to study was then taken as discussed in the next paragraph.

### Selection of product to study

A product was selected out of fifteen identified product categories for in-depth study using a checklist. The criteria for selection of products for case study was: social, cultural factors such as gender issues, potential for sustenance and livelihood, and environment issues; opportunities provided by trade and availability of natural resources; prospects for deepening value chains (opening way for other uses); available technological capability; strategies to exploit new opportunities, and the professional orientation of the MSc. student to carry out the scientific aspects of the case study. The product category of cassava production and processing using heap fermentation technique was selected because it is a critical income generating activity in the area, which is typically rooted in the traditional agricultural systems focusing on ensuring both food security and livelihoods sustainability. This category fulfilled all the requirements stipulated in the above criteria.

A case study approach was then used to get in depth information about modes of knowledge generation and determine how the communities innovate through product, process, organizational changes and services for their sustainability and trade. Previously, methodology used to generate information is described, and followed by presentation and discussion of the findings of the study.

[b] The District Chairperson also known as the Local Council Five Chairperson is the district political head

[c] The Resident District Commissioner is a government or presidential representative in the district

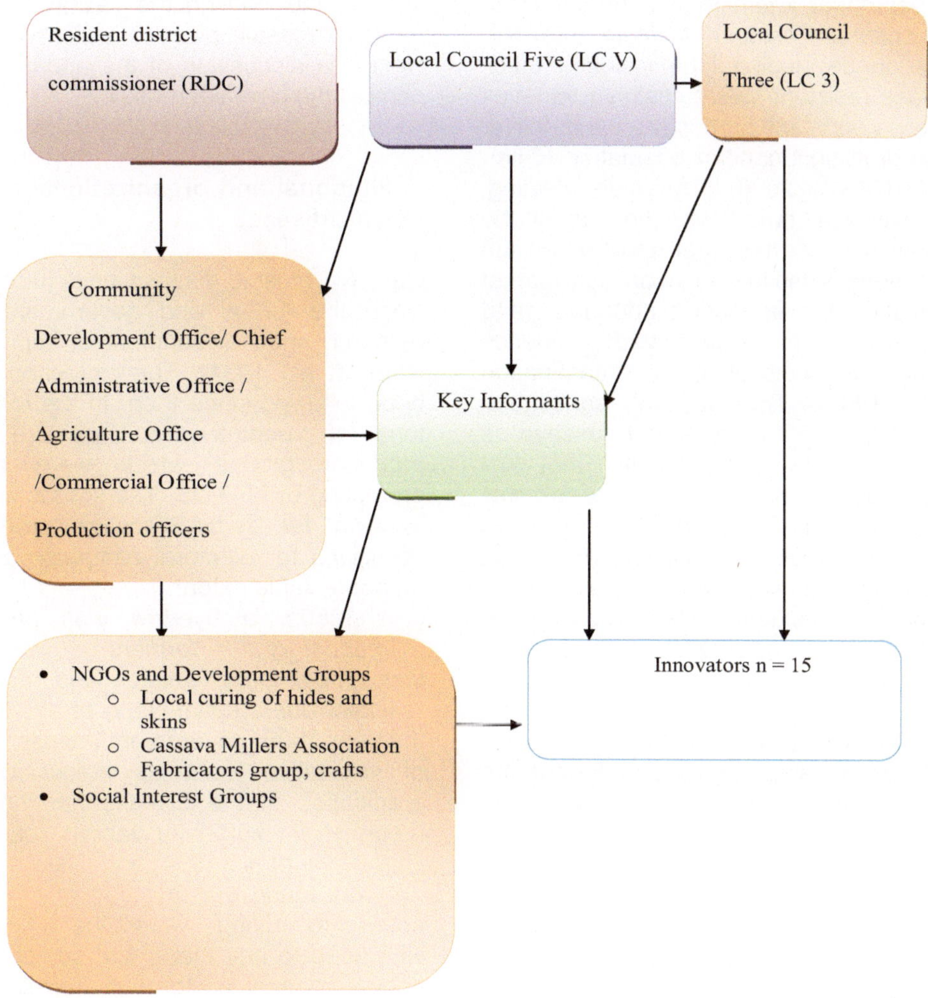


**Figure 1.** Sampling frame.

## RESULTS AND DISCUSSION

Here we report and discuss the findings of the study that was implemented to assess the traditional knowledge systems and identify innovations in the cassava production systems in Arua district of Uganda. It was found out that despite a lot of investment in biotechnology to increase productivity of cassava production systems (Abele et al. 2007; NARO, 2000) cassava production activities have continued to thrive on knowledge transactions within traditional knowledge system as further revealed in this research.

The findings are discussed under two themes starting with activities within the formal or modern innovation systems which are guided by government policy. The innovations in the traditional systems are further discussed, and in all scenarios highlighting benefits, novel attributes of the identified innovations, constraints and risks. The activities within the formal innovation systems are discussed as well.

### Activities within the formal innovation routes guided by the government

Information regarding the formal innovation systems was collected using literature search and key informant interviews. It was found out that innovation occurring in the formal system is planned (Friis-Hansen E, 2005; Friis-Hansen et al, 2004), structured to be responsive to macroeconomic conditions as devised by the policy makers at national level (Hakiza et al., 2004) as laid down in the strategy documents including the poverty eradication action plan (PEAP), plan for modernization of agriculture (PMA), and national agricultural advisory services (NAADS). The activities in cassava production are mainly driven by the National Agricultural Research Organization (NARO), an agency composed of highly trained and skilled personnel, and high tech infrastructure in Namulonge Agricultural Research Institute (NAARI). The main focus is on developing high yielding, quick maturing, and disease resistant sweet cassava varieties in order to fight disease, increase productivity; and deliver food security to the growing Ugandan population (Otim-Nape et al., 1994; NARO, 2000). The high yielding, disease resistant cassava sweet varieties must be harvested at maturity because they get spoilt when left for long in the soil. NARO capitalized upon this market oriented characteristic to enhance adoption and discouraged the growing of *'unimproved'*[d] varieties arguing that it causes problems of phyto-contamination (Otim-Nape et al., 1994). The national agricultural advisory services (NAADS) promoted sweet cassava as a way of increasing productivity, an objective that was *technically*[e] realized during the initial years of this program (Abele et al. 2007; NARO, 2000).

It was envisaged that increased production would lead to higher household incomes since farmers have plenty of sweet cassava that they had to market. It was reported (NARO, 2000) that there were plenty of cassava crops during the initial years of the biotechnology program, which report was also corroborated by the farmers during the research study. Due to the varietal attribute, however, the cassava had to be harvested at once and flooded the market. Increased total output led to very low prices (up to 50 shillings per kilogram in some areas! Bibangambah, 2002). This was complicated by the fact that the market for cassava changes with consumer income or availability of substitutes. Fresh cassava is only consumed as an inferior subsistence food for low income earners. Sweet cassava is moreover only eaten as a snack by some ethnic groups such as those in Arua, and has few alternative uses both locally and nationally. The local people claim that the sweet varieties do not make good fermented products, so do not fit in with their food preparation methods, and the local market requirements. Efforts to link production to manufacturing are dulled by the weak manufacturing base since there are few industries using cassava as their raw material.

At factory level moreover, processors decry the low volumes and high bulking cost in the context of small subsistence producers and few supplier Associations (Nkonya, 2002). It is still hard for farmers to access the market offered by the few factories. The farmers thus realize a lot of post-harvest loss (Mukadasi and Lusiba, 2006) and their aspirations for sustenance and improved livelihoods were thwarted. The years following the initial boom were bleak and cassava production plummeted (Otim-Nape et al., 1994). At that very time the innovation system was battling the cassava diseases such as the cassava mosaic disease which complicated the scene. Next is a discussion of the approach used by NARO to combat the cassava mosaic.

[d] Emphasis ours

[e] In technical terms there was increased cassava yield and more produce that was recorded and even observed in the markets, but farmers got a raw deal. Other reports show that this was due to increased use of more land not on increased per unit production ()

## Institutional and organizational response to cassava mosaic disease

Cassava mosaic disease epidemics have been reported since the 1920s and government intervened through research and development of improved varieties (Otim-Nape et al., 1994). The biotechnological approach has been followed since then. In the formal line, the seed in form of vegetative cuttings are distributed through producer groups which were developed following prompting by NAADS. The groups were required to pay a *'nominal'* fee by NARO which *'progressive farmers'* and politicians for electoral campaign purpose embraced to an appreciable extent.

In 1950s, a bye-law was passed where it was *"...mandatory for farmers to uproot all infected and susceptible local varieties and replace them with new resistant ones"* (Otim-Nape et al., 1994). Furthermore in the year 2009, government imposed a ban on movement of vegetative materials across districts. Success in eradication of the disease has however been elusive (Legg et al., 2002) for various reasons as reviewed by Otim-Nape et al. (1994) and Oleru et al. (2005).

It was found out that due to the structured nature of access to cuttings within the NARO program, coupled with need to buy them, and complicated by poverty and habits of access through familiar channels, farmers responded through parallel interventions applying traditional know-ledge and other social economic arrangements. These arrangements have not been acknowledged before, yet they have been central to ensuring people's livelihoods.

In essence, traditional knowledge has assuaged social, economic, science and technology development pro-blems as will be discussed next. We have discussed the approaches used in the modern innovation systems to manage cassava production and the outcome of these strategies on people's livelihoods and productivity, the producers in traditional knowledge systems in Arua district of Uganda developed innovations based on inherited experience gained over many generations and passed on verbally from generation to generation. These innovation processes are also discussed in this study.

## Innovations based on traditional knowledge

It has been revealed in the previous discussion that innovation within the formal sector is structured and responsive to macroeconomic conditions as devised by the policy makers at national level. It requires high outlay of inputs and often undertaken within research institutions that are removed from the mainstream network both in geographical and structural or institutional terms veiling them with an aura of superiority alien to the common person thus rendering them inaccessible. The innovation trail occurring in the informal system is spontaneous through selection of strains that are observed to possess required attributes, occurring within familiar context, and responsive to market needs and social, economic and cultural milieu.

In the traditional systems, seeds are distributed through family where an individual can enter the garden of a relative and pick cuttings in some cases even without prior consent. Clan linkages and neighborhood alliances transcend boarders and legal frameworks; and are often not through the monetary economy but bartering systems to the extent of exchanging different types of seeds.

This study reveals that while operating under tension of restrictive bye-laws, local people have achieved major breakthrough. In some cases, even accessing the international market and purposively seeking knowledge from the formal sector, showing that the system is viable. Next is an exemplification of the viable innovations that have been developed within the traditional innovations systems that could be nurtured.

### Selection of a mixture of cassava varieties

It was found out that regardless of the fact that NARO promotes sweet cassava varieties, local people continued to access bitter varieties even going as far as Congo to get them. A producer will plant a variety with its use and even ecological properties in mind.

It is believed that different varieties react differently to weather changes, soil conditions and cropping patterns even intercrop biotic factors. Seven '*varieties*' (which technical people referred to as strains of mainly three varieties) of bitter cassava each having its unique attributes were identified. These included varieties known by the following local names: *abiria, arua, omi, tongolo, mabulu, basimenge and mbazia.* The bitter varieties are preferred for fermentation. The variety locally called *abiria* can be dried in the shade. The producers say it is because it has high starch content, is hard and has low water content. If dried in the sun, it takes long to ferment and the product will not be good due to contaminants. When the cassava is matured, local people then select what to process into each specific product. The sweet varieties are preferred for cooking to eat as snack and as feed to domestic animals like pigs and goats.

### Control of cropping patterns

The NARO required that improved varieties should be planted on single stands to avoid phyto-contamination by the already infected local varieties. Producers' innovations take into consideration the varying ecological conditions in this rain fed agricultural system where farmers have little control and have learnt to adapt their cropping systems to weather and existing social economic and cultural needs. Local people continued to intercrop and all varieties of cassava are grown on same plot. The farmer plants cassava staggered all year round and harvesting is also done all year round. Where weeds become a problem, the varieties that form a canopy are grown to choke the weed. Intercropping is a main attribute of the farming system, whereby different varieties of cassava and other crops like ground nuts, beans, peas, maize, simsim, millet and sorghum are grown and each cassava variety serves a different purpose and occupies a specific niche in the ecological system. The bitter varieties are also planted at the fringes of the garden as a barrier to prevent domestic animals from browsing on the leaves of the more susceptible sweet varieties.

### Organizational innovations

Organizational innovations aimed at enhancing the productivity, marketing and ensuring food security were developed through social, economic and technological arrangements rooted in traditional STI and blended with modern approaches. Bitter cassava cuttings are sourced by the local people from neighboring regions, even Democratic Republic of Congo (DRC). A local producer organization known as the West Nile Cassava Millers Association (The Association) established according to the NAADS requirements, trains processors in improved processing methods, recruits suppliers, packages the flour and labels it for bulk transport.

The Association helps to develop products, standardize, and market them. The interesting finding is that, in order to ensure consistent product quality, the Association identified farmers and taught them how to ferment and dry the product; use racks or plastic sheets to spread out to dry on soil free surface and provided carpets for drying; wash with water instead of scrapping off the black spores with knives after fermentation is complete. The varieties that are preferred are the bitter ones since they fit in with the local food habits, food preparation and presentation.

It was also found out that market (effective demand) is a critical driver of uptake of new and improved technology. Through the Ministry of Disaster Preparedness in collaboration with the World Food Programme (WFP) of the World Bank and the NAADS, an initiative to get food from local suppliers to people in Karamoja region of Uganda who were affected by famine started. It is significant to note that the Association competed for and was awarded a contract to supply 250 tons of cassava flour. They also succeeded in servicing that contract through supply of cassava flour; which is a marketing innovation based on local bitter cassava detoxification technology. Through the incentive motivated by the market and skills acquired, the Association also got other outlets in Kampala; the product range includes: Anyafura or cassava flour composed of 50/50 fermented cassava and millet; millet flour; and gari porridge. A standard for the cassava flour was developed in collaboration with the Uganda National Bureau of Standards (UNBS). The Association even took samples of their products to Kampala (a distance of 500 kms away) for analysis at Makerere University then Department of Food Science and Technology to demonstrate quality, thus fulfilling the regulatory and trade requirement.

The producers are given a price which is less by 20% of the one offered by the buyer. This has benefitted not only individual farmers and households as a group of women respondents asserted that "...one can always get market for the product whether with the Association, Schools or individual consumers..." but has also benefited the government agencies like UNBS, technology generating institutions like Makerere University. Farmers as individuals and in groups are engaged in cassava growing and selling business. They sell to schools, restaurants, cafeteria, millers - the West Nile Cassava Millers Association (the Association) - and to individual consumers. This finding is fertile learning ground for policy and strategies for modernization of agriculture.

## The range of product innovations

Cassava is processed into several products for different markets. The popular products currently made out of cassava include starch, animal feed; heat gelatinized cassava gum, ethanol, and cassava flour either from fermented or unfermented cassava (Graffham et al., 2000).

The bitter cassava varieties are traditionally detoxified either through soaking, drying or aided spontaneous heap fermentation (Essers, 1996). Fermented cassava is a product with high demand in local urban centers. When the quality of the product became critical in terms of safety and standardization, particular skills and forms of knowledge were required to ensure high quality products and test methods to demonstrate safety from aflatoxins[f] as a trade and regulatory requirement. Producers sourced bitter cassava cuttings from neighboring regions. Quality in terms of consistence and safety in terms of aflatoxin content remain issues of concern. Thus, the experience of the Association is very informative.

## Response to biotic and environment factors

Producers carry out holistic innovations across a production system value chain. Innovative adaptations are made in the farming system. Where weeds become a problem, the varieties that form a canopy are grown to choke the weed, which has reduce the need and cost of weeding. The cropping system is adjusted to exploit varietal features relative to microenvironment, and adapt cropping pattern to weather patterns. A producer will plant a variety with its use in mind. With increase in population and density on arable land, and also competition for different uses, that is, tobacco and cassava, soil quality is affected due to heavy and continuous use. Depleted soils seem to support growth of certain weeds and neither favor cassava sweet varieties nor cereals.

Bitter cassava varieties often give equitable yields on depleted soils. It was also observed that after tobacco is harvested the soils have high nitrogen, phosphorus and potassium (NPK) fertilizer content; sweet varieties may become bitter and will rot very fast in such high salinity soils (Yang et al., 2011); an issue of biological significance. Accumulation of nitrogen in form of bitter substances called cyanogenic glycosides may be part of the nitrogen cycle in the ecosystem that plants use efficiently for other ecological needs like stress resistance and protection against pests and diseases. The cassava producers said that when the soils become depleted, they change the varieties often replacing sweet varieties, possibly because they require a more balanced soil profile, with the bitter ones. What is not clear is whether this shift is good for the soil; and why in two apparently dissimilar circumstances; when there is too much NPK and when the soils are depleted, cassava plants seem to respond similarly by accumulating bitter substances.

## Response to economic factors, sustenance and livelihoods

A dual strategic course is being followed by the local community. The NARO are implementing the biotechnology strategy to increase productivity. The farmers adopted the NARO varieties and also continued to plant the local varieties and developed innovative processing technologies, looked for new markets, added value to their products through standardization and UNBS underwriting. The Association managers looked for the required information.

The NARO strategy to increase productivity was deemed successful by national standards because there was a lot of cassava on the market and prices were low to consumers hence delivering food security. The low

[f] Aflatoxins are toxin produced by some molds in crops, especially peanuts and also fermented cassava

***Cassava heap fermentation technique***

*According to the European Commission (2007), the global market for bio-based products is estimated to grow to $ 250 billion by 2020. This knowledge led to selection of fermented cassava as a product for in-depth study. It was found out that detoxification of bitter cassava through aided spontaneous heap fermentation starts with harvesting the roots, which are peeled and sundried for one to three days depending on sun intensity. The producers gauge the dryness of the cassava by breaking several pieces, and based on skill determine the right stage to initiate the fermentation stage. It is known that adequate drying is necessary to avoid growth of spoilage microbes some of which are aflatoxin formers. Spontaneous inoculation occurs as a result of handling which includes peeling without washing and drying on the ground. After sun drying, the cassava is heaped in a corner usually in a hut and covered with sack cloth or dried banana leaves. Microorganisms will grow on it for a period of three to four days. The producer will determine the right stage to stop this process when she observes black/pink spores of moulds on the roots and also tests by pressing several roots between the thumb and middles fingers for softness. When ready, the roots crumble when pressed. The spores are then scrapped off using a knife. The roots are then sun dried and milled into flour. The roots can also be kept as sun dried chips until required for food or sale.*

**Box 1.** Cassava heap fermentation technique.

prices that ensued in response to market forces however meant that farmer's expectations were thwarted. The farmers became prey to national success. Due to the varietal attributes of the improved varieties, producers had no control over the crop and the market was small. There were no alternative outlets since the sweet cassava value chain is shallow in Uganda. Attempts to find alternative uses were not fruitful. The farmers resorted to their traditional varieties and processing technologies (Box 1), which later were boosted by the WFP market. However, this posed a demand for specifications of the product for which the farmers could not provide the necessary information. At this point, they needed more skill and knowledge to solve the technological and business related problems. The UNBS intervened to develop standards for the products. The extent of this intervention was only at the level of documentation, the Association played the role of technology development and product description. The Association managers looked for the required information. There is an opportunity for intervention through collaborative linkages with a research and development agency to give the required information for improved inoculums for use as starter cultures to enhance colonization of the mass before contaminants have chance to grow; and developing a diagnostic recipe for aflatoxin in locally processed fermented cyanogenic cassava flour, which issues are being studied.

## Social and cultural factors e.g. gender, community organization and relationships

The NAADS organized groups for development initiatives. Value chains are being developed by the Association through organizing individual farmers into processor suppliers. They are trained and skilled in quality assurance concepts. The availability of innovations for all to use, and continuous improvement is the essential characteristic of knowledge diffusion. In traditional systems, the users of such knowledge pay a form of royalty like direct payment by bartering with products, such as giving back some of the harvest later or bailing the person out when they are also in need. The cost depends on how close the relationship or friendship or the assumed value of the knowledge, product or innovation.

## Nutrition innovations and benefits

Fermentation technology can prolong shelf life and ensure food security and safety of the bitter cassava products. The product must be thoroughly processed. If not well done incidents of poisoning are not rare. The local people said that sweet varieties are not good for fermentation because the product can easily be contaminated with aflatoxin forming molds with black charcoal like spores which cause vomiting, headache and diarrhea when consumed. It is not clear whether presence of cyanogenic glycosides has an effect on the aflatoxin formers and what this effect is. It was further reported that the handling methods can lead to too many stones in the product and cases of appendicitis are many in the area.

### The heap fermentation technology

Farmers detoxify the bitter cassava by a heap fermentation process (Box 1). It is not clear where and who developed the heap fermentation technology. The heap fermentation technique improves nutritional and commercial value but it is not efficient. It requires a long time to completely detoxify the cassava, which is limiting during periods of food shortage. Since it is not standardized, it leads to products of variable quality and safety.

These improvements are however not enough to ensure a safe product of consistently high quality. The people training also lacked information on the biological aspects of the process. There was argument regarding signs of good quality such as which type of mold was the best-the pink or the black spores mold? Some processors argued that it is the process conditions that determine which type of spores are produced, but the best product is when spores are black. This observation is in tandem with known biology of some of the molds that occur in more than one morphological form (Perkins and Turner, 1988; Dettman et al., 2001). The need for improved product quality prompted a research study to develop a starter culture as an appropriate value addition technology to enhance quality of the product and later develop a process. With this as entry point other improvements such as formulation of composite flour for baking industry will be made. The innovations, benefits, novel attributes of the identified innovations, constraints and risks within both modern and traditional innovation systems that were identified have been previously defined. The findings have demonstrated that rooting innovation, technology development and industrial development in the traditional systems is inherently valuable and strategically pragmatic. When managed and coordinated, the bottom up and inclusive development process for long hyped will be achieved. Conclusions were drawn from the findings and synthesis.

## CONCLUSIONS AND RECOMMENDATIONS

### Familiar contexts for integration of traditional innovation systems into the modern systems

This paper describes novel contexts for development that is rooted in the traditional STI centered on knowledge and skills which originate in different bodies of knowledge and representations. A combination of practices, with biological and physical processes, social and economic contexts interacting at every level creates innovation. The need to reach beyond the biotic factors (e.g., mycological) and conservation approaches to integrate traditional innovation into agriculture, forestry, manufacturing and other land management practices, poverty reduction initiatives and development strategies is indicated. Ignorance of innovation activities within the traditional knowledge systems constitutes missed opportunities for intensive social and economic transformation. Subsistence agricultural production, processing and manufacturing systems draw on traditional knowledge during production activities and have made significant contribution to food security and livelihood.

The paper further reveals critical entry points for knowledge transactions for example variable product quality due to un-standardized and uncontrolled process conditions and safety risk due to aflatoxins are issues that require collaborative efforts. Likewise, critical stage for stopping the fermentation and techniques of process control needs to be developed. To demonstrate safety from aflatoxins, moreover, a simple diagnostic recipe is required as a trade and regulatory aid to show compliance to standards. The example of organizational innovations and how local people adopted new technologies within familiar contexts clearly shows how modernization of agriculture and industrial growth and development is possible starting at local level.

Critical policy and procedural issues were also isolated including: disjointed top down interventions that are not targeted to local people's needs and perception; top down research agenda setting that is not demand driven and fails to address community needs. It shows that improved productivity is not only key driver of modernization of agriculture, raising income and food security but a medley of factors and resources systematically structured synchronized in tandem. It also demonstrated that sustainable income from agriculture or other sources is necessary for food security and improved livelihoods which can only be assured if there is effective demand.

It is finally concluded that linkages between the modern and traditional knowledge systems should be enabling and based on mutual benefit and not intrusive and exclusive.

### Conflict of Interest

The authors have not declared any conflict of interest.

## ACKNOWLEDGEMENTS

The author would like to acknowledge support of the

Canadian International Development Research Council (IDRC) who funded the study. This publication is an output from a project (Grant No. 105359-005). Dr. Watu Wamae is specially acknowledged and thanked for her significant contribution to the success of this study through training on innovation systems framework. The author also thanks the district leaders and innovators involved for both their time and their information. The author would also like to acknowledge the Uganda Industrial Research Institute (UIRI) Administration and the University Lecturers who supervised the study.

## REFERENCES


Abele S, Twine E, Ntawuruhunga P, Baguma Y, Kanobe C, Bua A (2007). Development and Dissemination of Improved Cassava Varieties in Uganda: Analysis of Adoption Rates, Variety Attributes and Speed of Adoption AAAE Conference proceedings, pp. 479-482.

Asheim B (2007). Differentiated knowledge bases and varieties of regional innovation systems. Innovation 20 (3) Available at: http://www.nfa.org.ug/docs/forests_functions_and_classification.pdf. Accessed: 05/01/2012

Bell M (2006). Background discussion paper for the L20 workshop Communiqué. Science and Technology for Development Conference.

Bibangambah JR (2002). Review of Information on Marketing, Processing and Storage of Uganda's Agricultural Commodities'. Final report for PMA sub-committee on agro-processing and marketing.

Bolwig S, Omamo W, Kiiza B (2004). The effects of local market institutions on technology adoption and productivity among small holder farmers in Uganda. Department of Development Research Danish Institute for International Studies. A research Project Description, RUF/DANIDA, Project No. 91201

Borras S (2011). Policy learning and organizational capacities in innovation policies. Sci. Public Policy *38*(9):725-734.

Buyinza M (2008). Human ecology and household socioeconomic Determinants of community forestry projects in Uganda. Pak. J. Social Sci. 5: 280-288

Buyinza M (2010). Poverty alleviation and environment management. Res. J. Appl. Sci. 5(2)A:101-107.

Dettman JR, Harbinski FM, Taylor J (2001). Ascospore morphology is a poor predictor of the phylogenetic relationship of Neurospora and Gelasinospora, Fungal Genet. Biol. 34: 49-61. PMid:11567551

Essers AJA (1996). Removal of cyanogens from cassava roots studies on domestic sun – drying and solid – substrate fermentation in rural Africa. PhD – thesis at Wageningen Agricultural University, The Netherlands.

Friis-Hansen E (2005). Agricultural development among poor farmers in Soroti District, Uganda. Impact assessment of agricultural technology, farmer empowerment and changes in opportunity structures. In paper presented at Impact Assessment Workshop at CYMMYT 19:21.

Friis-Hansen E, Aben C, Kidoido M (2004). Small holder agricultural technologies development in Soroti district: synergy between NAADS and farmer field schools. Uganda J. Agric. Sci. 9:250-256.

Furman JL, Porter ME, Stern S (2002). The determinants of national innovative capacity. Res. Policy 31:899-933.

Gans J, Stern S (2003). Assessing Australia's Innovative Capacity in the 21st Century: Intellectual Property Research Institute of Australia.

Graffham A, Kleih U, Jagwe J, Nabawanuka J, Wanda K, Kalunda P , Ntibarikure G (2000). Industrial markets for starch-based products: An assessment of the potential for cassava in Uganda Natural Resources Institute (NRI), FOODNET / International Institute of Tropical Agriculture (IITA) and Kawanda Agricultural Research Institute (KARI) Exchange 6

Hakiza JJ, Odogola W, Mugisha J, AR S, Nalukwago J, Okoth J, Ekwamu A (2004). Challenges and prospects of disseminating technologies through farmer field schools: Lessons learnt based on experience from Uganda. Uganda J. Agric. Sci. 9(11)163-175.

Hall A, Nahdy S (1999). New Methods and Old Institutions: The systems Context of Farmer Participatory Research in National Agricultural Research Systems: the Case of Uganda. ODI.

Hall AJ, Yoganand B (2004). New institutional arrangements in agricultural research and development in Africa: concepts and case studies. Innovations in Innovation: reflections on partnership, institutions and learning, pp.105-131.

Legg JP, French R, Rogan D, Okao-Okuja G, Browns JK (2002). A distinct *Bemisia tabaci* (Gennadius) (Hemiptera: Sternorrhyncha: Aleyrodidae) genotype cluster is associated with the epidemic of severe cassava mosaic virus disease in Uganda. Mol. Ecol. J. 11:1219-1229.

MAAIF (Ministry of Agriculture, Animal Industry and Fisheries), (2000). The National Agriculture Advisory Services program: Master document of the NAADS taskforce and joint donor group. Kampala.

MAAIF and MoFPED (Ministry of Finance, Planning and Economic Development (2000). Plan for modernization of agriculture: Eradicating poverty in Uganda. Government strategy and operational framework. MAAIF and MoFPED. Kampala.

Mikkelsen JH, Langohr R (2004). Indigenous knowledge about soils and a sustainable crop production, a case study from the Guinea woodland savannah (Northern Region, Ghana). Danish J. Geogr. 104 (2):13-26.

Mukadasi B, Lusiba B (2006). Farmers perception of the relevance of agricultural technologies under Plan for Modernization of Agriculture in Uganda. Uganda J. Agric. Sci. 12(2):7-15.

NARO (2000). Nation Agricultural Research Organization Annual Report 1998-1999

Natario MMS, Couto JPA, Tiago MTB, Braga AMM (2011). Evaluating the determinants of national innovative capacity among European Countries; Global J. Manage. Bus. Res. 11(11):version 1.0.

National Agricultural Advisory Services (2001). Available at http://www.naads.or.ug/files/downloads/The%20NAADS%20Act%202001.pdf Accessed on 17.09.2014

Nkonya E (2002). Uganda crop market characteristics, constraints and opportunities. Strategic Criteria for Rural Investment in Productivity SCRIP report, IFPRI, Washington DC.

Oleru J, Semana AR, Bua A, Mutimba J, Obaa B (2005). Assessing the level of farmers' participation in the choice of enterprises promoted by NAADS in Arua district, northwestern Uganda. Afr. Crop Sci. Conf. Proc. 7:725-731.

Otim-Nape GW, Bua A, Baguma Y (1994). Accelerating the transfer of improved technologies: Controlling African Cassava Mosaic Virus Disease Epidemics in Uganda. Afr. Crop Sci. J. 2(4):479-495.

Perkins DD, Turner BC (1988). Neurospora from natural populations: Towards the population biology of a haploid eukaryote. Experimental mycology. 12: 91-131.

Porter ME, Stern S (2001). National innovative capacity. *The global competitiveness report, 2002*, 102-118.

Sharif N (2005). Contributions from the sociology of technology to the study of innovation systems. Knowl. Technol. Policy 17(3-4):83-105.

The World Bank (2006). Project appraisal document on a proposed credit in the amount of SDR 20.9 million (US $ 30.0 million equivalent) to the Republic of Uganda for a Millennium Science Initiative Project .

Wolfe DA (2000). Social capital and cluster development in learning regions. Adam J and Wolfe DA, Eds. Knowledge, clusters and learning regions ed. *J. Adam Holbrook and David A. Wolfe, Kingston: School of Policy Studies, Queen's University* (11-12).

Yang X, Wang X, Wei M, Hikosaka Sh, Goto E (2011). Changes of glutamine and asparagines content in cucumber seedlings in response to nitrate stress. Int. J. Plant Prod. 5(1):1- 7.

# Effect of locust bean fruit pulp (*parkia biglobosa*) flour on the quality of wheat biscuits

**Zakari U. M., Ajayi S. A. and Hassan A.**

Department of Food Technology Kaduna Polytechnic, P. M. B. 2021, Kaduna, Nigeria.

**Composite biscuits were produced from blends of wheat flour and African locust bean fruit pulp flour. The wheat/locust bean fruit pulp flours were mixed in the ratio of 100:0, 90:10, 85:15, 80:20 and 70:30%, respectively. The physical characteristics of the biscuits ranged in value, with diameter from 6.10 to 6.35cm, height from 1.08 to 1.20 cm, weight from 15.20 to 16.40 g, spread ratio from 4.90 to 5.88 and break strength from 530 to 820 g. Protein content ranged from 8.25 to 9.61%, fat from 0.62 to 1.19%, ash from 3.25 to 3.50%, crude fiber from 1.70 to 3.70% and carbohydrate from 69.93 to 71.75%. The vitamin C ranged from 400 to 460 mg/100 g. Protein content, fat and carbohydrate content decreased, while vitamin C, ash and crude fiber contents augmented with increased addition of locust bean fruit pulp flour. Sensory panelists rated the products highly for all the parameters investigated ($P < 0.05$). The biscuit with 10% substitution was rated next to the control (100% wheat biscuits) in the parameters tested. Acceptable biscuits can therefore be produced from blends of wheat/locust bean fruit pulp flour up to 20% substitution level.**

**Key words:** Biscuit, flour, locust bean, wheat.

## INTRODUCTION

Biscuits may be regarded as a form of confectionery, dried to a very low moisture content. Biscuits are usually produced from cereal flours (mainly wheat) and consumed extensively all over the world. According to Agu et al. (2007), a biscuit is defined as a small thin crisp cake made from unleavened dough. Okaka (1997) described the biscuits as a mixture of flour and water, often with the addition of fat, sugar and other ingredients, mixed together into dough which is rested for a period of time and then passed between rollers to make a sheet that is cut in smaller pieces and baked.

Biscuits are ideal for nutrient availability, palatability, compactness and convenience. They differ from other baked products like bread and cakes because of low moisture content, comparative freedom from microbial spoilage and long shelf-life (Mian et al., 2009). Biscuits may be classified either by the degree of enrichment and processing or by the method adopted in shaping them. Based on the enrichment criterion, biscuits may be produced from hard dough, soft dough or from batter (Fayemi, 1981).

The nutritional content however varies with the type of flour used. Soft wheat flour is the best flour for biscuit making, because of the composition of its storage proteins, gliadins and glutenins, which undergo hydration in the presence of water, salt and sugar. These proteins form a visco-elastic matrix known as gluten, which is responsible for the rising nature of dough and permits substantial increase in the volume of baked product with its gas retention capability (Agu et al., 2007).

Nigeria, a tropical country, cannot grow wheat in commercial quantity due to its hot climatic condition. Therefore, for the survival of biscuits and other confectionery products, the use of locally available grains or legumes to substitute wheat flour is essential (Kent, 1984). A lot of work involving the use of non-wheat flour

**Table 1.** Physical properties of the biscuits.

| Sample | Diameter (cm) | Height (cm) | Weight (g) | Spread ratio | Break strength (g) |
|---|---|---|---|---|---|
| 100% WF | 6.10±0.00 | 1.20±0.02 | 15.21±0.20 | 4.90±0.9 | 820±0.40 |
| 90WF:10LBFP | 6.22±0.50 | 1.10±0.00 | 15.20±0.40 | 5.65±0.00 | 620±0.30 |
| 85WF: 15LBFP | 6.27±0.30 | 1.10±0.00 | 15.20±0.40 | 5.70±0.01 | 540±0.60 |
| 80WF:20LBFP | 6.30±0.31 | 1.10±0.00 | 16.35±0.60 | 5.72±0.01 | 540±0.50 |
| 70WF:30LBFP | 6.35±0.41 | 1.08±0.01 | 16.40±0.50 | 5.88±0.80 | 530±0.51 |

NF = Wheat flour. LBFP = Locust bean fruit pulp flour.

from various cereals and legumes has been done to substitute wheat in baked products (Eneche, 1999, Nochera and Caldwell, 1992). The African locust bean (*Parkia biglobosa*) fruit pulp, which has been found to have good nutritional value (prot 6.56, fat 1.8, CHO 67.30, ash 4.18 and crude fibre 11.75%) and low anti-nutrients/toxins (phytic acid 60.00, crude saponin 17.80 and tannin 18.00 mg/100 g), has not been widely exploited as raw materials in confectionery products (Gernah et al., 2007; Akoma et al., 2002, Musa et al., 2005).

The aim of this study was to produce biscuits from various blends of wheat flour and African locust bean fruit pulp flour and to evaluate the physical, chemical and sensory properties of the developed products.

## MATERIALS AND METHODS

### Source of raw materials

This research was carried out at the department of Food Technology, Kaduna Polytechnic, Kaduna, Nigeria. The locust bean fruit pods were bought from Kasuwan Magani, a village in Chukun Local Government of Kaduna State. Wheat flour (Dangote brand), baking fat (margarine), granulated sugar, powdered milk, eggs, baking powder and salt were bought from a commercial store in Abubakar Gumi Central Market, Kaduna. All the equipment used was from the department of Food technology, Kaduna Polytechnic, Kaduna. The chemicals were of analytical grade and obtained from the laboratory of the department of Food Technology, Kaduna Polytechnic, Kaduna.

### Preparation of locust bean fruit pulp flour

The preparation of locust bean fruit pulp flour was done using the method of Gernah et al. (2007). The outer brown cover of the pods was manually stripped open and the yellow fruit pulp was separated from seeds embedded within the pulp. The yellow pulp was dried in a hot air oven (model T1211, Genlab Widnes, UK) at 60°C for 9 h to a moisture content of 10%. The dried powder was milled with a laboratory hammer mill (Christy Hunt, UK), and sieved through a 0.5 mm mesh screen to obtain a fine flour, which was packaged in low density polythene bags, and stored in air-tight container at room temperature.

### Preparation of biscuits

The ingredients used were wheat flour, locust bean fruit pulp flour, baking fat, granulated sugar, baking powder, milk powder, salt, egg (whole egg) and water. Wheat flour and the locust bean fruit pulp flour were the ingredients varied in the research.

To prepare blends of different proportions of wheat flour (WF) and locust bean fruit pulp (LBFP), the WF/LBFP flours were mixed in the ratio of 100:0, 90:10, 85:15, 80:20 and 70:30%, respectively. Control biscuits with wheat flour and experimental biscuits with flour blends were prepared using a modified method of Jane and Emma (1998). The ingredients were mixed for 10 min to form dough, kneaded into stiff dough. The dough was rolled out on a sheeting board to a sheet of uniform thickness of about 0.4 cm, and the sheet was stamped out in circular shapes of about 5.8 cm diameter, using a biscuit cutter. The biscuit cuts were placed on greased baking trays, covered, rested for about 15 minutes and baked for 20 min at 180°C. The biscuits were then removed and allowed to cool on a rack, packaged in low-density polyethylene bags and kept in an air-tight container.

### Analytical determinations

Physical properties of the biscuits such as the diameter, the weight and the height were determined by the AACC method (AACC, 1995). The spread ratio was determined by the method of Gomeaz, et al. (1997) and was calculated as diameter/height. The break strength was determined using break strenht device by the method of Okaka and Isieh (1990). A biscuit of known thickness (0.4 cm) was placed centrally between two parallel metal bars (3 cm apart), and weights were added on the biscuit until it snapped. The last weight that caused the breaking of the biscuit was regarded as the break strength of biscuit. Proximate composition was determined by AOAC methods (protein (micro-kjedahl, 2005); fat (Soxhlet method, 2005); ash (muffle furnace, 2005), fibre (acid digestion, 2005) and moisture content hot air oven, 2005).The carbohydrate content was determined by difference as described by Ihekoronye and Negoddy (1985). The vitamin C was analysed by Indophenol method (AOAC, 2005). Sensory evaluation was conducted on both the control and experimental biscuits. A total of fifteen untrained panelist drawn from Kaduna Polytechnic, Kaduna, based on their familiarity with the product, were used for the evaluation. The parameters investigated include, taste, colour, aroma, texture and general acceptability. A 7-point hedonic scale was used, in which 1 represents dislike extremely and 7 represents like extremely. The resulting data were tested using the analysis of variance (ANOVA) to detect significant differences among treatment. Tukey's test was used to separate the means, when significant differences existed (Iwe, 2002).

## RESULTS AND DISCUSSION

The results of the assessment of the physical properties of the biscuits are presented in Table 1. The diameter of the biscuits varied from 6.10 to 6.35 cm, the height ranged from 1.08 to 1.20 cm and the weight of the

**Table 2.** Chemical composition of the biscuits.

| Sample | Protein (%) | Moisture (%) | Fat (%) | Ash (%) | Crude fibre (%) | Carbohydrate (%) | Energy (Kcal) | Vit. C (Mg/100 g) |
|---|---|---|---|---|---|---|---|---|
| 100% WF | 9.61±0.16 | 12.5±0.05 | 1.19±0.00 | 3.25±0.02 | 1.70±0.00 | 71.75±0.01 | 336.15 | 400 |
| 90WF:10LBFP | 8.85±0.22 | 13.0±0.05 | 0.98±0.21 | 3.27±0.01 | 3.03±1.33 | 70.87±0.12 | 327.70 | 420 |
| 85WF:15LBFP | 8.63±0.23 | 13.5±0.12 | 0.85±0.13 | 3.28±0.01 | 3.32±0.29 | 70.42±0.06 | 323.85 | 440 |
| 80WF:20LBFP | 8.40±0.15 | 13.7±0.14 | 0.73±0.12 | 3.37±0.09 | 3.45±0.13 | 70.35±0.07 | 321.57 | 440 |
| 70WF:30LBFP | 8.25±0.10 | 14.0±0.06 | 0.62±0.11 | 3.50±0.13 | 3.70±0.25 | 69.93±0.09 | 308.30 | 460 |

**Table 3.** Sensory evaluation of the biscuits.

| Sample | Colour | Taste | Texture | Aroma | Acceptability |
|---|---|---|---|---|---|
| 100WF | $6.7^a$ | $6.5^a$ | $5.9^a$ | $6.2^a$ | $6.2^a$ |
| 90WF:10LBFP | $5.7^a$ | $6.1^a$ | $5.7^a$ | $5.9^a$ | $6.1^a$ |
| 85WF:15LBFP | $5.4^a$ | $6.0^a$ | $5.2^a$ | $5.2^a$ | $5.5^a$ |
| 80WF:20LBFP | $5.1^b$ | $5.7^a$ | $5.4^a$ | $5.4^a$ | $5.2^a$ |
| 70WF:30LBFP | $4.3^b$ | $5.4^a$ | $4.7^a$ | $4.7^b$ | $4.7^b$ |

Values with the same letter(s) in the same column are not significantly different at 5% level of significant.

biscuits ranged from 15.20 to 16.40 g, while the spread ratio ranged from 4.90 to 5.88. The wheat bread (100% wheat flour) had a lower spread ratio (4.90) than the other samples, which compares favorably with the work of Agu et al. (2007). This is because the starch polymer molecules are highly bound with granules and swelling is limited when heated. On cooling, the starch rapidly forms a rigid gel with capacity characteristics of large molecular aggregates. The formation of large molecular aggregates may not be unconnected with the gluten structure which is very strong resulting to high degree of contraction of the biscuit. The wheat flour biscuit (100% wheat flour) exhibited the highest break strength (820g), while all the composite biscuits fractured under a load of 700g. The high break strenght obseved for the 100% wheat flour biscuit compared favorably with work of Okaka and Isieh (1990). However, the biscuit products substituted with 10% LBFP flour were comparable to the control wheat flour biscuits in break strength.

The result of chemical composition analysis is shown in Table 2. The average protein contents of the biscuits ranged from 8.25 to 9.61%. A decrease in protein content was observed with increased addition of LBFP flour. The decrease could be attributed to substitution effect, since the LBFP flour has lower protein content. However, the biscuits made from the blends still maintained a reasonable protein level (8.25 to 8.85%). There were slight increases in moisture as the level of locust bean fruit pulp flour increased, which may be attributed to the initial high moisture content of locust bean fruit pulp flour. The increase in moisture content of the biscuits from the blends could also be due to high sugar content of locust bean fruit pulp flour, which made the biscuits more hygroscopic (Gernah et al., 2010).

The ash content (3.25 to 3.50%) increased with increased addition of LBFP flour, thus indicating high level of minerals in LBFP flour. Crude fibre of the products also increased. Since the fibre level was high, the biscuits produced from the blends could be considered as functional biscuits: their fibre will add bulk to the food and thereby facilitate bowel movement and help in preventing many gastrointestinal diseases in man (Sudha et al., 2007). There were slight decreases in the carbohydrate content of the biscuits, but the decrease was not significant. The carbohydrate ranged between 69.93 and 75% (70:30 composite and 100:0 wheat flour only, respectively). Vitamin C increased with increased addition of locust bean fruit pulp flour, as a result of the addition of LBFP, which is characterized by high vitamin C content. The result of sensory evaluation is shown in Table 3.

The results indicate that the products made from the blends were highly rated for all the parameters (colour, taste, texture, aroma and overall acceptability) investigated. There was significant difference ($P < 0.05$) in colour, possibly because of the high level of sugar of the LBFP flour, which caramelized during baking. The sensory results indicated that biscuits products could be highly accepted up to 20% substitution level.

## Conclusion

This research showed that acceptable biscuits could be produced with blends of locust bean fruit pulp flour and wheat four, up to 20% substitution level. Such biscuits could be regarded as functional biscuits because of the high fibre level.

### REFERENCES

AACC (1995). Approved methods of the American Association of Cereal Chemists. 10th ed. The Association: St. Paal, MIN.

Agu HO, Ayo JA, Paul AM, Folorunsho F (2007). Quality Characteristics of biscuits made from wheat and African breadfruit (*Treculia afriana*). Nig. Food J. 25(2):19-27.

Akoma O, Olawepo O, Ogunrinde BA (2002). The production of `tsamiya winie' from Tamarindus ladica. Nig. Food J. 27(1):17-19.

AOAC (2005). Official Methods of Analysis, 16th Ed. Association of Official Analytical Chemists, Arlington, V. A., pp. 806-842.

Eneche EH (1999). Biscuit making potential of millet/pigeon pea flour flour blends. Plant Food Hum. Nutr. 54:21-27.

Fayemi PO (1981). Home economics teachers guide, Ibadan. Macmillan Nig. Publisher Ltd. P. 201.

Gernah DI, Atolagbe MO, Echgwo CC (2007). Nutritional Composition of African Locust Bean (*Parkia biglobosa*) fruit pulp flour. Nig. J. Nut. Sci. 31(2):20-24.

Gernah DI, Akogwu AM, Sengeve AI (2010). Quality Evaluation of Cookies produced from blends of wheat flour and African locust bean pulp flour. Nig. J. Nut. Sci. 3(2):20-24.

Gomeaz MI, Obilantan AB, Martin D, Madvamase T, Manyo ES (1997). Grain quality evaluation in manual of laboratory procedure for quality evaluation of sorghum and pearl millet technology. International Crop Research Institute of Semi Arid Crops (ICRISAT).

Ihekoronye AJ, Ngoddy PO (1985). Integrated Food Science and Technology for the Tropics. Macmillan Publishers, Ltd. London, pp. 165-119.

Iwe MO (2002). Handbook of sensory Methods and Analysis. Rejoint Communications Services Ltd, Enugu, pp. 64-84.

Jane CO, Emma NI (1998) Composition and Characteristics of Cassava-soybean and wheat-soybean Biscuits. J. Food Sci. Technol. 35(2):128-131.

Kent NL (1984). Technology of Cereals. Pergamon Press Ltd, Oxford, United Kingdom, pp. 27-73.

Mian KS, Massod SB, Faqir MA, Haq N (2009). Preparation of Fiber and Mineral enriched defatted Rice Bran Supplemented Cookies. Pak. J. Nutr. 8(5):571-577.

Musa S, Yusuf YA Koleola AA, Akama O (2005). Chemical and Nutritional Characteristics of `Kinbwang' (a traditional fermented locust bean fruit pulp snack). Nig. Food J. 23:8-12.

Nochera CC, Caldwell M (1992). Nutritional evaluation of bread fruit containing composite flour product. J. Food Sci. 59(6):14-20.

Okaka JC and Isieh L (1990). Development and Quality Evaluation of Cowpea – Wheat Biscuits. Nig. Food J. 8:56-62.

Okaka JC (1997). Cereals and legumes Storage and processing Technology. Data and Micro-system publishes, Enugu, Nigeria, pp. 11-124.

Sudha MIR, Vetrimani K, Leclavathi K (2007). Influence of fibre from different cereals on the rheological characteristics of wheat flour dough on biscuit quality. J. Food Chem. 100:1365-1370.

# Handling, processing and utilization of milk and milk products in Ezha district of the Gurage zone, Southern Ethiopia

**Abebe Bereda[1], Zelalem Yilma[2] and Ajebu Nurfeta[3]**

[1]Department of Animal Science, Debre Berhan University P. O. Box 445, Debre Berhan, Ethiopia.
[2]Heifer International, East Africa Dairy Development, Country Program Mobilization Coordination Office, Addis Ababa, Ethiopia.
[3]School of Animal and Range Sciences, College of Agriculture, Hawassa University, P. O. Box 5, Hawassa, Ethiopia.

**A survey study was conducted in Ezha district of Gurage zone to understand the hygienic practices during production and further handling of milk and milk products; and their utilization. One hundred and twenty households were selected based on ownership of dairy cows, milk processing practice and willingness to participate in the study from two agro-ecologies (Dega and Woyna Dega) within the district. None of the respondents washed udder before milking. The majority of women washed the equipments (90.8%) and their hands (71.5%) before milking. *Olea africana* and *Hygenia abyssinica* plant leaves were the most commonly smoking and cleaning plant species in the district. The average volume of milk churned at a time was 6 L. Women preserve butter by mixing with spices such as *Nigella sativa*, *Aframomum angusti-folium*, *Trigonela fenum* and *Ocimum hardiense*, while Ayib is preserved with the use of *O. hardiense*. Out of the total monthly milk production (55 L), 13.5 L were consumed, whereas the remaining was accumulated for further processing. Among milk and milk products produced, only butter and Ayib were supplied to local markets. Lack of clean water for cleaning purpose; limited knowledge on hygienic handling of milk and milk products; and unimproved milk processing materials were the three major constraints reported by the respondents according to their importance. Recognizing the importance milk and milk products to the producing household nutrition, health and income, development interventions are required to boost production, improve the quality of the products and efficiency of the traditional milk processing equipment.**

**Key words:** Milk and milk products, hygiene, processing, utilization, Ezha, Ethiopia.

## INTRODUCTION

Ethiopia holds a substantial potential for dairy development mainly due to its large livestock population coupled with the relatively suitable environment for livestock production (Mohammed et al., 2004). In 2011, Ethiopia was home for an estimated 53.4 million cattle, 22.8 million goats, 25.5 million sheep, 49.3 million chicken and 1.1 million camels (CSA, 2011). However, the productivity of the livestock resources and the benefits obtained from the sector does not commensurate with the high livestock population. Given the considerable potential for smallholder income and employment generation from high-value dairy products, development of the dairy sector in Ethiopia can contribute significantly to poverty alleviation and nutrition in the country (Mohammed et al., 2004).

Smallholder farmers and pastoralists together produce and supply 98% of the total annual milk production of the country (YONAD, 2009).

According to CSA (2011), over 85% of the milk produced by rural households is consumed within the producer households with the proportion marketed being less than 7%. The small amount of milk produced by a large number of producers but the low marketable output in Ethiopia posses limitations on the possibilities of exploiting distant but rewarding markets due to high transaction costs arising from transportation and high opportunity costs of labor involved. As reported by Muriuki and Thorpe (2001) the vast majority `of milk produced outside urban centers in the country is processed into milk products at household level using traditional technologies.

In areas where the climate is hot and humid, the raw milk gets easily fermented and spoiled during storage unless it is refrigerated or preserved. However, such storage facilities are not readily available in rural areas and cooling systems are not feasible due to lack of the required dairy infrastructure and when available high cost of facilities such as refrigerator for resource poor smallholder producers (O'Mahony and Peters, 1987). In Ethiopia, in general and in the study areas in particular milk, milk products are Important for producer family consumption and as a source of income through sale of products such as butter and Ayib (Ethiopian cottage cheese).

Due to small volume of daily milk produced, producers keep milk produced over 3 to 4 days until sufficient amount is accumulated to process into the aforementioned more shelf stable products. In the study area, traditional milk production, processing and handling is a typical feature. Traditional milk products are generally reported to be of substandard quality mainly due to inadequate dairy infrastructure such as refrigeration facility and clean water and limited knowledge o the hygienic handling of milk and milk products. This necessitates better understanding of the traditional processes and handling of milk and milk by products, which is a prerequisite for development. Although, milk production represents an important part of the livelihood of the community in Ezha district of the Gurage zone, there is shortage of information with regard to milk processing, handling and utilization.

Understanding the prevailing traditional practices of milk production, processing and storage is of paramount importance to make future improvement interventions. The objectives of the current study were therefore to assess the hygienic conditions during handling and processing, and utilization of milk and milk products in Ezha district of the Gurage zone.

## MATERIALS AND METHODS

### Study area

The study was conducted between February and March 2010 in Ezha district of the Gurage zone, which is located at 200 km South west of Addis Ababa. The altitude of the district ranges from 1800 to 3098 m above sea level and receives an annual rainfall of 900 to 1600 mm with the mean minimum and maximum annual temperatures of 5 and 38°C, respectively.

### Sampling procedure

A random sampling procedure was employed to select sample kebeles and households for the study. The district was first stratified as Dega and Woina Dega agro-ecologies. From a total of 28 Kebeles located in the district 2 Kebeles were randomly selected from each of the two agro-ecologies. A total of 30 households per Kebele that own at least one local milking cow were randomly selected. Thus, a total of 120 (30 households × 4 kebeles) households were interviewed. A semi-structured questionnaire was used to gather the required information on the hygienic practices during handling and processing of milk and milk and milk products, and their utilization.

### Data analysis

Both qualitative and quantitative data collected on hygienic practices during milk production, processing and storage at household level were analyzed and summarized using both mean and frequency procedures of SPSS statistical package for social science version 13.

## RESULTS and DISCUSSION

### Handling practices during milking, milk processing and storage

Generally, proper sanitary milking practices were not followed by the majority of the respondents in the study area. Although, most of the respondents reported to wash their hands and milk vessels before milking their cows, washing of udders was not reported (Table 1). Moreover, milkers dip their fingers in the milking vessel and moistening teats of the cows with the intention of facilitating milking. However, such practice may cause microbial contamination of the milk from the milker's hand. The majority of the respondents (57.2%) had access to river water followed by tap water (28%) and hand dug well water (7.2%) (Table 1). However, the quality of both river and hand dug well waters used for cleaning may not be of the required standard thus can attribute to the poor quality of milk in the area. It is, therefore, important to heat treat water from river and hand dug wells intended for cleaning purpose.

The interviewed households used different utensils for milking, storage and processing. All of the respondents reported to use plastic jar for milking, while clay pot was the only material used for churning (butter-making). As reported by most of the respondents (92%), clay pot is also used for storage of milk until the desired volume is collected for processing. Wondu (2007) reported a similar result where 88% of small-scale producers in Southern Ethiopia used traditional clay pot for fermentation and butter-making. As indicated by the respondents, clay pot keeps milk well at the prevailing high ambient temperature as compared to

**Table 1.** Handling practices during milking by households in the two agro-ecological zones.

| Variable | Agro-ecology | |
|---|---|---|
| | Dega (n=60) | Woina Dega (n=60) |
| **Milk hygienic practice (%)** | | |
| Wash udder before milking | 0 | 0 |
| Wash hand after milking | 95.0 | 48.3 |
| Wash milking utensil with hot water | 80 | 95 |
| **Source of water for cleaning milk utensils (%)** | | |
| Tap water | 10 | 61 |
| Hand dug well water | 3.3 | 11 |
| River water | 86.7 | 28 |
| **Milk utensils used for milking (%)** | | |
| Plastic jar | 100 | 100 |
| Nickel | 10 | 0 |
| **Milk utensils used for storage (%)** | | |
| Clay pot | 88.3 | 96.7 |
| Plastic jar | 16.7 | 5 |
| **Milk utensil used for churning (%)** | | |
| Clay pot | 100 | 100 |

n= Number of observations.

plastic containers.

The production of milk of acceptable hygienic quality for consumers requires good hygienic practices. One of the major factors affecting the quality of dairy products is related to the practice of proper milking procedures and cleanness of the milking utensils (Almaz et al., 2001). About 30% of farmers reported to rub milk storage vessels with leaves of shrubs such as 0 (Kosso in Amharic), *Sida tenuicarpa* Vollesen (Ojaja in the local language), Odaterk *Conyza spinosa* (Odaterk), *Olea africana* (Weira), and Korekonda (maize cob) and wash with water before use. Washing is commonly followed by drying and smoking the containers with embers of wood sprinters of plants specifically used for this purpose (*Hygenia abyssinica*, *Sida tenuicarpa* Vollesen, *O. africana*, maize cob, *Thymus vulgaris*, Enset fiber, *Conyza spinosa* and *Thymus schimperi* are the most commonly used). Procedures of cleaning and disinfection of milking utensils prior to milk collection reported herein were similar to previous results from Ethiopia (Sintayehu et al., 2008).

Reasons for smoking milk containers using different parts of various plant species as reported by the respondents were similar to earlier reports in Ethiopia (Lemma et al., 2004; Yitaye, 2008). Smoking is also reported to give a good flavor to the product and disinfect the vessels, thereby reducing the microbial load and therefore extending the shelf life of the product. As reported by Mogessie and Fekadu (1993), smoked containers tend to lower the microbial load of milk as compared to unsmoked containers.

## Processing milk and shelf life of milk and milk products

As it is the case under smallholder setting in other rural areas of Ethiopia, naturally fermented milk is the basis of processing milk into more shelf stable fermented milk products in the present study area. The major milk products produced in the study area were Ergo (Ethiopian naturally fermented milk), traditional butter (Kibe), traditional ghee (Neter Kibe), cottage cheese (Ayib), sour defatted milk (Arrera), and whey (Aguat). According to most of the respondents, milk is fermented for the main reason that the daily production is small and to get the volume that justifies processing (primarily churning for butter-making), the daily milk yield should be accumulated over a few days. In this process, milk ferments naturally at the prevailing tropical ambient temperatures.

In the rural Ethiopia, in general and in the study area in particular, milk processing and other household activities are almost always the responsibility of women. In the Woina Dega area, the majority of the women (70%) process the milk twice per week followed by once per week (30%), while 53.4% the women around Dega area process their milk twice per week, whereas the remaining 38 and 8% of the respondents reported to process once per week and once fortnightly, respectively (Table 2).

Generally, high ambient temperature favors natural fermentation and therefore, due to the relatively high ambient temperature in the Woina Dega areas, producers in these areas processed milk at higher frequency as

**Table 2.** Milk processing practices (%) in the two agro-ecological zones.

| Variables | Agro-ecology | |
|---|---|---|
| | Dega (n=60) | Woina Dega (n=60) |
| **Processing of milk (%)** | 100 | 100 |
| **Processing frequency (%)** | | |
| Twice/week | 53.4 | 70 |
| Once/week | 38.3 | 30 |
| Once/two week | 8.3 | 0 |
| Volume of milk churned at a time (liter) | $6.4 \pm 2.1^{a}$ | $5.5 \pm 1.14^{b}$ |
| **Reasons for processing of milk (%)** | | |
| To generate more income | 85 | 68.3 |
| To increase shelf life | 88.3 | 96.7 |
| To purchase agricultural commodity | 91.7 | 61.7 |
| Cultural reluctance towards consumption of fresh milk | 7 | 5 |
| Consumer preference | 3.3 | 13.3 |

n= Number of observation.

**Table 3.** Milk products processing practice in the study area.

| Variables | Agro-ecology | |
|---|---|---|
| | Dega (n=60) | Woina Dega (n=60) |
| Amount of butter recovered in a single churn (kg) | 0.34±0.12 | 0.28±0.77 |
| Amount of *Ayib* in single churn (kg) | 1.07±0.11 | 0.97±0.05 |
| Volume of milk used to produce kg of butter (lit) | 21±1.6 | 23.07±1.2 |
| Volume of defatted sour milk used to produce kg of *Ayib* | 6± 0.56 | 6.7±0.54 |

compared to those in the Dega areas. As observed in the present study, about 93% of the respondents process milk to extend the shelf life of the product followed by generation of income to purchase agricultural commodities (76.7%) and to fulfill other day to day necessities (76.7%) (Table 2). Ayantu (2006) in her report in Delbo watershed area of the Wolayta zone also indicated that milk is processed in order to increase the family's income, diversify the products for consumption and to increase the shelf life of the products. Kassahun (2008), however, reported that milk is primarily processed whenever there is market constraint for fresh milk in the Ada'a district.

## Butter making

All of the respondents produced traditional butter from naturally fermented whole milk. While churning, traditional churns are covered with materials such as dry Enset pseudo stem sheath and/or a piece of plastic material, which is stretched over the opening of the churn and securely tied. The average milk churned at a time observed was about 6 L at household level though markedly (P<0.01) varied between the two agro-ecologies (Table 3). Zelalem and Ledin (1999) also reported a similar value (6.4 L) in the central highlands of Ethiopia. However, lower values of 2.1 L (Ayantu, 2006) and 1.97 L (Rahel, 2008) were reported in the Wolayta zone. The amount of butter obtained from a single churn in the current study (0.25 kg) (Table 3) was higher than that reported in the Wolayta zone (0.13 kg) (Rahel, 2008) and (0.16 kg) (Ayantu, 2006), which is likely to be attributed to the difference in the volume of milk churned at a time.

In the present study, the amount of milk required to produce one kg of butter was 22.32 L (Table 3). This value is comparable with that of earlier reports of 25 kg in Borena plateau (Coppock, 1994), 20 to 22 L in rural areas of Southern Ethiopia (Fekadu, 1994) and 21 L in the central highlands of Ethiopia (Zelalem and Ledin, 1999). However, lower values were also reported; 15 to 20 L in Awassa (Wondu, 2007), 15.79 L in the Wolayta zone (Rahel, 2008), 16.7 L in East Wollega (Alganesh, 2002), and 17.1 L in East Shewa zone (Lemma et al., 2004). Such variations are likely attributed to the fat content of the milk used which in turn is a function of breed. Milk of local zebu cattle breeds has higher fat content as compared with that of pure exotic and crossbred cows.

All of the interviewed farmers used different spices as means of preservation for traditional butter. The major spices used for this purpose were Koserete (*Ocimum hardiense*), Korerima (*Aframomum angusti-folium*), Tikur azmud (*Nigella sativa*) and Abish (*Trigonela fenum*). As

**Table 4.** Production and consumption of milk and milk product in the two agro-ecological zones.

| Variables | Agro-ecology | |
|---|---|---|
| | Dega (n=60) | Woina Dega (n=60) |
| **Product (%)** | | |
| Milk | 68.3 | 46.7 |
| Butter | 68.3 | 96.7 |
| Ergo | 3.3 | 5.0 |
| Ayib | 93.3 | 95 |
| Whey | 100 | 100 |
| **Average milk produced and consumed (lit) per month ($\pm$S.E)** | | |
| Produced | 57.6±3.6 | 52.2±3.2 |
| Consumed | 18.73±2.9$^{a}$ | 8.37± 2.6$^{b}$ |
| Daily milk consumption | 0.62±0.09$^{a}$ | 0.28±0.056$^{b}$ |
| **Average Ayib produced and consumed (kg) per month ($\pm$S.E)** | | |
| Ayib produced | 5.96±0.63 | 7.1±0.7 |
| Ayib consumed | 3.37±0.35$^{a}$ | 4.95±0.51$^{b}$ |
| **Average butter produced and consumed (kg) per month ($\pm$S.E)** | | |
| Butter produced | 1.9±0.18 | 2.05±0.15 |
| Butter consumed | 0.94±0.12$^{a}$ | 1.73±0.15$^{b}$ |

Row mean bearing with different superscripts letters are significantly different from each other (P<0.05). S.E = Standard error.

reported by the respondents, the traditional butter can be stored at ambient temperatures on average for about one year without losing its desired flavour and taste.

### Ethiopian cottage cheese (Ayib) making

The amount of Ayib obtained at a time in this finding was 1.05 kg, which is comparable between the two agro-ecologies. On average, 6.5 L of milk was required to produce one kg of Ayib in both agro-ecologies (Table 3). This value is in agreement with the 6 L reported by Zelalem and Ledin (1999) in the central highlands of Ethiopia. However, Rahel (2008) reported a higher value (8.69 L) in Wolayta zone. The spices mainly used as a preservatives to extend shelf life of Ayib were Koserete (*O. hardiense*), Tikure azmude (*N. sativa*), Korerima (*A. angustifolium*), Tena Adam (*Ruta chalepensis*) and Abish (*Trigonela fenum*) and in addition to these salt is also used to extend the shelf life of Ayib.

### Consumption and utilization of milk and milk products

Like in most other rural areas of Ethiopia, the common milk products manufactured and consumed in the current study areas include fresh whole milk, whey (Aguate), butter, Ayib and ghee (Niter Kibe). Out of the total monthly milk production (55 L) per household, about 13.5 L was used for consumption within the producing households, the rest being left to be processed in to more value added products mainly butter and Ayib that have better shelf life and fetch better market price (Table 4). In the present study, about 68.3 and 47% of the respondents in the Dega and Woina Dega areas, respectively, used whole milk for consumption. Contrary to this finding, consumption of raw milk in the Wolayta zone is not a usual practice rather in most cases the small amount of milk produced daily is accumulated over a few days to manufacture the butter that fetches a better price (Ayantu, 2006).

Butter of the Wolaita zone is one of the most preferred butter for its taste not only in the area of production but also all the way up to Addis Ababa. Out of the total butter production (1.9 kg) per household per month, about 1.33 kg was used for consumption (Table 4). About 68.3 and 96.7% of the respondents in the Dega and Woina Dega areas, respectively, used butter for producer household consumption.

Butter consumption in the Woina Dega area was 1.73 kg/month/household, which was significantly (P<0.01) higher than that consumed by households in the Dega area (0.94 kg/month/household). This could be due to the common practice of accompanied butter while consuming the homemade flat bread (*Kocho*) made from Enset (*Ensete ventricosum*) in the Woina Dega area.

The mean monthly production of Ayib per household in the district was 6.6 kg. Out of this total monthly production, 2.16 kg is consumed within the household. The average monthly production per household observed in this study was much higher than 0.31 kg reported in the Ada'a Woreda (Kassahun, 2008) and 1.2 kg in Wolayta zone (Rahel, 2008).

**Table 5.** Marketing of milk and milk products in the two agro-ecological zones.

| Variables | Agro-ecology | |
|---|---|---|
| | *Dega (n=60)* | *Woina ega(n=60)* |
| **Average butter and Ayib sold per month in kg (±S.E)** | | |
| Butter | 0.96±0.16[a] | 0.32±0.07[b] |
| Ayib | 2.54±0.38 | 2.23±0.44 |
| **Average price of butter and Ayib per kg (±S.E)** | | |
| Butter price | 57.8±10.5 | 44.42±1.08 |
| Ayib price | 15.11±0.28[a] | 17.46±0.35[b] |
| **Selling practices (%)** | | |
| Butter selling practice (%) | 60 | 35 |
| Ayib selling practice (%) | 76.7 | 65 |
| Milk selling practice (%) | 0 | 0 |
| **Reasons for not selling milk (%)** | | |
| Cultural restriction | 60 | 46.7 |
| Lack of market | 28.4 | 11.7 |
| Scarcity of milk | 85 | 71.7 |

Row mean bearing with different superscripts letters are significantly different from each other (P<0.05). S.E= Standard error.

## Marketing of milk and milk products

As observed during the current study, the sale of fresh whole milk was not a common practice (Table 5). Among different multiple reasons, inefficient milk production (78.3%) and cultural restrictions (taboos) toward selling fresh whole milk (53.3%) followed by lack of market (20%) are the most common reasons reported (Table 5). A similar situation was reported in Eastern Wollega where about 21.3 and 19% of the women did not sell fresh milk mainly due to scarcity and cultural reasons, respectively (Alganesh, 2002). On the other hand, about 60 and 35% of the respondents in the Dega area and Woina Dega area, respectively, stated that they sold butter, while the rest of the respondents never sold, while 29% of the respondents did not sell Ayib.

In the study areas, Ayib is the most important marketed product. Ayib sold per household per month was 2.4 kg (Table 5). This result is lower when comparable with results (3.4 kg) of Rahel (2008). The average weight of the butter sold per household per month in this district was 0.64 kg (Table 5). Significant difference (P<0.05) was observed between the Dega and Woina Dega areas in average monthly butter sold per household where households in the Dega area sold significantly high amount of butter. This difference could be attributed to the lower consumption of butter per household in the same (Table 5). The average butter sold per household per month in the current study is much less than the amount (7.2 kg) reported by Rahel (2008).

The average prices of Ayib and butter were 16 birr and 52 birr/kg, respectively (Table 5). The mean price of Ayib in the Woina Dega area was 17 birr, which is significantly higher (P<0.001) than that of the Dega area (14 birr) (Table 5). The price base of the product is subjective and depends on weight estimation by the customers who estimate by putting them on their hands. The price of Ayib reported in the current study is comparable with that (15 birr/kg) reported by Rahel (2008) and Zewide (2010) (16.5 birr/kg) in Sebeta. Kassahun (2008) and Zewdie (2010), however, reported lower prices in urban areas of Lume (12 birr/kg) and Zeway (13 birr/kg), respectively.

In Ethiopia, studies indicate that butter prices differ from place to place. Zewdie (2010) reported the price of butter in Sebeta, Ziway, Debre Berhan and Jimma to be 75, 66, 61 and 60 birr/kg, respectively, during the dry season. As indicated by Zelalem and Ledin (1999), such price differences may arise from factors such as access to better market in urban areas as compared to rural areas in the central high lands of Ethiopia. In addition, Zewdie (2010) indicated that price of butter and Ayib differences may arise due to season, highest price in the dry season than wet season. This could be due to the good supply of butter during the rainy season owing to better grass availability following the rain, which leads to high production of milk.

## Constraints of handling, processing and utilization of milk and milk products

The major constraints pertaining to milk handling, processing and utilization as reported by the respondents in the Ezha district of the Gurage zone are summarized in Table 6. Close to 80% of the respondents reported lack of clean water for cleaning purpose, limited knowledge on the hygienic handling of milk and milk products and unimproved milk processing materials to be the three major constraints listed according to their importance.

**Table 6.** Major constraints to milk handling, processing and utilization

| Variable | % of total respondents | Rank |
|---|---|---|
| Lack of clean water | 37 | 1st |
| Limited knowledge | 24 | 2nd |
| Unimproved milk processing materials | 17 | 3rd |
| Lack of access and high price of cooling facilities | 11 | 4th |
| Lack of electricity | 7 | 5th |
| Low milk production | 4 | 6th |

## Conclusion

Dairy production is an important and integral part of the crop-livestock mixed production system in Ethiopia in general and in the study area in particular; milk processing is a culture of the community of the Ezha district of the Gurage zone and milk products have various social, nutritional and economic values. The milk production system of the study areas can generally be characterized by small herd size with no improved crossbred cattle and poor productivity of cows with very low milk yield, and long calving interval and lactation length. Although milk and milk products are essential, the majority of the respondents did not practice recommended hygienic practices (such as hand and udder washing) during milking and further handling (processing, storage and marketing) of milk and milk products.

Producing milk and milk products of not only acceptable quality but also of high quality are important from consumer health point of view and may also lead to increased demand and income to producers. Among others, lack of clean water for cleaning purpose (milkers' hand, cows udder, and containers and equipment of milk and milk products); limited knowledge on the hygienic handling of milk and milk products; and unimproved milk processing materials in terms of capacity (volume), time elapsed to recover the intended milk product and the quantity of final product obtained from a given volume of raw material used (milk, naturally fermented milk, defatted sour milk) are the three major constraints reported by the respondents according to their importance.

Recognizing the importance of dairy farming to the livelihood of the community in question and milk and milk products to the producing household nutrition, health and income, development interventions are required to boost production, improve the quality of the products and efficiency of the traditional milk processing equipment. Such interventions coupled with linking producers with other important actors in the dairy value chain of the study district will definitely improve the benefits that producers should get from the dairy sub sector.

## ACKNOWLEDGEMENTS

The authors are grateful for the financial support provided by Debre Berhan University under Ministry of Education, Ethiopia.

## REFERENCES

Alganesh T (2002). Traditional milk and milk products handling practice and raw milk quality in Eastern Wollega, Ethiopia. M.Sc Thesis, Alemaya University, Ethiopia.

Almaz G, Howard AF, Wilhelm HH (2001). Field survey and literature review on traditional fermented milk products of Ethiopia. Int. J. Food Microbiol. 68:173-186.

Ayantu M (2006). Women's role on production, processing and marketing of milk and milk products in Delbo watershed of Wolayta. M.Sc Thesis, Hawassa University, Ethiopia.

Coppock DL (1994). The Borana Plateau of Southern Ethiopia: Synthesis of pastoral research, development and change, 1980-91. International Livestock Center for Africa (ILCA). System study 5, Addis Ababa, Ethiopia, p. 390.

Central Statistical Agency (CSA) (2011). Agricultural Sample Survey. Report on Livestock and Livestock Characteristics (Private Peasant Holdings). Volume II. Addis Ababa, Ethiopia, pp. 9-26.

Fekadu B (1994). Present situation and future aspects of milk production, milk handling and processing of dairy products in Southern Ethiopia. The case of Aneno, Bulbula and Sogora in Southern Ethiopia. PhD Dissertation. Agricultural University of Norway. Norway.

Kassahun M (2008). Characterization of milk products consumption patterns, preference and compositional quality of milk in Ada'a and Lume Woreda of East Shoa zone, Central Ethiopia. M.Sc. Thesis, Hawassa University, Ethiopia.

Lemma F, Fekadu B, Hogde PB (2004). Rural smallholder milk and dairy products production, utilization and marketing systems in East Showa zone of Oromia. In: Participatory innovation and research: Lesson for livestock development. Proceedings of the 12th Annual Conference of the Ethiopian Society of Animal Production (ESAP) held in Addis Ababa, Ethiopia, August 12-14. ESAP, Addis Ababa, Ethiopia, pp. 17-28.

Mogessie A, Fekadu B (1993). Effect of container smoking and cleaning on the micro flora and keeping quality of raw milk from a dairy farm in Awassa, Ethiopia. Trop. Sci. 33:368-376.

Mohammed AM, Simeon E, Yemesrach A (2004). Dairy development in Ethiopia. EPTD Discussion Paper No.123, (International Food Policy Research Institute), Washington DC, USA.

Muriuki HG, Thorpe W (2001). Smallholder Dairy Production and Marketing in Eastern and Southern Africa: Regional synthesis. Nairobi, Kenya. In:Rangnekar D. and Thorpe W. (eds), smallholder dairy production and marketing-Opportunties and constraints. Proceedings of a South-South Workshop held at NDDB, Anand, India, 13-16 March 2001.NNDB (National Dairy Development Board), Anand, India and ILRI (International Livestock Research Institute) Nairobi, Kenya. pp. 185-200.

O'Mahony F, Peters J (1987). Options for Smallholder Milk Processing in Sub-Saharan Africa. International Livestock Center for Africa (ILCA) Bulletin 27. Addis Ababa, Ethiopia, pp. 206-247.

Rahel N (2008). Traditional and improved milk and milk products handling practices and compositional and microbial quality of raw milk and butter in Delbo, Water shed of Wolyita Zone, Ethiopia. M.Sc Thesis, Hawassa University, Ethiopia.

Sintayehu Y, Fekadu B, Azage T, Berhanu GM (2008). Dairy production, processing and marketing systems of Shashemene - Dilla area, South Ethiopia. Improving Productivity and Market Success (IPMS) of Ethiopian farmer's project, International Livestock Research Institute (ILRI), Addis Ababa, Ethiopia.

Wondu M (2007). Composition, microbial quality and acceptability of butter produced from cow's milk in Awassa, Southern Ethiopia. M.Sc Thesis Hawassa University, Ethiopia.

Yitaye A (2008). Characterization and analysis of the urban and peri-urban dairy production systems in the North western Ethiopian highlands. M.Sc. Thesis, University of Natural Resources and Applied Life Sciences, Vienna, Austria.

YONAD (2009). Value chain Analysis of Milk and Milk products in Borana pastoralist area, unpublished manuscript.

Zelalem Y, Ledin I (1999). Milk production, processing, marketing and the role of milk and milk products on smallholder farmers' income in the central highlands of Ethiopia.

Zewdie W (2010). Livestock production systems in relation with feed availability in the highlands and central rift valley of Ethiopia. M.Sc. Thesis, Haramaya University, Ethiopia.

# Production, consumption and nutritional value of cassava (*Manihot esculenta,* Crantz) in Mozambique: An overview

**Salvador E. M.[1], Steenkamp V.[2] and McCrindle C. M. E.[1]**

[1]School of Health Systems and Public Health, Faculty of Health Sciences, University of Pretoria, HW Snyman Building Private Bag X323, Pretoria 0001, South Africa.
[2]Department of Pharmacology, Faculty of Health Sciences, Basic Medical Sciences, University of Pretoria (BMS) building, Private Bag X323, Arcadia, 0007 Pretoria, South Africa.

**Both soils and climate in Mozambique suit cassava cultivation and nine million tons fresh weight is produced annually, with a consumption of 85 kg per person per year. The roots are a staple carbohydrate and cooked leaves are served as a vegetable. Cassava is essential to food security, as it is a subsistence crop. Roots and leaves contain vitamin C and some minerals but are deficient in proteins and amino-acids. Although cassava is cultivated by about 63% of the population, cyanogenic glycosides and other anti-nutritional factors, threaten food safety. There are more than 100 varieties, but the more drought and insect resistant bitter types predominate. Traditional products made from cassava that rely on sun-drying, cooking or fermentation to reduce toxicity include "*rale*", "*xinguinha", "karakata"* "*mahewu*" and "*oteka*". Cassava flour has replaced up to 20% of wheat flour in bread, for economic reasons. An overview of the distribution, consumption patterns and nutritional value of cassava in Mozambique could contribute to knowledge, as much of the existing data has not been published. Food safety and nutritional value could be improved by commercializing the production of traditional products or fortifying the affordable staple carbohydrate. This could improve the health of vulnerable rural populations.**

**Key words:** Cassava production, cassava consumption, food security, traditional foods, cyanogenic glycosides, Mozambique.

## INTRODUCTION

Mozambique, one of the poorest countries in Africa (UNDP, 2010) is located in the south east of the continent. The country covers an area of about 799,380 $km^2$ which includes 36 million hectares of arable soil and a coastline of 2,740 km (FAO, 2011b). The climate in Mozambique is tropical and humid with two distinct seasons. Winter (April to September) is normally cold and dry, while the summer (October to March) period is warm and rainy. The annual ambient temperature is between 23 and 26°C for the coastal zones of southern and northern Mozambique with a mean annual rainfall of approximately 1200 mm (Hoguane, 2007).

The most recent census data was published in 2009 by the National Institute of Statistics (INE) following the last National Population Census, done in 2007. The population of Mozambique was recorded as 21 284 701, with more than 70% living in rural areas. Population density was 27 per km$^2$, growing at a rate of 1.7% per year. The gross birth ratio in the country was 42.2 per 1000 inhabitants, with an infant mortality of 94 deaths per 1000 live births and life expectancy of 49.4 years. About 80% of Mozambicans depended on agriculture for their livelihood and the agricultural sector remained the key for social and economic growth (INE, 2009).

Mozambique has been ranked 8$^{th}$ in the world and 5$^{th}$ among African countries with regards to cassava production (FAO, 2011a). It is an important staple food in Mozambique, second only to maize (Walker et al., 2006). Maize does not grow easily in certain regions of Mozambique (Tostao and Wade, 2005). The transportation of maize from high production areas to consumers in other areas is not cost effective (Nielson, 2009). Imported maize originates mainly from South Africa and it is significantly more expensive than cassava (Howard et al., 2003; Tschirley et al., 2006). Mozambique was involved in a protracted war which ended in 1992. This war impacted negatively on food production and bitter cassava was sometimes consumed without further processing, causing acute and chronic poisoning due to cyanogenic glycosides (Cliff, 1994). In the two decades since the war ended, efforts have been made to increase local crop production to reduce imports and improve foreign exchange (Mader, 2005; FAOSTAT, 2011b). Cassava is easily cultivated as a subsistence crop in Mozambique, with more than 90% being produced for human consumption (FAOSTAT, 2011b). Consumption has been estimated as 85kg per person per year (IOF, 2008/2009). Cassava is drought-tolerant and mature roots maintain their nutritional value for up to three years (Montagnac et al., 2009b; Gbadegesin et al., 2013). It is a multiple year crop and roots can be kept in the ground until needed, thus increasing food security (El-Sharkawy, 2004; Donovan et al., 2011; Gbadegesin et al., 2013). Cyanogenic glycosides and other anti-nutritional factors are a threat to human health that is well recognized in Mozambique (Ministry of Health Mozambique, 1984; Essers et al., 1992; Cliff, 1994; Cliff et al., 1997; Nhassico et al., 2008; Cliff et al., 2011). However, a less recognized threat is the poor nutrient quality, which because of the high level of consumption of this staple carbohydrate, threatens a large proportion of the population with malnutrition and poor quality of life.

Fermentation of cassava is known to reduce the toxicity of cyanogenic glycosides. It also results in higher levels of vitamins, especially the B group, and essential amino acids and improves digestibility of protein (Amoa-Awua and Frisvad, 1997; Oyewole, 2001; Boonnop et al., 2009). Traditional foods made from fermented cassava in Mozambique include *rale, karakata, mahewu, ottaca* and *oteka.* (Tivana et al., 2009; Donovan et al., 2011; Tivana, 2012; Hagglades et al., 2012; Salegua et al., 2012). This overview aims to consolidate knowledge on the production, consumption and nutritional value of cassava in Mozambique, with a view to stimulating research to improve food safety and nutrition, particularly in vulnerable rural populations.

## Cassava production

### *Area cultivated and yield*

The cropping season for cassava is described as flexible, because roots can be harvested between eight months and three years after planting (FAO/WFP, 2010). In 2011 the total cultivated area was estimated as 5 632 781 ha. The greater proportion of this was used by small scale farmers (96.4%). The remainder was cultivated by medium (2.3%) and larger scale (1.3%) farmers (INE, 2011). Approximately 2 425 240 ha (43%) of this total cultivated area was used for cassava production. Similar to other food crops in Mozambique, cassava is grown mainly (99.7%) by subsistence and small scale family farmers. Medium scale farmers use only 0.3% of the available land and large scale farmers only 0.01% (INE, 2011).

According to the Agriculture and Livestock Census 2009 to 2010, cassava is cultivated throughout the country. The three provinces with the highest production are Nampula Province, in the northern region (29.27%); Zambesia Province in the central region (26.76%); and Inhambane Province (8.80%) in the southern region (ine, 2011). The estimated production per region is shown in Figure 1. Cassava production increased between 2002 and 2008, when the average production was estimated as six million tons. It rose further to nine million tons in 2010 (FAOSTAT, 2011b). Overall, between 2005 and 2012, the yield of cassava in hectogram per hectare increased from 43.155 to 131.804 (Factfish, 2014). By 2012, the production of cassava had escalated to 10.05 million tons (Factfish, 2014).

### *Cassava varieties*

Varieties (also called cultivars) are classified according to morphological traits as well as taste, cyanide content, average yield, disease performance and pubescence (MIC, 2007; Gbadegesin et al., 2013). More than 5,000 cassava cultivars are recognized globally (Best, 1993; Bokanga, 1994; Gade, 2003; IFAD/FAO, 2005) and 100 varieties have been documented in Mozambique (INIA/IITA/SARRNET, 2002). Varieties grown in Mozambique are known to vary according to region; the

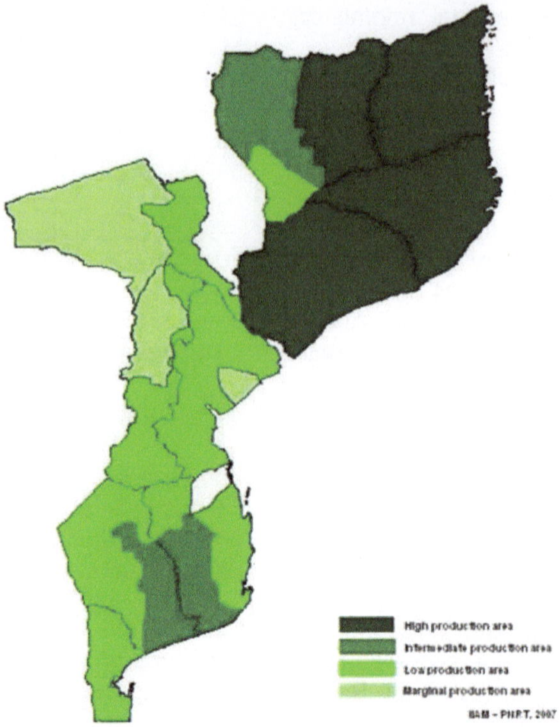


**Figure 1.** Cassava production area in Mozambique; Source: IIAM/PNRT, (2007).

*Munhaça* variety being most common in the Southern Region; *Inciricano* and *Bedo* in the Central Region and *Nikwaha*, *Tomo* and *Cororo* varieties, in the Northern Region (MIC, 2007). Consumer preferences influence the variety consumed and are based on the color of the plant, sharpness of the leaf lobe and the color of the root peel (MIC, 2007; Gbadegesin et al., 2013).

There are two major types of cassava: sweet and bitter (Chiwona-Karltun et al., 2004; Mkumbira et al., 2003; MIC, 2007). The flavor is influenced by the amount of cyanogenic glycoside present (McKey and Beckerman, 1993; Chiwona-Karltun et al., 2004). In the roots, cyanogenic glycosides range from 10 to 500 mg cyanide equivalent/kg dry matter (Arguedas, 1982; Siritunga and Sayre, 2003). Cassava leaves contain between 53 to 1,300 mg cyanide equivalents/kg of dry weight (Siritunga and Sayre, 2003; Wobeto et al., 2007). Bitter varieties contain more than 100 mg/kg fresh weight of hydrogen cyanide (Dufour, 1988; McKey et al., 2010).

These bitter varieties comprise more than 90% of production as they have a higher yield, are tolerant to pests and diseases and the potential for storage in the soil is greater than 12 months (Chiwona-Karltun et al, 1998; MIC/FAO/EC, 2004; Mader, 2005). It is essential to process bitter cassava to remove cyanogenic glycosides before consumption (Zvauya et al., 2002; Cardoso et al., 2005; Bradbury, 2006; Cumbana et al., 2007). Sweet varieties contain less than 100 mg/kg fresh weight of hydrogen cyanide, so are consumed fresh, sometimes even raw (Dufour, 1988; Mowat, 1989; Cardoso et al., 2005; Bradbury 2006; Cumbana et al., 2007; Donovan et al 2011).

### *Climate and production areas*

In Mozambique there are ten agro ecological zones (Figure 2). The classification of these zones is based on climate, vegetation, altitude, soil and farming systems. It can be seen from Table 1 that the majority of cassava is grown in the poor and marginal zones of Mozambique (INIA, 1994).

### Cassava consumption

The form in which cassava is consumed varies in the different regions of the country. In the Northern Region cassava flour (71% of production) is mainly boiled with water to make porridge. In this region the consumption of dried cassava accounts for 21% and fresh cassava for 8%, while the consumption of roasted cassava (*rale*) is not common (IOF, 2008 to 2009; Donovan et al, 2011; Hagglades et al., 2012).

In the Central Region, cassava is mainly consumed fresh (60%), cooked in different kinds of traditional

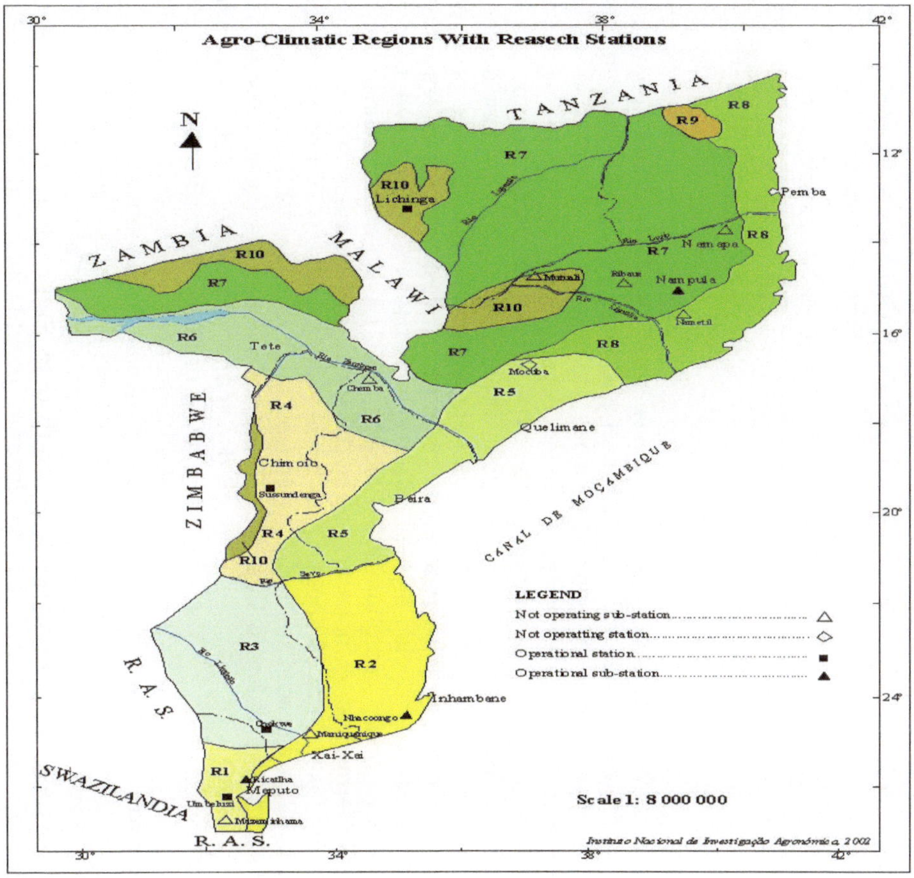


**Figure 2.** Agro-ecological zones of Mozambique (R1-R10); Source: INIA, (1994).

dishes. The consumption of flour and dried cassava accounts for only 31 and 9% of consumption respectively, with *rale* consumption being uncommon (Tivana et al., 2009; Donovan et al., 2011; Hagglades et al., 2012). Cassava consumption in the Southern Region is mainly in the fresh form (74%), followed by *rale* (19%). The flour and dried forms constitute 2 and 5%, respectively (IOF, 2008/09; Donovan et al, 2011; Hagglades et al., 2012).

Bitter cassava varieties must be processed to reduce or eliminate cyanogenic content (Chiwona-Karltun et al., 2000). The roots are boiled, roasted, sun dried, heap fermented or grated. Sweet varieties are mainly consumed fresh (Zvauya et al., 2002; Cardoso et al., 2005; Bradbury, 2006; Cumbana et al., 2007). Products that can be made from either sweet or bitter cassava roots include bread, biscuits, cakes and beverages.

### Cooked cassava

Approximately 10% of total consumption is cooked cassava, from mainly sweet varieties (Donovan et al., 2011). Due to the high temperature achieved during boiling, the enzyme linamarase can inactivate cyanide in bitter cassava roots (Nambisan and Sundareson, 1985, 1994). Traditionally, in Southern Mozambique, cassava is cooked with other vegetables and peanuts, in a dish known as *xinguinha*. Boiled cassava root, either sweet or bitter, is traditionally served with a green salad and tea. Cassava leaves are generally cooked to make a dark green vegetable that looks like spinach and can be flavored with different spices.

### Cassava flour

About 67% of cassava root harvested, is processed into flour (IOF, 2008 to 2009). The roots are first peeled and chipped or fermented, then sun dried before milling at community flour-mills. Direct sun-drying is commonly used for preparation of sweet varieties; whereas fermenting followed by sun drying is used in the preparation of bitter varieties (Zvauya et al., 2002; Tivana et al., 2007; Tivana et al., 2009). The inclusion of the fermentation process reduces or removes cyanogenic glycosides, as well as anti-nutritional factors, such as phytates and polyphenols (Montagnac et al., 2009a). In the Northern Region, cassava flour is mainly used to

**Table 1.** Agro ecological zones for cassava growth with their agro-climatic conditions.

| Zone | Altitude (m) | Precipitation (mm) | Temperature (°C) | Humidity index | Predominant soils |
|---|---|---|---|---|---|
| R2 | 0 - 200 | 800 - 1000 | 24 - 26 | Humid semi-arid, with some sub-humid spots in the littoral | Arenosol, fluvisols and manangas |
| R5 | 0 - 200 | 1000 - 1400 | 24 - 26 | Humid semi-arid | Fluvisols and arenosols |
| R7 | 200 - 500 | 1000 - 1200 | 20 - 25 | Humid semi-arid, with sub-humid | Feralsols, luvisols, acrisols |
| R8 | 0 - 200 | 800 - 1200 | 24 - 26 | Humid semi-arid, with spots sub-humid and an extensive of spots of dry semi-arid | Lixisols, leptosols, arenosols |

Source: INIA, (1994).

make a stiff porridge known as *karakata* (Tivana et al., 2007; Tivana, 20012), which is served with vegetables, fish or meat. Cassava flour is used to bake bread and has replaced between 10 and 20% of wheat flour in Mozambique (Salegua et al., 2012). It is also used to bake cakes, as it is more affordable and available than wheat.

### *Rale (roasted cassava roots)*

*Rale* is the traditional name for roasted fermented cassava root and it is a ready to eat food. It can be consumed as a snack with tea, or used as a basic staple food with cooked vegetables or meat. The vegetables or meat are often curried. Although, *rale* only accounts for approximately 1% of overall consumption of cassava products, (IOF, 2008 to 2009) it is almost 20% in the Southern Region (Donavan et al., 2011).

### *Cassava beverages*

*Oteka* and *impala* are alcoholic cassava beverages. These beverages are prepared by cooking cassava flour with water to make a thin starch porridge, which is then mixed with sorghum and malt. This form of cassava is consumed when there is a family celebration, or as a reward at the end of a working day (Donavan et al., 2011; Hagglades et al., 2012). *Mahewu* is a non-alcoholic fermented beverage made from sweet and bitter cassava. The preparation of cassava *mahewu* in Mozambique has not been documented. It was observed that, in Mozambique, fresh cassava root or flour was cooked to form a porridge, which was cooled and then fermented. Traditionally, bread, sorghum, millet malt or wheat flour is fermented to make a starter culture. Fermentation takes 24 to 36 hours at room temperature. Generally sugar was seen to be added to sweeten the *mahewu* before it was consumed.

## Nutritional value of cassava

The nutritional content of cassava depends on the specific plant part (root or leaves), geographic location, variety, age of the plant and environmental conditions (FAOSTAT, 2011a). The cassava root is composed of carbohydrates and is therefore mainly a source of energy. The starch content varies between 32 to 35% of the mass of fresh roots and 80 and 90% of the mass of dried roots (Montagnac et al., 2009b). Amylopectin comprises 83% of the carbohydrate content of cassava roots while amylose comprises 17% (Rawel and Kroll, 2003). The amount of sucrose, glucose, fructose and maltose is usually low (Tewe, 2004), however, in sweet varieties more than 17% of the root content is sucrose (Okigbo, 1980; Charles et al., 2005). Fiber content differs according to the variety and stage of development of the root. In fresh roots it is less than 1.7%, while it comprises4% of cassava flour (Gil and Buitrago, 2002). Lipid content varies between 0.1 and 0.3% of the fresh mass of cassava roots and is present as nonpolar (45%) or glycolipids (52%) (Hudson and Ogunsua, 1974).

The protein content is trivial, between 1 and 3% of dry matter (Buitrago, 1990) and about 1.5 mg/100 g of fresh mass (Bradbury and Holloway, 1988). Essential amino acids are present in low quantities, with the exception of arginine, glutamic acid and aspartic acid (Gil and Buitrago, 2002). The mineral content of the roots varies (Table 2). According to Burns at al. (2012) in a study in Mozambique (Maputo and Nampula Provinces) iron content of cassava roots, varied from 8 to 24 mg/kg whereas zinc was between 3 to 140 pm (Burns et al., 2012). The vitamin content of the roots is low, except for vitamin C, which is found at relatively high levels of between 15 to 45 mg/100g per edible portion (Okigbo, 1980; Gil and Buitrago, 2002; Charles et al., 2004). The content

**Table 2.** Proximate nutrient composition of cassava roots and leaves.

| Variable | Unit | Raw cassava[a] | Cassava roots[b,c,d] | Cassava leaves[b,c] |
|---|---|---|---|---|
| Proximate composition 100 (g) | | | | |
| Food energy | kcal | 160 | 110 - 149 | 91 |
| Food energy | kj | 667 | 526 - 611 | 209 - 251 |
| Moisture | g | 59.68 | 45.9 - 85.3 | 64.8 - 88.6 |
| Dry weight | g | 40.32 | 29.8 - 39.3 | 19 - 28.3 |
| Protein | g | 1.36 | 0.3 - 3.5 | 1.0 - 10.0 |
| Lipid | g | 0.28 | 0.03 - 0.5 | 0.2 - 2.9 |
| Total carbohydrate | g | 39.06 | 25.3 - 35.7 | 7 - 18.3 |
| Dietary fiber | g | 1.8 | 0.1 - 3.7 | 0.5 - 10.0 |
| Ash | g | 0.62 | 0.4 - 1.7 | 0.7 - 4.5 |
| | | | | |
| **Vitamins** | | | | |
| Thiamin | mg | 0.087 | 0.03 - 0.28 | 0.06 - 0.31 |
| Riboflavin | mg | 0.048 | 0.03 - 0.06 | 0.021 - 0.74 |
| Niacin | mg | 0.854 | 0.6 - 1.09 | 1.3 - 2.8 |
| Ascorbic acid | mg | 20.6 | 14.9 - 50 | 60 - 370 |
| Vitamin A | µg | - | 5.0 - 35.0 | 8300 - 11800[e] |
| | | | | |
| **Minerals** | | | | |
| Calcium | mg | 16 | 19 - 176 | 34 - 708 |
| Total phosphorus | mg | 27 | 6 - 152 | 27 - 211 |
| Ca/P | | 0.6 | 1.6 – 5.48 | 2.5 |
| Iron | mg | 0.27 | 0.3 - 14.0 | 0.4 - 8.3 |
| Potassium | % | - | 0.25 - 0.72 | 0.35 - 1.23 |
| Magnesium | % | - | 0.03 - 0.08 | 0.12 - 0.42 |
| Copper | ppm | - | 2.00 - 6.00 | 3.00 - 12.0 |
| Zinc | ppm | - | 14.00 - 41.00 | 71.00 - 249.0 |
| Sodium | ppm | - | 76.00 - 213.00 | 51.0 - 177.0 |
| Manganese | ppm | - | 3.00 - 10.00 | 72.0 - 252.0 |

Source: [a]Values were obtained from the USDA Natl. Nutrient database for standard references (http://www.nal.usda.gov/fnic/foodcomp/search/). Nutrient values and weights are for the edible portion; [b]Bradbury and Holloway (1988); [c]Woot-Tsuen et al., (1968); [d]Favier (1977); [e]Lancaster et al., (1982).

of B vitamins of cassava roots is also found to be trivial (Gil and Buitrago, 2002).

The nutrient density of cassava leaves differs in quantity and quality according to cassava variety, stage of development of the shrub and the relative size of the leaves and stem (Gil and Buitrago, 2002). As can be seen from Table 2, cassava leaves contain a higher amount of protein, essential amino acids, vitamins $B_1$, $B_2$, and C and carotenoids than roots (Adewusi and Bradbury, 1993; Okigbo 1980; West at al., 1988) and the starch content varies from 7 to 18 g/100g (Gil and Buitrago, 2002). They also contain more minerals such as calcium, iron, manganese, and magnesium and zinc (Webeto at al., 2006). In Mozambique (Maputo and Nampula Provinces) iron content in cassava leaves ranged from 284 to 93 mg/kg and the zinc concentrations varies between 47,8 and 27,0 mg/kg (Burns et al., 2012).

### *Anti-nutritional factors in cassava*

Cassava contains anti-nutritional compounds that affect the digestibility and absorption of nutrients. Processing techniques are used to counteract or denature anti-nutritional factors and increase nutritional value of the leaves and roots (Montagnac et al., 2009b). These anti-nutritional factors include phytates, polyphenols, oxalates, nitrates and saponins. The quantity depends on the variety of cassava and plant maturity (Nambisan and Sundareson, 1994; Montagnac et al., 2009a; Bandna et al., 2012 - 2013).

Phytate content of roots is about 624 mg/100 g in the roots (Marfo et al., 1990). It interferes with the uptake and utilisation of minerals such as calcium, iron, molybdenum and zinc in consumers (Hambidge at al., 2008). Phytate has also been reported to interfere with the enzymatic process of protein digestion in the gut (Singh and Krikorian, 1982).

Polyphenols or tannins, mainly found in cassava leaves, range from 2.1 to 120 mg/100g dry weight (Wobeto et al., 2007). Studies have reported that tannins interfere with the uptake of nonheme-iron in consumers and decrease thiamin absorption and the digestibility of starch (Silva and Silva, 1999). They also interfere with lipid and protein digestibility (Bravo, 1998). Oxalates are found in cassava leaves and it is well known that oxalates interfere with calcium uptake (Wobeto et al., 2007). Oxalate a in cassava leaves varies from 1.35 to 2.88 g/100 g dry weight (Correa, 2000; Wobeto et al., 2007).

### *Cassava toxicity*

The most toxic substances found in both cassava roots and leaves are cyanogenic glycosides. The level depends on the genotype as well as environmental factors, for instance drought. Acute poisoning and death following consumption of cassava is rare. Cases that have been reported are mainly associated with undernourishment, marasmus and kwashiorkor (Cock, 1973; FAO/WHO, 2005). In Mozambique, acute poisoning from cassava consumption has, however, been reported in Nampula Province (Essers et al., 1992). To avoid acute toxicity in humans the level of cyanide should be less than 10 mg equivalent/kg dry matter (FAO/WHO, 1991a). Unfortunately, it is estimated that the level of cyanide from cassava, consumed per person in Mozambique, is 14 to 70 times higher based on bodyweight (Burns et al., 2012), than the safety limit proposed by the World Health Organization (FAO/WHO, 1991b).

Chronic poisoning follows long-term consumption of cassava with high cyanogenic glycoside content. This has been reported in countries where cassava is used as staple food, including Mozambique (Dufour, 1988; Mckey and Beckerman, 1993; Ministry of Health Mozambique, 1984; Cliff at al., 1997; Cliff at al., 2011). Chronic intoxication from cassava manifests as tropical neuropathy (Osuntokun, 1994), glucose intolerance and konzo (Ernesto et al., 2002, Cliff at al., 1997; Cliff at al., 2011), goitre and cretinism (Delange et al., 1994).

In Mozambique, konzo is the most common cassava intoxication reported. The first case of konzo was reported in 1981 (Ministry of Health Mozambique, 1984). Population groups found to be most vulnerable are women of child bearing age and children up to two years of age (Cliff et al., 2011). All cases reported in the country have been found to be related to consumption of bitter cassava which was poorly processed, collected during a famine period, during the civil war, or to a diet deficient in sulfur amino acids (Ministry of Health Mozambique, 1984; Essers at al., 1992; Tylleskar er al., 1993; Cliff at al., 1997; Ernesto et al., 2002; Cliff at al., 2011). Patients with konzo have been found to have a high level of thiocyanete in their urine (Ministry of Health Mozambique, 1984; Ernesto et al., 2002; Cliff at al., 2011). Thiocyanate remains in the body as a result of the detoxification of cyanide. It is stored in the stomach and patients with konzo are also at risk to develop stomach cancer (Mirvish, 1983; Maduagwu and Umoh, 1988).

Like anti-nutritional factors, cyanogenic glycosides can be reduced or removed by processing. Processing methods include drying, roasting, boiling, soaking, wetting, and fermentation (Zvauya et al., 2002; Cardoso et al., 2005; Tivana et al., 2007; Cumbana et al., 2007; Tivana et al., 2009). Detoxification processes can be used singularly or in combination. In most cases the combined use of processes results in better nutrient quality (Ikemefuna et al., 1991). A long term approach to the removal of cyanogenic glycosides would be by genetic manipulation to breed less toxic cassava plants (Soetan, 2008). However, this would not solve the problem completely, as environmental factors also play a role.

## CONCLUSIONS

This overview consolidates recent publications on the production, consumption and nutritional value of cassava in Mozambique. Many rural parts of Mozambique are remote and lack infrastructure. It is mainly in these deep rural areas that cassava plays an essential role in food security. It is well recognized that the toxic and anti-nutritional factors in cassava threaten food safety in vulnerable populations, particularly during famine or civil unrest in Mozambique. Yet there are affordable traditional ways of reducing and eliminating these toxic compounds and improving the health of rural populations, using traditional cassava products such as *karakata, rale, mahewu* and *oteka*. On a positive note, cassava provides an affordable and accessible source of carbohydrates in Mozambique, as it is easily cultivated, even at subsistence level and is resistant to drought and insects.
The macro-economic impact of cassava flour is particularly important to make bread more affordable, as imported wheat is expensive in a country with a very low Gross Domestic Product and a high level of poverty. More research is required on ways to enrich cassava products, to meet the need for essential nutrients in vulnerable populations, particularly growing children and women of childbearing age, possibly including fortification with essential nutrients such as iron.

## ACKNOWLEDGEMENTS

Financial assistance from the South African National Research Foundation and the Graca Machel Scholarship is gratefully acknowledged. Permission to undertake this study was granted by the Research Ethics Committee of the University of Pretoria (67/2013).

## Conflict of interest

The author(s) have not declared any conflict of interest.

## REFERENCES


Adewusi SRA, Bradbury JH (1993). Carotenoid in cassava: comparison of open column and HPLC methods of analysis. J. Sci. Food Agric. 62:375-83.

Amoa-Awua W, Frisvad J (1997). Sefa-Dedeh S, Jakobsen M: The contribution of moulds and yeasts to the fermentation of "agbelima" cassava dough. J. Appl. Microbiol. 83:288-296.

Arguedas PCR (1982). Residual cyanide concentrations during the extraction of cassava starch. Food Technol. 17:251-262.

Bandna C (2012/2013). Effect of processing on the cyanide content of cassava products in Fiji. JMBFS 2(3):947-958.

Best RTRH (1993). Cassava: The latest facts about an ancient crop. CIAT publication, Cali, Colombia. Bokanga M (1994). Distribution of cyanogenic potential in the cassava germoplasm. Acta Hortic. 375:117-123.

Boonnop K, Wanapat M, Nontaso N, Wanapat S (2009). Enriching nutritive value of cassava root by yeast fermentation. Sci. Agric (Piracicaba, Braz), 66(5):629-633.

Bradbury JH, Holloway WD (1988). Chemistry of tropical root crops: significance for the nutrition and agriculture in the Pacific. Austr. Centre. Int. Agric. Res. (ACIAR). 6:1-20.

Bradbury JH (2006). Simple wetting method to reduce cyanogen content of cassava flour. J. Food Compost. Anal. 19:388-393.

Bravo L (1998). Polyphenols: chemistry, dietary sources, metabolism, and nutritional significance. Nutr. Rev. 56:317-33. PMid:9838798

Buitrago A (1990). La yuga en la alimentacion animal. Centro Internacional de Agricultura Tropical (CIAT). Cali, Colombia. pp. 10-18.

Burns AE, Gleadow RW, Zacarias AM, Cumbana CE, Miller RE, Cavagnaro TR (2012). Variation in the chemical composition of cassava (*Manihont esculenta*, Crantz) leaves and roots as affected by genotypic and environmental variation. J. Agric. Food Chem. 60:4946-4956. PMid:22515684

Cardoso AP, Mirione E, Ernesto M, Massaza F, Cliff J, Haque MR, Bradbury JH (2005). Processing of cassava roots to remove cyanogens. J. Food Compost Anal. 18:451-460.

Charles AL, Chang YH, Ko WC, Sriroth K, Huang TC (2004). Some physical and chemical properties of starch isolates of cassava genotypes. Starch-Starke. 56(56):413-418.

Charles AL, Sriroth K, Huang TC (2005). Proximate composition, mineral contents, hydrogen cyanide and phytic acid of five cassava genotypes. Food Chem. 92:615-620.

Chiwona-Karltun L, Mkumbira J, Saka J, Bovin M, Mahungu NM, Rosling H (1998). The importance of being bitter – a qualitative study on cassava cultivar preference in Malawi. Ecol. Food Nutr. 37:219-245.

Chiwona-Karltun L, Tylleskar T, Mkumbira J, Gebre-Medhin M, Rosling H (2000). Low dietary cyanogen exposure from frequent consumption of potentially toxic cassava in Malawi. Int. J. Food Sci. Nutr. 51:33-43. PMid:10746103

Chiwona-karltun L, Brimer L, Kalenga SJD, Mhone AR, Mkumbira J, Johansson L, Bokanga M (2004). Bitter taste in cassava roots correlates with cyanogenic glucoside levels. J. Sci. Food Agric. 84:581-590.

Cliff J (1994). Cassava safety in times of war and drought in Mozambique. Acta Hortic. (ISHS) 375:373-378.

Cliff J, Nicala D, Saute F, Givragy R, Azambuja G, Taela A (1997). Konzo associated with war in Mozambique. Trop. Med. Int. Health. 2:1068-1074. PMid:9391509

Cliff J, Muquingue H, Nhassico D, Nzwalo H, Bradbury JH (2011). Konzo and continuing cyanide intoxication from cassava in Mozambique. Food Chem. Toxicol. 49:631-635. PMid:20654676

Cock JH (1973). Cyanide toxicity in relation to the cassava research program of CIAT in Colombia. In ME Nestel B (eds), Chronic cassava toxicity. Proceedings of the interdisciplinary workshop edn. Ottawa: International Development Research Centre. pp. 37-40.

Correa AD (2000). Farinha de folhas de mandioca (*Manihot esculenta* Crantz cv. Baiana) efeito de processamentos sobre alguns nutrients e antinutrientes. 108f. Tese (Doutorado em Ciências de Alimentos). Lavras, Brazil: Univ. Federal de Lavras.

Cumbana A, Mirione E, Cliff J, Bradbury JH (2007). Redution of cyanide contents of cassava flour in Mozambique by wetting method. Food Chem. 101:894-897.

Delange F, Ekpechi LO, Rosling H (1994). Cassava cyanogenesis and iodine dificiency disorders. Acta Hortic. 375:289-293.

Donovan C, Haggblade S, Salrgua AV, Cuambe C, Mudema J, Tomo A (2011). Cassava commercialization in Mozambique. MSU International development working . 120:1-59.

Dufour DL (1988). Cyanide content of cassava (*Manihot esculenta*, Euphorbiaceae) cultivars used by Tukanoan Indians in Northwest Amazonia. Econ. Bot. 42:255-266.

El-Sharkawy MA (2004). Cassava biology and physiology. Plant Mol. Biol. 56:481-501. PMid:15669146

Ernesto M, Cardoso AP, Nicala D, Mirione E, Massaza F, Cliff J, Haque Mr, Bradbury JH (2002). Persistent Konzo and cyanide toxicity from cassava in Northern Mozambique. Acta Trop. 82:357-362.

Essers AJA, Alsen P, Rosling H (1992). Insufficient processing of cassava induced acute intoxications and the paralytic disease konzo in rural area of Mozambique. Ecol. Food Nutr. 27:172-177.

Factfish (2014) Mozambique cassava. Available online: http://www.factfish.com/statistic-country/mozambique/cassava Accessed on 04 Jun 2014.

FAO/WHO (1991a). Joint FAO/WHO food standard programme. Codex Alimentarius commission. XII Suppl. 4 Ed. Rome: FAO. pp. 1-42.

FAO/WHO (1991b). Joint FAO/WHO food standard programme: In: Codex Alimentarius Commission XII.

FAO/WHO (2005). Codex standard for sweet cassava. In Codex standard, (eds), Joint FAO/WHO Food standard programmer: Codex Alimentarius Commission. Rome: FAO/WHO. 238-2003.

FAO/WFP (2010). Crop and food security assessment to Mozambique.

FAO (2011a). Food outlook. Global market analysis, global information and early warning system on food and agriculture. Rome: FAO. 1-109.

FAO (2011b). Nutrition country profile of Mozambique. Rome: FAO.

FAOSTAT (2011a). Food balance sheet. Available online: http://www.faostat.fao.org Accessed on 20 Jun 2012.

FAOSTAT (2011b). Mozambique country profile. Available online: http://wwwfao.org/corp/statistics/en/ Accessed on 20 November 2013.

Favier JC (1977). Valeur alimentaire de deux aliments de base Africains: le manioc et le sorgho. Paris, France: ORSTOM (editions de l'Office de la Recherche Scientifique et Technique Outre-mer). Travaux et documents nr 67. Available from: http://www.congoforum.be/upldocs/manioc.pdf. Accessed on 07 December 2013.

Gade D (2003). Names for Manihot esculenta: Geographical variation and lexical clarification. JLAG. 1:43-57.

Gbadegesin MA, Olaiya CO, Beeching JR (2013). African cassava:

Biotechnology and molecular breeding to the rescue. Br Biotechnol J. 3(3):305-317.
Gil JL, Buitrago AJA (2002). La yuga en la alimentacion animal. In: Osopina B, Ceballos H, (Eds). La yuga en el tercer milénio: sistemas modernos de produccion, procesamiento, utilizacion y comercializacion. Cali, Colombia: CIAT: pp. 527-569. Available online: http://www.clayuca.org/PDF/libro-yuca/capitulo28.pdf Accessed on 29 Jun 2012.
Hagglades S, Cumbana C, Donovan C, Droppimunn K, Jirstrom M (2012). Assava commercialization in Southeastern Africa. JADEE. 2(1):4-40.
Hambidge KM, Miller LV, Westcott JL, Krebs NF (2008). Dietary reference intakes for zinc may require adjustment for phytate intake based upon model predictions. J. Nutr. 138:2363-6.

PMCid:PMC2635502
Hoguane AM (2007). Diagnosis of Mozambique coastal zone. JICZM. 7(1):69-82.
Howard J, Crawford C, Kelly V, Demeke M, Jeje JJ (2003). Promoting high-input maize technologies in Africa: the Sasakawa-Global 2000 experience in Ethiopia and Mozambique. Food Policy. 28:335-348.

Hudson BJF, Ogunsua OA (1974). Lipids of cassava tubers (*Manihot esculenta,* Crantz). J. Sci. Food Agric. 25:1503-1508. PMid:4437153
IFAD/FAO (2005). Proceedings of the validation forum on the global cassava development strategy: A review of cassava in Africa with country case studies on Nigeria, Ghana, the United Republic of Tanzania, Uganda and Benin. 2:1-66.
Ikemefuna C, Obizoba J, Atii JV (1991). Effects of soakin, sprouting, fermentation and cooking on nutrient composition and some antinutritional factors of sorghum (Guinesia) seeds. Plant. Food Hum. Nutr. 41:203-212.
IIAM/PNRT (2007). Potential production areas of cassava in Mozambique. Maputo, Mozambique: MINAG.
INE (Instituto Nacional de Estatísticas) (2009). Apresentação dos resultados definitivos do censo 2007 (Presentation of the final results of the census 2007). Available online: http://www.ine.gov.mz/censo2007/censo2007 Accessed on 12 May 2012.
INE (Instituto Nacional de Estatíscas) National institute of Statistics. Censo Agro-pecuário (Agriculture and Livestock Census) 2009-2010: Resultados definitivos (Definitive results) (2011). Available online: http://www.ine.gov.mz/censos_dir/agro-pecuaria/CAP_VF.pdf. Accessed on 03 Jun 2014.
INIA (Instituto Nacional de Investigação Agrária) (1994). Zonas agras ecológicas de Moçambique (Agro-ecological zones of Mozambique). Maputo, Mozambique: INIA.
INIA (Instituto Nacional de Investigação Agrária)/IITA/SARRNET (2002). Cassava and sweet potato production, processing and markiting in Mozambique. Maputo, Mozambique: INIA.
IOF (2008/09). Mozambique household budget and expenditure survey.
Lancaster PA, Ingram JS, Lim MY, Coursey DG (1982). Traditional cassava-based foods: survey of processing techniques. Econ. Bot. 36:12–45.
Mader (Ministério de Agricultura e Desenvolvimento Rural) (2005). Trabalho de inquérito agrícola 2005 (National Agricultural Survey 2005). National Agricultural survey Maputo, Mozambique: Mader.
Maduagwu EN, Umoh IB (1988). Dietary thiocyanate and N-nitrosation in vivo in wistar rat. Ann. Nutr. Metab. 32(1):30-37. PMid:3355109
Marfo EK, Simpson BK, Idowu JS, Oke OL (1990). Effect of local food processing on phytate levels in cassava, cocoyam, yam, maize, sorghum, rice, cowpea, and soybean. J. Agric .Food Chem. 38:1580-1585.
Mckey D, Beckerman S (1993). Chemical ecology, plant evolution and traditional manioc cultivation systems. *In* Hladik CM, Hladik A, Linares OF, Pagezy H, Semple A, Hadley M [eds], Tropical forests, people and food: Biocultural interactions and applications to development, pp. 83–112. UNESCO, Paris, France, and Parthenon, Canforth, UK.
Mckey D, Cavagnaro TR, Cliff J, Gleadow R (2010). Chemical ecology in coupled human and natural systems: People, manioc, multitrophic interactions and global change. Chemoecology 20:109-133.

Mkumbira J, Chiwona-Karltun L, Lagercrantz U, Mahungu NM, Saka J, Mhone A, Bokanga M, BrimerL, Gullberg U, Rosling H (2003). Classification of cassava into 'bitter' and 'cool' in Malawi: from farmers' perception to characterization by molecular markers. Euphytica. 132:7-22.
MIC (Ministério do Comercio) /FAO/EC (2004). External market task force: Report on cassava. 3(1):1-90.
MIC (Ministério de Comercio) (2007). Cassava development strategy for Mozambique 2008-2012. Maputo, Mozambique: MIC.
Ministry of Health Mozambique (1984). Mantakassa: an epidemic of spastic paraparesis associated with chronic cyanide intoxication in a cassava staple area of Mozambique: epidemiology and clinical and laboratory findings in patients. Bull. World Health Org. 62:477-484.
Mirvish SS (1983). The etiology of gastric cancer, intr-gastric nitrosamide formation and other theories. J. Nat. Cancer. Inst. 71:630-636.
Montagnac JA, Davis CR, Tanumihardjo SA (2009a) Reduce toxicity and anti-nutrients of cassava for use as a staple food. Compr. Rev. Food Sci. Saf. 8:17-27.
Montagnac JA, Davis CR, Tanumihardjo SA (2009b) Nutritional value of cassava for use as a staple food and recent advances for important. Compr. Rev. Food Sci. Saf. 8:181-194.

Mowat L (1989). Cassava and chicha: Bread and beer of the Amazonian Indians. Shire Publications, Aylesbury, UK.
Nambisan B, Sundareson S (1985). Effect of processing on the cyanoglucosidr content of cassava. J. Sci. Food Agric. 36:1197-1103.

Nambisan B, Sundareson S (1994). Distribution of linamarin and its metabolizing anzimes in cassava tissues. J. Sci. Food Agric. 66:503-507.
Nhassico D, Muquingue H, Cliff J, Cumbana A, Bradbury JH (2008). Review Rising African cassava production, diseases due to high cyanide intake and control measures. J. Sci. Food Agric. 88:2043-2049.
Nielson HH (2009). The role of cassava in smallholder maize marketing in Zambia and Mozambique. Michigan State University.
Okigbo BN (1980). Nutritional implication of projects giving high priority of the production of staples of low nutritive quality. In the case of cassava (Manihont esculenta, Crantz) In the humid tropics West Africa. Food Nutr. Bull. 2:1-10.
Osuntokun BO (1994). Chronic cyanide intoxication of dietary origin and a degenerative neuropathy in Nigerians. Acta Hortic. 375:311-321.
Oyewole OB (2001). Characteristics and significance of yeast involvement in cassava fermentation for '*fufu*' production. Int. J. Food Microbiol. 65:213-218.

Rawel HM, Kroll J (2003). Importance of cassava (*Manihot esculenta* Crantz) as the main staple food in tropical countries. DLR. 99:102-110.
Salegua V, Donavan C, Haggblades S, Cuabe C, Nhatumbo A (2012). Dinâmica da cadeia de valores da mandioca no Norte de Moçambique (Dynamics of the cassava value chain in the North Mozambique) IIAM Boletim Nordeste. 1:1-7.
Silva MR, Silva MAAPd (1999). Nutritional aspects of phytates and tannins. Revista de Nutrição. 12:21-32.
Singh M, Krikorian AD (1982). Inhibition of trypsin activity *in vitro* by phytate. J. Agric. Food Chem. 30:799-800.

Siritunga D, Sayre RT (2003). Generation of cyanogen-free transgenic cassava. Planta. 217:367-373. PMid:14520563
Soetan KO (2008). Pharmacological and other beneficial effects of antinutritional factorsin plants. Afr. J. Biotechnol. 7(25):4713-4721.
Tewe OO (2004). Cassava for livestock feed in sub-Saharan Africa. FAO/FAD. 1255:1-64.
Tivana LD, Bvochora JM, Mutukumira AN, Owens JD (2007). A study of heap fermentation processing of cassava roots in Nampula Province, Mozambique. J. Root Crops. 33(2):118-128.

Tivana LD, Da Cruz FJ, Bergenstahl B, Dejmek P (2009). Cyanogenc potential of roasted cassava (*Manihot esculenta* Crantz) roots *rale* from Inhambane Province, Mozambique. Czech. J. Food Sci. 27:S375-S378.

Tivana LD (2012). Cassava processing: Safety and protein fortification, Lund University.

Tostao E, Wade BB (2005). Spatial efficiency in Mozambique's post-reform maize markets. Agricultural Economics, Int. Assoc. Agric. Econ. (IAAE). 33(2):205-214.

Tschirley D, Abdula D, Weber MT (2006). Toward improved maize marketing and trade policies to promote household food security in Central and Southern Mozambique. Research Report 60E, Ministry of Agriculture, Republic of Mozambique.

Tylleskar T, Howlett WP, Rwiza HT, Aquilonius SM, Stalberg E, Linden B (1993). Konzo: a distinct disease entity with selective upper motor neuron damage. J. Neurol. Neurosurg. Psychiatry 56:638-643. PMid:8509777 PMCid:PMC489613

UNDP (2010) Human development report 2010. The real wealth of nations: pathways to human development. Palgrave MacMillan, Basingstoke.

USDA National Nutrient Database for Standard Reference. Available from: http://www.nal.usda.gov/fnic/foodcomp/search/. Accessed on 10 July 2013.

Walker T, Pitoro R, Tomo A, Sitoe I, Salência C, Mahanzule RC. Donovan C, Mazuze F (2006). Priority setting for public-sector. Agricultural Research in Mozambique with the National Agricultural Survey Data. Research report No. 3E. Institute of Agricultural Research of Mozambique. Maputo: Mozambique.

West CE, Pepping F, Temalilwa CR (1988). The composition of foods commonly eaten in East Africa. The Netherlands: Wageningen Agricultural University.

Wobeto C, Correa AD, de Abreu CMP, dos Santos CD, de Abreu JR (2006). Nutrients in the cassava (*Manihot esculenta*, Crantz) leaf meal at three ages of the plant. Cienc. Technol Aliment. 26:865-869.

Wobeto C, Correa AD, de Abreu CMP, dos Santos CD, Pereira HV (2007). Antinutrients in the cassava (*Manihot esculenta* Crantz) leaf powder at three ages of the plant. Cienc Technol Aliment. 27:108-112.

Woot-Tsuen WL, Busson F, Jardin C (1968). Food composition table for use in Africa. FAO corporate document repository. Rome, Italy. Available online: http://www.fao.org/docrep/003/X6877E/X6877E00.htm#TOC. Accessed on 20 November 2013.

Zvauya R, Ernesto M, Bvochora T, Tivana LD, Da Cruz FJ (2002). Effect of village processing methods on cyanigenic potential of cassava flour collected from selected districts of Nampula Province in Mozambique. Int. J. Food Sci. Technol. 37:463-469.

# Effect of ridge height and planting orientation on *Ipomea batatas* (sweet potato) production

**C. Parwada*, C. T. Gadzirayi and A. B. Sithole**

Department of Agriculture, Bindura University of Science Education, P. Bag 1020, Bindura, Zimbabwe.

***Ipomea batatas*** **(sweet potato) yield vary widely among farmers due to improper ridging and planting orientation. This study was to establish the proper ridging and planting orientation, so as to enhance constant reliable yields among sweet potato producing farmers. The objectives of the study were to compare vine length and to determine root yield, central root diameter and root length among different ridge heights and planting orientations. A 3 x 3 factorial in a completely randomised block design with 3 replications was used. Ridge height had levels; 30, 40 and 50 cm while planting angle had levels; horizontal 180°, inclined 45° and vertical 90°. Vine length was measured at 3, 6, 9 and 12 weeks after planting and root yield, root diameter and root length were measured at harvesting. Data was analysed using M STAT C for variance between treatment means. The results showed interaction effect ($p = 0.05$) of ridge height and planting angle on the vine length, root diameter, length and yield. Lower ridges produced longer vines than higher ridges. Decreasing inclination of cuttings also increases vine length. Root diameter decreases with increasing ridge height while root length increased as ridge height increased. Medium ridge height (40 cm) with inclined planting angle may be recommended for higher root yields and horizontal planting angle on low ridges can be used to produce rounded swollen roots.**

**Key words:** Ridge, height, planting orientation, sweet potato, vine, yield.

## INTRODUCTION

Sweet potato tolerates a wide range of soils and even on poor acid soils, it gives satisfactory yields (Chipangura and Jackson, 2003). Though grown in areas with relatively high rainfall, it cannot withstand water logging conditions; hence, why they are grown on ridges and mounds (Gomes, 1999). It also has good drought tolerance ability. In Zimbabwe, its production is concentrated in natural regions I, II and III which have high rainfall and good soils (Chipangura and Jackson, 2003). Sweet potato was discovered to be potentially high yielding such that, it can yield up to 50 tonnes per hectare under minimum management (Coertze and Van den Berg, 1995). The area under sweet potato production in Zimbabwe has recently increased; however, most communal farmers have learnt by experience to grow sweet potatoes. Variable planting methods are being used by sweet potato farmers in Zimbabwe. Many farmers believe that, high yields are produced from very high ridges, yet Dhliwayo and Chiunzi (2004) reiterate that, small to medium sized ridges that are easy to make may produce good yields as long as fertility is present. Ridges should also be high enough to prevent water logging (Gomes, 1999). On the planting angle, various authors have diverged. Dhliwayo and Chiunzi (2004) stipulate that, planting at an angle or horizontally produce more yields while Onwuene (1999) recommends vertical orientation. It was the purpose of this study to explore and examine these different lines of thoughts and get the most appropriate recommendation to farmers. This was achieved by determining vine length on different ridge heights under different planting angles and root diameter, length and yield on different planting angles under varying ridge heights. For this study, Brondal variety which takes about 135 days to mature was used. The variety was developed in South Africa and is now popular in Zimbabwean rural areas. Vine cuttings were used as

*Corresponding author. E-mail: crparwada@yahoo.com

**Table 1.** Factors, factor levels and treatments.

| Factor A: Planting angle | Factor B: Ridge height (cm) | | |
|---|---|---|---|
| | (1) 30 | (2) 40 | (3) 50 |
| 1. Horizontal (H) | 30H (1) | 40H (2) | 50 H (3) |
| 2. Inclined (I) | 30 I (4) | 40I (5) | 50 I (6) |
| 3. Vertical (V) | 30 V (7) | 40V (8) | 50V (9) |

planting material.

## MATERIALS AND METHODS

### Study site description

The experiment was carried out in Natural Farming Region IIa of Zimbabwe. The region receives high annual rainfall of 700 to 1000 mm with warm summer (18 to 22°C) and cool winter (16 to 18°C). The site is characterised by loam soil type with pH 5.8 ($CaCl_2$ scale) and local vegetation consists of veld grasses and Musasa trees. The area is located 1248 m above sea level. The site is gently sloping and well drained. The plot was previously occupied by organically produced green mealies.

### Experimental design

A 3 x 3 factorial experiment in a Randomised Complete Block Design (CRBD) with 3 replicates was used. Slope was the blocking factor.

### Land preparation

The trial plot was ploughed to a depth of 30 cm using an ox drawn plough, after which decomposed cattle manure was broadcast at a rate of 20 tonnes per hectare and incorporated into the soil. 27 ridges, each 1.2 m long and 50 cm wide were made. Ridges were spaced 1 m between blocks and 50 cm within blocks. Ridges differed in height among treatments.

### Planting

The vine cuttings were cut into 30 cm long and planted with two thirds of their lengths buried into the soil uniformly on all treatments, making sure an equal number of nodes were buried into the soil. Cuttings were spaced 30 cm apart along the 1.2 m long ridges in two rows, leaving 15 cm spaces at both edges. Therefore, a total of 6 cuttings were planted per ridge. A wooden campus was placed on the levelled ridges to obtain a planting angle. The horizontal, vertical and inclined angles were measured 180°, 90° and 45° respectively (Table 1).

### Data collection

Vine measurements (in centimetres) were taken at a three week interval from week 3 after planting up to week 12. Total vine length per ridge was obtained and average length recorded. At 16 weeks post planting, the yield from each ridge was weighed in kilograms and converted into tonnes per hectare. Mean root length and mean central root diameter were also calculated and recorded in cm.

### Data analysis

A computer package M STAT C was used to analyse the variance between treatment means at 95% significance level using the F-test. The means were separated using the least significant difference ($LSD_{0.05}$) test at 5% to determine difference between the treatment means.

## RESULTS

### The effect of ridge height and cutting planting angle on vine length

Vine length was determined by the interaction ($p = 0.05$) of ridge height and planting angle as from 3 to 12 weeks after planting (Table 2). At 3 weeks after planting, treatments 40H and 40I had the highest mean vine lengths while 50H had the least. At 6, 9 and 12 weeks after planting, treatment 30H had the highest mean vine length while 50V had the least (Table 2).

### The effect of ridge height and planting angle on root diameter, root length and yield

Central root diameter at harvesting was determined by both the ridge height and planting angle ($p = 0.05$) (Table 3). The thickest roots were recorded from 30H (6.3 cm) while the thinnest roots were recorded from 50I and 50V (4.4 cm). Ridge height had a significant effect ($p = 0.05$) on root length at harvesting (Table 3). The longest roots were obtained from 50 cm ridges while the shortest roots were obtained from 30 cm ridges. There was interaction effects of ridge height and planting angle on final root yield ($p = 0.05$). The highest mean yield recorded was from 40I treatments (14.08 t/ha) while the lowest mean yield recorded was from 50V treatments with 9.73 t/ha (Table 3).

**Table 2.** Mean vine lengths in cm at 3, 6, 9 and 12 weeks after planting.

| Treatment combinations | 3 weeks | 6 weeks | 9 weeks | 12 weeks |
|---|---|---|---|---|
| 1. Horizontal + 30 cm | $24.10^{c}$ | $96.33^{a}$ | $216.3^{a}$ | $341.0^{a}$ |
| 2. Horizontal + 40 cm | $25.33^{a}$ | $80.67^{e}$ | $201.0^{c}$ | $330.3^{d}$ |
| 3. Horizontal + 50 cm | $24.00^{c}$ | $72.33^{g}$ | $191.7^{d}$ | $310.3^{fg}$ |
| 4. Inclined + 30 cm | $24.77^{b}$ | $90.33^{b}$ | $209.7^{b}$ | $336.0^{b}$ |
| 5. Inclined + 40 cm | $25.33^{a}$ | $86.00^{c}$ | $207.7^{b}$ | $332.7^{c}$ |
| 6. Inclined + 50 cm | $24.47^{bc}$ | $67.00^{h}$ | $190.0^{d}$ | $308.7^{g}$ |
| 7. Vertical + 30 cm | $24.27^{bc}$ | $83.33^{d}$ | $207.3^{b}$ | $326.7^{e}$ |
| 8. Vertical + 40 cm | $24.17^{c}$ | $75.00^{f}$ | $192.0^{d}$ | $312.3^{f}$ |
| 9. Vertical + 50 cm | $24.23^{bc}$ | $65.00^{i}$ | $189.3^{d}$ | $304.3^{h}$ |
| Coefficient of variation | 1.27% | 1.19% | 0.90% | 0.40% |
| $LSD_{0.05}$ | 0.5391 | 1.645 | 3.134 | 2.250 |
| Grand mean | 24.52 | 79.56 | 200.6 | 322.5 |
| P Values: Planting angle | 0.0020 | <0.001 | <0.001 | <0.001 |
| Ridge height | <0.001 | <0.001 | <0.001 | 0.0004 |
| Interaction | 0.0083 | <0.001 | <0.001 | <0.001 |

Means with same letter(s) are statistically insignificant at p = 0.05.

**Table 3.** Mean root diameter (cm), mean root length (cm) and mean root yield (t/ha) at harvesting.

| Treatment combinations | Root diameter | Root length | Root yield |
|---|---|---|---|
| 1. Horizontal + 30 cm | $6.3^{a}$ | $11.6^{c}$ | $12.64^{c}$ |
| 2. Horizontal + 40 cm | $5.4^{c}$ | $12.7^{b}$ | $13.50^{b}$ |
| 3. Horizontal + 50 cm | $4.8^{d}$ | $16.2^{a}$ | $10.66^{f}$ |
| 4. Inclined + 30 cm | $6.1^{ab}$ | $11.6^{c}$ | $11.85^{d}$ |
| 5. Inclined + 40 cm | $5.8^{b}$ | $12.4^{b}$ | $14.08^{a}$ |
| 6. Inclined + 50 cm | $4.4^{e}$ | $16.2^{a}$ | $11.05^{e}$ |
| 7. Vertical + 30 cm | $6.2^{a}$ | $11.3^{c}$ | $11.01^{e}$ |
| 8. Vertical + 40 cm | $5.2^{c}$ | $12.7^{b}$ | $13.53^{b}$ |
| 9. Vertical + 50 cm | $4.4^{e}$ | $16.1^{a}$ | $9.73^{g}$ |
| Coefficient of variation | 3.61% | 2.09% | 0.95% |
| $LSD_{0.05}$ | 0.3374 | 0.4865 | 0.1974 |
| Grand mean | 5.381 | 13.422 | 12.007 |
| P Values: Planting angle | 0.0471 | | <0.001 |
| Ridge height | <0.001 | <0.001 | <0.001 |
| Interaction | 0.0178 | | <0.001 |

Means with same letter(s) are statistically insignificant at p = 0.05

## DISCUSSION

### Effect of ridge height and planting angle on mean vine length

The vine length was determined by the interaction of ridge height and planting angle. Too high (50 cm) or low (30 cm) resulted into shorter vines, as compared to the intermediary ridge height (40 cm). This could be that plant available water and organic matter rich top soil were at a thicker depth of the planting zone on the 40 cm ridges which is necessary for rapid early vine growth. Whilst on the highest and lowest ridges, these could have been inadequate to support a vigorous increase in length. This tallies well with Edmond (2001) who noted that, ridge height and planting angle affect vine length of sweet potatoes at all growth stages.

However, Gomes (1999) denoted that sweet potato early vine growth does not differ significantly but only at its mid and late stages. He argued that this difference at later stages after planting might be due to the different adaptive growth rates as the cuttings established differently on the various treatments. Since cuttings roots were well established at 6 after planting, longer vines on 30 cm ridges might be justified as the cuttings roots could easily tap capillary water and nutrients over a short distance as compared to higher ridges, where cutting roots were far from capillary water. The increased root growth to reach capillary water levels might have compensated reduced vine growth on 40 and 50 cm ridges.

These results concur with Dhliwayo and Chiunzi (2004) but contradicts with Gomes (1999) who postulated that, vine growth tends to be higher on high ridges where there is increased root penetration and ample soil from which nutrients are extracted.

On all ridge heights, horizontally planted cuttings had the highest mean vine lengths, followed by inclined cuttings and lastly vertical cuttings. This could be due to root orientation; as sweet potato roots grow downwards following positive geotropism rather than growing laterally (Bose et al., 2003). This might have resulted in roots of horizontally planted cuttings being evenly spaced and having a larger area from which to tap water and nutrients towards vine growth. On the other hand, roots of vertically planted cuttings grew closer together; having a limited area from which water and nutrients could be tapped for photosynthesis, hence reduced vine growth. These results are also consistent with Gomes (1999).

### Effect of ridge height and planting angle on root diameter at harvesting

Root thickness was proportional to ridge height. The lower the ridge height, the thicker the roots such that, the 30 cm ridges had the thickest roots, followed by 40 cm ridges while 50 cm ridges had the thinnest roots. The thicker roots on lower ridges as compared to higher ridges might be due to the fact that, tuberous roots form after adventitious roots hit hard soil and thickens upwards; so 30 cm ridges' roots hit hard soil earlier than the 40 and 50 cm ridges' roots, resulting in more time for tuber swelling on 30 cm ridges than on 50 cm ridges. This is supported by Dhliwayo and Chiunzi (2004) and Gomes (1999).

The thicker roots on horizontally planted cuttings as compared to vertically planted cuttings, might be due to positive geotropism of roots which resulted in horizontally planted cuttings, having ample space for free swelling, as compared to vertically planted cuttings where roots grew closer together with limited space for expansion (Gomes, 1999).

### Effect of ridge height and planting angle on root length at harvesting

The significant difference on mean root length ($p = 0.05$) due to ridge height where higher ridge produced longer roots, could be as a result of increased root penetration through the soil on higher ridges before hitting hard soil, to initiate tuber formation, resulting in thin elongated roots unlike on low ridges where adventitious roots quickly hit hard soil to initiate root thickening. These results are consistent with Gomes (1999).

### Effect of ridge height and planting angle on root yield at harvesting

Ridge height and planting angle determined the final root yield where the 40 cm ridges had the highest yield. This could be due to the fact that, at this ridge height all the necessary conditions for an ideal tuber formation were satisfied. At this ridge height, both the root diameter and length were optimal and so resulted to highest yield. This is in agreement with Chipangura and Jackson (2003) who noted that, neither too high nor low ridge heights results to high tuber yield.

## CONCLUSION AND RECOMMENDATIONS

Planting sweet potato at a horizontal angle on 30 cm ridge height resulted to longer vines than at vertical angles with high ridges (50 cm). Lower ridges therefore produce longer vines than higher ridges. Decreasing inclination of cuttings also increases vine length. Inclined planting angle on 40 cm ridge heights had higher root yield. Root diameter decreased with increasing ridge height while root length increased as ridge height increased. From the results it can also be concluded that, increasing inclination of cuttings reduces root diameter while root length increases as cutting inclination increases.

Since the results were based on one season, it is recommended for the research to be repeated for the next two or more seasons, ideally in multiple sites of different climate and soil type to increase the scope of validity of findings. Medium ridge height (40 cm) with inclined planting angle may be recommended for higher root yields. To produce rounded swollen roots, horizontally planted cuttings on low ridges (30 cm) may be recommended. Farmers can establish nurseries and fodder sweet potato on low ridges with horizontally planted cuttings to obtain maximum vine yields.

## REFERENCES


Bose TK, Kabir J, Maity TK, Parthasarathy VA, Som MG (2003). Vegetable Crops: NAYA UDYOG, Kolkata, 2: 3.

Chipangura B, Jackson J (2003). Sweet Potato for high yield in Zimbabwe, Farming World, Harare, pp. 5-12.

Coertze AF, Van den Berg AA (1995). Sweet Potato Cultivars, Vegetable and Ornament Plant Institute, Agric. Res. Council, South

Africa. pp. 18.
Dhliwayo P, Chiunzi PD (2004). A Guide to profitable sweet potato production, Harare, Biotechnology Trust of Zimbabwe, pp. 3-9.
Edmond JB (2001). Sweet Potatoes. The AVI Publishing Company, New Delhi, p. 13.
Gomes G (1999). Sweet Potato Growth Characteristics, (First Edition), Academic Press, New York, pp. 4-7.
Onwuene IC (1999). The Tropical Tuber Crops, John Wiley and Sons Ltd, London, p. 43.

# Survey on honey production system, challenges and opportunities in selected areas of Hadya Zone, Ethiopia

**Haftu Kebede**[1] **and Gezu Tadesse**[2]

Department of Animal Science Wachemo University, Hossana Ethiopia.

**The study was carried out in Hadiya zone to asses honey production systems, challenges and opportunities in the study area. A total of 196 farmers were selected randomly and interviewed using semi-structured questionnaires. The result shows most (90.7%) of the beekeepers in the study area have owned only traditional beehives. The average honey yield from traditional, transitional and frame hive was 3.04 ±0.92, 4.9±1.12 and 8.2±2.62 kg/year/hive respectively. There was significance difference (P<0.05) in honey yield per hive/year from traditional hives while there was no significant difference observed (P>0.05) transitional and moveable frame hive between Woredas. Pesticides and herbicides application, shortage of bee forage, presence of pests and predators and shortage of water were found to be the most important constraints of beekeeping in the study area. Despite all the constraints, there is a great potential to increase the production and quality of honey in the district. Strong extension and technical intervention, proper pesticides application, establishing beekeepers cooperative and enhancing bee forage plantation through introducing multi-purpose trees should be encouraged to increase honey production and income of beekeepers in the study area.**

**Key words:** Beekeepers, colony, challenges, honey production.

## INTRODUCTION

Beekeeping is an important agricultural activity in Ethiopia. Owing to its varied ecological and climatic conditions, the country is home to some of the most diverse flora and fauna in Africa. This diversity makes it highly suitable for sustaining a large number of bee colonies (Adgaba, 2007). The country has the largest bee population in Africa. According to CSA (2012/2013) about 5.21 million hives is estimated to be found in the rural sedentary areas of the country. From this total hives, 870,544 and 65,961 beehives are from Southern Nations, Nationalities and hadya zone respectively.

This makes the country one of the largest honey producers and the third largest beeswax producer worldwide. Ethiopian honey production accounts for approximately 2.5% of world production and 21.7% of African honey production (MoARD, 2007). Beekeeping is an environment friendly activity that can be integrated with agricultural practices like crop production, animal husbandry, horticultural crops, conservation of natural resources and it is non-farm business activity that has immense contribution to the economies of segments of the society and to a national economy as a whole (Beyene and Phillips, 2007).

Beekeeping research carried out in the country is though hopeful but did not cover to describe and document the apicultural resources and associated constraints of the sector for its proper intervention and utilization to specific potential regions (Chala et al., 2012). Although, Hadya zone is endowed with diversified type of vegetation and horticultural crops and expected to be potential for beekeeping activities so far there is no researchable information on honey production system, opportunities and constraints in the area. Therefore; production system study is important to identify problems and come up with appropriate development plan for an area (Edessa, 2002). Therefore, this study was designed to assess honey production systems, challenges and opportunities of the study area.

## MATERIALS AND METHODS

### Description of the study area

The study was conducted in two woredas of Hadya zone (Lemo and Analemo), Southern Nations, Nationalities and People Regional State (SNNPRS). Hadya Zone is located at a distance of 232 km from Addis Ababa capital city and 212 km from Hawassa regional city. The astronomical location of Hossana is 7°30' longitude to the North and 37°48' latitude to the East. The average annual temperature and rainfall ranges from 15.1 to 20°C and 1001 to 1200 mm respectively.

### Sampling techniques

A purposive sampling procedure was applied for the study woredas. Based on honeybee colonies potential of the study areas 16 representative kebeles, 8 from each woredas were selected using random sampling techniques.

The households were selected by giving equal chance from a list of farmers which participate in beekeeping activity of the rural kebele in their respective area. Single household respondent was used as sampling unit in this study. Finally, to compromise the representative-ness of the sample for the study 15% of the households among the total bee keeper in each of the sample kebeles were selected using random sampling method for interview and successive data collection. A total of 196 sample households were interviewed by well trained enumerators using a semi-structured questionnaire.

### Methods of data collection

The study used both primary and secondary data to come up with reliable information on honey production system, opportunities and major constraints of the selected potential areas. Primary data were collected on the following aspects: socio-economic demographic data, current practices and placement of bee colony, types of hives, honey yield, inspection of honeybee colonies, honey harvesting, major honeybee flora, honey storage practices, opportunities and major constraints of beekeeping. Secondary data were used to select potential localities based on number of honeybee colonies and honey production acquired from woredas Agricultural Offices and zonal office of agriculture.

### Statistical analyses

All collected data were entered into Microsoft Excel 2007 and descriptive statistics such as mean, frequency, percentage and one way ANOVA were used to analyze the data using SPSS (version 16.0 software of statistical analysis).

## RESULTS AND DISCUSSION

### General characteristics of respondents

The general characteristics associated with beekeeping households respondents are distributed by sex, age, marital status, and educational status are presented in Table 1. The majority of the respondents (93%) were male while the remaining (7%) were female. This is in line with Addis and Malede (2014) who reported that 97.5% of the interviewed beekeepers were male and only 2.5% were female. Moreover, this very limited number of female participation agrees with Tessega (2009) who states that only 1.7% were female. Similarly Hartmann (2004) reported that in Ethiopia traditionally beekeeping is men's job. So, we can say that few women are participated in the beekeeping job in the area. As a result, women are not economically empowered through beekeeping.

The majority age of the beekeepers in the study area ranges between 40 to 49 years (37.6%). This result showed that people in the most productive age is actively engaged in beekeeping activities. This result agrees with Chala et al. (2013) who report that mean age of the respondents was 40.47 years. Of the total households interviewed, 96.8% are married. This agrees with Tessega (2009) who stated high percentages of the respondents (97.5%) were married.

Concerning to level of education, the highest percentage (51.9%) of the respondents had higher primary level of education and 27.2% of those interviewed beekeepers had not attended any formal or informal education. The rest were at different stages of education ranging from reading and writing skills to completion of secondary school (Table 1). Gichora (2003) noted that for more advanced beekeeping, one should have a good grasp of bee biology and behavior of bees for better colony management. Moreover, for illiterate people there is a need of intensive training and persuading of beekeepers before distributing movable frame hives. Therefore, according to the result of this study the (27.2%) in the district limits the effectiveness of formal training programs and requires more emphasis to be placed on practical demonstration of essential concepts especially in improved beekeeping. This shows that traditional beekeeping practices are based on informal opportunities and an individual's level of formal education does not matter as most of the beekeepers in this study are uneducated people. This is in line with Gichora (2003) who noted the insignificant role of level of education in the traditional beekeeping.

### Current practices and placement of bee colony

Most respondents (60.3%) replied that they have got their colonies by catching swarms and the rest obtained from their parents and buying (Table 2). This is in line with

**Table 1.** Characteristics of the Sample Households in bee keeping.

| Category | Variables | Anlemo woreda (n=75) | Lemo woreda (n=121) | Overall (N =196) |
|---|---|---|---|---|
| Sex | Male | 89.3% | 96.7% | 93% |
| | Female | 10.7% | 3.3% | 7% |
| Age (yrs) | < 30 | 2.7 | 2.5 | 2.6 |
| | 30 -39 | 45.3 | 23.1 | 34.2 |
| | 40-49 | 38.7 | 36.4 | 37.55 |
| | >50 | 13.3 | 38 | 25.65 |
| Marital status | Married | 98.7 | 95 | 96.85 |
| | Single | 1.3 | 0.8 | 1.05 |
| | Widowed | - | 0.8 | 0.4 |
| | Divorced | - | 3.3 | 1.66 |
| Education | Cannot read and write | 42.7 | 11.6 | 27.15 |
| | Can read | 2.7 | 19.0 | 10.85 |
| | Primary | 46.7 | 57 | 51.85 |
| | Secondary edu and above | 8.0 | 12.4 | 10.2 |

**Table 2**. Source and placement of bee colonies.

| Variables | Anlemo woreda (%) N= 75 | Lemo woreda (%) N= 121 | Overall (%) N= 196 |
|---|---|---|---|
| From parent | 1.3 | - | 0.65 |
| Catching swarms | 38.7 | 81.8 | 60.25 |
| Buying | 13.3 | 9.1 | 11.2 |
| From parent and catching swarms | 9.3 | 5.0 | 7.15 |
| Catching swarms and buying | 36.0 | - | 18 |
| Parent and buying | 1.3 | 4.0 | 2.65 |
| Placement of hive | | | |
| Back yard | 80.0 | 44.6 | 62.3 |
| Under the eaves | 18.7 | 35.5 | 27.1 |
| Back yard and under the eaves | 1.3 | 13.2 | 7.25 |
| under the eaves and hanging near homestead | - | 5.8 | 2.9 |
| under the eaves and hanging on forest | - | 0.8 | 0.4 |

Addis and Malede (2014), who noted that 49.2% of the beekeeper started by catching swarms. Moreover, Tesfaye and Tesfaye (2007) reported that about 70% of respondents have got their bee colonies by trapping swarms. From this result, one can conclude that catching swarm is the main sources of honeybee colonies in the study areas.

Most (62.3%) beekeepers in the study area kept their hives around their homestead (back yard) (Table 2) and 90.7% owned traditional hives (Table 3). This indicates that the adoption rate of improved technology (modern bee hive) is very low. This might be because of the cost of constructing and purchasing of modern hives and due to lack of harvesting and processing equipment's to use modern and improved hives.

Similarly, Mahari (2007) in eastern Tigray reported that modern beekeeping productions require more expensive establishment cost, accessories and skill training although yield better quality and quantity honey. Moreover, Tesfaye and Tesfaye (2007) reported almost all beekeeping practices are traditional and only little intervention was made with improved beekeeping practices due to lack of appropriate honey processing materials, Lack of bee equipments (like modern hives, casting mold, frame wires) and lack of skilled manpower. However, the present result disagree with Tesfaye and

**Table 3.** Types of bee hive.

| Types of hive | Anlemo woreda (%) | Lemo woreda (%) | Overall (%) |
|---|---|---|---|
| Traditional only | 92.0 | 85.1 | 90.7 |
| Traditional and transitional | 1.3 | 1.7 | 0.4 |
| Traditional and movable frame | 6.7 | 12.4 | 8.5 |
| All types of hives | - | 0.8 | 0.8 |

**Table 4.** Beehive inspection frequency of beekeepers.

| Inspection frequency | Anlemo woreda (%) | Lemo woreda (%) | Overall (%) |
|---|---|---|---|
| Frequently | 37.3 | 50.4 | 43.9 |
| Some times | 61.3 | 46.3 | 53.8 |
| Rarely | 1.3 | 3.3 | 2.3 |

Tesfaye (2007) who reported that about 97.6% of the respondents in Adami Tulu put their hives on a branch of tree and the rest at back yard.

### Inspection of honeybee colonies

The majority (53.8%) of the respondents replied that they inspect hives some times (Table 4).This is in line with Tessega (2009) who reported that 46.7, 20.6 and 7.5% of respondents inspect internally rarely, every month and every fifteen days respectively. However, Tesfaye and Tesfaye (2007) reported that 53.5% of respondents (beekeepers) visit their bees every day while 30.2% of them visit their bees at every three days and the rest visit their bees to check if the hive was occupied with bees and at least during honey harvesting seasons. About 60.2% of the interviewed beekeepers in study area believe that visiting the apiary and hive during rainy season causes diseases. For this reason, during rainy seasons the apiary is covered with grasses which may intern serve as a hiding place of pests of honey bee.

### Honey bee forage of the study area

According to this study some important local honeybee plants (trees, shrubs, herbs and cultivated crops are known as a source of nectar and pollen in the study area, namely Girawa *(Vernonia amygdalina)*, Wanza (*Cordia africaca)*, Wyira *(Olea Africana)*, Meche *(Guizotia scabra)*, Bahirzaf *(Eucalptus spp)*, dogma (*Syzygium guineese)*, Bisana (*Croton machrostachyus)* are some examples.

Even though there are different types of bee plants available in the area, there is still shortage of bee food in the district especially during the dry seasons. 60 % of the beekeepers pointed out that bee forages become declining as compared with the past due to deforestation and expansion of cultivated lands in the area. Karan et al. (2013) reported that depleting floral resources has reduced the beekeeping potentials.

### Honey yield and management

In the present study there was significance difference ($P<0.0015$) in honey yield per hive/year from traditional hives while there was no significant difference ($P>0.05$) in transitional hives and moveable frame hive between the two woredas (Table 5). The present result was below the result reported by Addis and Malede (2014) and Chala et al. (2013) that states the average honey yield per year/colony was 7.20±0.23, 14.70±0.62 and 23.38±0.73 kg for traditional, transitional and moveable frame hives, respectively. Furthermore, the results of the current study, also below the average amount of honey harvested from traditional, top bar and frame hive were 8.94, 10.66 and 15.56 kg per hive respectively Tessega (2009).

This indicates that the studied area is below the line of productivity what the beekeeping industry can perform. Honey yield decreasing in the present study might be due to lack of using improved bee hives, pests and predators and poor management. Similarly, a result reported by Tessega (2009) in Bure district indicated that honeybee products production was in a decreasing trend due to shortage of bee forages, drought, pesticides and herbicide application, lack of water and poor management.

Honey can be harvested once or twice, while in some cases even three times in a year largely depending on the availability of bee forage. It was reported that any production obtained in the remaining periods of the year would be left as food for the colony to strengthen it for the next harvest.

**Table 5.** Honey yield and price in the study area.

| Variables | Woreda | Mean+SE | Overall mean | P- value |
|---|---|---|---|---|
| Honey yield /hive/year from traditional hive | Lemo<br>Anlemo | 3.46 ± 0.9<br>2.62 ± 0.099 | 3.04 ± 0.915 | 0.000*** |
| Honey yield /hive/ year from transitional hive | Lemo<br>Anlemo | 5.33 ± 0.88<br>4.5 ± 0.5 | 4.9 ± 1.12 | 0.537NS |
| Honey yield /hive/ year from modern hive | Lemo<br>Anlemo | 9.86 ± 1.01<br>6.5 ± 0.65 | 8.2 ± 2.61 | 0.192NS |
| Price of crude honey/kg | Lemo<br>Anlemo | 70 ±1.02<br>81.27±2.91 | 75.6 ± 1.8 | 0.000*** |
| Honey lose due to chemical(kg/beekeeper/ year) | Lemo<br>Anlemo | 8.08 ± 0.81<br>5.26 ±0.37 | 6.67 ± 0.59 | 0.007** |
| honey lose due to chemical(ETB/ beekeeper/ year) | Lemo<br>Anlemo | 507.83 ± 51.10<br>540.62 ±39.66 | 524.23± 45.38 | 0.644 |

*** Significant at P<0.001, ** Significant at P < 0.01, NS = Non- significant at P>0.05.

This research result was in line with Tessega (2009) reported honey harvested once or twice, and in some cases even three times in Burie district. None of the beekeepers of the study area collect crude beeswax. According to the respondents all interviewed beekeepers did not harvest beeswax because of lack of knowledge or awareness (39.2%), about the product, lack of beeswax market (21.5%), lack of processing skill (20.8%) and lack materials (18.5%). As a result no one has benefited from the product except using it for oiling 'elee' (materials used for local injera making).

## Storage (packing) practices of honey

Even though the majority (75%) of the households do not store honey primarily because of high demand for cash but some farmers keep some amount for different purposes. Beekeepers sell the largest proportion of their honey during harvest at low price mainly to meet their demand for cash for social obligations (Beyene and Phillips, 2007).

As reported by the sample respondents, 59% of the sample beekeepers used plastic containers (Table 6). Other beekeepers still use traditional containers such as gourd and tin to store honey. However, Rivera et al. (2007) reported that most of the farmers (90.7%) store honey in clay pots until consumption or sale. Farmers use traditional containers which are technically not appropriate storage facilities as they result in quick crystallization, fermentation of honey, changing of general appearance and taste of honey (Tesfaye and Tesfaye, 2007).

## Trends of honeybee products production in the study area

The majority of the respondents (86.4%) observed decreasing trend of bee products. On the other hand, 10.25 and 3.4% of the respondents replied that bee products in the given years have constant and increased trends respectively (Table 7). This result agrees with Tessega (2009) who stated that honeybee products production in Bure district was in a decreasing trend due to shortage of bee forages, drought, pesticides and herbicide application, lack of water and poor management. In the present study the decreasing trend in honeybee products could be due to pesticides and herbicide application, drought and lack of bee forage, due to decrease in the number of bee colonies and lack of using improved bee hives.

## Constraints for beekeeping in the study areas

Each beekeeper involved in the study was requested to prioritize the challenges of beekeeping. Accordingly, the most important constraints of beekeeping in the study areas were shortage of bee forage, pesticides and herbicides application, pests and predators and shortage of water were found to be the top challenges for beekeeping in the area, and the detailed result is summarized in Table 8. This result agrees with Yirga et al. (2012) reported that bee pests and predators, absconding and shortage of bee forage were the major constraints affecting the honey sub sector in asgede

**Table 6.** Containers used to store honey.

| Variables | Anlemo woreda (%) | Lemo woreda (%) | Overall (%) |
|---|---|---|---|
| Earthen pot | 16 | 7.4 | 11.7 |
| Tin | 2.7 | - | 1.4 |
| Plastic container | 56 | 62 | 59 |
| Earthen pot and tin | - | 0.8 | 0.4 |
| Gourd and animal skin | 1.3 | - | 0.7 |
| Earthen pot and plastic container | 22.7 | 13.2 | 18 |
| Earthen pot, tin and plastic container | - | 16.5 | 8.3 |

**Table 7.** Trend of honey yield in percent.

| Trend of honey yield | Anlemo woreda (%) | Lemo woreda (%) | Overall (%) |
|---|---|---|---|
| Increase | 2.7 | 4.1 | 3.4 |
| Stable | 14.7 | 6.6 | 10.65 |
| Decrease | 82.7 | 89.3 | 86 |

**Table 8.** Major Constraints identified by respondent beekeepers in the study area.

| Constraints | Lemo woreda (%) | Rank | Anlemo woreda (%) | Rank |
|---|---|---|---|---|
| Shortage of bee forage | 30.6 | 3 | 53.3 | 1 |
| Shortage of water | 35.5 | 4 | 38.7 | 4 |
| Absconding | 15.7 | 8 | 34.9 | 6 |
| Pests and predators | 38.8 | 1 | 46.7 | 2 |
| Pesticides and herbicides application | 52.1 | 2 | 33.3 | 3 |
| Death of colony | 32.2 | 6 | 29.3 | 7 |
| Migration | 26.4 | 7 | 32.0 | 8 |
| Swarming | 36.4 | 5 | 22.7 | 5 |
| Storage facilities | 66.9 | 9 | 73.3 | 9 |

tsimbla district, Northern Ethiopia. SOS-Sahel-Ethiopia (2006) reported that the main constraints in Ethiopia are lack of beekeeping knowledge, shortage of trained manpower, shortage of beekeeping equipment, pests and predators and inadequate research and extension services to support apiculture development programmes.

## Opportunities for beekeeping

**Some of the opportunities for beekeeping in the study area include:**

(i) Availability of many numbers of local bee hives and suitable environment with different agro ecology.
(ii) Currently the government is promoting self-contained plan to create modern hives with low cost by using locally available materials and organizing beekeepers co-operatives at grass-root level and connect them with local carpenters who produce modern bee hive with low cost.
(iii) Farmers willingness to improve beekeeping practices in the area,
(iv) Although there is shortage of bee food during dry season, there are many varieties of flowering plants during wet seasons which used as bee food,
(v) The current high market demand for bee products

## Conclusions

Generally the most widely used type of beekeeping in the study area is traditional using local hives. The most important constraints of beekeeping in the study area were found Pesticides and herbicides application, shortage of bee forage, Pests and predators and shortage of water. In the area despite the presence of different constraints and challenges, there are high potentials and opportunities to maximize the out puts of the resource to improve the livelihoods of the communities in a sustainable ways, as the current government plan to develop apiculture as one of the strategies to reduce poverty, high demand for hive products and the establishments of beekeepers co-operatives at grass-root level.

## Conflicts of Interest

The authors have not declared any conflict of interest.

## ACKNOWLDEGEMENTS

Authors would like to thank Wachemo University for providing us fund. We would like to express our thanks to the Lemo and Anlemo Woreda Agricultural offices for their assistance during data collection. Lastly we would also like to express our thanks to beekeepers in the study area for their willingness to give us all the valuable information.

## REFERENCES

Addis G, Malede B (2014). Chemical analysis of honey and major honey production challenges in and around gondar, Ethiopia. Aca. J. Nut. 3(1).

Adgaba N (2007). Atlas of pollen grains of major honeybee flora of Ethiopia, Holetta Bee Research Centre, Holetta, Ethiopia.

Beyene T, Phillips D (2007). Ensuring small scale producers in Ethiopia to achieve sustainable and fair access to honey markets. P. 64.

CSA (Central Statistical Agency) (2013). agricultural sample survey 2012/2013, Volume II, Report On Livestock and Livestock Characteristics. Addis Ababa. Statistical Bulletin, P. 570.

Chala K, Taye T Kebede D (2013). Production and marketing system in Gomma District, South Western Ethiopia, G. J. Bus. Mana. Stu. 3(3):99-107.

Edessa N (2002). Survey on honey production system in West Shoa Zone (unpublished). Holeta Bee Research Center (HBRC), Ethiopia. P. 15.

Gichora M (2003). Towards realization of kenya's full beekeeping potential: A Case Study of Baringo district. Ecology and Development series No.6, 2003. Cuvillier Verlag Gottingen, Gottingen, Germany. P. 157.

Karan B, Yashwant S, Parampal S (2013). Constraints of apiculture in India. Inter. J. Life Sci. Res. 1(1):(1-4).

MoARD (2007). Livestock development master plan study phase I report - data collection and analysis, Volume N- apiculture, Ministry of Agriculture and Rural Development (MoARD), Addis Ababa, Ethiopia.

Mahari G (2007). Impact of beekeeping on household income and food security: The Case of Atsbi Wemberta and Kilte Awlailo Woredas of Eastern Tigray, Ethiopia M.Sc. Thesis, Mekelle University.

Rivera J, Losada H, Lopez M, Cortes J, Vieyra JY, Grande D (2007). Production system of honey bees in the periurban area of the south east of Mexico city. Liv. Res. for Rural Dev. 19(2). http://www.cipav.org.co/lrrd/lrrd19/2/rive19029.html

SOS–SAHEL-ETHIOPIA (2006). Smallholders apiculture development and trade promotion project terminal report. ANRS Food Security Program Coordination and Disaster Prevention Office, Addis Ababa, Ethiopia.

Tessega B (2009). Honeybee production and marketing systems, constraints and opportunities in Burie District of Amahara Region, A Thesis Submitted to the Department of Animal Science and Technology, School of Graduate Studies Bahir Dar University, Ethiopia.

Tesfaye K, Tesfaye L (2007). Study of honey production system in Adami Tulu Jido Kombolcha district in mid rift valley of Ethiopia. Adami Tulu Agri. Res. Center.

Gidey Y, Bethelhem K, Dawit K, Alem M (2012). Assessment of beekeeping practices in Asgede Tiembla district, Northern Ethiopia: Absconding, bee forage and bee pests. Afr. J. Agric. Res. 7(1):1-5.

# Effect of fermentation on proximate composition, physicochemical and microbial characteristics of pearl millet (*Pennisetum glaucum* (L.) R. Br.) and Acha (*Digitaria exilis* (Kippist) Stapf) flour blends

A. O. Ojokoh[1], O. E. Fayemi[1], F. C. K. Ocloo[2] and F. I. Nwokolo[1]

[1]Department of Microbiology, Federal University of Technology, P. M. B. 704, Akure, Ondo State, Nigeria.
[2]Radiation Technology Centre, Biotechnology and Nuclear Agriculture Research Institute. Ghana Atomic Energy Commission, P. O. Box LG 80, Legon, Accra, Ghana.

**The objective of this study was to determine the effect of spontaneous fermentation on proximate composition, physicochemical and microbial characteristics of sprouted millet-acha blends with the aim of producing fermented flour blends for traditional beverage. Pearl millet and acha grains were sprouted, milled and blended into different ratios. The blends were mixed with water, fermented for 72 h and dried in an air-oven. Proximate composition, pH, titratable acidity and microbial profile were determined using suitable methods. Moisture contents ranged from 5.60 to 5.94% and 10.44 to 13.33% for unfermented and fermented flour blends, respectively. Increase in pearl millet proportions significantly decreased ($p \leq 0.05$) crude fat, ash, fibre and protein contents of the blends. In contrast, carbohydrate contents increased with increase in pearl millet proportion. Fermentation decreased significantly ($p \leq 0.05$) the fat, fibre and carbohydrate contents of the flour blends; whereas ash and protein contents were significantly increased ($p \leq 0.05$) with fermentation. pH values decreased with simultaneous increase in titratable acidity during fermentation. A total of 13 bacteria were isolated during fermentation, namely, *Lactobacillus acidophilus, Lactobacillus plantarum, Lactobacillus brevis, Lactobacillus leichmonnii, Lactobacillus cellobiosus, Lactobacillus casei, Pediococcus* sp., *Streptococcus thermophilus, Micrococcus luteus, Bacillus subtillis, Pseudomonas aeruginosa, Staphylococcus aureus* and *Proteus vulgaris.* Fermentation has a potential to improve protein and mineral contents of pearl millet-acha flour blends for beverage preparation.**

**Key words:** Bacteria, beverage, fermentation, pearl millet, acha, flour, physicochemical, microbial characteristic.

## INTRODUCTION

Pearl millet (*Pennisetum glaucum* (L.) R. Br.) is considered as one of the most important drought tolerant crops of the tropical and subtropical regions in the world (Zakari et al., 2010). It is cultivated on more than 28. million ha of land in the semi-arid tropical regions of Asia (USDA, 2005). Nigeria is reported to produce about 6.7

million tons of pearl millet on 5.8 million hectares (USDA, 2005). Pearl millet contains about 1.6 to 2.4% ash, 2.6 to 4.0% crude fibre, 2.7 to 7.1% fat, 8.5 to 15.1% crude protein, 58 to 70% carbohydrate and 354 to 796 mg/g phytic acid (Abdalla et al., 1998). It is rich in minerals, such as calcium (10 to 80 mg/g), magnesium (180 to 270 mg/g) and phosphorus (450 to 990 mg/g) (Abdalla et al., 1998). Pearl millet also contains, potassium (70 to 110 μg/g) sodium (4 to 13 μg/g), zinc (53 to 70 μg/g), manganese (18 to 23 μg/g), copper (10 to 18 μg/g) and iron (70 to 180 μg/g) (Abdalla et al., 1998). In Nigeria and other parts of West Africa, pearl millet is used traditionally to produce fermented cereal based porridge ("Akamu" or "Ogi") (Zakari et al., 2010) and indigenous fermented beverage, such as "kunu-zaki" (Oluwajoba et al., 2013).

Acha (*Digitaria exilis* (Kippist) Stapf), also called "fonio" or "hungry rice" is reported as the oldest African cereal. It is an indigenous annual crop in West Africa (Chukwu and Abdul-kadir, 2008). Acha is reported to supply food to about 3 to 4 million people (Jideani, 1990; NRC, 1996). Acha contains about 1.6 to 3.6% ash, 0.48 to 1.90% crude fibre (Irving and Jideani, 1997; Chukwu and Abdul-kadir, 2008), 2.1 to 2.5% fat, 6.9 to 10.6% crude protein (Chukwu and Abdul-kadir, 2008; Irving and Jideani, 1997; Jideani, 1999) and 81.8 to 87.8% carbohydrate (Irving and Jideani, 1997; Jideani, 1999). It also contains minerals, such as calcium, magnesium, phosphorus, potassium, sodium, manganese, iron, zinc and copper (Chukwu and Abdul-kadir, 2008). Acha finds its application in traditional foods, such as porridge, couscous, alcoholic and non-alcoholic beverages (Jideani, 1999).

The use of acha and millet composite grains in the preparation of traditional fermented beverage (kunu-zaki) has been reported (Oluwajoba et al., 2013). However, in the study by these authors, proximate composition of the acha-millet grains composites were not reported, instead, the proximate composition of the kunu-zaki produced was reported. The present research sought to formulate pearl millet-acha flour blends for possible utilization in traditional beverage production. The objective of this study was to evaluate the effect of fermentation on the proximate composition, physicochemical and microbial characteristics of pearl millet and acha flour blends.

## MATERIALS AND METHODS

### Source of materials

Dry pearl millet (*P. glaucum* (L.) R. Br) and acha (*Digitaria exilis* (Kippist) Stapf) grains were purchased from a local market in Akure, Nigeria. The samples were transported to the laboratory in clean low density polyethylene bags.

### Preparation of samples

Stones and other grits were sorted out from the grains, after which they were separately steeped in water for 5 h. The grains were then drained and separately spread on a clean tray and allowed to sprout at 25 ± 2°C for 24 h. The sprout grains were separately milled into flour using Hammer Mill (Brook Crompton, Huddersfield, England). The sprout pearl millet flour and sprout acha flour were mixed in sterile containers with appropriate labels in the following proportions (in grams): 200:100 (2:1) pearl millet to acha (A), 225:75 (3:1) pearl millet to acha (B), 240:60 (4:1) pearl millet to acha (C), 250:50 (5:1) pearl millet to acha (D) and 300:0 pearl millet to acha (E). Each of the samples was mixed with 1000 ml distilled water. The containers with the mixtures were covered, taped at the edges and the contents fermented for 72 h (Figure 1). At the end of fermentation, samples were dried in an air-oven at 65°C for 24 h and then packaged in low density polyethylene pouches and stored at 8°C prior to analyses. The fermentation process was carried out in three batches.

### Proximate composition

The moisture, crude protein (N x 6.25), crude fibre, crude fat and total ash contents of samples were analysed before and after 72 hrs of fermentation using the method described by Association of Official Analytical Chemists' (AOAC, 1990) approved methods 925.10, 920.87, 920.86, 920.39 and 923.03 respectively. Total carbohydrate content of the samples was calculated by difference method (subtracting the sum of percent moisture, crude protein, crude fibre, crude fat, and ash from 100%).

### Physico-chemical properties

#### *pH and titratable acidity*

The method described by AOAC (1990) was used to determine pH and titratable acidity of the fermenting medium. Samples were taken every 24 h during the fermentation period according to the procedure described by Fayemi and Ojokoh (2014). The pH of the samples was determined using an Orion pH meter (Model 310, Orion Research Inc., Beverly, MA) equipped with glass electrode. The titratable acidity (TTA) was determined by titrating 10 ml of thoroughly mixed sample against 0.1 M NaOH using phenolphthalein as an indicator. Values obtained were expressed as percent lactic acid. All analyses were carried out in triplicate.

### *Microbial characteristics*

The microbial profile of the raw and fermenting (at 24 h interval) blend samples were determined at 24 h interval. The changes in microbial population (cfu/g) of the total aerobic bacteria were determined using nutrient agar (NA) (Merck, Darmstadt, Germany) while De Man, Rogosa and Sharpe (MRS), (Merck) and M17 agar media (Oxoid, Basingstoke, Hampshire, England, UK) was used for the isolation of lactic acid bacteria (LAB). Four different colonies were randomly picked following visual assessment from the highest dilution factor of MRS and M17 agar plates to determine the dominant bacteria during the fermentation of the blends. Samples were analysed by homogenizing 1 g of the fermenting blend with 9 ml sterile 0.1% buffered peptone water (BPW) (Merck) followed by appropriate dilutions, spread plating and incubation at required temperatures. The NA agar plates were incubated at 37°C for 24 h while MRS agar plates were incubated anaerobically using anaerobic jar together with anaerocult system (Merck) at 37°C for 48 h. Colonies were selected randomly and purified before biochemical identification of the bacterial isolates (Collins et al., 1989; Fayemi and Ojokoh, 2014).

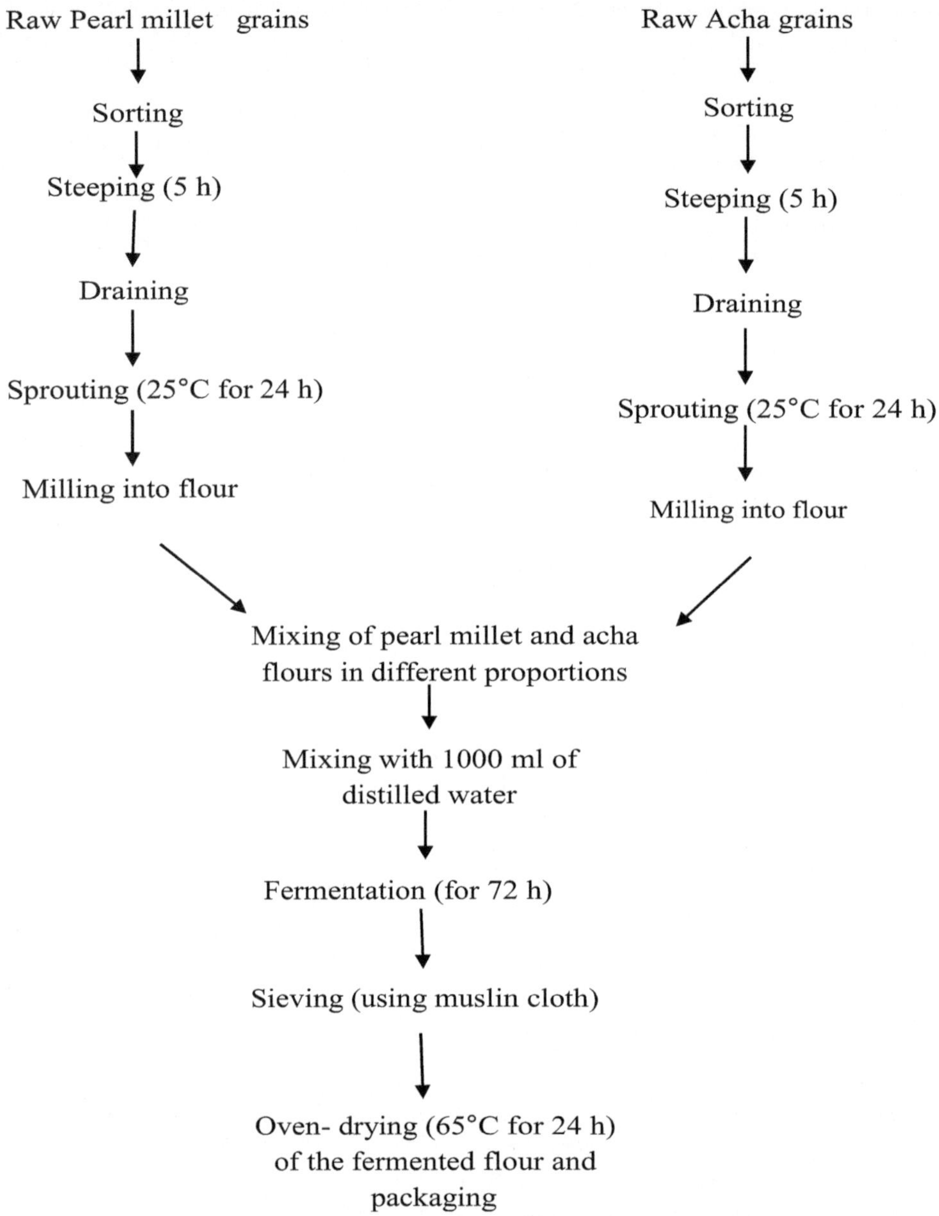


**Figure 1.** Flow chart showing the preparation of fermented pearl millet- acha flour blends.

#### Statistical analyses

Analysis of variance (ANOVA) was performed on the data at $p \leq 0.05$ using MINITAB statistical software (Minitab® Release 14.13, Minitab Inc., USA). Significant means were separated using the least significant difference (LSD) at $p \leq 0.05$.

## RESULTS AND DISCUSSION

### Proximate composition

The proximate composition of fermented and unfermented pearl millet-acha flour blends is shown in Table 1. Moisture contents of flour blends before fermentation (unfermented blends) ranged from 5.60 to 5.94%. Unfermented pearl millet flour (control, E) had significant ($p \leq 0.05$) low moisture content compared to flour blends and acha flour. After fermentation and drying, the flour blends had moisture contents ranging from 10.44 to 13.23% (Table 1). The moisture contents of all the samples were significantly ($p \leq 0.05$) different from each other. It could be seen from the results that moisture content of fermented pearl-millet flour blends were significantly higher than the unfermented samples. This could be attributed to the temperature and the duration of drying of the fermented samples. Moisture

**Table 1.** Proximate composition of fermented and unfermented sprout pearl millet-acha flour blends.

| Proximate | A | | B | | C | | D | | E | | Acha* |
|---|---|---|---|---|---|---|---|---|---|---|---|
| (%) | Unfermented | Fermented | Unfermented | Fermented | Unfermented | Fermented | Unfermented | Fermented | Unfermented | Fermented | Unfermented |
| Moisture | $5.94^{Cb} \pm 0.03$ | $13.23^{g} \pm 0.10$ | $5.88^{Bb} \pm 0.00$ | $12.78^{f} \pm 0.03$ | $5.88^{Bb} \pm 0.01$ | $11.89^{e} \pm 0.00$ | $5.85^{Bb} \pm 0.01$ | $11.26^{d} \pm 0.06$ | $5.60^{Aa} \pm 0.00$ | $10.44^{c} \pm 0.08$ | $7.21^{D} \pm 0.01$ |
| Fat | $7.31^{Dg} \pm 0.01$ | $6.27^{f} \pm 0.03$ | $6.22^{Ce} \pm 0.00$ | $6.11^{d} \pm 0.01$ | $6.18^{Ce} \pm 0.01$ | $5.95^{c} \pm 0.00$ | $6.10^{Bd} \pm 0.00$ | $5.87^{b} \pm 0.04$ | $5.94^{Ac} \pm 0.03$ | $5.31^{a} \pm 0.00$ | $8.40^{E} \pm 0.06$ |
| Ash | $1.96^{Dc} \pm 0.01$ | $3.68^{d} \pm 0.03$ | $1.90^{Cbc} \pm 0.01$ | $4.83^{e} \pm 0.00$ | $1.84^{Bb} \pm 0.00$ | $4.91^{f} \pm 0.03$ | $1.69^{Aa} \pm 0.01$ | $4.94^{fg} \pm 0.01$ | $1.67^{Aa} \pm 0.04$ | $5.00^{g} \pm 0.07$ | $2.22^{E} \pm 0.00$ |
| Fibre | $2.28^{Dh} \pm 0.01$ | $0.65^{a} \pm 0.04$ | $2.20^{Cg} \pm 0.01$ | $0.79^{b} \pm 0.01$ | $2.14^{Bf} \pm 0.00$ | $0.93^{c} \pm 0.04$ | $2.10^{Bf} \pm 0.00$ | $1.39^{d} \pm 0.01$ | $1.97^{Ae} \pm 0.03$ | $1.43^{d} \pm 0.00$ | $2.80^{E} \pm 0.03$ |
| Protein | $11.52^{Ee} \pm 0.03$ | $15.07^{f} \pm 0.10$ | $11.00^{Dd} \pm 0.00$ | $15.76^{g} \pm 0.14$ | $9.40^{Cc} \pm 0.08$ | $18.52^{i} \pm 0.00$ | $9.10^{Bb} \pm 0.14$ | $17.43^{h} \pm 0.15$ | $8.00^{Aa} \pm 0.00$ | $9.52^{c} \pm 0.03$ | $20.15^{F} \pm 0.07$ |
| Carbohydrate | $70.99^{Be} \pm 0.28$ | $61.10^{c} \pm 0.14$ | $72.80^{Cf} \pm 1.13$ | $59.73^{bc} \pm 1.03$ | $74.56^{Dg} \pm 0.79$ | $57.80^{a} \pm 0.99$ | $75.16^{DEg} \pm .23$ | $59.11^{ab} \pm 0.16$ | $76.82^{Eh} \pm 0.88$ | $68.30^{d} \pm 0.28$ | $59.22^{A} \pm 0.31$ |

Values are means ± standard deviation of duplicates. Values in the same row with different subscripts and or superscripts are significantly different at $p < 0.05$. Subscripts A – F compares all unfermented samples. * Acha was not fermented. A = 2:1 Pearl millet to acha blend, B = 3:1 Pearl millet to acha blend, C = 4:1Pearl millet to acha blend, D = 5:1 Pearl millet to acha blend, E = 100 % Pearl millet (Control).

content values ranging from 10.16 to 10.33% and 9.25 to 10.35% have been reported for acha and fermented acha, respectively (Echendu et al., 2009). Zakari et al. (2010) reported moisture content values ranging from 12.0 to 12.9% have been reported for pearl millet-bambara groundnut blend flours.

Crude fat content of the unfermented pearl millet-acha flour blends ranged from 5.94 to 7.31% (Table 1). The fat content of acha flour was significantly ($p \leq 0.05$) higher (8.40%) compared to the blend samples and pearl millet flour (control). As the proportion of pearl millet flour in the flour blends increases, the crude fat content of the flour blends significantly decreased (Table 1). The fat content of the fermented pearl millet-acha flour blends ranged from 5.31 to 6.27% (Table 1). Fermentation also decreased the fat content of the pearl millet-acha flour blends (Table 1). Fat content ranging from 2.7 to 7.1% has been reported for some pearl millet genotypes flours (Abdalla et al., 1998). Fat content values of 5.7, 5.6, 7.2 and 2.4% for raw, germinated, roasted and fermented millet flours, respectively, have been reported (Sade, 2009). Crude fat contents ranging from 2.1 to 2.5% has been reported for acha (*D. exilis*) (Chukwu and Abdul-kadir, 2008; Irving and Jideani, 1997; Jideani, 1999). The reduction in fat contents of pearl millet-acha flour blends with increasing proportions of pearl millet flour could be due to the low fat composition of the pearl millet grains (Table 1). Decreased fat content has been reported for fermented millet flour (Sade, 2009).

The ash content of unfermented acha flour was 2.22 %, and that for unfermented pearl-millet flour blends ranged from 1.67 to 1.96% (Table 1). Increasing the proportion of pearl millet flour significantly ($p \leq 0.05$) decreased the ash contents of the flour blends. However, no significant ($p > 0.05$) differences were observed between the ash content of unfermented 100 % pearl millet (E, control) and 5:1 pearl millet to acha flour blend (D) samples. Similarly, the ash content of 3:1 pearl millet to acha flour blend (B) and 4:1 pearl millet to acha flour blend (C) were not significantly different. The ash content of the fermented pearl millet-acha flour blends ranged from 3.68 – 5.00 % (Table 1). In contrast to the unfermented pearl millet-acha flour blends, the ash content in the fermented pearl millet-acha flour blends significantly ($p \leq 0.05$%) increased with increasing proportion of pearl millet flour. Fermentation significantly increased the ash content of the pearl millet-acha flour blends (Table 1). Abdalla et al. (1998) reported ash content values ranging from 1.8 to 2.4% for flours from pearl millet genotypes. Ash content of 2.1, 2.1, 2.7 and 1.9% for raw, germinated, roasted and fermented millet flours, respectively, have been reported (Sade, 2009). Ash content values of 1.4% (Irving and Jideani, 1997) and 3.6% (Chukwu and Abdul-kadir, 2008) have been reported for acha (*D. exilis*). The reduction in ash content of pearl millet-acha flour blends with increasing proportions of pearl millet flour could be due to the low ash content of the pearl millet grains (Table 1). Decreased ash content has been reported for fermented millet flour (Sade, 2009). The fibre content of unfermented pearl millet-acha flour blends ranged from 1.97 to 2.28% compared to 2.80 % for acha flour (Table 1). Increased proportion of pearl millet flour significantly ($p \leq 0.05$) decreased the fibre content of the pearl millet- acha flour blends. However, the fibre content between 4:1 pearl millet to acha flour blend (C) and 5:1 pearl millet to acha flour blend (D) were not significantly different. The fibre content of fermented pearl millet-acha flour blends ranged from 0.65 to 1.43% (Table 1). Fibre content of fermented pearl millet-acha flour blends significantly increased with increased proportion of pearl millet flour. However, no significant differences were observed between fermented flour blends from 100% pearl millet (E, control) and 5:1 pearl millet to acha flour blend (D). Fermentation significantly ($p \leq 0.05$) decreased the fibre content of the pearl millet-acha flour blends (Table 1). Fibre content ranging from 2.6 to 4.0% have been reported for pearl millet genotypes flours (Abdalla et al., 1998). Fibre content of 2.0, 1.8, 1.8 and 1.8%, respectively, have been reported for flours from raw, germinated, roasted and fermented millet (Sade, 2009). Fibre content of 0.48% (Irving and Jideani, 1997) and 1.9% (Chukwu and Abdul-kadir, 2008) have been reported for acha (*D. exilis*). The decrease in fibre content of pearl millet-acha flour blends with increasing proportions of pearl millet flour could be due to the low fibre content of the pearl millet grains (Table 1). The reduction in fibre contens of fermented flour blends could be attributed to enzymatic breakdown of fibre during fermentation by the lactic acid bacteria (Ojokoh et al., 2013; Ojokoh et al., 2014). Decreased fibre content has been reported for fermented millet flour (Sade, 2009). Amankwah et al. (2009) reported increased crude fibre content in fermented maize flour.

The protein content of unfermented pearl millet-acha flour blends ranged from 8.00 to 11.52% compared to 20.15% for acha flour (Table 1). Increasing the proportions of the pearl millet flour significantly ($p \leq 0.05$) decreased the protein content of the pearl millet-acha flour blends. The protein content of the fermented pearl millet-acha flour blends ranged from 9.52 to 18.52% (Table 1). As the proportion of pearl millet flour increased, the protein content of the fermented pearl millet-acha flour blends significantly ($p \leq 0.05$) increased and then drop after 4:1 pearl millet to acha flour blend (C) sample. Fermentation significantly ($p \leq 0.05$) increased the protein content of the pearl millet- acha flour blends. Protein content values ranging from 8.5 to 15.1% have been reported for pearl millet genotypes flours (Abdalla et al., 1998). Crude protein content values of 14.0, 19.4, 15.7 and 17.5% have been reported for flours from raw, germinated, roasted and fermented millet, respectively (Sade, 2009). Protein content values ranging from 6.9 to 10.6% have been reported for acha (*D. exilis*) (Chukwu and Abdul-kadir, 2008; Irving and Jideani, 1997; Jideani, 1999). The reduction in protein content of the blends with increasing proportions of pearl millet could be explained by the low content of protein in pearl millet (Table 1). The increases in crude protein content recorded with fermentation could be due to the activities and increase in number of microorganisms (Ojokoh et al., 2013; Ojokoh et al., 2014). Increase in protein content has been reported in fermented maize flour (Amankwah et al., 2009) and fermented millet flour (Sade, 2009). Increase in protein content has been attributed to proteolytic activities of enzymes produced by microorganisms during fermentation (Amankwah et al., 2009).

Carbohydrate content of unfermented pearl millet-acha flour blends ranged from 70.99 to 76.82% compared to 59.22% for acha flour (Table 1). Increased proportions of pearl millet flour contributed to significant ($p \leq 0.05$) increase in carbohydrate content of pearl millet-acha flour blends. However, no significant ($p > 0.05$) differences were observed between the carbohydrate contents of 4:1 pearl millet to acha flour blend (C) and 5:1 pearl millet to acha flour blend (D). Carbohydrate content of fermented pearl millet-acha flour blends ranged from 57.80 to 68.30% (Table 1). As the proportion of pearl millet flour increased, the carbohydrate content of the fermented pearl millet-acha flour blends significantly ($p \leq 0.05$) decreased and then increased after 5:1 pearl millet to acha flour blend (D) sample. Fermentation significantly ($p \leq 0.05$) decreased the carbohydrate content of the pearl millet-acha flour blends. Carbohydrate content of 76.3, 71.1, 72.6 and 76.5% have been reported for flours from raw, germinated, roasted and fermented millet, respectively (Sade, 2009). Carbohydrate content ranging from 81.8 to 87.8% has been reported for acha (*D. exilis*) (Irving and Jideani, 1997; Jideani, 1999). The increase in carbohydrate content with increasing pearl millet proportion could be due to high carbohydrate composition of the pearl millet (Table 1). The reduction in carbohydrate content with fermentation could be attributed to utilization of fermentable sugars by lactic acid bacteria for growth and other metabolic activities (Ojokoh et al., 2013).

## Physico-chemical properties

### *pH and titratable acidity*

The changes in pH and titratable acidity of pearl millet-acha flour blends during fermentation are represented in Figure 1. pH values for all the samples before fermentation (time 0) ranged from 4.55 to 4.82. The pH values for pearl millet-acha flour blends before fermentation were significantly ($p \leq 0.05$) different from each other. During the fermentation period, pH values decreased significantly for all the samples (Figure 2A). However, pH values for 4:1 pearl millet-acha flour blend fermenting medium (C) and 5:1 pearl millet-acha flour blend fermenting medium (D) at 48 h were not significantly different from each other. Similarly, pH values for 2:1 pearl millet-acha flour blend fermenting medium (A) and 100% pearl millet flour blend fermenting medium (E) at 72 h were not significantly different from each other.

Decreases in pH of the fermenting pearl millet-acha flour

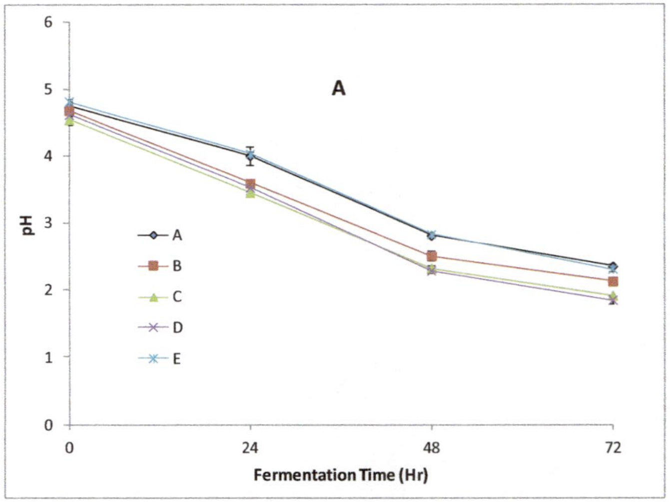

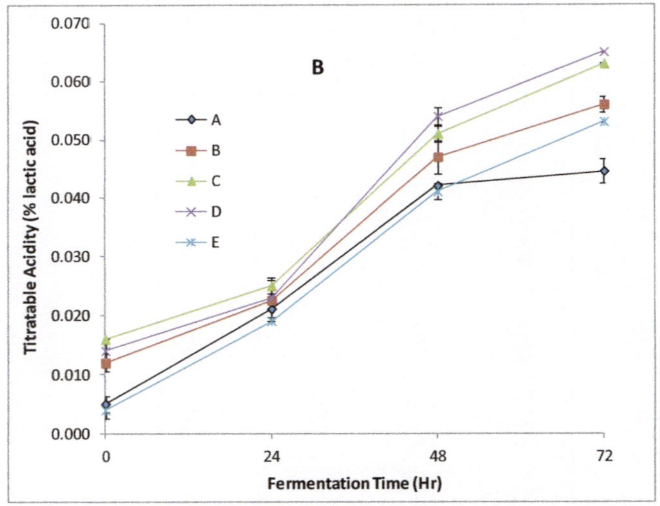


**Figure 2.** Changes in pH (**A**) and titratable acidity (**B**) of sprout pearl millet-acha flour blends during fermentation period. Error bars = ± standard deviations. A = 2:1 Pearl millet to acha blend, B = 3:1 Pearl millet to acha blend, C = 4:1Pearl millet to acha blend, D = 5:1 Pearl millet to acha blend, E = 100 % Pearl millet (Control).

**Table 2.** Morphological and biochemical characteristics of lactic acid bacteria isolated during the spontaneous fermentation of pearl millet-acha flour blends.

| Cellular morphology | Gram reaction | Catalase | Oxidase | Motility test | Growth at 4°C | Growth at 45°C | Growth at pH 3.9 | $NH_3$ from Arginine | $H_2S$ production | Growth at 4% Nacl | Homo/Hetero-fermentation | Methyl Red | Voges Proskaeur | Glucose | Xylose | Raffinose | Sucrose | Fructose | Mannose | Maltose | Mannitol | Trehalose | Arabinose | Rhamnose | Lactose | Galactose | Probable organism |
|---|---|---|---|---|---|---|---|---|---|---|---|---|---|---|---|---|---|---|---|---|---|---|---|---|---|---|---|
| Rods | + | - | - | - | - | + | + | - | - | - | HM | + | - | + | - | - | + | + | - | - | - | + | - | + | + | + | *Lactobacillus acidophilus* |
| Rods | + | - | - | - | + | - | + | - | + | + | HM | + | - | + | + | + | + | + | - | + | + | + | + | + | + | + | *Lactobacillus plantarum* |
| Rods | + | - | - | - | + | - | + | + | - | - | HE | + | - | + | + | + | + | - | - | + | + | - | + | - | + | + | *Lactobacillus brevis* |
| Rods | + | - | - | - | + | + | + | + | - | + | HM | + | - | + | - | - | + | + | - | + | - | - | - | - | + | + | *Lactobacillus leichmannii* |
| Rods | + | - | - | - | - | - | + | + | - | - | HE | + | - | + | - | + | + | - | + | + | - | + | + | - | + | + | *Lactobacillus cellobiosus* |
| Cocci | + | - | - | - | + | + | + | - | - | + | HM | + | - | + | + | - | + | - | - | - | - | - | - | - | + | + | *Pediococcus* sp |
| Rods | + | - | - | - | + | - | + | - | - | + | HM | + | - | + | - | - | + | + | + | + | + | - | + | + | + | + | *Lactobacillus casei* |
| Cocci | + | - | - | - | + | + | + | - | - | + | HE | + | - | + | - | - | + | + | + | + | - | - | + | - | + | + | *Streptococcus thermophillus* |

+ = positive test; - = negative test; HE = Heterofermentative HM = Homofermentative.

**Table 3.** Morphological and biochemical characteristics of bacterial isolates from raw and during fermentation of pearl millet–acha flour blends.

| Morphological Characteristics | Gram reaction | Spore test | Motility | Catalase | Oxidase | Coagulase | Urease | Glucose | Sucrose | Mannitol | Lactose | Starch hydrolysis | Probable bacteria |
|---|---|---|---|---|---|---|---|---|---|---|---|---|---|
| Creamy, punctiforms, Umbonate, entire and wrinkled cocci | + | - | - | + | + | - | + | NAG | NAG | NAG | NAG | - | *Micrococcus luteus* |
| Creamy white, irregular, Raised, undulate and rough surface | + | + | + | + | + | - | - | AG | NAG | AG | NAG | + | *Bacillus subtilis* |
| Milky, irregular, Raised, entire, opague and wet surface | - | - | + | + | + | - | - | NAG | NAG | NAG | NAG | - | *Pseudomonas aeruginosa* |
| Circular, convex, creamy, entire and smooth surface | + | - | - | + | - | + | - | ANG | ANG | AG | ANG | - | *Staphylococcus aureus* |
| Cream, irregular, raised, entire and rough surface | + | - | + | + | + | - | + | AG | AG | NAG | NAG | - | *Proteus vulgaris* |

+ = positive test; - = negative test; AG = Acid and Gas production; NAG = No Acid and Gas production; ANG = Acid production and no Gas.

blend medium were accompanied by significant ($p \leq 0.05$) increases in titratable acidity (Figure 2B). The decrease in pH with simultaneous increase in titratable acidity during fermentation has been reported for millet flour (Khetarpaul and Chauhan, 1989), millet-acha based *kunun zaki* (beverage) (Ayo, 2004), breadfruit-cowpea blend flours (Ojokoh et al., 2013, 2014) and maize flour (Amankwah et al., 2009). The reduction in pH and simultaneous increase in TTA during fermentation could be attributed to the activities of lactic acid bacteria (Ojokoh et al., 2013). Lactic acid bacteria have been reported to degrade carbohydrates (Table 1) resulting in acidification (Ojokoh et al., 2013).

### *Microbial characteristics*

A total of 13 bacteria were identified both in raw and during fermentation of the blend samples (Tables 2 and 3). *Micrococcus luteus, Bacillus subtilis, Proteus vulgaris, Staphylococcus aureus* and *Pseudomonas aeruginosa* were isolated from the raw sample and at the early stage of the fermentation of the blends followed by disappearance towards the end of fermentation process. Studies have reported the presence of *Bacillus subtilis, Staphylococcus aureus* and *Pseudomonas aeruginosa* in similar products and were attributed to inadequate precautionary measures during processing (Adeyemi and Umar 1994; Osuntogun and Aboaba, 2004). However, the disappearance of these organisms as the fermentation progressed had been reported in a similar product known as 'koozh' an Indian fermented millet beverage (Ilango and Antony, 2014). Several studies had shown that fermentation of cereals helps in altering the pH to levels that do not favour growth of pathogenic bacteria due to the production of antimicrobial compounds by the LAB during fermentation. (Steinkraus, 2002; Hernández-Ledesma et al., 2004; Adams and Nicolaides, 1997; Odumodu and Inyang, 2006; Oliveira et al., 2014). Lactic acid bacteria (LAB) were the prevalent bacteria during the fermentation of pearl millet and acha flour blends with the *L. plantarum* been the most dominant LAB. This is similar to the findings of Adeyemi and Umar (1994) who reported LAB to be the most prevalent bacteria associated with the fermentation and production of "kunu-zaki" made from sorghum and millet blends.

In conclusion, the findings from this study suggest that fermentation of pearl millet-acha flour blends with the purpose of preparing traditional beverages can improve their nutritional composition.

## Conflict of Interest

The authors have not declared any conflict of interest.

## REFERENCES

Abdalla A, El Tinay A, Mohamed B, Abdalla A (1998). Proximate composition, starch, phytate and mineral contents of 10 pearl millet genotypes. Food Chem. 63:243-246.

http://dx.doi.org/10.1016/S0308-8146(97)00228-8

Adams MR, Nicolaides L (1997). Review of the sensitivity of different foodborne pathogens to fermentation. Food Contr. 8:227-239. http://dx.doi.org/10.1016/S0956-7135(97)00016-9

Adeyemi IA, Umar S (1994). Effect of method of manufacture on quality characteristics of kunun zaki, a millet based beverage. Nig. Food J. 12:34-41.

Amankwah E, Barimah J, Acheampong R, Addai L, Nnaji C (2009). Effect of fermentation and malting on the viscosity of maize-soyabean weaning blends. Pak. J. Nutr. 8:1671-1675. http://dx.doi.org/10.3923/pjn.2009.1671.1675

AOAC (1990). Official methods of analysis. Washington D.C.: Association of official analytica chemists.

Ayo J (2004). Effect of acha (*Digitaria exilis* staph) and millet (*Pennisetum typhodium*) grain on kunun zaki. Brit. Food J. 106:512-519. http://dx.doi.org/10.1108/00070700410545719

Chukwu O, Abdul-kadir AJ (2008). Proximate chemical composition of Acha (*Digitaria exilis* and *Digitaria iburua*) Grains. J. Food Technol. 6:214-216.

Collins CH, Lyne PM, Grange JM (1989). Collins and Lyne's Microbiological Methods. 6th Ed., Butterworth & Co. (publisher) Ltd., U.K., ISBN 0-68301117-0.

Echendu CA, Obizoba IC, Anyika JU, Ojimelukwe PC (2009). Changes in chemical composition of treated and untreated hungry rice acha (Digitaria exilis), Pak. J. Nutr. 8:1779-1785. http://dx.doi.org/10.3923/pjn.2009.1779.1785

Fayemi OE, Ojokoh AO (2014). The Effect of different fermentation techniques on the nutritional quality of the cassava product (fufu). J. Food Proc. Preserv. 38:183-192. http://dx.doi.org/10.1111/j.1745-4549.2012.00763.x

Hernández-Ledesma B, Amigo L, Ramos M, Recio I (2004). Angiotensin converting enzyme inhibitory activity in commercial fermented products. Formation of peptides under simulated gastrointestinal digestion. J. Agric. Food Chem. 52:1504-1510. http://dx.doi.org/10.1021/jf034997b PMid:15030202

Ilango S, Antony U (2014). Assessment of the microbiological quality of koozh, a fermented millet beverage. Afr. J. Microbiol. Res. 8:308-312. http://dx.doi.org/10.5897/AJMR2013.6482

Irving D, Jideani I (1997). Microstructure and composition of *Digitaria exilis* Stapf (acha): a potential crop. Cereal Chem. 74:224-228. http://dx.doi.org/10.1094/CCHEM.1997.74.3.224

Jideani IA (1990). Acha - *Digitaria exilis* - The neglected cereal. Agric. Int. 42:132-134.

Jideani IA (1999). Traditional and possible technological uses of Digitaria exilis (acha) and Digitaria iburua (iburu): A review. Plant Foods Hum. Nutr. 54:363-374.
http://dx.doi.org/10.1023/A:1008193503688 PMid:10798347

Khetarpaul N, Chauhan BM (1989). Effect of fermentation on protein, fat, minerals and thiamine content of pearl millet. Plant Food Hum. Nutr. 39:169-177. http://dx.doi.org/10.1007/BF01091897 PMid:2762245

National Research Council (NRC) (1996). Lost crops of Africa. Grains National Academy Press, Washington, DC, USA Vol 1.

Odumodu CU, Inyang CU (2006). Effects of fermentation on microbial loads of formulated complementary food. Ann. Microbiol. 56:331-334. http://dx.doi.org/10.1007/BF03175026

Ojokoh AO, Daramola MK, Oluoti OJ (2013). Effect of fermentation on nutrient and anti-nutrient composition of breadfruit (*Treculia africana*) and cowpea (*Vigna unguiculata*) blend flours. Afr. J. Agric. Res. 8:3566-3570.

Ojokoh AO, Fayemi EO, Ocloo FCK, Alakija O (2014). Proximate composition, antinutritional contents and physicochemical properties of breadfruit (*Treculia africana*) and cowpea (*Vigna unguiculata*) flour blends fermented with Lactobacillus plantarum. Afr. J. Microbiol. Res. 8:1352-1359. http://dx.doi.org/10.5897/AJMR2013.6469

Oliveira PM, Zannini E, Arendt EK (2014). Cereal fungal infection, mycotoxins, and lactic acid bacteria mediated bioprotection: From crop farming to cereal products. Food Microbiol. 37:78-95. http://dx.doi.org/10.1016/j.fm.2013.06.003 PMid:24230476

Oluwajoba SO, Akinyosoye FA, Oyetayo OV (2013). Comparative sensory and proximate evaluation of spontaneously fermenting kunu-zaki made from germinated and ungerminated composite cereal grains. Food Sci. Nutr. 337-349. http://dx.doi.org/10.1002/fsn3.45

Osuntogun B, Aboaba OO (2004). Microbiological and physico-chemical evaluation of some non-alcoholic beverages. Pak. J. Nutr. 3:188-192.http://dx.doi.org/10.3923/pjn.2004.188.192

Sade FO (2009). Proximate, antinutritional factors and functional properties of processed pearl millet (*Pennisetum glaucum*). J. Food Technol. 7:92-97.

Steinkraus KH (2002). Fermentations in world food processing. Comprehensive Rev. Food Sci. Food Safety. 1:23-32. http://dx.doi.org/10.1111/j.1541-4337.2002.tb00004.x

USDA (2005). Millet- Production, consumption, exports and imports statistics. United States Department of Agriculture.

Zakari UM, Hassan A, Abbo ES (2010). Physico-chemical and sensory properties of "Agigi" from pearl millet (*Pennisetum glaucum*) and bambara groundnut (*Vigna subterranean*) flour blends. Afr. J. Food Sci. 4:662-667.

# Comparative effects of organic and inorganic soil amendments on the growth of cashew nut (*Anacardium occidentale l.)* seedlings

**Awodun M. A.[1], Osundare O. T.[2], Oyelekan S. A.[1] and Okonji C. J.[3]**

[1]Department of Crop, Soil and Pest Management, Federal University of Technology,Akure, Ondo State, Nigeria.
[2]Teaching and Research Farm, Federal University Oye, P.M.B. 373, Ekiti State, Nigeria.
[3]Department of Crop Science and Horticulture, Federal University Oye, P.M.B. 373, Ekiti State, Nigeria.

**This experiment was established to investigate the comparative effect of organic and inorganic soil amendments on the growth of cashew seedlings. Cashew nuts were planted in potting medium filled with well drained loamy soil and placed in a screenhouse. Cashew seeds with 8 to 9 g weight were selected and viability test was carried out using Tetrazolium test (TZ) viability test to ascertain how viable the seeds were before sown in a potting medium. After two weeks of germination, amended with the following treatments at various levels; NPK 15:15:15, composted cow dung, composted poultry dropping, sunshine organo-mineral, pig manure and control (No fertilizer). Complete randomized design (CRD) was used for the experimental layout with three replicates. Agronomic parameters in the study include; stem girth, plant height, number of leaves, leaf length and width. Cashew seedlings were more responsive to the animal sourced fertilizers (organic manure) compared to the inorganic (NPK 15, 15, 15) and control (No fertilizer). Composted poultry droppings fertilizer (PDF) showed higher significant differences on stem girth, plant height and cow dung gave significant difference on number of leaves from 2 to 8 weeks of growth after treatments application, sunshine organomineral and composted poultry droppings fertilizer gave higher significant differences at 8 weeks on leaf length and width respectively.**

**Key words:** Cashew nut (*Anacardium occidentale L.)*, seedlings, organo-mineral, soil amendments.

## INTRODUCTION

Cashew tree, *Anacardium occidentale*, belongs to the Anacardiaceae family of plants which also includes the mango, pistachio, and poison ivy. A medium-sized, spreading evergreen tree native to Brazil, but now grown widely in the tropics for its edible nuts and the oil contained in the shells. The fruit consists of a fleshy, red or yellow, termed the *apple*, at the end of which is borne a hard-shelled, kidney-shaped nut. Although, cashew trees are spread throughout the tropics, commercial production is centered in India, which handles 90% of the world trade. The cashew tree grows with minimum of attention and is easily cultivated. Cashew nuts are edible when roasted and used extensively in the confectionery and baking trade. The cashew nuts shell liquid is a valuable by-product, containing 90% anacardic acid and 10% cardol, and is used in the varnish and plastic industries. The cashew nuts and apples are astringent for eating without being processed, but when processed may be used for jams, chutney, pickles, and wine. It is a nut crop of considerable economic importance to Nigeria.

Unlike most fruits where the seed is found inside the flesh, the cashew seed hangs from the bottom of the cashew apple. Although, fresh cashew fruit (apple) is not only edible but delicious. Cashew apples begin to ferment as soon as they are picked and will barely last 24 h. Cashew apples are highly prized in their growing locale, where they are sometimes found canned in jams, or used to make liquors. The cashew tree is a tropical ever-green, resistant to drought unexacting soil. The tree has extensive root systems, which helps it to tolerate a wide range of moisture levels and soil types. Cashew nut processing allows for the development of an important by-product, which can increase its added value. The liquid inside the shell represents 15 percent of the gross weight and has some attractive possible medicinal and industrial uses. Cashew Nut Shell Liquid (CNSL) is one of the few natural resins that is highly heat resistant and is used in braking systems and in paint manufacture. It contains a compound known as *anacardium*, which is used to treat dermatological disorders. Successful cashew cultivation depends on the selection of the best varieties suited to agro climatic condition.

. Cashew is becoming one of the most important cash crops for farmers in Nigeria where there is a great potential for increased production for the local and export market (Olife et al., 2013). Cashew has assumed greater importance due to its special qualities and consumer preference that resulted from its good taste, supplement of plant protein in human diet and as supplement for soybeans in livestock mashes (Ojewola et al., 2004). Africa accounts for about 36% of world cashew production and Nigeria is the world's 6th largest producer of cashew, with annual production of about 120,000 tonnes (FAO, 2000). Prior to this, production was relatively static at 25,000 tonnes over a 25 year period from 1965. As in the case of other developing countries, Nigeria has recognized the potential economic value of cashew and has made a concerted effort to improve production of the crop (Olife et al., 2013).

Cashew is usually planted as mono-crop in plantations for the nuts that have pleasant taste and flavor when roasted (Yahaya, 2003). Generally, the fleshy part of the whole fruit is eaten raw or the apple juice is being extracted, bottled and sold, while jam and jelly are produced from the pulp following juice extraction and the pulp is a useful ingredient in livestock feedstuff (Yahaya et al., 2010). This research came up due to low usage of fertilizers (inorganic) to fruit crops such as cashew, either at the seedling stage or fruiting stage and the reason may be due to high cost and scarcity of inorganic fertilizers which most farmers cannot afford; so this necessitated into low cost, internally sourced, cheap and adoptable organic materials that can serve as fertilizer and it was also observed that cashew has not received much research attention in terms of fertilizer requirement in Nigeria but research so far indicated that it is needed for optimum cashew seeding growth (Ibiremo, 2008). Low establishment of cashew seedlings to be developed into full grown cashew plantation and un-rehabilitated moribund cashew plantation that has led to drastic reduction in production, hence, the urgent need to use soil amendment to assess their performance and at the same time making recommendation for the best fertilizer types that will have positive effect on cashew growth, compare to the control (with no fertilizer) (Hammed and Adeyemi, 2005).

The objectives of this study are:

1. Identify local and available sources of organic fertilizers that promote fruit crop growth;
2. Identify other fertilizer sources that could cushion the effect of high costing inorganic fertilizers.

## MATERIALS AND METHODS

This research work was conducted January to April, 2012 and May to August, 2013 at the Screenhouse of Federal University of Technology, Akure, Ondo State, South western, Nigeria {54° 30$^1$ N and 70° 30$^1$ E}. Screenhouse was built with translucent plastic to allow penetration of daylight. The study area has an average temperature of 27°C at room temperature and total mean annual rainfall of 1600 mm.

Viable seeds of cashew with 8 to 9 g were selected for the study using Tetrazolium test (TZ) for seed viability and planted in 30 by 10 cm (length and width) nursery pot filled with non-disinfected, very fine dried loam soil. After two weeks of germination (a seed was considered germinated when the radical protrusion was approximately 2 mm), treatments were applied. Plastic bags (Pots) were placed on polythene of 0.5 mm thickness to cover the ground surface. Seeds were sown 3 cm depth in each pot. Weeds were controlled manually by hand pulling. Water was introduced through watering can to supplement rainfall and provide adequate moisture. Complete randomized design (CRD) was used for the experimental layout with each of the six treatments having three replicates. After two weeks of transplanting, treatments were applied in ring form at the following rates (Table 2):

1. N.P.K 15:15:15 (NPKF) = 100 g/stand
2. Sunshine organo-mineral (S.ORGM) = 300 g/stand
3. Composted cow dung (CDGF) = 300 g/stand
4. Composted pig dung (PDF) = 300 g/stand
5. Composted poultry droppings (PDG) = 300 g/stand
6. Control (No fertilizer) = 0 g/stand

### Data collection

After two weeks of applying the treatments, data on growth parameters were collected on forth night bases (two weeks interval) which included plant height (PH), stem girth (SG), number of leaves (NL), leaf length (LL) and leaf width (LW) of the leaves. The seedlings height were measured from ground level in pot (soil base to the shoot tip) using meter rule (cm), girth of stem at the plant - soil base level using analogue veneer caliper (cm), number of leaves by visual count and leaf length and width using leaf area

**Table 1.** Soil analysis of pretreatment collected from the soil sample used.

| Textural class | Percentage (%) |
|---|---|
| Sandy clay loam | 67 sand, 12 silt, 21 clay |
| pH | 5.4 |
| **Nutrient/organic composition** | |
| Organic matter | 1.25 |
| Nitrogen | 0.12 |
| Available P | 9.5 ppm |
| K | 0.19 cmol/kg |

meter; data were collected Four times.

#### Data analysis

Data collected for the experiment were averaged and analyzed. The data obtained were statistically analyzed with SPSS (version 15) through the analysis of variance (ANOVA) and significant mean differences were separated by Tukey at $p \leq 0.05$ level.

## RESULTS AND DISCUSSION

Composted poultry droppings fertilizer (PDF) gave higher significant difference (0.76) compare with other organic fertilizer sources used in this experiment on stem girth of seedlings, this may be attributed to higher concentration of Nitrogen (N) in PDF than composted cowdung (CDGF) and pig dung (PDG), CDGF and PDG also gave higher significant values (0.75 and 0.73) respectively in comparism with sunshine organomineral, NPK and control. Composted poultry droppings also gave significant difference (16.44) on plant height of seedlings, this may be attributed to its high N and P concentration compared to cowdung and pig manure as stated in the treatment analysis above. Cow dung gave significant difference (7.41) on the number of leaves produced by seedlings at 2 to 8 weeks, this reiterates the fact that organic fertilizers steadily release available nutrients overtime and not like inorganic fertilizers that release nutrients within a short period to plant, then leaches away from the root zones (Carl and Roger, 2005). Sunshine organomineral (S. ORGM) and composted poultry droppings fertilizers (PDF) gave significant differences on leaf length and width (9.01 and 4.37) respectively, this also reiterates the fact that two separate mixture of fertilizers, based on the combination of Nitrogen (N), phosphorus (P) and potassium (K), manure and fertilizers promote growth of plants (Ojeniyi et al., 2007). These organic fertilizers have higher nutrient concentration compared with pig manure, cow dung and control, hence enhance more growth of seedlings. The positive growth response of the cashew seedlings to organic and inorganic soil amendments than control indicated that the soil collected for this research is inherently low in some essential nutrients (Table 1), hence the level of many essential nutrients in these soils is lower than critical value which was readily available and supplied from the organic and inorganic amendments (Tables 3 to 7). The cashew seedlings were more responsive to the animal sourced (organic manure) amendments compare to the inorganic fertilizer, thus indicating there were some important nutrients that were being supplied to the cashew seedlings by the organic manures which were not available in the inorganic NPK (15:15:15), thus, cashew seedlings were more responsive to organic amendments for stem girth (SG), plant height (PH) in pig dung and cow dung (Tables 3 to 7). Inorganic and organic sourced fertilizers were competitive in the growth pattern of cashew seedlings, and after four weeks of their application, they resulted to a general significantly higher plant height than control (Table 3). This trend was maintained throughout the observation period. Plant stem girth (SG) was generally higher for the fertilizer treatments compared to the control (Table 3).

Plant number of leaves, plant height and stem girth were significantly ($p \leq 0.05$) different for cashew seedlings treated with organic manures compared to the control from 2 to 8 weeks after its application (Tables 3 to 5). It was observed from this experiment that the amendments, especially the organic manures have further enhanced better growth performance of the seedlings in all the parameters measured compared to the inorganic (NPK) and organo-mineral nutrients source that is noted to be fast in nutrients release and prone to leaching, losses with the accompany environmental hazards (Giles, 2005). Better growth performance due to organic fertilizers than inorganic nutrient sources have been reported for *Amarantus cruentus* (Ipinmoroti et al., 2003), *Camellia sinensis* (Adeoye et al., 2007), coffee plants (Daniel and Obi, 2006), coffee seedlings (Ipinmoroti et al., 2006) and tomato (Adetunji et al., *2003*). The addition of organic matter content resulting from organic fertilizer application helps to improve nutrient availability to plants, especially in tropical soils that are generally low in soil organic matter and clay (Ogunwale et al., 2002).

### Conclusion

It could be deduced from this study that application of nutrient supplements as soil amendments enhanced the growth performance of cashew seedlings resulting in rapid plant height, leaf production and increased plant stem girth that are inevitable for plant vigorous growth. Composted poultry droppings fertilizer (PDF) showed higher significant differences on stem girth, plant height and cow dung gave significant difference on number of leaves from 2 to 8 weeks of growth after treatments application. The use of organic fertilizers from animal source was significant in this experiment and hence, capable to cushion the hardship faced by farmer's on high costing and scarcity of inorganic fertilizers in Nigeria.

**Table 2.** Approximate major nutrient composition of organic and inorganic fertilizers used as treatments.

| Organic | N (% dry weight basis) | $P_2O_5$ | $K_2O$ |
|---|---|---|---|
| Composted poultry droppings | 3.5 | 5.2 | 2.3 |
| Pig manure | 2.9 | 2.3 | 1.5 |
| Composted cow dung | 1.4 | 2.2 | 2.4 |
| Sunshine organomineral | 4.5 | 2.0 | 2.2 |
| **Inorganic** | | | |
| NPK | 15 | 15 | 15 |
| **Control** | 0 | 0 | 0 |

Sunshine organomineral is a fortified organomineral fertilizer made up of plant and animal manure manufactured by Ondo State government, Nigeria.

**Table 3.** Effect of soil amendments on stem girth (cm).

| Treatments | Weeks after application | | | |
|---|---|---|---|---|
| | 2 | 4 | 6 | 8 |
| S.ORGM | $0.42^{b}$ | $0.42^{b}$ | 0.53d | $0.73^{c}$ |
| NPKF | $0.42^{b}$ | $0.42^{b}$ | 0.53d | $0.73^{c}$ |
| PDG | $0.43^{ab}$ | $0.43^{b}$ | $0.54^{c}$ | $0.73^{bc}$ |
| PDF | $0.44^{ab}$ | $0.45^{a}$ | $0.55^{b}$ | $0.76^{a}$ |
| CDGF | $0.45^{a}$ | $0.46^{a}$ | $0.56^{a}$ | $0.75^{ab}$ |
| CONTROL | $0.42^{b}$ | $0.42^{b}$ | $0.52^{e}$ | $0.69^{d}$ |

**Table 4.** Effect of soil amendments on plant height (cm).

| Treatments | Weeks after application | | | |
|---|---|---|---|---|
| | 2 | 4 | 6 | 8 |
| S.ORGM | $7.04^{d}$ | $9.08^{d}$ | $13.09^{d}$ | $15.14^{d}$ |
| NPKF | $6.82^{f}$ | $8.84^{f}$ | $12.88^{f}$ | $15.03^{e}$ |
| PDG | $7.32^{c}$ | $9.45^{c}$ | $13.79^{b}$ | $15.82^{b}$ |
| PDF | $7.53^{b}$ | $9.55^{b}$ | $14.39^{a}$ | $16.44^{a}$ |
| CDGF | $8.34^{a}$ | $10.35^{a}$ | $13.61^{c}$ | $15.68^{c}$ |
| Control | $6.94^{e}$ | $8.95^{e}$ | $12.98^{e}$ | $14.99^{e}$ |

Means on the same column with same prescript are not significantly different by Tukey at 0.05% probability level.

**Table 5.** Effect of soil amendments on number of leaves (visual counting).

| Treatments | Weeks after application | | | |
|---|---|---|---|---|
| | 2 | 4 | 6 | 8 |
| S.ORGM | $2.77^{c}$ | $2.82^{c}$ | $4.87^{c}$ | $6.94^{c}$ |
| NPKF | $2.69^{d}$ | $2.72^{d}$ | $4.75^{d}$ | $6.77^{d}$ |
| PDG | $3.03^{b}$ | $3.04^{b}$ | $5.07^{b}$ | $7.09^{b}$ |
| PDF | $3.04^{b}$ | $3.05^{b}$ | $5.10^{b}$ | $7.11^{b}$ |
| CDGF | $3.33^{a}$ | $3.35^{a}$ | $5.37^{a}$ | $7.41^{a}$ |
| CONTROL | $2.83^{c}$ | $2.85^{c}$ | $4.85^{c}$ | $6.87^{c}$ |

Means on the same column with same prescript are not significantly different by Tukey at 0.05% probability level.

**Table 6.** Effect of soil amendments on leaf length (cm).

| Treatments | Weeks after application | | | |
|---|---|---|---|---|
| | 2 | 4 | 6 | 8 |
| S.ORGM | $3.00^{a}$ | $4.01^{a}$ | $7.01^{a}$ | $9.01^{a}$ |
| NPKF | $2.96^{a}$ | $2.99^{cd}$ | $5.96^{d}$ | $5.97^{b}$ |
| PDG | $2.96^{a}$ | $3.03^{bc}$ | $6.84^{b}$ | $6.99^{ab}$ |
| PDF | $2.97^{a}$ | $2.98^{d}$ | $6.86^{b}$ | $7.03^{ab}$ |
| CDGF | $2.98^{a}$ | $3.00^{c}$ | $6.85^{b}$ | $7.02^{ab}$ |
| CONTROL | $2.98^{a}$ | $3.05^{b}$ | $6.06^{c}$ | $6.07^{b}$ |

Means on the same column with same prescript are not significantly different by Tukey at 0.05% probability level.

**Table 7.** Effect of soil amendments on leaf width (cm).

| Treatments | Weeks after application | | | |
|---|---|---|---|---|
| | 2 | 4 | 6 | 8 |
| S.ORGM | $1.17^{c}$ | $2.24^{ab}$ | $3.26^{bc}$ | $4.30^{b}$ |
| NPKF | $1.18^{bc}$ | $2.32^{a}$ | $3.34^{a}$ | $4.35^{ab}$ |
| PDG | $1.27^{abc}$ | $2.27^{a}$ | $3.32^{ab}$ | $4.34^{ab}$ |
| PDF | $1.29^{ab}$ | $2.32^{a}$ | $3.34^{a}$ | $4.37^{a}$ |
| CDGF | $1.30^{a}$ | $2.32^{a}$ | $3.32^{ab}$ | $4.34^{ab}$ |
| CONTROL | $1.20^{abc}$ | $2.24^{ab}$ | $3.25^{bc}$ | $4.27^{b}$ |

Means on the same column with same prescript are not significantly different by Tukey at 0.05% probability level.

Composted organic manure used in this experiment can be incorporated on large scale as against the filthy and bulky nature believed to be its limiting factor of use.

## Conflict of Interest

The authors have not declared any conflict of interest.

### REFERENCES


Adeoye IB, Denton OA, Oladapo MO, Olufunmi OO, Okafor BN, Ajetunmobi T (2007). Consumer preference and awareness for some exotic vegetables in Ibadan, Oyo state. In: proceedings of 25th annual conference of the horticultural society of Nigeria, pp. 228-233

Adetunji LA, Odeleye GO, Adelaja BO (2003). Effect of combined manual sources and inorganic phosphorus on growth and yield of *Lycopersicum esculentum* var. Ife 1, in: proceeding of 21st Annual Conference of Horticultural Society of Nigeria, Lagos State Polytechnic, Ikorodu, Lagos State, November 10-13, pp. 80-83.

Carl R, Roger E (2005). Nutrient management for commercial fruit and vegetable Crops in Minnesota, published by University of Minnesota, extension service, pp 6-11.

Daniel MA, Obi OA (2006). The effect of three organic fertilizers on the growth Coffee Canephora seedlings. *Proceedings of the 21st international conference on coffee Science 11-15 September, 2006*, Montepella, France, pp. 1206 1208, 1209-1211.

FAO (2000). Cashew production in Africa, 1961-2000;Food and Agriculture Organisation of the United Nations. Production database, *http://apps.fao.org/page/collections*

Giles J, (2005). Nitrogen study fertilizers fears of pollution, nature *International weekly Journal of science* online search *http://www.nature.com/nature/journal/v433/n7028/full/433791a.html.*

Hammed LA, Adeyemi EA (2005). Germination and Seedlings Performance of Cashew (*Anacardium occidentale*) as affected by nut sowing orientations and cotyledon removal. Nig. J. Hort. Res. 10:59-64.

Ibiremo OF (2008) Effect of phosphorus fertilizers and abuscular mycorrhizal fungi inoculation on the growth of cashew seedlings in two soils in Nigeria. Nig. J. Soil Sci. 18:138-146.

Ipinmoroti RR, Adeoye GO, Daniel MA (2003). The comparison of locally blended organic fertilizers on Amaranthus cruentus production at Ibadan, Southwester Nigeria. *Proceedings, Horticultural Society of Nigeria (HORTSON) Conference held at Ikorodu, Lagos*. November 10-13, 2003, pp. 58-61.

Ipinmoroti RR, Adebowale LA, Ogunlade MO, Iremiren GO, Adeoye GO (2006). Effect of inorganic and organic nutrient sources on growth, dry matter yield and nutrient uptake of coffee seedlings. Proceedings of international coffee (ASIC) conference, France, 13-19 September, 2006, pp. 1196-1198.

Ogunwale JA, Olaniyan JO, Aduloju MO. (2002). Morphological, physio-chemical and clay mineralogical properties of soils overlaying basement complex rocks in Ilorin East, Nigeria. Moor J. Agric. Res. 3.2:147-154.

Ojeniyi SO, Awodun MA, Odedina SA (2007). Effect of Animal Manure Ammended Spent Grain and Cocoa Husk on Nutrient Status Growth and Yield of Tomato. Int. J. Agric. Res. 2:406-410.

Ojewola GS, Okoye and I.Agbakuru (2004).Replacement Value of Cashew-nut Meal for Soyabean Meal in Finishing Broiler Chickens. Int. J. Poult. Sci. 3(8):513-516

Olife IC, Jolaoso MA, Onwualu AP (2013). Cashew processing for economic development in Nigeria. Agric. J. 8(1):45-50

Yahaya LE, Aroyeun SO, Ajao AA, Igbinadolor RO (2010). Antibacterial efficacy of CSNL-modified liquid soap. Int. J. Biosci. *5(2)*:104-108.

Yahaya LE (2003). Synthesis and characterization of Cashew nut shell liquid-modified resin Coating. Nig. J. Polym. Sci. Tech. 3(1):202-205.

# Permissions

All chapters in this book were first published in JABSD, by Academic Journals; hereby published with permission under the Creative Commons Attribution License or equivalent. Every chapter published in this book has been scrutinized by our experts. Their significance has been extensively debated. The topics covered herein carry significant findings which will fuel the growth of the discipline. They may even be implemented as practical applications or may be referred to as a beginning point for another development.

The contributors of this book come from diverse backgrounds, making this book a truly international effort. This book will bring forth new frontiers with its revolutionizing research information and detailed analysis of the nascent developments around the world.

We would like to thank all the contributing authors for lending their expertise to make the book truly unique. They have played a crucial role in the development of this book. Without their invaluable contributions this book wouldn't have been possible. They have made vital efforts to compile up to date information on the varied aspects of this subject to make this book a valuable addition to the collection of many professionals and students.

This book was conceptualized with the vision of imparting up-to-date information and advanced data in this field. To ensure the same, a matchless editorial board was set up. Every individual on the board went through rigorous rounds of assessment to prove their worth. After which they invested a large part of their time researching and compiling the most relevant data for our readers.

The editorial board has been involved in producing this book since its inception. They have spent rigorous hours researching and exploring the diverse topics which have resulted in the successful publishing of this book. They have passed on their knowledge of decades through this book. To expedite this challenging task, the publisher supported the team at every step. A small team of assistant editors was also appointed to further simplify the editing procedure and attain best results for the readers.

Apart from the editorial board, the designing team has also invested a significant amount of their time in understanding the subject and creating the most relevant covers. They scrutinized every image to scout for the most suitable representation of the subject and create an appropriate cover for the book.

The publishing team has been an ardent support to the editorial, designing and production team. Their endless efforts to recruit the best for this project, has resulted in the accomplishment of this book. They are a veteran in the field of academics and their pool of knowledge is as vast as their experience in printing. Their expertise and guidance has proved useful at every step. Their uncompromising quality standards have made this book an exceptional effort. Their encouragement from time to time has been an inspiration for everyone.

The publisher and the editorial board hope that this book will prove to be a valuable piece of knowledge for researchers, students, practitioners and scholars across the globe.

# List of Contributors


**O. A. Omotesho**
Department Agricultural Economics and Farm Management, University of Ilorin, P. M. B. 1515, Ilorin, Nigeria

**A.Muhammad-Lawal**
Department Agricultural Economics and Farm Management, University of Ilorin, P. M. B. 1515, Ilorin, Nigeria

**E. A. Oruade-Dimaro**
Nigerian Institute for Oil Palm Research (NIFOR), Plant Pathology Division, P. M. B. 1030 Benin City,Edo State, Nigeria

**I. B. Omamor**
Nigerian Institute for Oil Palm Research (NIFOR), Plant Pathology Division, P. M. B. 1030 Benin City,Edo State, Nigeria

**E. I. Eziashi**
Nigerian Institute for Oil Palm Research (NIFOR), Plant Pathology Division, P. M. B. 1030 Benin City,Edo State, Nigeria

**T. O. Adejumo**
Department of Microbiology, Adekunle Ajasin University, P. M. B. 01, Akungba-Akoko, Nigeria

**Ahmed Abed Gatea**
Department of Agricultural mechanization, College of Agriculture, University of Baghdad, Iraq

**Maria Cristina Silva**
Departamento de Química, Universidade Federal de Lavras, 37200-000, Lavras, MG, Brazil

**Angelita Duarte Corrêa**
Departamento de Química, Universidade Federal de Lavras, 37200-000, Lavras, MG, Brazil

**Custódio Donizete dos Santos**
Departamento de Química, Universidade Federal de Lavras, 37200-000, Lavras, MG, Brazil

**Flávia Cristina**
Departamento de Química, Universidade Federal de Lavras, 37200-000, Lavras, MG, Brazil

**Almeida Marcos**
Departamento de Química, Universidade Federal de Lavras, 37200-000, Lavras, MG, Brazil

**Celeste Maria Patto de Abreu**
Departamento de Química, Universidade Federal de Lavras, 37200-000, Lavras, MG, Brazil

**H. D. Zakpaa**
Department of Biochemistry and Biotechnology, Faculty of Biosciences, College of Science, Kwame Nkrumah University of Science and Technology, Private Mail Bag, Kumasi, Ghana

**E. E. Mak-Mensah**
Department of Biochemistry and Biotechnology, Faculty of Biosciences, College of Science, Kwame Nkrumah University of Science and Technology, Private Mail Bag, Kumasi, Ghana

**J. Adubofour**
Department of Biochemistry and Biotechnology, Faculty of Biosciences, College of Science, Kwame Nkrumah University of Science and Technology, Private Mail Bag, Kumasi, Ghana

**A.A. Ogunjobi**
Department of Botany and Microbiology, University of Ibadan, Nigeria
International Institute of Tropical Agriculture (IITA), P. M. B. 5320, Ibadan, Nigeria

**O. E. Fagade**
Department of Botany and Microbiology, University of Ibadan, Nigeria

**A.G. O. Dixon**
International Institute of Tropical Agriculture (IITA), P. M. B. 5320, Ibadan, Nigeria

**R. Bandyopadhyay**
International Institute of Tropical Agriculture (IITA), P. M. B. 5320, Ibadan, Nigeria

**Sayed Alireza Valadabadi**
Islamic Azad University, Shahr-e-Qods Branch, Iran

**Hossein Aliabadi Farahani**
Islamic Azad University, Takestan Branch, Iran

**B. A. Senjobi**
College of Agricultural Sciences, Olabisi Onabanjo University, Ogun State, Nigeria

**O. T. Ande**
College of Agricultural Sciences, Olabisi Onabanjo University, Ogun State, Nigeria

**C. T. Senjobi**
College of Agricultural Sciences, Olabisi Onabanjo University, Ogun State, Nigeria

**O. A. Odusanya**
College of Agricultural Sciences, Olabisi Onabanjo University, Ogun State, Nigeria

**O. D. Ajilore**
College of Agricultural Sciences, Olabisi Onabanjo University, Ogun State, Nigeria

**M. Fofana**
Africa Rice Center (AfricaRice); 01 B.P. 2031 Cotonou, Benin

**M. Cherif**
Université de Cocody-Abidjan, 22 BP 582 Abidjan

**B. Kone**
Université de Cocody-Abidjan, 22 BP 582 Abidjan

**K. Futakuchi**
Africa Rice Center (AfricaRice); 01 B.P. 2031 Cotonou, Benin

**A.Audebert**
CIRAD, Département cultures annuelles, B.P. 5085, 34032, Montpellier

**E. D. Fagbohun**
Department of Microbiology, University of Ado-Ekiti, P. M. B. 5363, Ado-Ekiti, Ekiti State, Nigeria

**O. K. Abegunde**
Department of Microbiology, University of Ado-Ekiti, P. M. B. 5363, Ado-Ekiti, Ekiti State, Nigeria

**O. M. David**
Department of Microbiology, University of Ado-Ekiti, P. M. B. 5363, Ado-Ekiti, Ekiti State, Nigeria

**P. G Udofia**
Department of Hotel and Catering Management, Akwa Ibom State Polytechnic, Ikot Ekpene, Nigeria

**P. J Udoudo**
Department of Hotel and Catering Management, Akwa Ibom State Polytechnic, Ikot Ekpene, Nigeria

**N. O Eyen**
Department of Hotel and Catering Management, Akwa Ibom State Polytechnic, Ikot Ekpene, Nigeria

**N. S Udoekong**
Department of Science Technology, Akwa Ibom State Polytechnic, Ikot Ekpene, Nigeria

**B. A. Senjobi**
Department of Soil Science and Land Management, University of Agriculture, Abeokuta, Nigeria

**A. O. Ogunkunle**
Department of Agronomy, University of Ibadan, Ibadan, Nigeria

**ML Adriano-Anaya**
Centro de Biociencias. Universidad Autónoma de Chiapas. Carretera a Puerto Madero Km 2.0 Tapachula, 30700,Chiapas, México

**FA Gutiérrez-Miceli**
Departamento de Ingeniería Química y Bioquímica. Instituto Tecnológico de Tuxtla-Gutiérrez, Tuxtla-Gutiérrez,Chiapas, México

**L Dendooven**
Laboratory of Soil Ecology, GIB, Dept. Biotechnology and Bioengineering, Cinvestav, México

**M Salvador-Figueroa**
Centro de Biociencias. Universidad Autónoma de Chiapas. Carretera a Puerto Madero Km 2.0 Tapachula, 30700,Chiapas, México

**E. O. Imoloame**
Regional Agricultural Training School, Ngala, Borno State, Nigeria

**S. D. Joshua**
Department of Crop Production, Faculty of Agriculture, University of Maiduguri, Borno State, Nigeria

**N. A. Gworgwor**
Department of Crop Production, Faculty of Agriculture, University of Maiduguri, Borno State, Nigeria

**E. O. Imoloame**
Regional Agricultural Training School, Ngala, Borno State, Nigeria

**S. D. Joshua**
Department of Crop Production, Faculty of Agriculture, University of Maiduguri, Borno State, Nigeria

**N. A. Gworgwor**
Department of Crop Production, Faculty of Agriculture, University of Maiduguri, Borno State, Nigeria

**Ahmed Abed Gatea**
Agricultural Mechanization College of Agriculture, University of Baghdad, Iraq

**A.R. Heidarian**
M.Sc. Graduate, Department of Agronomy, Islamic Azad University, Mashhad Branch, Iran

**H. Kord**
M.Sc. Graduate, Department of Agronomy, Islamic Azad University, Mashhad Branch, Iran

**Khodadad Mostafavi**
Department of Agronomy and Plant Breeding, Karaj Branch, Islamic Azad University, Karaj, Iran

**Amir Parviz Lak**
Department of Agronomy and Plant Breeding, Karaj Branch, Islamic Azad University, Karaj, Iran

**Faezeh Amini Mashhadi**
Department of Statistics, Khayyam Institute of High Education, Mashhad, Iran

**S. Nwodo Chinedu**
Department of Biological Sciences, School of Natural and Applied Sciences College of Science and Technology, Covenant University, KM 10 Idiroko Road, Canaan Land, PMB 1023, Ota, Ogun State, Nigeria

**C. Obinna Nwinyi**
Department of Biological Sciences, School of Natural and Applied Sciences College of Science and Technology, Covenant University, KM 10 Idiroko Road, Canaan Land, PMB 1023, Ota, Ogun State, Nigeria

**Tuğba Kiper**
Department of Landscape Architecture, Faculty of Agriculture, Namık Kemal University, Tekirdağ/Turkey

**Lotfie A.Yousif**
Agricultural Research Corporation, Agricultural Engineering Research Programme, Gedarif, Sudan

**Mohamed H. Dahab**
Department of Agricultural Engineering, Faculty of Agriculture, University of Khartoum, Shmbat, Sudan

**Haitham R. El Ramlawi**
Center of Dry Land Research and Studies, Faculty of Agriculture and Environmental Sciences, University of Gedarif, Gedarif, Sudan

**I. W Nyakudya**
Department of Agriculture, Bindura University of Science Education, P. Bag 1020, Bindura, Zimbabwe

**L. Jimu**
Department of Environmental Science, Bindura University of Science Education, P. Bag 1020, Bindura, Zimbabwe

**V.Muripira**
Department of Agriculture, Bindura University of Science Education, P. Bag 1020, Bindura, Zimbabwe

**T. J. Chikuvire**
Department of Agriculture, Bindura University of Science Education, P. Bag 1020, Bindura, Zimbabwe

**J. O Othira**
Department of Biochemistry and Molecular Biology, Egerton University, P. O. Box 536 - 20115, Egerton, Kenya

**J. O Omolo**
Department of Chemistry, Egerton University, P. O. Box 536 - 20115, Egerton, Kenya

**F. N Wachira**
Tea Research Foundation of Kenya, P. O. Box 820 - 20200, Kericho, Kenya

**L. A. Onek**
Juba University, Private Bag, Juba, South Sudan

**Sulayman Adeniran Shittu**
Department of Arabic and Islamic Studies, University of Ibadan, Nigeria

**K. S. Subramanian**
Department of Soil Science and Agricultural Chemistry, Tamil Nadu Agricultural University, Coimbatore 641 003, India

**V. Tenshia**
Department of Soil Science and Agricultural Chemistry, Tamil Nadu Agricultural University, Coimbatore 641 003, India

**K. Jayalakshmi**
Department of Soil Science and Agricultural Chemistry, Tamil Nadu Agricultural University, Coimbatore 641 003, India

**V. Ramachandran**
NA & BTD, Bhabha Atomic Research Centre, Trombay, Mumbai 220 085, India

**M. T. Lahai**
Department of Crop Science, Njala Campus, Njala University, PMB, Freetown, Sierra Leone

**I.J. Ekanayake**
Crop Improvement Division, International Institute of Tropical Agriculture (IITA), Oyo Road, PMB 5320, Ibadan, Nigeria
IBRD, 1850, I-street, Washington DC 20043, USA

**J. P. C. Koroma**
Department of Crop Science, Njala Campus, Njala University, PMB, Freetown, Sierra Leone

**Bertukan Mekonnen**
Department of Plant Sciences, Haramaya University, P. O. Box: 138, Dire Dawa, Ethiopia

**Kebede Woldetsadik**
Department of Plant Sciences, Haramaya University, P. O. Box: 138, Dire Dawa, Ethiopia

**Tamado Tana**
Department of Plant Sciences, Haramaya University, P. O. Box: 138, Dire Dawa, Ethiopia

**Asha Yahya**
Department of Crop Production Ecology, Swedish University of Agricultural Sciences, Sweden

**Deborah Wendiro**
Microbiology and Biotechnology Centre of Excellence, Department of Product Development,Uganda Industrial Research Institute, P. O. Box 7086, Kampala, Uganda

**Watu Wamae**
The Open University Faculty of Mathematics, Computing and Technology, TheOpen University, Milton Keynes,MK7 6AA, UK

**Anne Kingiri**
African Centre for Technology Studies P. O. Box 45917 – 00100 Nairobi, Kenya

**Margaret Dhabangi**
Microbiology and Biotechnology Centre of Excellence, Department of Product Development,Uganda Industrial Research Institute, P. O. Box 7086, Kampala, Uganda

**Paul Alex Wacoo**
Microbiology and Biotechnology Centre of Excellence, Department of Product Development,Uganda Industrial Research Institute, P. O. Box 7086, Kampala, Uganda

**U. M Zakari**
Department of Food Technology Kaduna Polytechnic, P. M. B. 2021, Kaduna, Nigeria

**S. A Ajayi**
Department of Food Technology Kaduna Polytechnic, P. M. B. 2021, Kaduna, Nigeria

**A Hassan**
Department of Food Technology Kaduna Polytechnic, P. M. B. 2021, Kaduna, Nigeria

**Abebe Bereda**
Department of Animal Science, Debre Berhan University P. O. Box 445, Debre Berhan, Ethiopia

**Zelalem Yilma**
Heifer International, East Africa Dairy Development, Country Program Mobilization Coordination Office, Addis Ababa, Ethiopia

**Ajebu Nurfeta**
School of Animal and Range Sciences, College of Agriculture, Hawassa University, P. O. Box 5, Hawassa, Ethiopia

**E. M Salvador**
School of Health Systems and Public Health, Faculty of Health Sciences, University of Pretoria, HW Snyman Building Private Bag X323, Pretoria 0001, South Africa.

**V Steenkamp**
Department of Pharmacology, Faculty of Health Sciences, Basic Medical Sciences, University of Pretoria (BMS) building, Private Bag X323, Arcadia, 0007 Pretoria, South Africa

**C. M. E McCrindle**
School of Health Systems and Public Health, Faculty of Health Sciences, University of Pretoria, HW Snyman Building Private Bag X323, Pretoria 0001, South Africa

**C. Parwada**
Department of Agriculture, Bindura University of Science Education, P. Bag 1020, Bindura, Zimbabwe

**C. T. Gadzirayi**
Department of Agriculture, Bindura University of Science Education, P. Bag 1020, Bindura, Zimbabwe

**A.B. Sithole**
Department of Agriculture, Bindura University of Science Education, P. Bag 1020, Bindura, Zimbabwe

**Haftu Kebede**
Department of Animal Science Wachemo University, Hossana Ethiopia

**Gezu Tadesse**
Department of Animal Science Wachemo University, Hossana Ethiopia

**A.O. Ojokoh**
Department of Microbiology, Federal University of Technology, P. M. B. 704, Akure, Ondo State, Nigeria

**O. E. Fayemi**
Radiation Technology Centre, Biotechnology and Nuclear Agriculture Research Institute. Ghana Atomic Energy Commission, P. O. Box LG 80, Legon, Accra, Ghana

**F. C. K. Ocloo**
Department of Microbiology, Federal University of Technology, P. M. B. 704, Akure, Ondo State, Nigeria

**F. I. Nwokolo**
Department of Microbiology, Federal University of Technology, P. M. B. 704, Akure, Ondo State, Nigeria

**M. A Awodun**
Department of Crop, Soil and Pest Management, Federal University of Technology, Akure, Ondo State, Nigeria

**O. T Osundare**
Teaching and Research Farm, Federal University Oye, P.M.B. 373, Ekiti State, Nigeria

**S. A Oyelekan**
Department of Crop, Soil and Pest Management, Federal University of Technology, Akure, Ondo State, Nigeria

**C. J Okonji**
Department of Crop Science and Horticulture, Federal University Oye, P.M.B. 373, Ekiti State, Nigeria